DOMESTIC ELECTRONIC FUEL INJECTION AND COMPUTER SYSTEMS

Frederick D. Allen, M.Ed

Assistant Professor, Automotive Technology
Brookdale Community College
Lincroft, New Jersey

DOMESTIC ELECTRONIC FUEL INJECTION AND COMPUTER SYSTEMS

PRENTICE HALL, Englewood Cliffs, New Jersey 07632

Library of Congress Cataloging-in-Publication Data

ALLEN, FREDERICK D.
Domestic electronic fuel injection and computer systems

Includes index.
1. Automobiles—Motors—Fuel injection systems. 2. Automobiles—Motors—Fuel injection system—Maintenance and repair. I. Title.
TL214.F78A45 1988 629.2′53 87-11411
ISBN 0-13-218256-4

Editorial/production supervision and interior design: Tom Aloisi
Cover design: Ben Santora
Manufacturing buyer: Lorraine Fumoso

A Division of Simon & Schuster
Englewood Cliffs, New Jersey 07632

Printed in the United States of America

10 9 8 7 6 5 4 3 2 1

ISBN 0-13-218256-4 025

Prentice-Hall International (UK) Limited, *London*
Prentice-Hall of Australia Pty. Limited, *Sydney*
Prentice-Hall Canada Inc., *Toronto*
Prentice-Hall Hispanoamericana, S.A., *Mexico*
Prentice-Hall of India Private Limited, *New Delhi*
Prentice-Hall of Japan, Inc., *Tokyo*
Simon & Schuster Asia Pte. Ltd., *Singapore*
Editora Prentice-Hall do Brasil, Ltda., *Rio de Janeiros*

Contents

Preface

The knowledge, repair, and diagnosis abilities of any automotive technician or student can be developed or improved through good training, study, and experience. This text contains information that will help you understand the operation, servicing, and diagnosing of the domestic electronic fuel injection and computer systems that are available on today's cars.

Included for each system is its description, operation, components, basic testing and servicing, and diagnostic aids and procedures. The servicing and diagnostic information given here is not intended to supersede that given by the automotive manufacturers, but it should prepare you to know when to seek more detailed instructions.

Each chapter covers a specific type of fuel injection system and includes computers, sensors, electronic control devices, and modules, which are all described in detail. A glossary is not included as each term is defined as it is used in the descriptive material. Each chapter ends with a listing of the abbreviations used.

CAUTION: — SAFETY NOTICE: To reduce the possibility of personal injury or damage to property or your vehicle, the following instructions must be observed:

1. Service and diagnostic procedures must be followed explicitly as provided by the repair manual. Correct service and repair are important to your safety as well as the safe, reliable operation of the vehicle being serviced or repaired.
2. If part replacement is necessary, replace it with one of the same part number or one that is equivalent. Do *not* use a replacement part of lesser quality.
3. This text provides recommended service information that is introductory in nature. The appropriate service manual will provide effective service and repair information and specify the tools and equipment necessary to do the best possible job.
4. Anyone who uses a replacement part, service procedure, or tool that is not recommended should first ascertain that neither his or her safety nor the safe operation of the vehicle will be jeopardized as a result.
5. These rules are general and not exhaustive. It is virtually impossible to warn of all possible hazardous conditions that might result from not following manufacturers' instructions.
6. The service information contained in this text is written to assist in the correct understanding, diagnosis, and repair of electronic fuel injection and computer systems.

Frederick D. Allen

Acknowledgments

My appreciation is expressed to everyone that encouraged me through the development and writing of this text.

Special thanks are due to the technical reviewers for having done an excellent job and for helping with technical details. Peggy Hynes of General Motors Product Service Training, John Greenhalgh of Chrysler Training, and Fran Treichler of Ford Parts and Service Division did an outstanding review.

Special thanks also to the administration of Brookdale Community College for their support and to Doug Welsh for his review and recommendations concerning this text.

The following companies deserve special recognition as they all cooperated with the aim of providing accurate and complete information that would be useful in the training of automotive technicians by granting permission for the use of illustrations and instructional material adapted from their training booklets and service manuals:

Buick Motor Division, General Motors Corporation
Cadillac Motor Car Division, General Motors Corporation
Chevrolet Motor Division, General Motors Corporation
Chrysler Corporation
Ford Motor Company
General Motors Product Service Training, General Motors Corporation
Oldsmobile Division, General Motors Corporation
Pontiac Motor Division, General Motors Corporation

Last, but certainly not least, a special thanks to Holly, Alice, Matt, Tim, Chuck, and everyone else at Prentice Hall that encouraged, guided, and assisted in making this a reality.

Introduction— Electronic Fuel Injection Background

The first American production car to use a solid-state fuel injection system as a standard item was Cadillac in 1975. This system was used, with minor modifications, up to and including 1980.

There are distinct advantages in using electronic fuel injection systems over carburetors and even feedback computer-controlled carburetor systems. The fuel-wetted area in the intake manifold is reduced, fuel is metered (injected) with more accuracy, the engine starts easier and quicker even in cold weather, and response to driver demands is practically instantaneous. There is no need for a choke, driveability problems are minimized, and the system is altitude compensated. All these lead to improved efficiency, in terms of both fuel economy and emissions levels, and better performance.

In a carbureted engine, air enters the air horn of the carburetor and causes fuel to be drawn from the float bowl as it rushes through the venturi; the resulting air/fuel mixture passes the throttle blades and flows through the intake manifold and past the open intake valve into the cylinder. The passages or runners through the intake manifold are not of equal length or temperature. As a result, equal amounts of mixture do not reach each cylinder. There can be as much as a 15% difference between the leanest and richest cylinders. This encourages such problems as rough idle, emissions, driveability symptoms, and sacrifice of fuel economy and performance.

The underlying cause of this unequal mixture problem is related to the laws of gas-flow theory, vaporization, and condensation. Air can easily flow around corners and through the changing shape of the intake manifold passages. When fuel is mixed with the air, it has a tendency to "settle out" or condense as it flows through the passages. This happens because it is heavier, and slows quicker, giving it opportunity to condense. A fuel-wetted area results inside the intake manifold, which encourages rough idle, emissions, and poor economy and performance.

The problems of the carbureted engine can be minimized through the use of fuel injection. With fuel injection, the intake manifold may carry only air, as the fuel can be injected directly into the intake valve ports in the cylinder head. This eliminates the possibility of a fuel-wetted area as previously described.

Additional problems of carbureted engines include the following:

1. Restrictions to the air/fuel flow through the carburetor and intake manifold, limiting the engine's ability to "breathe in."
2. Heating of the entering air/fuel mixture, causing expansion before the mixture can enter the cylinder.
3. Slow response of fuel delivery to throttle blade

position changes in the carburetor due to fuel surface tension and inertia.

Existing fuel injection systems are designed for gasoline only although they can be made adaptable for use with gasohol and other synthetic fuels.

The fuel injection system must meet at least six requirements if it is to prove advantageous:

1. It must provide precise control of air/fuel mixtures to meet all operating conditions.
2. It must provide equal fuel distribution to all cylinders.
3. It must minimize the fuel-wetted area in the intake manifold.
4. It must not be so complicated as to create maintenance problems or reduce system reliability or vehicle life expectancy.
5. It must be reasonably priced, cost-comparable to carbureted systems, and designed so as to be produced on a mass-production basis.
6. It must provide atomization of fuel for consistent, homogeneous mixtures.

Work with fuel injection systems dates back to the late 1800s. The invention of the airplane further encouraged experimentation as carburetors were not capable of delivering fuel under all flying conditions. Later carburetor developments overcame these problems and fuel injection suffered a setback.

The development of the diesel engine revived fuel injection experimentation around 1920, and by 1930 a system was perfected and ready for mass production.

By 1949, a port-injection system, providing a continuous spray of fuel within a throttle body at each intake port, had been developed for a car to be raced at the Indianapolis Motor Speedway. By 1952 almost all race cars entering the track were equipped with this system developed by Stuart Hilborn. Later, in 1955, when Chevrolet was developing the V-8 Corvette, Zora Arkus-Duntov and Chevrolet engineers developed a system that was put into production by the Rochester Products Division of General Motors Corporation.

Other manufacturers, such as Bendix, Borg-Warner, American Bosch, Lucas, were developing systems at the same time. However, in 1976, Holley unveiled a new carburetor with electronic fuel metering. This proved to be both a setback and a stepping stone for fuel injection. Existing carburetors were easier to change in order to allow use of electronic controls and computer systems. The developing technology did not stop with carburetors, but was also applied to the older mechanical fuel injection systems. So it is electronic fuel injection that has become the state-of-the-art system to use.

Injectors, instead of being opened by pressure, are operated by computer-controlled solenoids. Engine sensors feed information data, either analog or digital, to the computer, which in turn calculates the pulse width (injection duration), depending on operating conditions such as manifold pressure, engine speed, acceleration, cold startup, ambient temperature, altitude, and idling.

Electronic fuel injection studies began in the early 1950s when a Bendix engineer, Robert W. Sutton, developed a system for an experimental 1953 Buick V-8. The system, known as Electrojector, was used by American Motors Corporation as an option for the 1957 Rambler Rebel. Because of service problems and lack of interest from the public, no production cars equipped with the system were sold. Chrysler offered the system as an option in 1958 on the Chrysler 3000-D and the De Soto Adventurer. The system still needed improvement but was, overall, satisfactory.

In 1965, Bendix sold patent ideas to American Bosch, which resulted in Volkswagen's use of the Bosch electronic fuel injection system known as D-Jetronic.

As the years progressed, new developments such as emissions regulations and fuel economy standards made Cadillac Chief Engineer Carlton Rasmussen take a more serious look at fuel injection. Bendix had not worked much with the Electrojector system since 1965, but began to take a new look at the potential of the system. John Campbell, general manager of the Motor Components Division of Bendix, felt that the accuracy of fuel mixture control could help automotive manufacturers meet emissions levels. An Electronic Fuel Injection Division was formed in 1972.

By the same time, integrated circuits, microprocessors, and manufacturing methods had been developed, making mass production of electronic fuel injection a viable possibility. A contract with Cadillac was drawn and signed. The system was regarded as one of the most advanced fuel control systems available at the time anywhere in the world.

Using this background as a foundation, it should be easier to see and understand the rationale and development of newer electronic fuel injection systems as they are covered. The first system to be covered in depth is the Cadillac electronic fuel injection system offered as standard equipment in 1975 on the advance model 1976 Seville and as an option on full-sized Cadillac models.

DOMESTIC ELECTRONIC FUEL INJECTION AND COMPUTER SYSTEMS

UNIT I
General Motors
Corporation Systems

1 Cadillac Electronic Fuel Injection, 1975–1980

OVERVIEW

One problem encountered with carbureted engines is related to the fuel-wetted area inside the intake manifold. This happens because the manifold passages must carry an air/fuel mixture (Fig. 1-1).

The Cadillac electronic fuel injection (EFI) system is a port injection system. This means the passages in the manifold carry only air, while the fuel is injected at the intake valve ports in the cylinder heads (Fig. 1-2). This system provides a method of accurately controlling and adjusting the air/fuel ratio to satisfy engine operating demands and yet maintain emissions levels within limits set by the Environmental Protection Agency (EPA). To accomplish this, certain engine operating conditions must be monitored and fuel requirements must be controlled in accordance with those operating conditions.

The Cadillac EFI system uses eight individual fuel injectors, one for each cylinder, which are electrically operated or actuated fuel metering valves. The injectors are mounted on the intake manifold with their tips inserted into the manifold and aimed at the intake valve ports in the cylinder heads (Fig. 1-2). The injector is sealed to the intake manifold with a special rubber O-ring to prevent vacuum leaks. Four injectors are opened simultaneously in two groups and are timed so that the fuel will be injected before the intake stroke for the cylinder. The injectors are divided into two groups as follows: group I, cylinders 1, 2, 7, and 8; and group II, cylinders 3, 4, 5, and 6.

Fuel is supplied to the inlets of the injectors by a fuel rail. The injectors are sealed to the fuel rail by a special rubber O-ring to prevent fuel leaks. To prevent air bubble

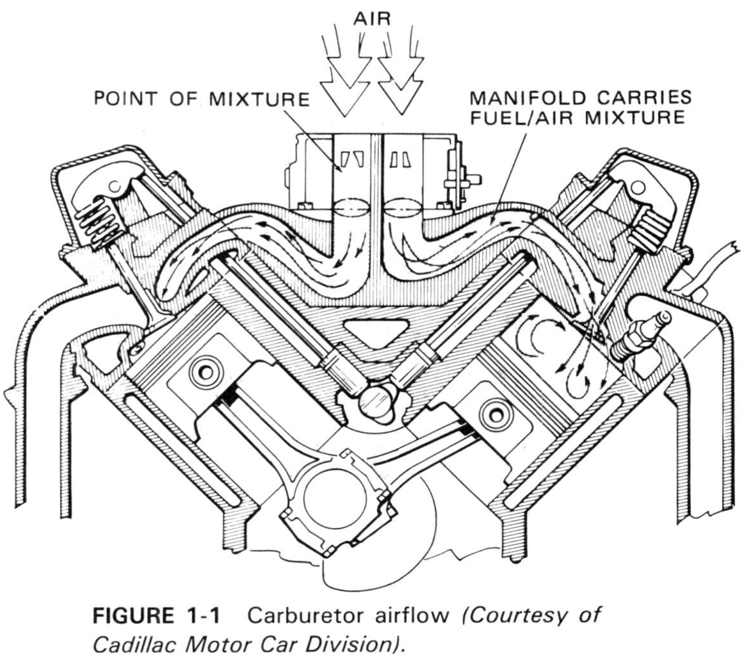

FIGURE 1-1 Carburetor airflow *(Courtesy of Cadillac Motor Car Division).*

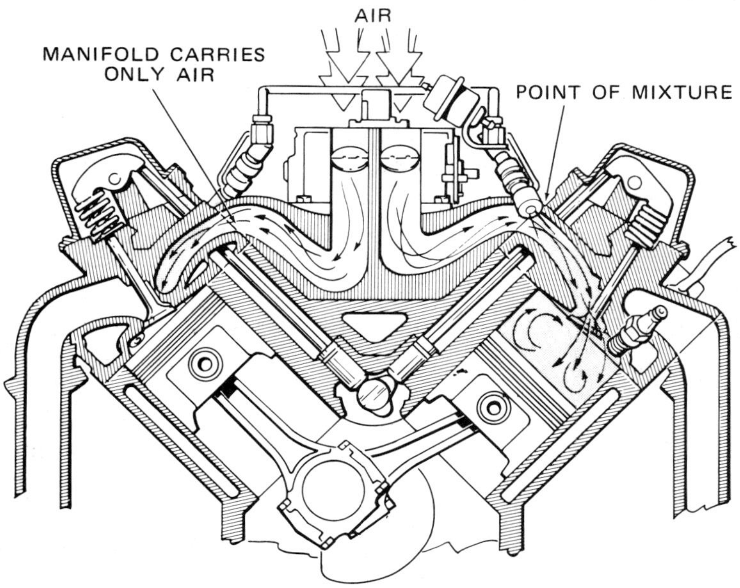

FIGURE 1-2 Fuel injection airflow *(Courtesy of Cadillac Motor Car Division).*

formation (vapor lock) problems in the fuel lines, fuel pressure is kept high (29 to 35 psi). Because the pressure is kept constant, by varying the amount of time the injectors are opened the fuel quantity can be controlled.

The amount of air entering the engine must be known, along with air temperature, engine speed and load, engine temperature, throttle position, and manifold pressure. This information, along with information such as engine displacement, volumetric efficiency, and other critical inputs allow the computer (ECU), to regulate fuel delivery through the injectors.

When compared to a carbureted vehicle, the EFI component functions become clearer:

Carburetor components	*EFI components*
Accelerator pump and power valve	Throttle position switch
Fast idle cam	Electric fast idle valve
Float	Fuel pressure regulator
Metering rods	Manifold absolute pressure sensor
Metering jets and idle circuit	Injectors, electronic control unit
Choke	Coolant sensor
Thermac air temperature	Air sensor

These comparisons may not be interchangeable when diagnosing problems but should aid in understanding operating principles. The function of the electronic control unit (ECU) is to calculate fuel requirements according to all possible input signals from various sensors and to adjust pulse width or injector open duration to provide the appropriate air/fuel ratio.

Pulse width is measured in milliseconds. One millisecond is equal to one-thousandth of a second. For example, an engine at idle and operating temperature would require a pulse width of about 3 milliseconds, whereas the same engine at wide-open throttle (WOT) would require a pulse width of about 13 milliseconds.

Figure 1-3 shows the sensors (input signals), the ECU, and the controlled devices (output signals).

Engine operating conditions are monitored by the ECU through the input sensors. The ECU then uses the input signals to determine pulse width for the injector groups and whether or not to command the fuel pumps, fast idle valve, and EGR solenoid.

The Cadillac EFI system can be broken down into four subsystems: (1) fuel delivery, (2) air induction, (3)

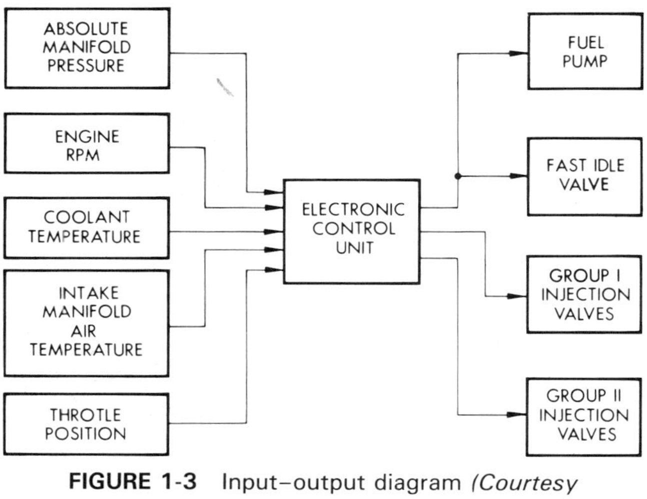

FIGURE 1-3 Input–output diagram *(Courtesy of Cadillac Motor Car Division).*

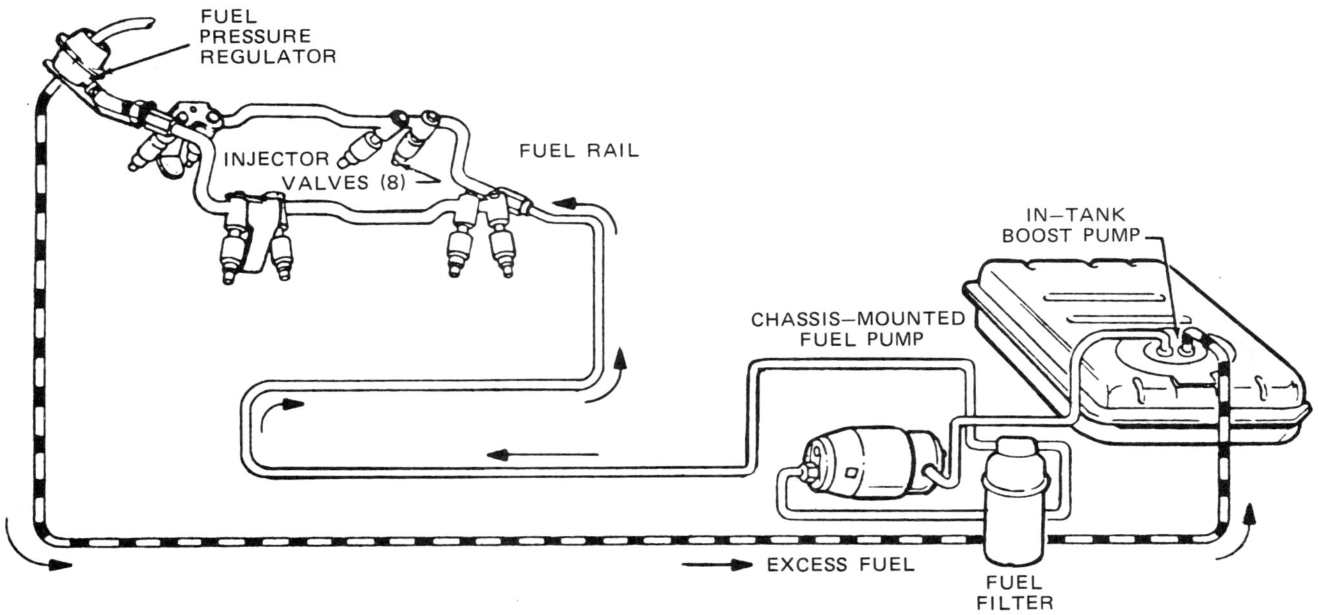

FIGURE 1-4 Fuel delivery system *(Courtesy of Cadillac Motor Car Division).*

sensors, and (4) electronic control unit (ECU). A detailed description of each will provide a better understanding of the overall system.

FUEL DELIVERY SYSTEM

The fuel delivery system consists of the fuel tank and cap, an in-tank electric fuel pump, a chassis-mounted high-pressure electric fuel pump, a fuel filter, fuel lines, fuel rail, a fuel pressure regulator, and return fuel lines (Fig. 1-4). An evaporative emission control (EEC) system is also used but will not be discussed at this time.

Fuel Tank

The fuel tank is different from those in carbureted systems in that it has a fuel reservoir or sump directly beneath the fuel gauge sending unit. The sump ensures a constant supply of fuel to the in-tank electric fuel pump during low-fuel, severe maneuvering conditions that might cause the fuel to "run" to one side of the tank. This is critical as the electric fuel pumps draw fuel at a rapid rate because the system circulates fuel from the tank to the injectors and back to the tank. The fuel return line delivers fuel to the tank from the fuel pressure regulator. The returned fuel is delivered into the reservoir area as an additional measure to prevent the pumps from "running dry" and damaging pump internal parts. The fuel tank cap is a standard pressure-vacuum relief emission control cap.

In-tank Electric Fuel Pump

The in-tank electric fuel pump is part of the fuel gauge sending unit and cannot be serviced separately (Fig. 1-5). It acts as a supply or boost pump to the chassis-mounted pump. It also prevents vapor lock on the intake or suction side of the chassis pump. Both pumps are activated by the ECU when the ignition key is turned to ON. The pumps are automatically deactivated by the ECU if the engine stalls (no rpm signal) or if there is no cranking signal (no rpm signal), within 1 second of the key switch being turned ON. The two pumps are connected in parallel, electrically,

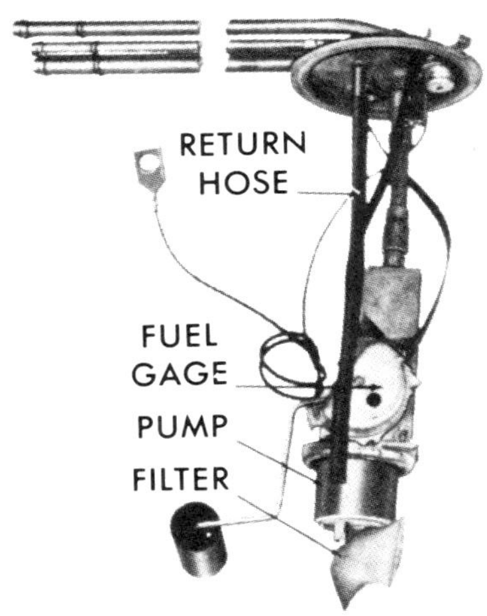

FIGURE 1-5 In-tank pump *(Courtesy of Cadillac Motor Car Division).*

and are protected by a 10-ampere in-line fuse in the wiring harness near the ECU connectors under the instrument panel.

Chassis Fuel Pump

The chassis-mounted high-pressure electric fuel pump is a 12-volt, roller-vane, constant-displacement-type pump. It has a check valve to prevent backflow of fuel from the high-pressure (output) side. The pump has a 33 gallons/minute flow rate at a normal operating pressure of 39 psi. Maximum pressure is limited to 55 to 95 psi by an internal relief valve. The pump is mounted on the chassis (frame) ahead of the left rear wheel on DeVille and Seville cars and ahead of the right rear wheel on Eldorados (Fig. 1-6).

Fuel Filter

The fuel filter is an in-line, paper element type located near the chassis-mounted pump on Seville and at the lower-left front engine bracket on DeVille and Eldorado (Fig. 1-7). This filter can remove particles as small as 20 microns or twenty-thousandths of a millimeter. It is disposable and should be replaced every 22,500 miles (Fig. 1-8).

Fuel Rail

High-pressure fuel is delivered from the chassis-mounted pump through standard rubber fuel lines and standard steel fuel lines to the fuel rail. The fuel rail feeds fuel to the injectors. It is mounted on top of the intake manifold and is a two-piece assembly with connecting fittings at the rear left corner and front right corner (Fig. 1-9).

Fuel Injectors

The fuel injectors are electrically operated solenoids that operate a pintle valve inside each injector (Fig. 1-10). When the injector is activated, based on a pulsed signal from the ECU, the solenoid causes the pintle valve to rise, allowing fuel to pass and spray into the intake valve port. The duration or time interval that the injector is open is called its *pulse width.* When the injector is deactivated, it prevents fuel flow from the pintle valve due to a return spring. Han-

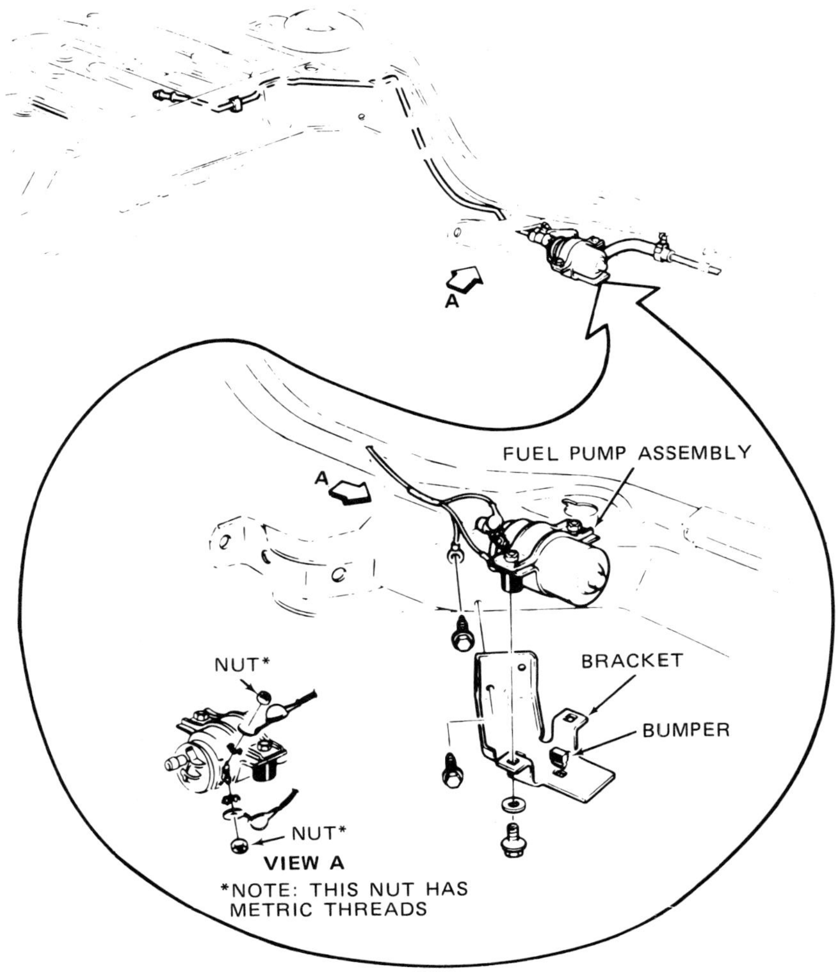

FIGURE 1-6 Chassis fuel pump *(Courtesy of Cadillac Motor Car Division).*

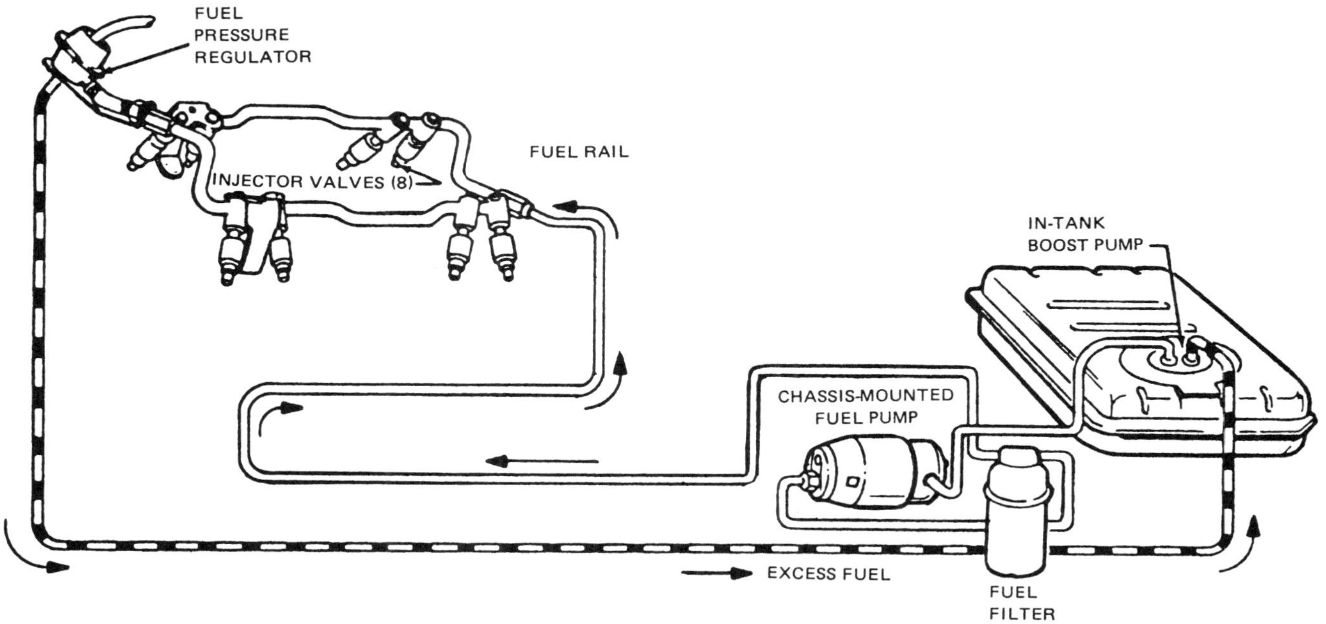

FIGURE 1-7 Fuel delivery system *(Courtesy of Cadillac Motor Car Division).*

FIGURE 1-8 Fuel filter *(Courtesy of Cadillac Motor Car Division).*

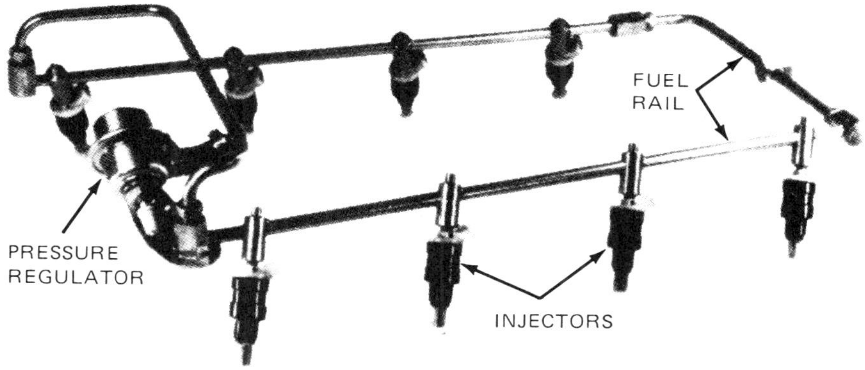

FIGURE 1-9 Fuel rail and injectors *(Courtesy of Cadillac Motor Car Division).*

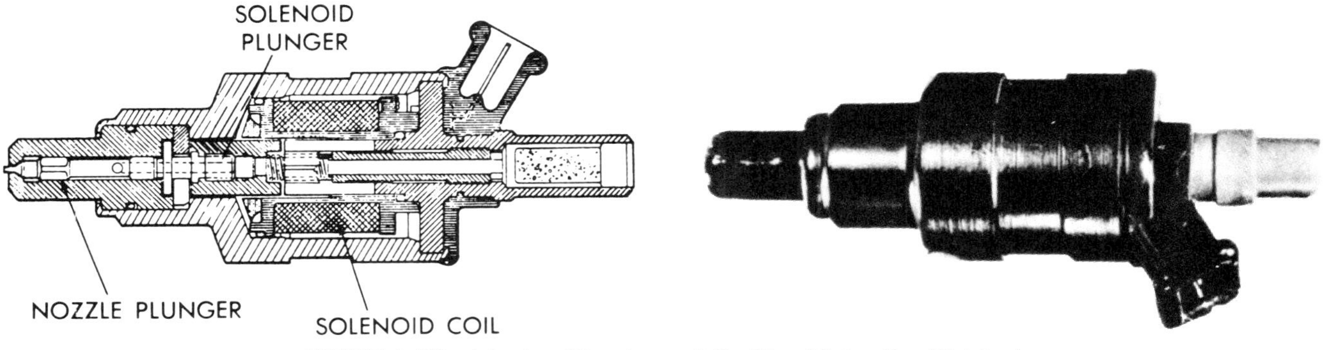

FIGURE 1-10 Injector *(Courtesy of Cadillac Motor Car Division).*

FIGURE 1-11 Fuel pressure regulator *(Courtesy of Cadillac Motor Car Division).*

dling and servicing of the injectors must be done carefully to avoid damaging the tips.

Pressure in the fuel rail is controlled by the fuel pressure regulator (Fig. 1-11). It contains an air chamber and a fuel chamber divided by a spring-loaded diaphragm. The air chamber is connected to engine vacuum by a hose between the regulator and a vacuum tap on the throttle body. The regulator is subjected to full manifold vacuum. Changing vacuum and the spring tension act on the diaphragm to cause changes in the orifice opening inside the regulator. When the orifice opens wider, more fuel is allowed to return to the tank, causing a drop in rail pressure. The orifice opens wider whenever manifold vacuum gets higher. If the orifice gets smaller, less fuel returns to the tank and rail pressure rises. Whenever manifold vacuum drops, the orifice gets smaller. The fuel pressure regulator is designed to maintain approximately 39 psi in the fuel rail at 0 inches of mercury (Hg) manifold vacuum. The operation of the pressure regulator also compensates or corrects for changes in altitude automatically. It is located at the right or left front corner of the fuel rail and is connected to it by fuel line fittings.

The fuel delivery system has similarities to but is different from carbureted systems. When servicing of the system is covered, these differences will be more apparent.

At this point, there should be a sufficient supply of fuel. Air is also needed for mixing an ideal air/fuel mixture, called a *stoichiometric* or *ideal mixture,* of about 14.7 to 1. The air induction system provides the needed oxygen for combustion.

AIR INDUCTION SYSTEM

The air induction system includes the air cleaner housing and element, a throttle body, a fast idle valve, and an intake manifold (Fig. 1-12). Vacuum connections can also be considered as each may affect the operation of the air induction system. Outside air is directed through the air

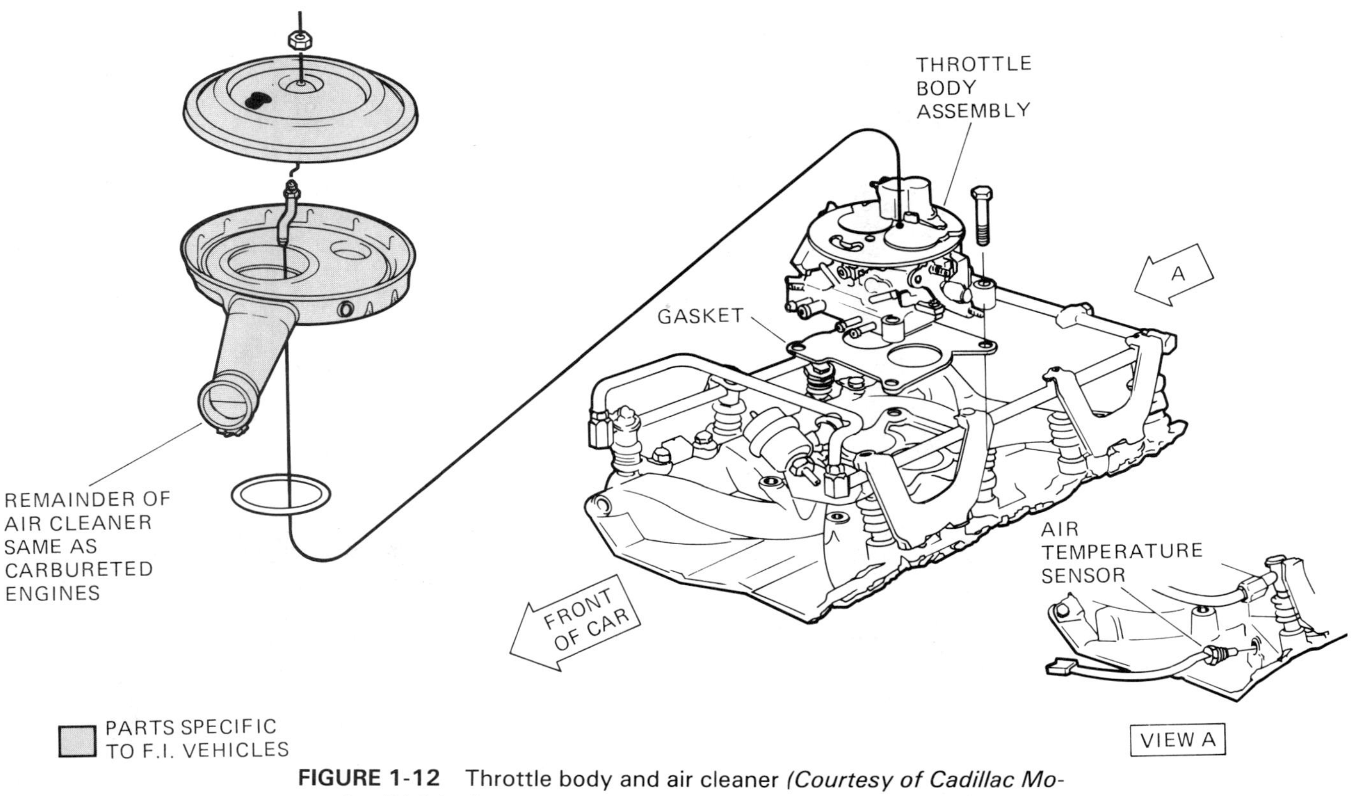

FIGURE 1-12 Throttle body and air cleaner *(Courtesy of Cadillac Motor Car Division).*

cleaner housing and filtering element, which is a disposable paper type. Air then passes over the throttle plates and into the intake manifold for distribution to the cylinders. When idling, air enters a bypass passage above the throttle plates and is allowed to bypass the plates. An adjusting screw in the passage is used to regulate the amount of air allowed to bypass and, therefore, adjust the curb idle speed. Primary airflow is controlled by the throttle plates.

Intake Manifold

The intake manifold used on DeVille and Eldorado is a dual-plane type. Dual plane means there are two levels of passages inside the manifold. The Seville uses a single-plane manifold in which the passages are on one level (Fig. 1-13). It is important to remember that these manifolds carry only air; have holes into which the injectors are mounted with O-ring seals; have a threaded hole for the manifold air temperature sensor; and do not have an exhaust crossover passage (sometimes referred to as a heat riser passage or early fuel evaporation passage); all vacuum taps are on the throttle body.

FIGURE 1-13 Intake manifold *(Courtesy of Cadillac Motor Car Division).*

Throttle Body

The throttle body is a two-piece assembly (Fig. 1-14). The upper body contains two bores, the fast idle valve, and the curb idle adjusting screw. The lower body contains the throttle shaft and plates, linkage for the accelerator pedal, and the vacuum taps. The right side also has mounts for fastening the throttle position switch. The accelerator linkage on the left side, when at idle, comes to rest against a factory-preset adjusting screw. The adjustment is made so the throttle blades are slightly open. It should *not* be changed.

The screw and locknut at the front of the throttle body are used to adjust curb idle at operating temperature. When the screw is turned clockwise (in), the air bypass passage is restricted and causes the idle speed to lower. Turning the screw counterclockwise, (out) allows more air to bypass and raise the idle speed (Fig. 1-15).

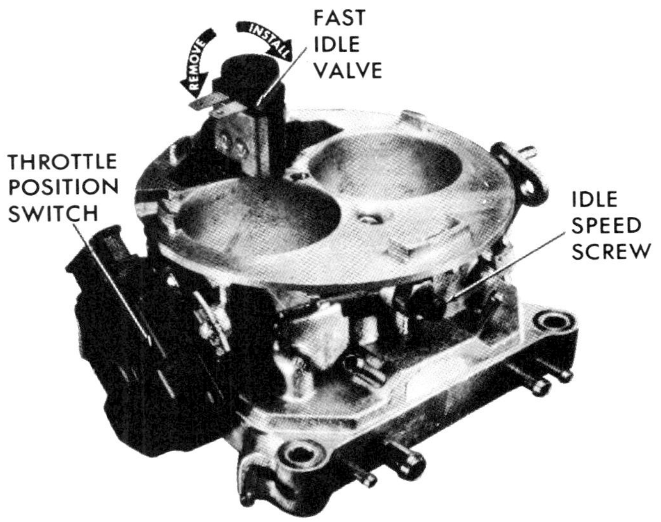

FIGURE 1-14 Throttle body *(Courtesy of Cadillac Motor Car Division).*

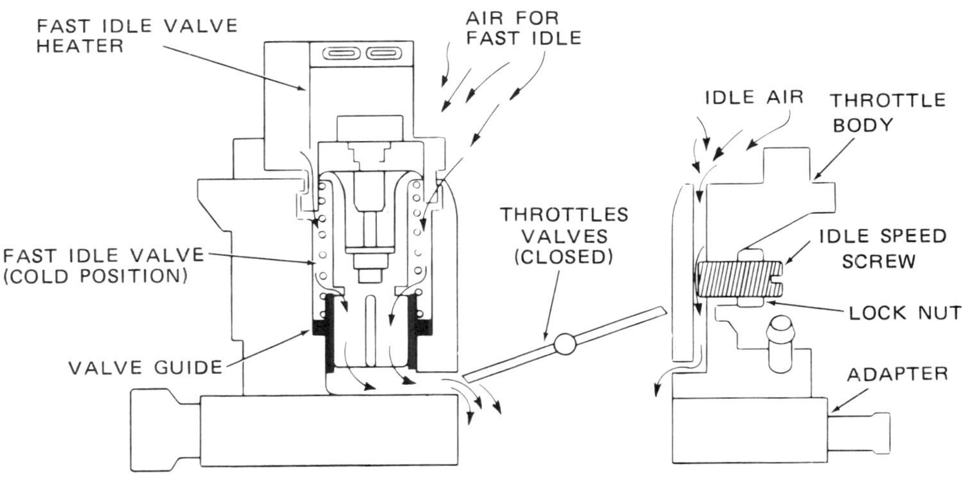

FIGURE 1-15 Throttle body cutaway *(Courtesy of Cadillac Motor Car Division).*

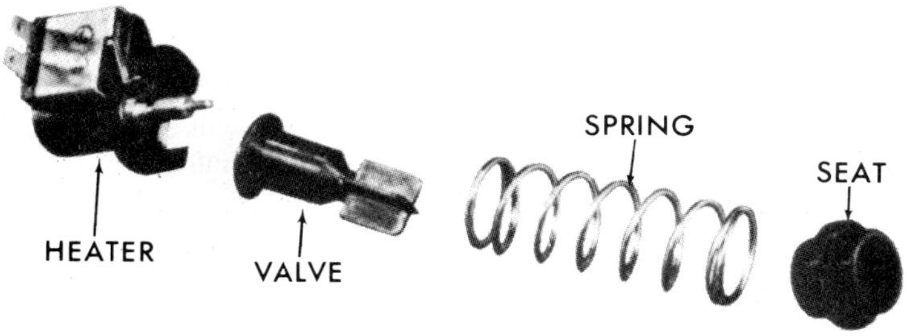

FIGURE 1-16 Fast idle valve *(Courtesy of Cadillac Motor Car Division).*

Fast Idle Valve

A fast idle valve is housed in the top rear section of the throttle body. It has a two-stage electric heater, a microswitch, a plunger and spring, and a temperature-sensing element (Fig. 1-15). It is connected electrically to the fuel pump circuit through the electronic control unit. When the engine is off and cool, the spring pushes the plunger up, lessening the restriction in the fast idle valve bore (Fig. 1-16). This allows more air to bypass the throttle plates when the engine is started. The extra airflow provides a fast idle for cold operation. At the same time, the electric heater begins to warm the temperature-sensing element, which expands and pushes the plunger down against spring tension and gradually restricts air bypass flow. The engine fast idle speed will gradually lower until normal curb idle speed is reached. This valve does not fully close; however, the leakage is calibrated into the system. Once the temperature-sensing element reaches 140°F, the fast idle valve will not affect engine operation. Ambient temperature (outside air temperatures) determine the rate at which the fast idle valve operates. For example, at 68°F, the valve should close within approximately 90 seconds; at −20°F, the valve should close within approximately 5 minutes.

Vacuum hose connections are located on the throttle body. Refer to the vacuum hose diagrams for correct connection and routing. The individual manufacturer shop manual should be used to ensure correct connection and routing as there are subtle changes with year and model.

ENGINE SENSORS

Engine sensors are used to provide information to the electronic control unit so it has a basis upon which to make calculated adjustments in pulse width. There are five separate engine sensors: manifold absolute pressure, engine speed (rpm), manifold air temperature, throttle position, and engine coolant temperature. Each operates independently and is connected electrically to the electronic control unit. The sensors provide a continuous flow of data to the electronic control unit, indicating to it changes in engine operation. The ECU can then vary pulse width to the injectors to meet engine needs.

Manifold Absolute Pressure Sensor

The manifold absolute pressure (MAP) sensor is mounted inside the ECU under the instrument panel (Fig. 1-17). A hose connects to a fitting at the rear lower-left corner of the ECU (Seville) and in turn connects to plastic tubing that is incorporated with the EFI wiring harness. On the engine, the plastic tubing again connects to a rubber hose that is clamped to a vacuum connection on the throttle body. This sensor monitors changes in intake manifold vacuum and provides an indication of engine load, barometric pressure, altitude and road speed changes. The sensor changes these pressure variations into electrical signals that are interpreted by the ECU.

When engine load is increased, the throttle plate opening must be increased in an attempt to maintain road

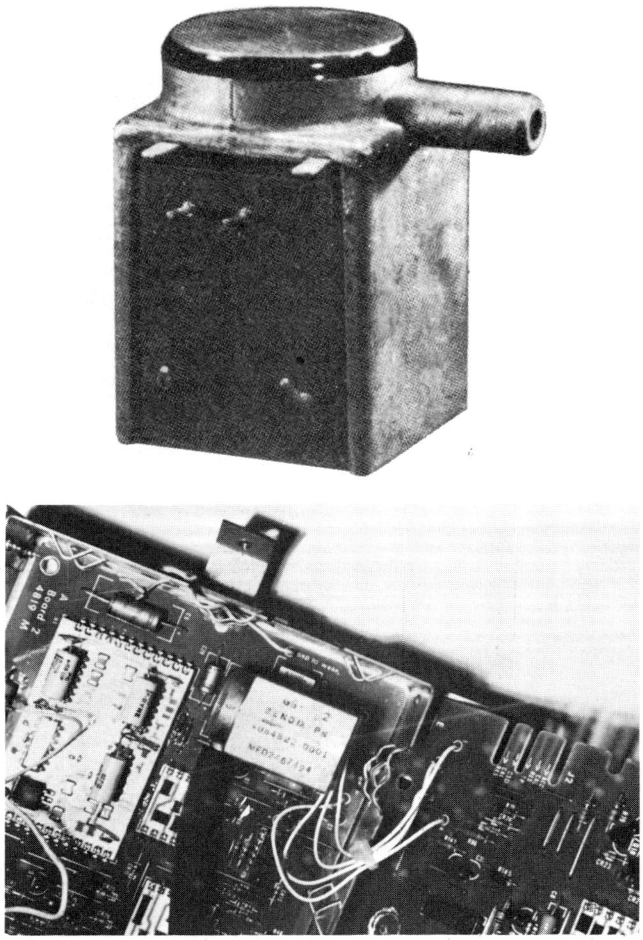

FIGURE 1-17 Manifold absolute pressure sensor *(Courtesy of Cadillac Motor Car Division).*

speed. Engine manifold vacuum decreases (pressure increases), and the MAP sensor sends an appropriate signal to the ECU that more fuel is needed. The ECU then increases pulse width. As the engine load decreases, the throttle plate opening decreases as an attempt is made to again maintain road speed. Engine manifold vacuum increases (pressure decreases), and the MAP sensor sends an appropriate signal to the ECU for a reduced pulse width. The MAP sensor is part of the ECU and is not serviceable separately.

Throttle Position Switch

The throttle position switch (TPS) is mounted on the right side of the throttle body and is connected to the end of the throttle shaft (Fig. 1-18). It is connected electrically to the electronic control unit. Movement of the accelerator pedal (opening or closing of the throttle plates) causes the throttle shaft to rotate. The throttle position switch senses this movement and electrically sends signals to the ECU. As a result, the ECU is able to vary pulse width in accordance with driver demands. The three active functions of the throttle position switch are the closed throttle signal, acceleration enrichment pulses, and wide open throttle signal (Fig. 1-19). A closed throttle signal indicates to the ECU that the engine is idling and needs a pulse width to match. The acceleration enrichment pulses indicate to the ECU that the driver wishes to increase speed. The ECU delivers additional pulses to the injectors to meet this demand. As the throttle plates are opened, the throttle position switch sends a signal to the ECU for every three degrees of opening. The last function of the TPS is to provide enrichment pulses through the ECU for wide-open throttle operation. The ECU increases pulse width to about 13 milliseconds under this operating condition.

There is a fourth function that the TPS serves, but it is not an active function. Although a flooded fuel-injected engine is a rare occurrence, it is a possibility. To clear the engine (with carbureted engines it is called "unloading"), the accelerator pedal should be held to the floor. At this point, when the engine is being cranked, the ECU receives a wide-open throttle signal and a cranking signal at the same time. The ECU responds by deactivating the injectors, enabling the flood condition to be cleared.

The throttle position switch consists of the following:

1. A printed circuit board with five copper *runs*.
2. Four wiper arms that contact and travel the runs.
3. A pair of leaf springs that may or may not contact two brass pins depending on throttle position (Fig. 1-20).

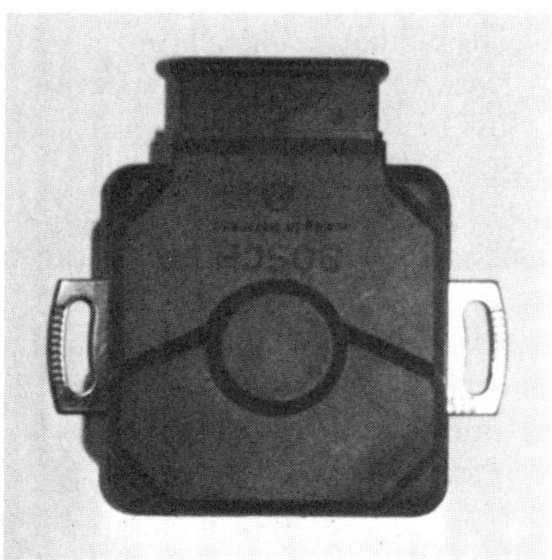

FIGURE 1-18 Throttle position switch *(Courtesy of Cadillac Motor Car Division).*

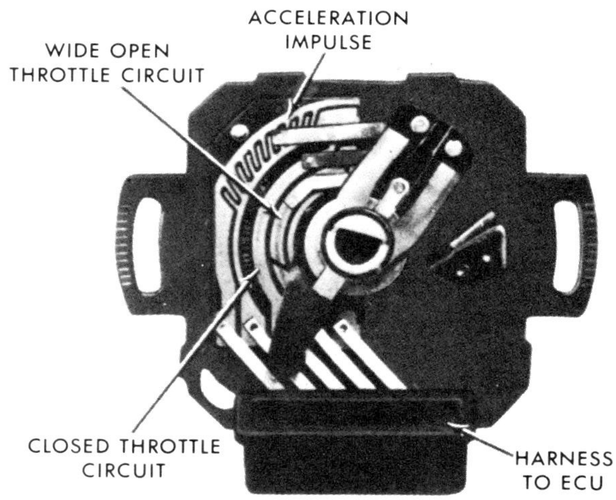

FIGURE 1-19 Throttle position switch *(Courtesy of Cadillac Motor Car Division).*

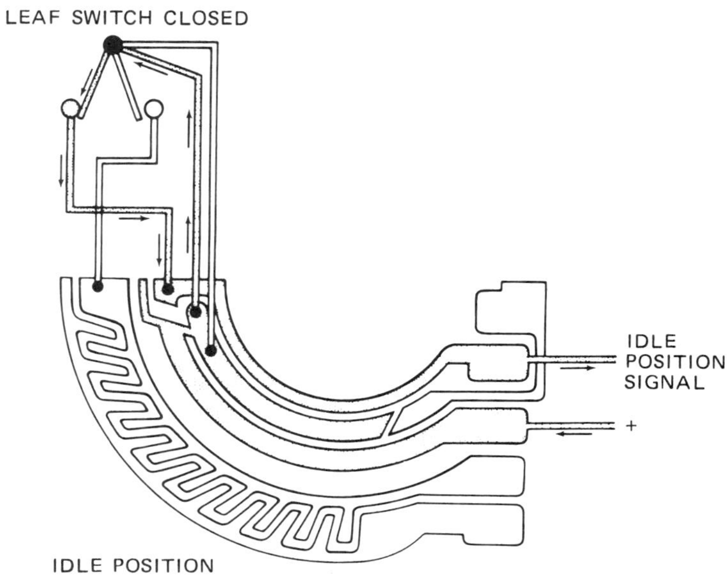

FIGURE 1-20 Throttle position switch-idle position *(Courtesy of Cadillac Motor Car Division).*

One wiper arm provides accelerator enrichment signals and contacts both accelerator runs. A second wiper arm provides an idle signal. The third and fourth wiper arms provide the wide-open throttle signal and travel over the same run. All wiper arms travel in unison. A plastic cover houses the arms and runs. A five-terminal connector links the TPS to the wiring harness and ECU. The center wire of the connector (light green) provides a positive voltage to the TPS from the ECU. The other four wires provide paths through which the TPS can transmit signals to the ECU.

The TPS is adjustable, but must be replaced if it malfunctions as it is not a serviceable part.

Engine Coolant and Manifold Air Temperature Sensors

Two temperature sensors are used, engine coolant and manifold air, which are made up of a coil of high-temperature nickel wire embedded in an epoxy case (Fig. 1-21). This is molded into a brass housing with two wires extending from the housing. The resistance of the coil increases as temperature increases. The ECU measures the voltage drop across the sensor. The manifold air temperature sensor is located in the rear of the intake manifold and is connected to the ECU by means of the engine EFI harness. The engine coolant temperature sensor is located in the left front intake manifold on Seville and in the heater hose fitting at the rear of the right cylinder head on DeVille and Eldorado.

At 40°F, the sensor should have an approximate resistance of 670 ohms, whereas at 260°F the sensor's resistance should rise to about 1625 ohms. Refer to the sensor resistance chart to be accurate (Fig. 1-22). Both sensors

FIGURE 1-21 Temperature sensors *(Courtesy of Cadillac Motor Car Division).*

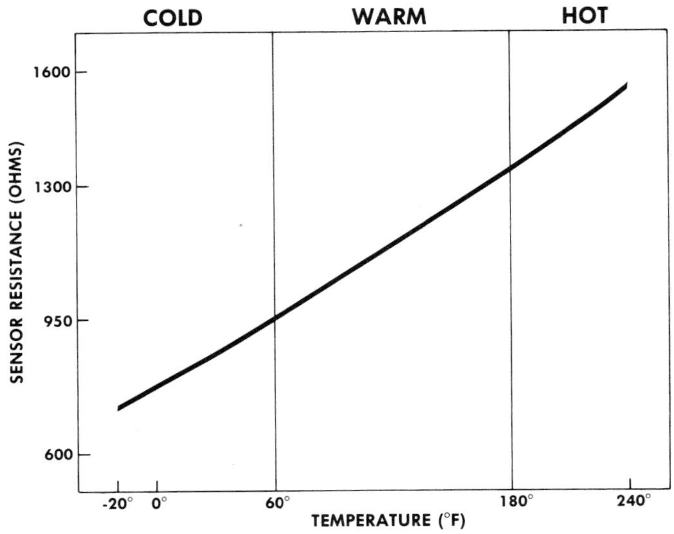

FIGURE 1-22 Sensor resistance characteristics *(Courtesy of Cadillac Motor Car Division).*

are identical and are interchangeable. Low resistance when cold allows higher current flow to the ECU. This results in a wider pulse width or more fuel. As the resistance increases with temperature, the pulse width from the ECU becomes shorter, leaning the fuel mixture. The coolant temperature sensor has the greatest effect on pulse width of the two sensors. It affects the pulse width over a wide range of temperatures; however, when coolant temperature reaches approximately 170°F, its affect on pulse width is greatly diminished. Neither sensor is serviceable or adjustable and must be replaced if out of specification.

Engine Speed Sensor

The engine speed sensor provides an indication of engine rpm. It is built into the housing of the high-energy ignition distributor housing beneath the casting on which the distributor cap is mounted; it contains two reed switches and a rotor with two rotating permanent magnets (Fig. 1-23). The reed switches are enclosed in a plastic casing, are opened and closed by the rotating magnets, and are connected to the ECU by three wires leading to the engine EFI wiring harness. Both permanent magnets are embedded in the rotor, which is fastened to the distributor shaft by a roll pin. The magnets are 180° apart (Fig. 1-24). As the distributor shaft spins, the magnets will close and open both reed switches once for each shaft revolution. Since the engine crankshaft turns twice for each turn of the distributor shaft, the two reed switches compensate for differences in shaft speeds.

This sensor provides data that enables the ECU to synchronize the activation of each injector group with intake valve timing and also enables the ECU to pulse the

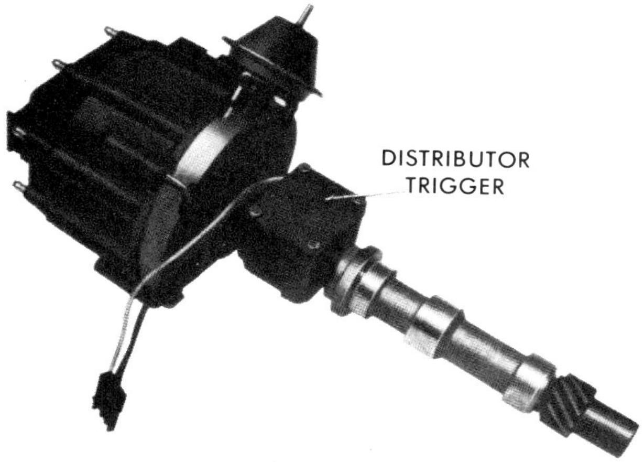

FIGURE 1-23 Engine speed sensor *(Courtesy of Cadillac Motor Car Division).*

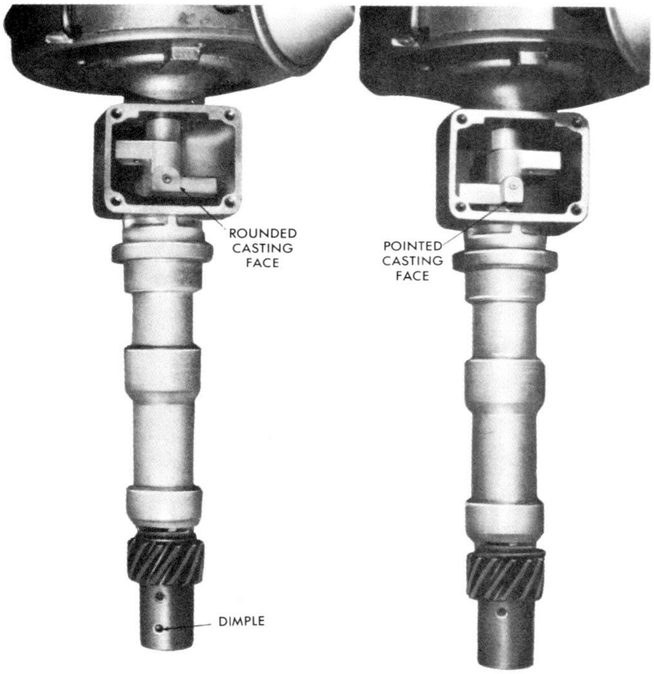

FIGURE 1-24 Speed sensor location *(Courtesy of Cadillac Motor Car Division).*

injectors in conjunction with engine rpm. When engine speed is below 900 rpm, the sensor enables the ECU to make a fuel metering correction. The basic pulse width is reduced 10% for better idle quality and lower emissions.

Finally, the speed sensor directs the ECU to shut off both fuel pumps and the fast idle valve when necessary. When the key switch is turned on, the electric fuel pumps and the fast idle valve heater come on for only 1 second if there is no engine speed signal from the speed sensor. Otherwise, both circuits would remain on, causing a potential hazard due to the pressurization of the fuel system when the engine is not running.

Electronic Control Unit

An electronic control unit (ECU) is used on all Cadillac EFI systems (Fig. 1-25). This preprogrammed computer consists of two circuit boards mounted in a steel casing. It is located beneath the instrument panel and is connected by three connectors (red, blue, and black) to the EFI harness. The DeVille and Eldorado models use a jumper harness between the EFI harness and the ECU because of its location above the glove box, while the Seville models are connected directly to the ECU as its location is below the radio.

It is very important that the correct ECU be used in the corresponding model car. Each is labeled according to its application. Refer to the appropriate parts manual for accuracy and best results. When the ignition switch is turned to ON or START, the ECU receives power from the battery. At this time, the following signals are received from the EFI sensors:

1. Engine coolant temperature
2. Manifold air temperature
3. Manifold absolute pressure
4. Engine speed and firing position (crank only)
5. Throttle position or wide-open throttle

As a result, the following commands are given by the ECU:

1. Electric fuel pump activation
2. Fast idle valve activation
3. Injection valve activation (crank only)
4. EGR solenoid deactivation (at operating temperature and above idle speed only)

The desirable air/fuel ratios for different operating conditions are programmed into the ECU. The ECU calculates injector pulse width based on its program and input

FIGURE 1-25 Electronic control unit *(Courtesy of Cadillac Motor Car Division).*

sensor signal data. It then issues commands in the form of electrical signals to the injectors to open for a specific duration. This process occurs continuously at the speed of electricity or approximately 186,000 miles per second. Changes in command pulse width happen at this speed. It is easy, then, to see why electronic fuel injection is replacing carburetion, whether it is vacuum or electronically controlled.

Summary

To summarize system operation, the EFI system is energized when the ignition switch is turned to the ON position. The following events take place:

1. The ECU receives battery voltage.
2. The electric fuel pumps are activated for 1 second unless the engine is cranking or running.
3. All engine sensors are activated and are sending signal data to the ECU.
4. The fast idle valve heater is activated for 1 second unless the engine is cranking or running. Then it will continue to operate.
5. The exhaust gas recirculation valve solenoid is activated and prevents a vacuum signal to the EGR valve until the engine coolant temperature reaches 130°F.

When the engine is running, the following activities take place:

1. The electric fuel pumps run continuously.
2. The fast idle valve heater is operating continuously.
3. The air bypass (curb idle adjusting screw) in the throttle body controls airflow into the intake manifold when the throttle plates are closed. For off-idle operation, the throttle plates control airflow.
4. The fuel pressure regulator on the fuel rail maintains fuel pressure at about 39 psi and returns excess fuel to the fuel tank.
5. Signals from the engine coolant and manifold air temperature sensors, the manifold absolute pressure sensor, the engine speed sensor, and the throttle position switch are continuously sent to the ECU.
6. The ECU sends out electrical pulses to each group of injectors that precisely controls pulse width for delivery of fuel to the engine.

SYSTEM SERVICE

Fuel Pressure Relief

NOTE: Any service work to be performed on the fuel delivery system should not be done until fuel pressure is relieved. The following procedure should be used to safely relieve fuel pressure.

1. The fuel rail is equipped with a Schrader-type fitting at the right or left front corner.
2. Remove the protective cap from the fitting and carefully install a valve depressor on the fitting.
3. Some fuel may spray out. Keep a rag under the fitting to absorb the gasoline and use a special pressure-relieving tool.
4. Tighten the valve depressor slowly until fuel pressure is relieved.
5. Dispose of the gasoline correctly.
6. Remove the valve depressor and proceed with system service.

NOTE: Pressure may build again slightly due to heat. When loosening fittings, use caution.

In-tank Electric Fuel Pump Service

The in-tank electric-boost fuel pump can be tested by installing a pressure gauge in its outlet line and turning the ignition switch to the ON position, OFF position, ON position two to three times. This energizes the pump long enough to build up pressure. Pressure should be between 7 and 10 psi. To remove the pump, the fuel tank must be drained and removed, and the fuel pump/sending unit assembly can be removed from the tank. The assembly is replaced as such and cannot be taken apart. When reassembling, be sure to connect wires correctly.

NOTE: A ''dry'' fuel system may require substantial cranking before starting.

Fuel Filter Service

The in-line fuel filter element is replaced by unscrewing the steel can and removing the element and large O-ring seal. Do not remove a ''plastic bubble'' in the bottom of the can if so equipped. This is a sound absorber and prevents a thump when the fuel system is cycled on and off. The bottom cover or can should be hand tightened only.

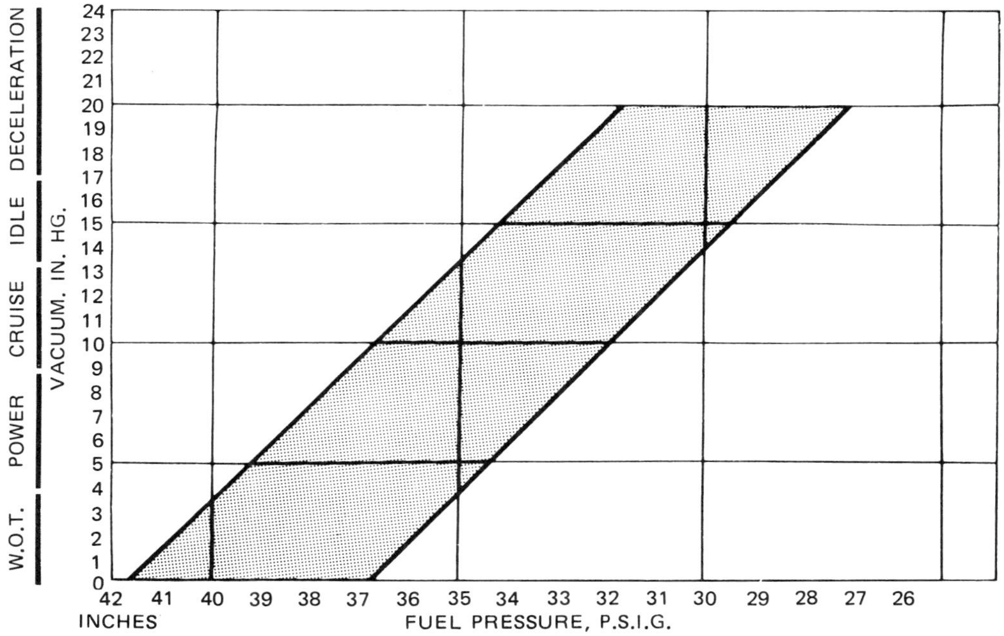

FIGURE 1-26 Normal fuel pressure *(Courtesy of Cadillac Motor Car Division).*

Chassis Electric Fuel Pump Service

Servicing the chassis-mounted electric fuel pump involves complete replacement only, as the unit is nonserviceable. The pump can be pressure tested at the fuel rail Schrader fitting with a fuel injection pressure gauge by connecting the gauge and cycling the ignition switch on and off to allow pressure to build. Refer to the fuel pressure chart for specifications (Fig. 1-26).

Fuel Injector Service

The fuel rails must be removed to service the injectors. They can be removed by disconnecting the rail inlet fitting at the left rear corner and by disconnecting the return line from the pressure regulator at the right front corner. Brackets that secure the rail and injectors to the intake manifold must be removed. After disconnecting wiring harness leads, vacuum hoses, throttle linkage, and any other parts necessary, the rail and injectors can be removed as an assembly. On some models the two parts of the rail can be separated to make removal easier. Do not bend or kink the rail as the pressure will be affected and fuel flow restricted. A conical copper gasket should be used whenever servicing the fuel rail fittings to ensure tight, leakproof connections. The fuel rail brackets should be torqued when replacing to minimize vacuum, or fuel leaks, and distortion.

Fuel Injector Cleaning

Fuel injector service should be done very carefully. Once the fuel rails have been removed, the injectors can be pulled from the rails. (Some injectors may have stuck in the intake manifold when the rail was lifted from the engine.) New O-rings should be installed at both ends of the injectors when reassembling, but first clean the injector tips and inspect them critically. The tip may appear as shown in Figs. 1-27 and 1-28. If the tips have a depositlike buildup on them, they should be soaked and cleaned as follows:

1. Remove the plastic cap carefully from the injector tip. Do not crack or distort it as this will necessitate replacement of the injector.
2. Soak only the tips in metal conditioner, such as Du Pont 5717-S or equivalent, mixed according to the instructions.
3. Carefully put only the tips in the conditioner so the injector coils will not be damaged.
4. Soak the injector tips for 30 minutes.
5. Remove the injectors and rinse the tips with alcohol and dry.
6. Visually inspect the tips. If still contaminated, soak them in conditioner an additional 15 minutes. **NOTE:** Under no circumstances should total soak time exceed 1 hour.

FIGURE 1-27 Contaminated injectors *(Courtesy of Cadillac Motor Car Division).*

FIGURE 1-28 Damaged injectors *(Courtesy of Cadillac Motor Car Division).*

7. Reinstall the plastic cap after rinsing and drying and install new O-rings.

Fuel Pressure Regulator Service

The fuel pressure regulator can be tested by connecting an EFI pressure gauge to the Schrader fitting on the fuel rail. Cycle the ignition switch and listen for a characteristic whine of the chassis-mounted fuel pump. Fuel pressure should develop and should hold steady (not drop more than 5 psi in 2 minutes). Less drop indicates a tight system. If the drop is more, the pressure regulator is a possible cause. If no external leaks are visible, the leak is occurring either at the pressure regulator, the injectors, or the chassis-mounted pump.

Pinch the fuel return hose coming out of the pressure regulator and cycle the ignition switch. If pressure holds, the pressure regulator is at fault and should be replaced. After releasing the pinched hose, check pressure again. If pressure is now normal, it is possible that foreign particles were preventing the pressure regulator from seating and have been dislodged. If rail pressure drops when the hose is pinched, the internal leak is either in the injectors or the chassis-mounted pump.

The fuel pressure regulator calibration can also be checked by connecting an EFI fuel pressure gauge to the fuel rail Schrader fitting, starting the engine, connecting an external vacuum source to the pressure regulator vacuum tap, and checking the pressure at specified vacuum amounts in the fuel pressure chart (Fig. 1-25). Removing and replacing the fuel pressure regulator is a matter of disconnecting the vacuum hose and fuel lines and removing the metric thread nut that fastens it to the fuel rail bracket. Be sure to use a conical copper gasket in the fuel rail fitting when reassembling.

Throttle Body Service

The Cadillac EFI throttle body should be cleaned periodically with a spray carburetor-type solvent. The cleanliness of the throttle plate area is very important as dirt accumulation affects airflow and vacuum signals. The throttle body can be removed, disassembled, and cleaned similarly to a carburetor. All passages should be free of accumulation. There are two rectangular ports in the base of the throttle body that lead from a passage cast into the intake manifold into the throttle bore beneath the throttle plates (Fig. 1-13). These are the EGR ports. They should be clean and open, and the passage in the manifold should also be unrestricted. If these passages are clogged, the result may be hesitation or stumble as the ECU is programmed with the EGR signal operational.

Fast Idle Valve Service

Cleaning the fast idle valve is easy. A clean rag and spray carburetor-type solvent are all that are necessary. The fast idle valve is removed by disconnecting the wiring harness connector, pushing down on the valve, and turning it counterclockwise at the same time to unlatch it from the throttle body bore (Fig. 1-29). The valve and seat are the most

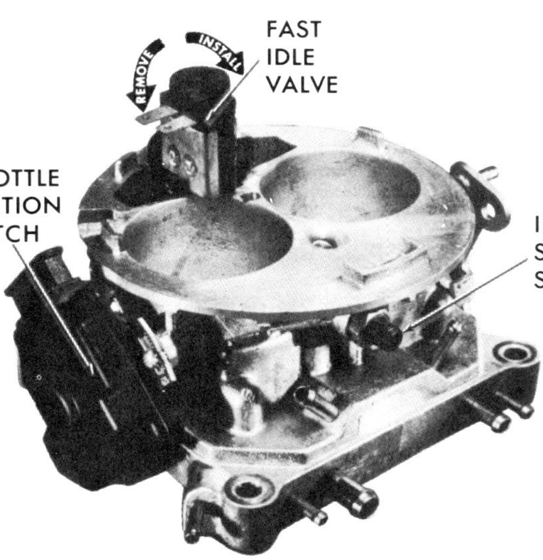

FIGURE 1-29 Throttle body *(Courtesy of Cadillac Motor Car Division).*

important parts to clean. A retainer clip is used on the valve and may be removed and discarded if it causes the valve to become *latched* or stick in the seat (Fig. 1-30). In cold weather, the fast idle speed may be excessive. The speed can be reduced by installing a shim between the electric heater plunger and the valve. Too thick a shim will cause the engine to run excessively rich and load up during warmup. When reinstalling the fast idle valve seat, be sure it is not upside down. The large-diameter shoulder has a square and a rounded corner. The square corner should go down against the bottom of the throttle body bore.

MAP Sensor Service

Inspection of the manifold absolute pressure sensor vacuum line is the only service that can be done as the sensor itself cannot be tested. Connect a hand vacuum pump to the MAP sensor line at the throttle body and develop about 15 inches of Hg of vacuum. Check for leak down. A damaged line can be repaired by splicing in a section of equal diameter (ID), hose that conforms to GM Specification 6107M.

Throttle Position Switch Adjustment

Throttle position switch adjustment is very important to eliminate driveability complaints. To check its adjustment, cycle the ignition switch to ON and slowly open the throttle plates. As the plates open, a clicking sound will be heard. There should be 21 clicks from closed to wide-open throttle. Just as wide-open throttle is reached, a click will be heard that should not be included with the count. If less than 21 clicks are heard, the TPS mounting screws should be loosened and the TPS should be rotated. Check for 21 clicks again and repeat these steps as necessary. When 21 clicks are heard, check to ensure the throttle plates are going fully closed or against the throttle stop screw (factory-preset screw). Now take a 0.020-inch feeler gauge and open the throttle until the first click is heard. The 0.020-inch feeler gauge should not fit between the factory-preset throttle stop screw and the throttle lever. If it does, readjust the TPS slightly clockwise. Recheck all steps. If the switch cannot be adjusted, it should be replaced.

Coolant and Air Temperature Sensor Service

The engine coolant temperature sensor and the manifold air temperature sensor can be tested in the same way. Disconnect the sensor from the EFI wiring harness and probe the sensor leads with an ohmmeter. Its resistance is dependent on the ambient (coolant or air) temperature and should fall within the specifications on the sensor resistance chart (Fig. 1-22).

Engine Speed Sensor Service

Testing the engine speed sensor requires an ohmmeter and disconnecting of the sensor from the EFI wiring harness. With the ohmmeter leads connected first between the black and white sensor terminals, crank the engine. The ohmmeter needle should pulse slightly. Connect the ohmmeter to the red and white terminals and crank the engine. The same results should be observed. This indicates a correctly operating speed sensor. If these results are not obtained, replace the sensor.

ECU Service

To test the electronic control unit, a special tester, Kent Moore J-25400, and the appropriate adapter harness are required. Any electrical testing of sensors can and sometimes should also be done at the ECU connectors inside the vehicle. Use of the wiring diagrams is essential. This will test the wiring harness and any connectors in the circuit.

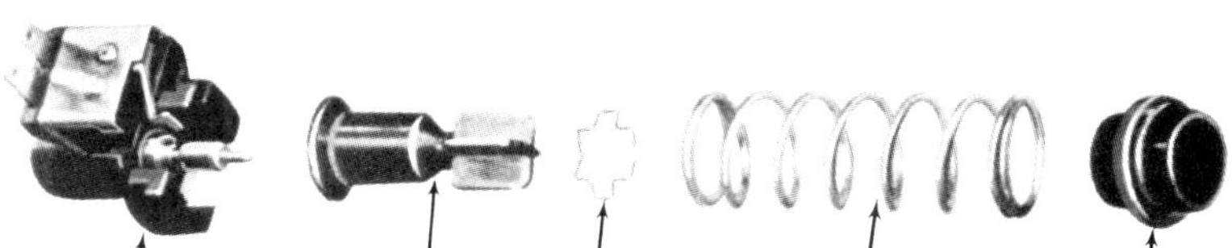

FIGURE 1-30 Fast idle valve *(Courtesy of Cadillac Motor Car Division).*

SYSTEM DIAGNOSIS

Decision trees are used to help the technician troubleshoot the Cadillac EFI system. Each chart is accompanied by a detailed description of what should be done, step by step. Each chart is for a different driveability symptom.

A diagnostic card is also available as a troubleshooting aid (Fig. 1-31). It lists the driveability symptoms, possible causes, and which of those causes to check in numerical order. It also lists repair manual reference pages for more information.

Some symptom definitions that apply to this system are as follows:

1. Stall after start (cold): engine is cold and, when started, stalls either immediately or as a driveaway is attempted.
2. Stall after start (hot): engine stalls immediately or as a driveaway is attempted with a hot to very hot engine, right after starting.
3. No start: engine will not start at all.
4. Hard start (cold): an unusually long cranking period, which eventually leads to startup. Cranking speed is normal.
5. Hard start (hot): unusually long cranking period, which eventually leads to startup. Cranking speed is normal.
6. Stalls when driving, immediate restart (intermittent): occasionally stalls during driving but immediately restarts.
7. Stalls while driving, no immediate restart (intermittent): occasionally stalls during driving but does not immediately restart.
8. Tip-in stumble: a delay in engine response when accelerating from idle with vehicle at a standstill.
9. Surge (lean operation): engine power variation under steady throttle or cruise. Vehicle feels like it is speeding up and slowing down with no change of acceleration pedal position. Can happen at any speed.
10. Rich operation (black smoke from tailpipe): excessive fuel causing engine to load up and emit black smoke from the tailpipe.
11. Hesitation: a delay in engine response when accelerating from steady throttle or cruise.
12. Rough idle: uneven engine running at idle and, if severe enough, may make vehicle shake.
13. Idle speed change: variations in idle speed under different operating conditions such as temperature or "hunting" of engine as idle speed rises and lowers.
14. Misfire on light acceleration: an engine miss or series of light hesitations that occurs under light acceleration. Moderate acceleration clears up this symptom.
15. No fast idle: lack of engine fast idle when starting cold; engine tends to load up.
16. Poor cold driveability: driving operation during engine warmup lacks response. Fuel economy is poor.
17. Poor hot driveability (sag): engine initially responds, then flattens out or slows down, then recovers. Severe cases may cause stall.
18. Cuts out at wide-open throttle—temporary complete loss of power at wide-open throttle. Engine quits until throttle is returned.
19. Cuts out while driving (no stall): engine quits at sharp, irregular intervals. Condition may occur repeatedly or intermittently. It is usually worse under heavy acceleration.
20. Sluggish performance: engine delivers limited power under load or at higher speeds. Will not accelerate as fast as normal, loses too much speed going uphill, or has less top speed than normal.
21. Poor fuel economy: fuel economy is significantly less than the average for the vehicle and drive-train combination.
22. Misfire: rough engine operation due to lack of combustion in one or more cylinders.

ABBREVIATIONS

The abbreviations and meanings given at the end of each chapter may help when reading this text or the manufacturers' shop manuals:

AE Acceleration enrichment
ATS, MAT Air temperature sensor, manifold air temperature (sensor)
BPEGR Back-pressure EGR
CTS Coolant temperature sensor
ECU Electronic control unit
EEC Evaporative emission control
EFI Electronic fuel injection
EGR Exhaust gas recirculation
ESS Electronic spark selection
FIV Fast idle valve
HEI High-energy ignition
MAP Manifold absolute pressure (sensor)
OCT Off closed throttle
rpm Revolutions per minute, engine speed, (sensor or distributor trigger)
TPS Throttle position switch, or acceleration enrichment switch
WOT Wide-open throttle

CAUSES ↓ / COMPLAINTS →	Poor Fuel Economy	Sluggish Performance	Intermittent Cut-Out No Stall	Poor Driveability Hot or Sag	Poor Driveability Cold	No Fast Idle	Cuts Out at WOT	Misfire on Light Acceleration	Idle Speed Change	Rough Idle	Hesitation	Rich Operation	Surge - Lean Operation	Tip-in Stumble	Intermittent Stall - No Restart	Intermittent Stall - Restart	Hard Start Hot	Hard Start Cold	Stall After Start - Hot	Stall After Start - Cold	No Start
Electrical Connections			1					1		3		1	2	4		2					1
Cranking Signal Wire Open							1														7
TPS WOT Circuit																					8
TPS Adjustment and Operation			2						3		3			2							
ECU	9	6	4	11			2	10		15	4	6	6	14	5	3	8	3	7	8	9
Speed Sensor Malfunction			3												2	1	5				6
Speed Sensor Magnet And Sensor Orientation														11							
Coolant Sensor - High Resistance Cold					1													1		1	5
Coolant Sensor - Low Resistance Hot																	4				
Air Temperature Sensor					3															7	
FIV Inoperative						2															
FIV Latched					5	1														3	
Intank Pump															1		6		4		
Fuel Pressure Low Or Drops Off		3		6	2			7		10			4				2	2	3	2	2
Fuel Pressure High	5									11		3									
Fuel Pressure Low At Speed		7																			
Restricted Filter		5											1								
Damaged Fuel Lines		4		7																	
Evaporative Canister Excessive Purge									2	2							3		2		
Low Idle Speed										1				1			1		1	4	
High Idle Speed	3																				
MAP Hose Leak Or Obstruction	4			1						8	1	2		3					5	6	
Dirty or Sticking Injectors				10				9		14		5	5						6		
Injector Tip Damage								8		13											
Sticking Throttle Linkage														12							
Vacuum Leaks at Throttle Body	6							6		5		4		13							
Throttle Body Bore Contamination									1												
Restricted EGR Port In Throttle Body				5				11						10							
Restricted EGR Passage	8			4				12						9							
Wrong EGR Orifice	7			2				13						8							
EGR Solenoid Inoperative					4			14													
EGR Transducer				3				15													
EGR Operating At Idle										9					3						
No EGR								16					3	7							
EGR Delay								17						6							
HEI										12					4		7				3
Spark Plugs Fouled								3		6								4			4
Spark Plugs Cracked								5		7								5			
Initial Timing	1	1		8						4				5						5	
Ignition Advance	2	2		9				2			2										
Plug Wires - Crossfire								4													

FIGURE 1-31 Diagnosis chart *(Courtesy of Cadillac Motor Car Division).*

UNIT I
General Motors Corporation Systems

2 Throttle Body Fuel Injection and Crossfire Fuel Injection, 1982–1985

OVERVIEW

The throttle body fuel injection system (TBI, and now referred to as EFI) was designed as an improvement over the carburetor and the electronically controlled carburetor. Exhaust emissions and fuel distribution can be precisely controlled to attain federally legislated standards. This is accomplished, in part, by using an electronic control module (ECM), which receives input signals from various sensors and sends out command signals to controlled devices. An air/fuel ratio of 14.7 to 1 is desired, and the ECM is programmed to try to provide this ratio as nearly as possible under almost all engine operating conditions.

The ECM monitors engine, and sometimes other, sensors, which send signals to the ECM's microprocessor. Then, in accordance with its program, the ECM can make adjustments in the amount of fuel injected by the fuel injector. This adjustment is accomplished by changing the electrical signal sent to the injector, which is called the *injector pulse width.*

The pulse width represents the amount of time the injector is energized or ON. The pulse width is varied by the ECM in response to changes in engine operation and driving conditions. For example, pulse width changes with conditions such as cranking, starting, warmup, acceleration, deceleration, and altitude. In addition, the pulse width will be varied to attain, as closely as possible, a 14.7 to 1 air/fuel ratio for lowest emissions and optimum fuel economy. When the pulse width is decreased, the mixture is leaned. The ECM sends pulses to the injector in two modes of operation, synchronized and nonsynchronized.

In the synchronized mode, the injector is pulsed once for each distributor reference pulse from the electronic spark timing module. The nonsynchronized mode pulses the injector either every 12.5 milliseconds or every 6.25 milliseconds, depending on the calibration of the program in the ECM. These pulses are not dependent on the distributor reference pulses. Nonsynchronized-mode pulses are issued by the ECM to the injector during acceleration, deceleration, and cranking just before the engine starts. These pulses are also issued by the ECM if the desired pulse width is too small to operate the injector accurately, or approximately 1.5 milliseconds.

When two fuel injectors are used, as in the crossfire fuel injection system, the injectors are pulsed alternately when in the synchronized mode. They are pulsed simultaneously when in the nonsynchronized mode.

The ECM relies on sensors for electrical input measurements of engine operating conditions. An oxygen (O_2) sensor in the exhaust stream provides feedback information to the ECM to reflect results of its efforts to enrich or lean the air/fuel mixture. This feature, in combination with a catalytic converter, achieves the lowest emissions and optimum fuel economy. A detailed description of the sensors and fuel injection system follows.

CROSSFIRE FUEL INJECTION

The crossfire fuel injection (CFI) system uses two single-injector throttle body assemblies mounted on a common intake manifold (Fig. 2-1). These units are mounted diagonally, front and rear. Two injectors and two air induction bores provide improved distribution in a new-design intake manifold. Each throttle body unit provides an air/fuel mixture through long tuned runners in the intake manifold to cylinders on the opposite side, giving the system its name, crossfire. A swirl plate is located beneath the throttle plate to aid in mixture distribution. The throttle body assemblies are model 400 units, which are different from other TBI models.

The two throttle bodies are connected by a common throttle rod and bearing assembly and a common fuel tube. Each has an idle air control motor but only the rear unit has a throttle position sensor. The throttle rod and bearing assembly has a synchronizing screw to maintain correct throttle lever relationship on both units.

The fuel pump provides fuel, under pressure, to the front throttle body. Its accumulator provides fuel, under pressure, to the rear throttle body through the common fuel tube. Excess fuel is returned through the rear pressure regulator to the fuel tank.

Servicing these throttle bodies is the same as for other TBI units except for the synchronizing screw adjustment. This procedure is provided in the service manual. The operation of this system is very similar to the throttle body fuel injection system.

SYSTEM DESCRIPTION

Several subsystems make up the throttle body fuel injection system. The fuel supply system is concerned with the supply, delivery, and return of gasoline. The throttle body assembly is concerned with pressure regulation, air control, and fuel injection. The idle air control functions to control idle speed and to prevent stalling. The electronic control module contains program information to sense vehicle operation and to control the fuel injection system in response to that operation within the guidelines provided by its program. Data sensors transmit information to the ECM to aid its operation. The electronic spark timing system provides ignition system operation that is coordinated with engine and fuel system operation. The emission control system reduces emission levels to within federally mandated standards. Each subsystem will be described in detail.

FUEL SUPPLY SYSTEM

The fuel supply system consists of the fuel tank and cap, an in-tank electric fuel pump, an in-line fuel filter, and a throttle body. A fuel line connects these components together, and a fuel return line runs from the throttle body back to the tank.

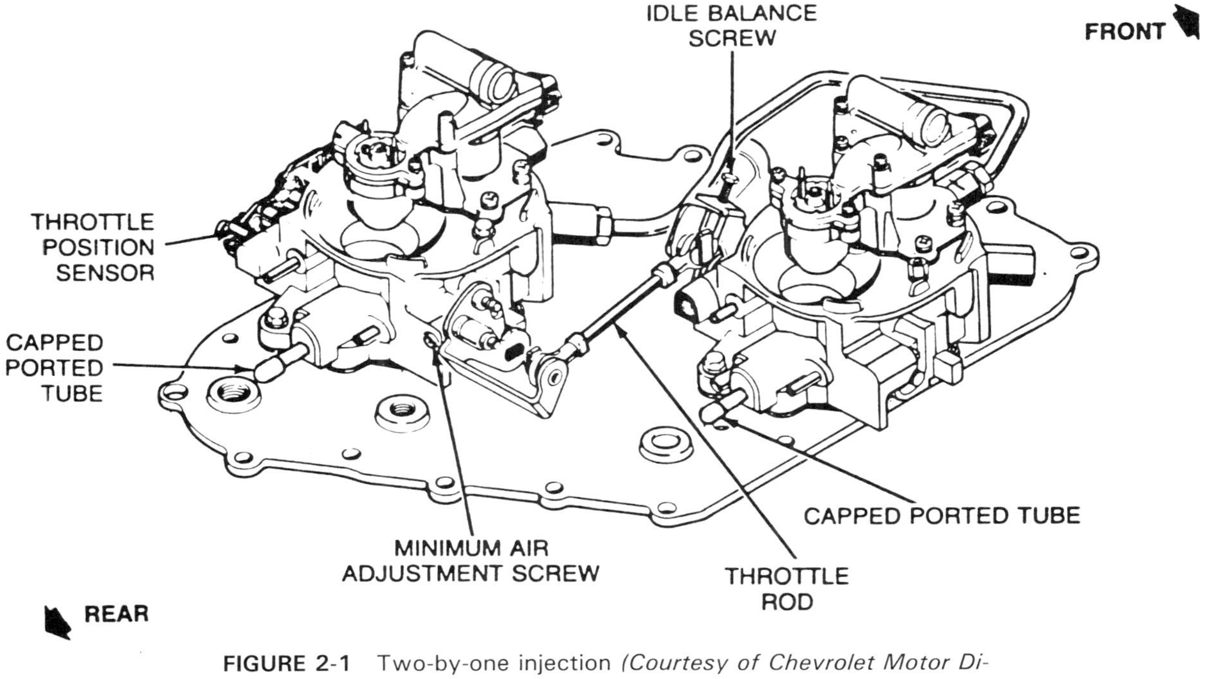

FIGURE 2-1 Two-by-one injection *(Courtesy of Chevrolet Motor Division).*

Fuel Tank

The fuel tank contains a sump area into which the fuel pump/sending unit assembly is installed. The sump contains fuel and prevents it from "running away" during low-fuel, severe maneuvering conditions. The fuel tank cap is an emission control, pressure vacuum type of cap. It prevents excessive pressure and vacuum buildup in the tank and otherwise seals the tank from the atmosphere. The fuel pump is attached to the fuel gauge sending unit and float arm assembly, but is serviceable separate from it (Fig. 2-2). The pump is sealed and is not rebuildable.

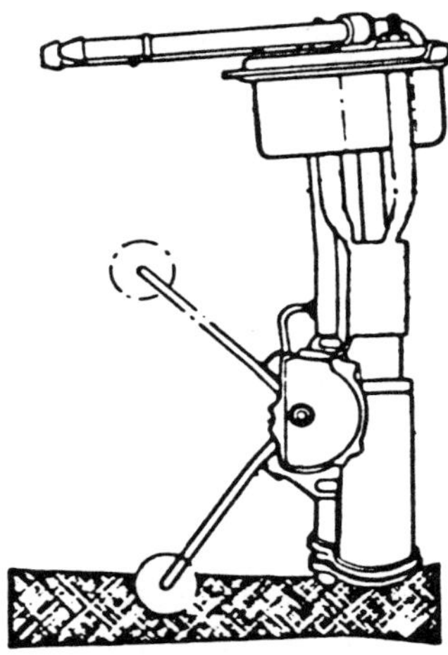

FIGURE 2-2 In-tank fuel pump (typical) *(Courtesy of Pontiac Motor Division).*

Fuel Pump

Fuel pump operation is controlled by a fuel pump relay located in the engine compartment on some engines. When the ignition switch is turned to ON, the ECM energizes the relay and the fuel pump will run. If the ECM does not receive distributor reference pulses to indicate engine cranking, it deactivates the relay within 1.5 to 2.0 seconds. This not only primes the fuel injector but also prevents the fuel pump from running when not needed. This also serves as a safety measure in accidents to prevent the pump from emptying the fuel tank, thereby reducing chances of fires. When distributor reference pulses are received by the ECM, it reactivates the relay and the fuel pump.

Some models use an oil pressure sending unit as part of the control circuit for the electric fuel pump. Initial activation of the pump is the same as described previously. However, if the fuel pump relay is defective or malfunctioning, the engine will not start. To avoid this condition, the oil pressure sending unit can complete the fuel pump circuit when oil pressure reaches about 4 psi. This sending unit has two internal circuits; one for the oil pressure telltale light or the gauge in the instrument panel and one for the fuel pump circuit.

Fuel Filter

The fuel filter is located under the hood or on the frame rail near the fuel tank in the fuel supply line to the throttle body (Fig. 2-3). A rubber fuel line hose connects to one end and a screw coupler fastens the filter to the steel fuel line leading to the throttle body. A rubber O-ring is used to seal the screw coupler. Another filter is located on the end of the fuel pickup tube in the fuel tank. It is part of the fuel pump/sending unit assembly, but can be replaced separately if necessary. This filter is woven plastic and prevents dirt and water from entering the pump. However, it passes water if it is completely submerged in water.

FIGURE 2-3 Fuel filter, 2.5 liter *(Courtesy of Pontiac Motor Division).*

Fuel Lines

The fuel lines are either steel double-flared lines or a specially manufactured fluoroelastomer rubber hose. Do not replace them with any other material. The rubber fuel hoses should not be closer than 4 inches to the exhaust system nor closer than 10 inches to the catalytic converter.

THROTTLE BODY ASSEMBLY

The throttle body assembly replaces the carburetor and consists of two main castings (Fig. 2-4). The throttle body casting contains a throttle plate for regulating airflow; ports or vacuum taps are located above or below the throttle plate for operating the exhaust gas recirculation (EGR) valve, the manifold absolute pressure sensor (MAP), and the canister purge, which is part of the evaporative emission control system (EEC).

1 FUEL METER ASSEMBLY

2 GASKET – FUEL METER BODY

3 SCREW & WASHER ASSY– ATTACH. (3)

4 FUEL INJECTOR KIT

5 FILTER – FUEL INJECTOR NOZZLE

6 SEAL – SMALL "O" RING

7 SEAL – LARGE "O" RING

8 BACK-UP WASHER – FUEL INJECTOR

9 GASKET – FUEL METER COVER

10 DUST SEAL – PRESS, REGULATOR

11 GASKET – FUEL METER OUTLET

12 SCREW & WASHER ASSY – LONG (3)

13 SCREW & WASHER ASSY – SHORT (2)

14 NUT – FUEL INLET

15 GASKET – FUEL INLET NUT

16 NUT – FUEL OUTLET

17 GASKET – FUEL OUTLET NUT

18 FUEL METER BODY ASSEMBLY

19 THROTTLE BODY ASSEMBLY

20 SCREW – IDLE STOP

21 SPRING – IDLE STOP SCREW

22 LEVER – TPS

23 SCREW – TPS LEVER ATTACHING

24 SENSOR – THROTTLE POSITION KIT

25 SCREW – TPS ATTACHING (2)

26 IDLE AIR CONTROL ASSY-

27 GASKET – CONTROL ASSY-TO T.B.

28 GASKET – FLANGE MOUNTING

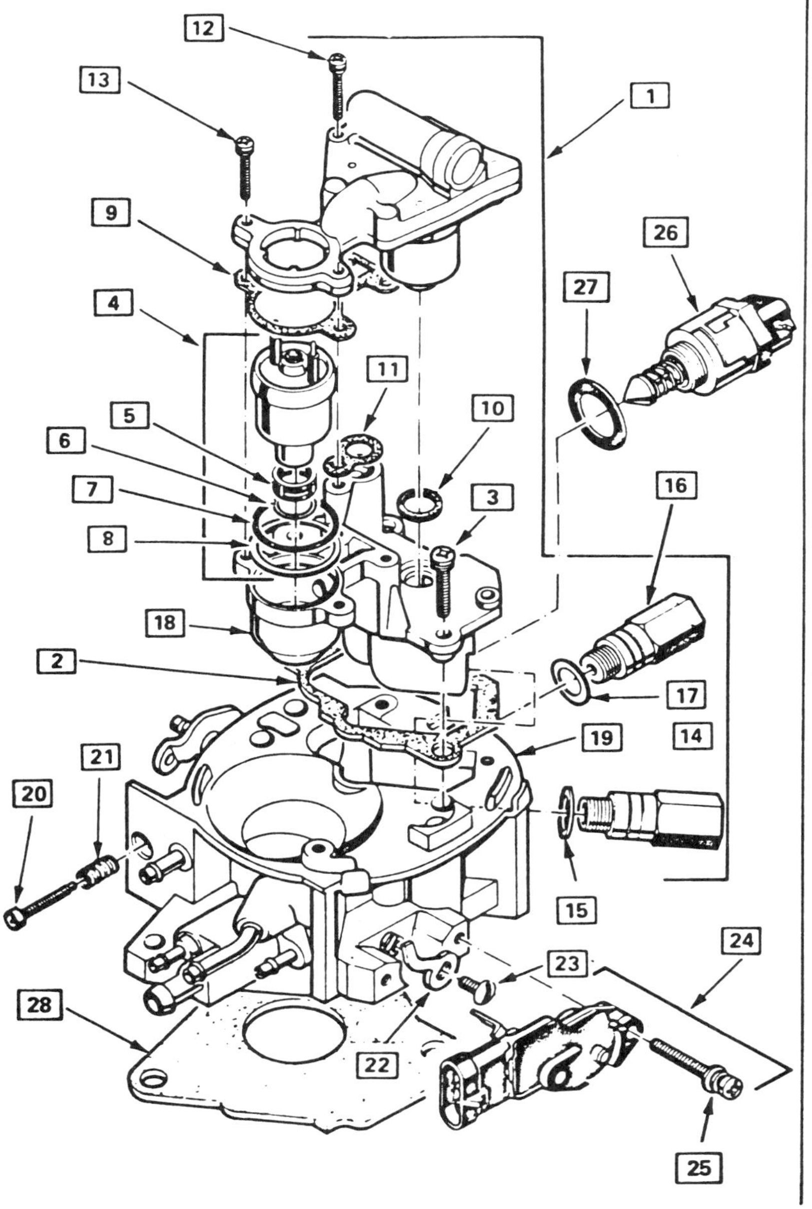

FIGURE 2-4 Model 300 throttle body injection *(Courtesy of Oldsmobile Division).*

Fuel Body

The fuel body casting has an integral pressure regulator and a fuel injector. The fuel injector is solenoid operated and is controlled by the ECM. Supply fuel is provided to the lower end of the injector assembly, where a 10-micron fine-mesh screen filter is located. When the ECM pulses the injector, a normally closed (spring loaded) ball valve is raised, and a conical spray of atomized fuel is directed against the walls of the throttle body bore above the throttle plate. The injector, when energized, has full system voltage applied to its solenoid coil until the current rises to a level of 4 amperes. At this point, current regulation begins and a holding current of 1 ampere is maintained. This design allows a fast opening response during turn-on and a low-current holding condition, which prevents overheating of the solenoid winding. Excess system fuel passes the injector and is routed to the pressure regulator.

Pressure Regulator

The pressure regulator has a diaphragm with injector pressure on one side and air cleaner pressure on the other side (Fig. 2-5). The diaphragm movement acts as a relief valve to maintain a constant pressure to the injector regardless of the operating load or speed range of the engine. Relieved fuel is returned to the fuel tank through the fuel return line. The end of the fuel return line inside the tank empties into the sump area. Injector pressure is maintained at between 10 and 12 psi by the regulator.

At this pressure, when ambient temperatures rise to about 100°F, vapor bubbles form in the fuel. If vapor bubbles, instead of plain fuel, are delivered to the injector, fuel metering is affected and the mixture is leaned. Since the system is exposed to this level of temperature, a way of separating vapor bubbles was designed into the fuel-metering body.

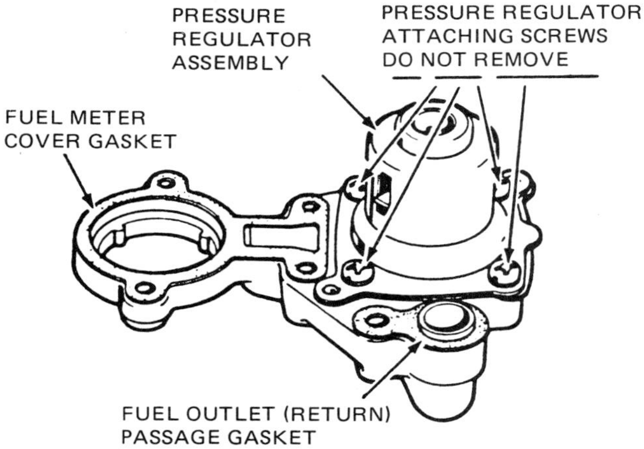

FIGURE 2-5 Fuel meter cover, bottom view *Courtesy of Pontiac Motor Division).*

The passage leading to the metering tip of the fuel injector presents little resistance to the low-velocity fuel (Fig. 2-6). As a result, the buoyancy of the vapor bubbles causes them to rise in the fuel-filled chamber surrounding the injector. Only fuel will enter the injector to be sprayed. Excess fuel not needed by the injector bypasses the metering tip, enters the pressure regulator, and, at the same time, purges collected vapor bubbles from the system.

Mounted on the throttle body assembly are two electronic devices, one to send information to the ECM and one that is controlled by the ECM.

Throttle Position Sensor

The throttle position sensor (TPS) sends varying signals to the ECM to indicate changes in throttle plate positions. This will be described in detail later.

Idle Air Control Valve

The idle air control (IAC) is controlled by the ECM to regulate or adjust engine idle speed and to prevent stalling. The installation of the IAC will help identify the model of throttle body assembly being serviced. If the IAC slants downward into the throttle body, it is a model 500 TBI unit. The model 500 also has a ''banjo''-type fuel line fitting (Fig. 2-7). If the IAC is horizontally mounted in the throttle body, it is a model 300 TBI unit. The dual TBI units used in the crossfire fuel injection system are model 400.

The idle air control is an electric motor (sometimes referred to as a stepper motor) operated valve that controls the amount of air allowed to bypass the throttle plate (Fig. 2-8). A conical valve can be retracted or extended within an air bypass passage in the throttle body to regulate idle speed. If engine rpm is too low (the ECM monitors engine rpm by the distributor reference pulses), the valve can be retracted to allow more air to bypass the throttle plate. If engine rpm is too high, the IAC can be extended to reduce bypass air. These adjustments do not affect air/fuel mixtures because the oxygen sensor indicates to the ECM richness or leanness, and the ECM compensates by changing the pulse width to the fuel injector.

When engine rpm drops below a specified level and the throttle plate is closed (indicated to the ECM by the throttle position sensor), the ECM determines a new IAC valve position based on barometric pressure (indicated to the ECM by the manifold absolute pressure sensor) to prevent a stalling condition.

There are currently three IAC valve designs in use; the single taper, dual taper, and blunt pintle designs (Fig. 2-9). When replacing an IAC, it is critical that it be replaced with the same design.

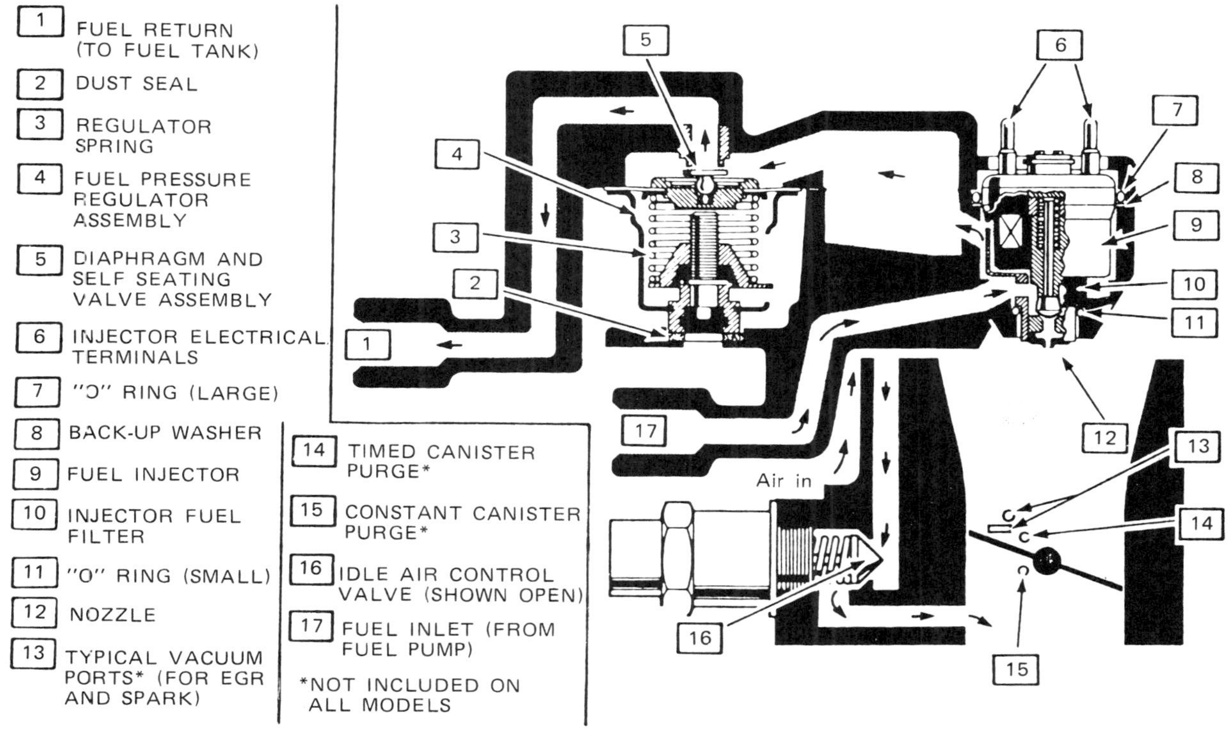

FIGURE 2-6 Fuel metering schematic *(Courtesy of General Motors Product Service Training).*

FIGURE 2-7 Fuel feed and return line attachment *(Courtesy of Oldsmobile Division).*

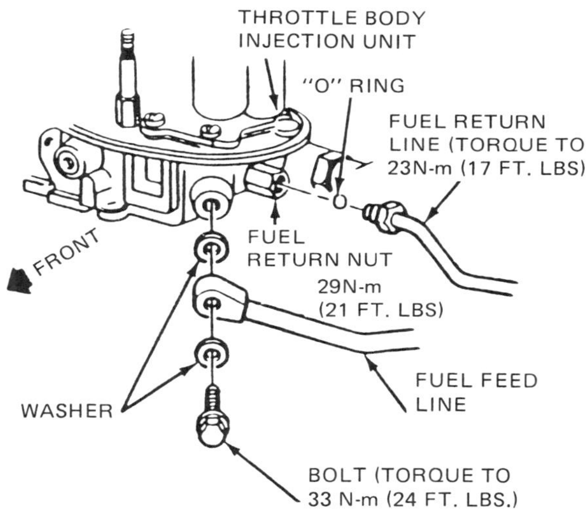

FIGURE 2-8 Idle air control *(Courtesy of General Motors Product Service Training).*

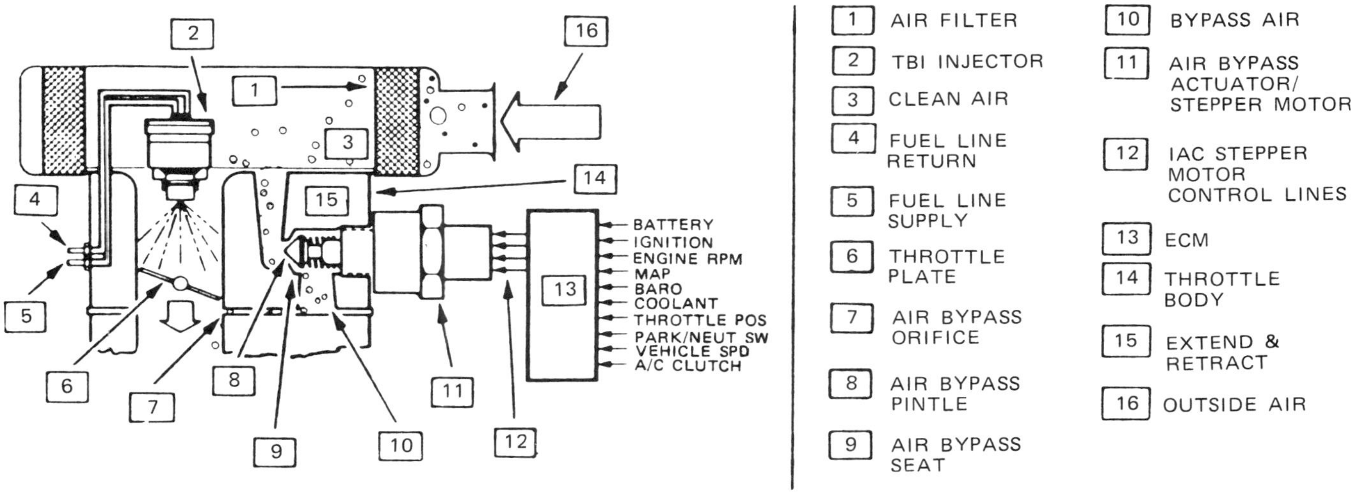

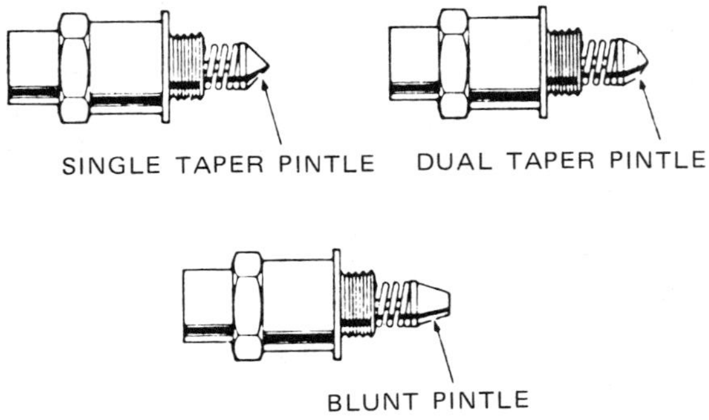

PINTLE DESIGNS

FIGURE 2-9 Pintle designs *(Courtesy of General Motors Product Service Training).*

ELECTRONIC CONTROL MODULE

The electronic control module is located in the passenger compartment (Fig. 2-10). It monitors input signals or parameters from various sensors and/or switches. Parameters sensed include vehicle speed, transmission/transaxle gear indication, time, throttle position, park/neutral mode, manifold absolute pressure, system voltage, exhaust oxygen content, engine detonation, engine cranking mode, engine crankshaft position, engine coolant temperature, barometric pressure, and air-conditioning system engagement. Some of these parameters may not be monitored on some engines.

The ECM continually monitors this input information, makes calculations based on it, and then produces output commands to its controlled actuators. The controlled actuators include the air control valve, air switching valve, canister purge control valve, exhaust gas recirculation control, electronic spark timing, idle air control motor, throttle body injector(s), transmission/transaxle torque converter clutch, air-conditioning compressor clutch, air door control, and engine cooling fan control.

The ECM contains a serviceable program that matches the ECM processes to the vehicle into which it is installed. This program is called a PROM (programmable read only memory) and is a chip. It contains information and specifications that pertain to a specific vehicle. Engine, transmission/transaxle, final drive ratio, and vehicle weight are examples of the kind of information found in the PROM. The combination of an ECM and a PROM is tailored to a specific vehicle and should not be interchanged with that from another vehicle.

The ECM also has a self-diagnostic function in that it can indicate to the driver that a problem exists by lighting a CHECK ENGINE light on the instrument panel. At the same time, a code is stored in a memory that can be entered by a service technician and the code can be revealed. This assists the technician in diagnosing the system and making repairs.

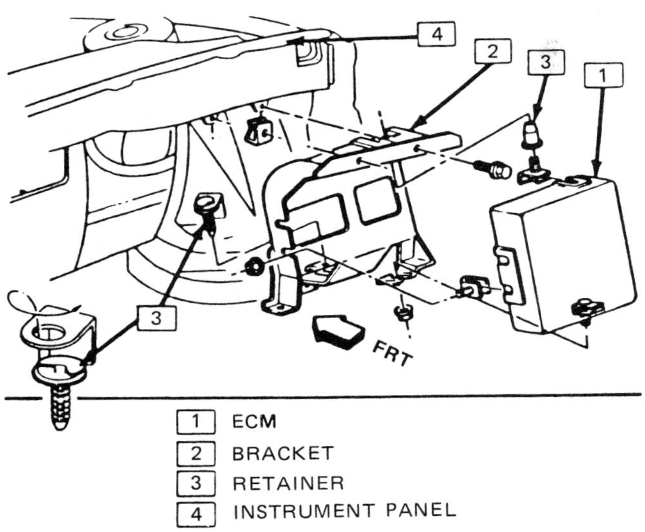

FIGURE 2-10 Electronic control module mounting, J series, with air conditioning *(Courtesy of Oldsmobile Division).*

DATA SENSORS

Data sensors are used to feed information to the ECM so that it can accurately function to control other systems.

Vehicle Speed Sensor

The vehicle speed sensor (VSS) is located behind the instrument panel and is attached to the back of the speedometer head (Fig. 2-11). It consists of a light-emitting diode whose light beam strikes the speedometer cup. This cup is painted black except for one shiny section and is mounted on and spins with the speedometer cable. As the shiny section passes under the infrared light beam, the light beam is reflected back to a photo-transistor. The photo-transistor produces an on-off voltage signal that is amplified by an electronic circuit called a *speed buffer.* The speed or frequency of the reflections and amplified signals from the speed buffer is interpreted by the ECM as vehicle speed. The speed buffer is also located under the instrument panel.

Third/Fourth Gear Switch

Another data sensor, the transmission/transaxle gear indication, is a pressure-sensitive switch that indicates to the ECM when a shift to high gear occurs. In some cases, there are two pressure sensors to indicate third- or fourth-gear engagement (automatic overdrive). The signal is an off–on signal or, in terms of voltage, a high–low signal.

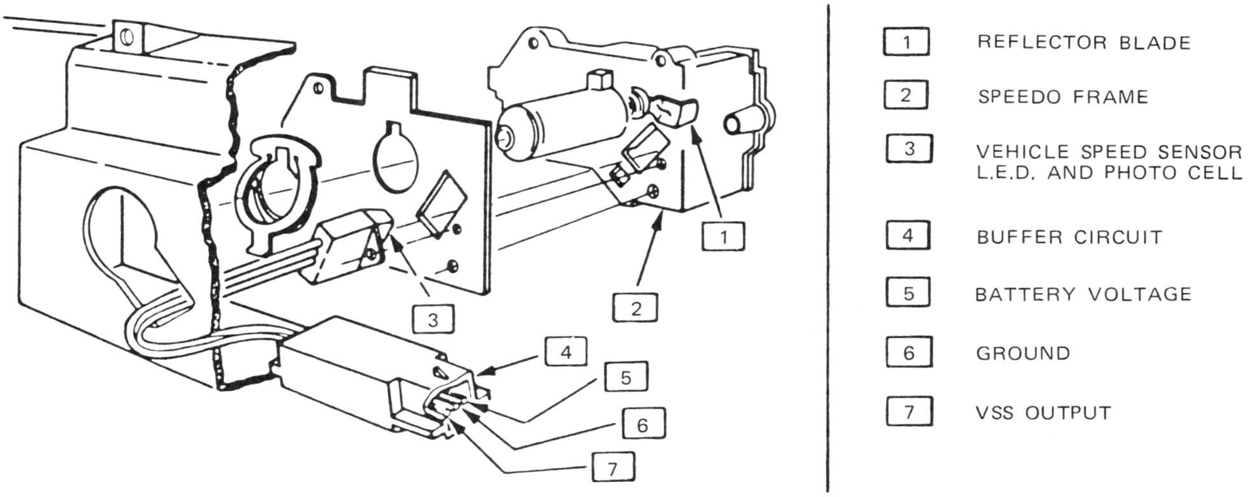

FIGURE 2-11 Vehicle speed sensor (typical) *(Courtesy of Oldsmobile Division).*

Time

Passage of time is measured by the ECM through an internal circuit. This, for example, is important for the ECM to be able to energize the fuel pump relay to prime the injector and then shut the relay off if no distributor reference pulses are received.

Throttle Position Sensor

Throttle position is transmitted to the ECM from a sensor mounted on the side of the throttle body and on the end of the throttle shaft (Fig. 2-12). The throttle position sensor is a variable resistor whose signal to the ECM changes as a result of changes in the throttle plate opening angle. This sensor is adjustable, so it can be calibrated to work with the circuit into which it is installed. It indicates to the ECM the demand of the driver on the accelerator pedal.

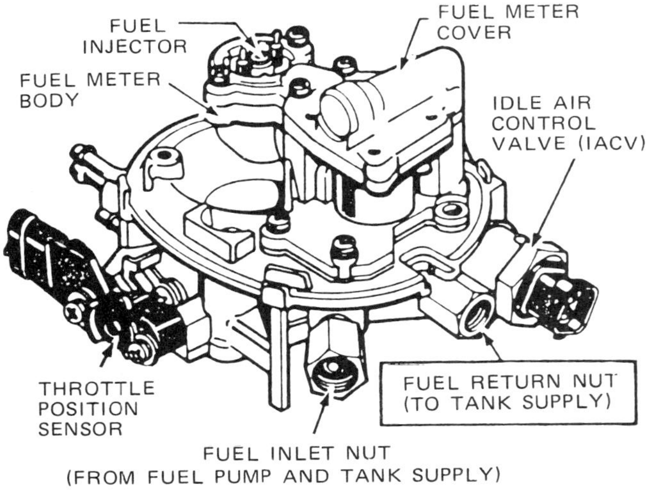

FIGURE 2-12 Fuel injection assembly *(Courtesy of General Motors Product Service Training).*

Park/Neutral Switch

To prevent a stalling condition when the driver shifts from park or neutral into another gear, the ECM needs to anticipate the load of the transmission/transaxle. The ECM monitors the PARK/NEUTRAL switch and senses its movement. In response, the ECM maintains the idle speed through the IAC as the shift from park or neutral occurs.

Manifold Absolute Pressure Sensor

The manifold absolute pressure (MAP) sensor provides a signal to the ECM that represents engine load (Fig. 2-13). The MAP sensor has a pressure/vacuum-sensitive diaphragm, connected to the manifold vacuum, that operates an internal transducer. The transducer interprets changes in vacuum into electrical signals, which are sent to the ECM. The ECM, knowing what load is placed on the engine, can

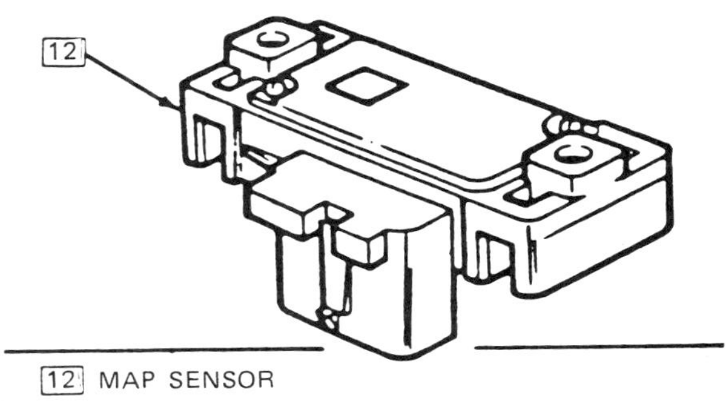

FIGURE 2-13 Manifold absolute pressure sensor *(Courtesy of Chevrolet Motor Division).*

control ignition timing and torque converter clutch application. These two functions will be discussed later in this chapter.

System Voltage

System voltage is monitored because it affects the operation of the overall computer system. When the ECM senses lower voltage, it tries to compensate by raising the idle speed with the IAC. Low system voltage changes all signal values to the ECM and also all output signals, resulting in driveability symptoms.

Oxygen Sensor

Exhaust oxygen concentration is measured by the oxygen (O_2) sensor (Fig. 2-14). It is a closed-end zirconia sensor located in the exhaust stream. Exhaust gases pass through louvers on the bottom end and oxygen in the exhaust acts on the zirconia. Fresh air is admitted into the sensor at the top end, and the sensor can then compare oxygen content in the exhaust stream to that in the surrounding air. The zirconia, when exposed to oxygen, produces an electrical signal. In the sensor, because of the fresh air versus exhaust oxygen comparison, the electrical signal produced will be proportional to the oxygen content in the exhaust stream. A rich exhaust mixture (low oxygen content) causes the oxygen sensor to produce a high voltage of approximately 450 to 900 millivolts. A lean exhaust mixture (high oxygen content) causes the oxygen sensor to produce a low voltage of approximately 100 to 450 millivolts. The output voltage of a normally operating oxygen sensor varies between 100 and 900 millivolts. The oxygen sensor will not generate until it reaches its operating temperature, generally around 600°F.

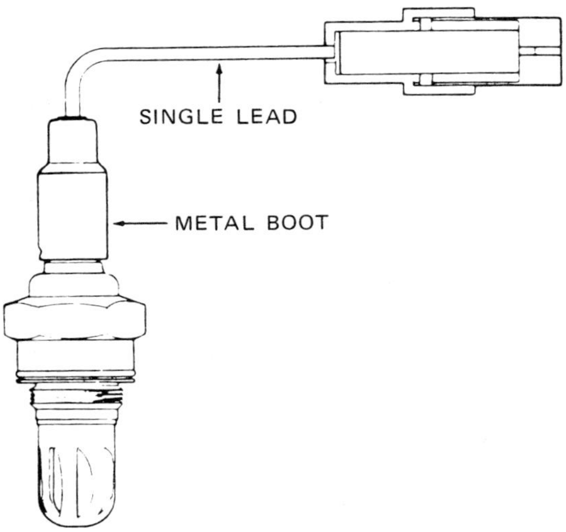

FIGURE 2-14 Oxygen sensor assembly *(Courtesy of Oldsmobile Division)*.

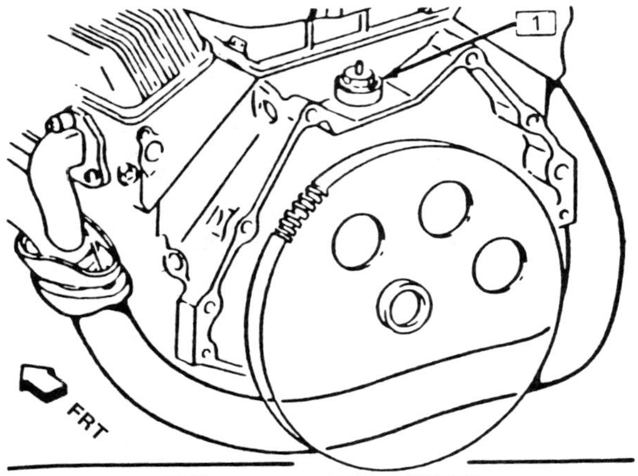

FIGURE 2-15 Knock sensor, 3.0 and 3.8 liter (typical) *(Courtesy of Oldsmobile Division)*.

Detonation Sensor

Engine detonation is a sensed parameter on engines utilizing the crossfire fuel injection system. A detonation sensor screwed into the engine block senses vibrations in the metal (Fig. 2-15). This piezoelectric sensor converts the vibrations into electrical signals that are sent to an electronic spark control (ESC) module. This module indicates to the ECM the need to retard the ignition timing through the electronic spark timing (EST) system. The ECM can then retard the ignition timing approximately 4° each second until the detonation stops. The detonation sensor then stops sending signals to the ESC module, and the ECM advances the ignition timing. After the detonation stops, the timing is advanced at a slower rate than it was retarded.

Cranking Mode

The engine crank mode is monitored by the ECM to control the electric fuel pump and the air/fuel mixture for starting. For each distributor reference pulse, the ECM issues a fuel injector pulse (synchronized). This cranking air/fuel ratio is provided as long as the throttle plate is less than 80% open. The ratio otherwise is determined by the engine coolant temperature. It can range from 1.5 to 1 at −40°F to about 14.7 to 1 at 200°F. The lower the coolant temperature is, the richer (longer pulse width) the mixture. The higher the coolant temperature is, the leaner (shorter pulse width) the mixture.

Crankshaft Position

Engine crankshaft position is monitored through the EST module. The ECM uses these data to regulate ignition timing. The pickup coil in the distributor sends signals or pulses to the EST module to fire spark plugs. These same pulses are used as an indication of the position of the crankshaft. This establishes a reference point for the ECM for purposes of controlling ignition timing.

Coolant Temperature Sensor

The coolant temperature sensor (CTS) is a thermistor whose resistance changes with temperature changes. As the temperature rises, the CTS resistance drops. At about −40°F, the CTS resistance is approximately 100,000 ohms, whereas at about 266°F its resistance is approximately 70 ohms. The ECM monitors these changes and varies pulse width, ignition timing, idle speed, EGR operation, canister purge, torque converter clutch engagement, and other engine controls.

Barometric Pressure

Barometric pressure is measured by the manifold absolute pressure sensor (MAP) when the ignition switch is first turned to ON. The atmospheric or barometric pressure acts on a diaphragm-operated transducer that produces data signals for the ECM. As a result, compensations can be made for changes in altitude and weather.

Air-conditioning System

An air-conditioning (A/C) circuit indicates to the ECM that the system has been turned on. The ECM can raise idle speed to counteract the load of the A/C system on the engine. A low idle speed, potential stalling condition, and low system voltage can be avoided by sensing this parameter.

The data sensors are used by the ECM to monitor engine operation. The ECM sends a 5-volt signal to many of these sensors and then measures what is returned. For example, a 5-volt reference is sent to the coolant temperature sensor, MAP sensor, and TPS. Each of these sensors modifies the 5-volt reference and allows either most or very little of that voltage to return to the ECM, which will then "see" what is happening through the returned high or low voltage.

A low return voltage from the coolant temperature sensor is interpreted by the ECM to mean low coolant temperature. A low return voltage from the MAP sensor is interpreted by the ECM to mean high manifold vacuum or closed throttle.

The ECM also has the ability to "learn." If its power source is disconnected, it must begin the learning process over again. This may cause a noticeable change in the vehicle's performance for a short period of time. In later models, the learning process is started over again with each cycle of the ignition switch if the switch is off for longer than 10 seconds.

EMISSION CONTROLS

Air Management Valves

Parameters that are controlled by the ECM include the air management control valves. These valves, sometimes called an electric air control (EAC) valve, and an electric air switching (EAS) valve are used to direct air from the air injection reaction pump to the exhaust manifold or to the air cleaner (divert) or to the catalytic converter, depending on the input parameters. This serves the same purpose as the older air injection system except that it is now electronically controlled.

The control of these valves enables the ECM to further reduce and control emission levels. Generally, the air control valve reduces hydrocarbons (HC) and carbon monoxide, (CO), while the air switching valve allows the front bed of the catalytic converter to reduce oxides of nitrogen (NO_x). Their actual functions may vary by model.

Canister Purge Control Valve

The canister purge control valve, when energized by the ECM, blocks purging of the evaporative emission control canister. The ECM de-energizes the purge control valve solenoid to allow purging to begin. This operation is similar to the older system except it is electronically controlled for better emission control.

Exhaust Gas Recirculation Valve

The ECM also controls the EGR valve on some engines by means of an EGR solenoid. When the solenoid is energized on some models, it blocks vacuum to the EGR valve, rendering it inoperative. On other models, vacuum is allowed to operate the EGR valve when the solenoid is energized. When EGR valve operation is desirable, the ECM de-energizes the solenoid and the valve can then operate in a similar manner to previous models.

Electronic Spark Timing

The electronic spark timing (EST) parameter is used on all engines. A new seven-terminal module is used in the distributor to perform the same functions as the older high-energy ignition module (four terminals) and in addition provides a reference pulse indicating engine rpm and crankshaft position to the ECM. The ECM considers this pulse and other input parameters, calculates the proper spark advance or retard for best engine operation, and then sends an EST pulse to the distributor.

Idle Speed Control

The idle control signal can be influenced by the following factors. Air-conditioning compressor clutch engagement causes the ECM to raise the idle speed to compensate for additional load on the engine. This is done by means of the A/C enable switch, which is monitored by the ECM. The idle control signal can also be influenced by the operation of the power steering system. A power steering cutoff switch, which is a normally closed switch, opens and causes the ECM to disengage the A/C compressor clutch when power steering pressure builds to approximately 300 psi. The ECM maintains idle speed to prevent stalling. Some models use an idle speed module located in the wiring harness next to the ECM to raise idle speed when power steering pressure exceeds a set value (Fig. 2-16a). Low system voltage causes the ECM to raise idle speed in an effort to compensate for low charge rate.

The throttle body injector control signal is delivered by the ECM in the form of a pulse width, as discussed earlier. Factors influencing pulse width include engine load, altitude, and weather, (MAP), engine rpm (HEI module), coolant temperature (CTS), exhaust oxygen content (O_2); throttle position (TPS); and system voltage (PROM).

Torque Converter Clutch

A torque converter clutch is used to improve fuel economy. This clutch provides a direct mechanical connection between the engine and the transmission/transaxle if certain conditions are met. Without this clutch, there would always be a loss of power transfer due to the design of the torque converter, resulting in a loss of fuel economy. Application of the torque converter clutch is controlled by a solenoid located in the transmission/transaxle pan (Fig. 2-16b). The solenoid is energized by the ECM to apply the clutch when the engine reaches operating temperature, vehicle speed is approximately 45 mph or higher, engine load is not too high or too low, the brake pedal is not being applied, and the transmission/transaxle is in high gear, which is third gear in most cases (fourth gear or overdrive in others). If any one of these conditions is not met, the torque converter clutch will not engage or, if it is already engaged, it will disengage.

1 CONN. (E.C.M. HARNESS)
2 IDLE SPEED MODULE ASM.
3 ELECTRONIC CONTROL MODULE

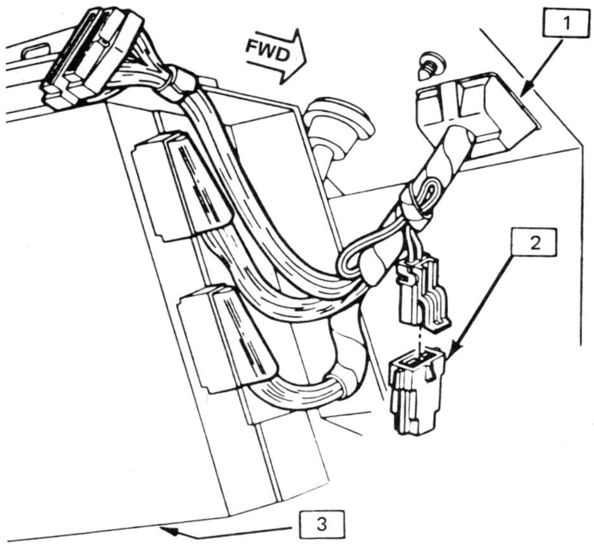

(a)

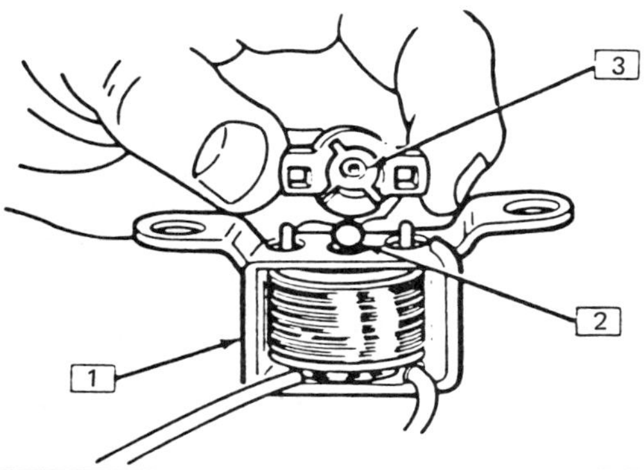

1 TCC SOLENOID
2 CHECK BALL
3 CHECK BALL SEAT

(b)

FIGURE 2-16 (a) Idle speed module, 2.0 liter *(Courtesy of Oldsmobile Division)*. (b) Torque converter clutch solenoid *(Courtesy of Chevrolet Motor Division)*.

A/C Clutch Control

The air-conditioning clutch control signal can be interrupted by the ECM if high pressure is present in the power steering system. This reduces engine load during peak power steering loads, which are especially important at idle speed because of the potential stalling condition. Other factors that cause the ECM to disengage the A/C compressor clutch are wide-open throttle, low idle speed, and low air temperature.

Air Door Control

The air door control signal from the ECM operates the hood louver solenoid relay on the crossfire fuel injection equipped 5.0-liter engines. When the engine is at operating temperature and is running or is at wide-open throttle the hood louver should be open. This allows for improved air induction.

Engine Cooling Fan

The cooling fan control is influenced by the ECM on A/C-equipped vehicles. All coolant fans operate when the coolant temperature rises to about 238°F. When the A/C mode lever is moved to the A/C mode, the control head energizes the electric coolant fan relay. This causes the fan to run constantly, providing cooling air across the condenser. On some vehicles, the ECM controls the ground side of the fan relay. When vehicle speed reaches 35 mph, the ECM can open the ground path and interrupt the operation of the coolant fan. At this speed, the airflow through the condenser and radiator should be sufficient to reduce A/C high-side pressure and also cool engine coolant. Better fuel economy is the end result. If coolant temperature rises above 239°F, the fan temperature switch will provide a ground to run the coolant fan.

THROTTLE BODY INJECTION EMISSION CONTROLS

The emission controls used in conjunction with throttle body injection systems include exhaust gas recirculation (EGR); early fuel evaporation (EFE); positive crankcase ventilation (PCV); thermostatic air cleaner (THERMAC); evaporative emission control (EEC); air injection reaction (AIR); and a catalytic converter.

Exhaust Gas Recirculation Valve

The EGR valve controls oxides of nitrogen (NO_x) and is either a negative or positive back-pressure transducer modulated design. It operates similarly to those on previous non-EFI cars; however, some models use an EGR solenoid to control the vacuum signal to the EGR valve. If the engine is at idle or if the coolant temperature is below 140°F, the ECM energizes the solenoid and blocks vacuum to the EGR valve. At any other time, the ECM de-energizes the solenoid and allows vacuum to operate the EGR valve. On some models, the ECM energizes the EGR solenoid to allow vacuum to operate the EGR valve.

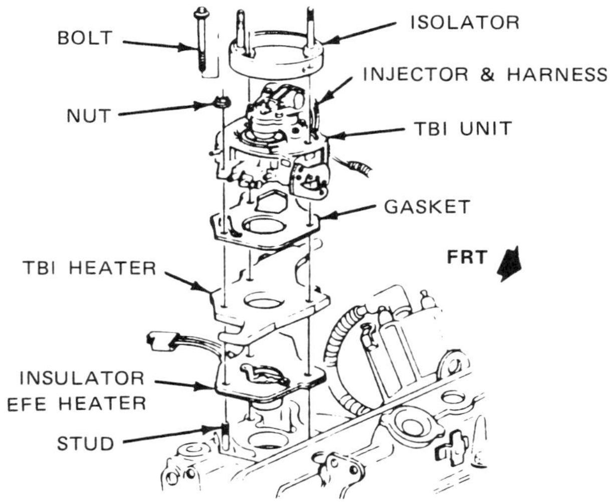

FIGURE 2-17 Throttle body injection unit replacement *(Courtesy of Oldsmobile Division).*

Early Fuel Evaporation

The EFE, when used, is an electric ceramic heater grid installed under the throttle body (Fig. 2-17). This heats the air/fuel mixture entering the intake manifold and improves cold driveaway conditions. The EFE grid is controlled by the ignition switch, self-regulates in terms of temperature, and is shut off by a snap action switch when coolant temperature reaches a predetermined level.

Positive Crankcase Ventilation

The PCV system is used to control crankcase emissions. It operates the same as in previous models.

Thermal Air Control

The THERMAC system preheats air entering the throttle body. This improves vaporization and driveability. The system operates in the same manner as previous models.

Evaporative Emission Control

The EEC system collects, stores, and meters fuel vapors into the intake air/fuel mixture. Most models operate the same as in the past; however, some have a controlled metering or purge rate. The ECM can regulate when the purge action takes place by controlling the EEC purge valve solenoid. If the computer system is in an open-loop mode of operation (the oxygen sensor has not yet begun to generate a voltage and the coolant temperature sensor is below a specified value), or if a specified time has not lapsed, or if engine rpm is lower than a specified amount, the ECM will energize the solenoid, which blocks vacuum to the purge valve. No fuel vapors will be metered into the throttle body.

When the preceding conditions are met, the ECM de-energizes the solenoid, allowing vacuum to the purge valve for recycling the fuel vapors.

Air Management

The AIR system reduces emissions (HC and CO) by providing fresh air to the exhaust manifold, catalytic converter, or air cleaner. This system is controlled by the ECM by energizing or de-energizing an air switching valve. Input data that affect the ECM control of this valve include coolant temperature, engine load, acceleration, and deceleration. The system operates similarly to previous computer command-control-equipped models. A modification of this system is also used on some models and is known as PULSAIR (pulsed air injection reaction). This system utilizes exhaust system pulsations to draw fresh air into the exhaust stream through check valves. The operation of this system is similar to carbureted models.

Catalytic Converter

A catalytic converter is used on all models. The catalytic converter used in combination with the PULSAIR system is a single-bed, bead-type converter. If the engine is equipped with an AIR system, it will have a dual-bed, monolith converter or a dual-bed, bead-type converter. The operation of these converters is the same as on carbureted models.

A brief discussion of some of the system modes of operation may provide a clearer overall picture.

SYSTEM OPERATION

Cranking Mode

During the cranking or startup mode, the ECM measures or senses several data inputs. When the ignition is turned to ON, engine cold, the ECM senses coolant temperature, MAP sensor pressure, TPS position, and engine rpm, and issues a wider-than-normal pulse to the fuel injector for enriched air/fuel mixture when cranking begins.

Run Mode

When the engine starts and runs, the computer system operates in the open-loop mode until the coolant temperature reaches a specified temperature, the oxygen sensor produces a recognizable signal to the ECM, and a specified amount of time has lapsed. At this point, the system goes into a closed-loop mode of operation, and the ECM monitors the oxygen sensor voltage as an indication of exhaust gas oxygen content. The ECM makes corrections by changing the pulse width to the fuel injector. As the coolant temperature rises, the ECM makes additional adjustments to the injector pulse width. Once the engine is at operating temperature, pulse width adjustments for the injector are dependent on the input parameters discussed earlier. During open-loop operation, the ECM controls the fuel injector based on engine data sensor information except from the oxygen sensor. The oxygen sensor does not initially provide a signal recognizable to the ECM and cannot provide feedback information related to the results of the commanded fuel injector pulse width. When the oxygen sensor signal is great enough to be recognized by the ECM, the computer system goes into closed loop. The ECM can now measure the results of its commanded fuel injector pulse width directly through the feedback information of the oxygen sensor.

Clear Flood Mode

The computer system also has a mode of operation for a flooded condition. When a flooded condition occurs, the driver must depress the accelerator pedal completely. This causes the TPS to send a wide-open throttle signal to the ECM. When the engine is now cranked, the ECM reduces injector pulse width enough to provide a 20 to 1 air/fuel ratio. As long as a wide-open throttle signal and cranking rpm signal (less than 600 rpm) are sent to the ECM, it remains in the clear-flood mode.

Acceleration Enrichment Mode

The acceleration enrichment mode is active when the throttle blade is opening. This causes a changing TPS signal and an increase in MAP sensor pressure (decrease in vacuum), and the ECM lengthens the injector pulse width. This mode prevents stumble and hesitation and allows acceleration. Injector pulses are provided in a nonsynchronized mode as discussed earlier. Any reduction in the throttle blade angle stops the injector enrichment pulses, and, as a result, any fast movements of the accelerator pedal do not provide an overrich mixture.

Deceleration Enleanment Mode

The deceleration enleanment mode provides a means of controlling emissions. As the throttle plate is closing, the TPS signal is changing and the MAP sensor pressure is decreasing (vacuum increasing). The ECM uses these inputs to decrease the pulse width in order to try to maintain a 14.7 to 1 air/fuel ratio.

In cases of severe deceleration, the ECM operates in its decel fuel cutoff mode. This mode overrides the deceler-

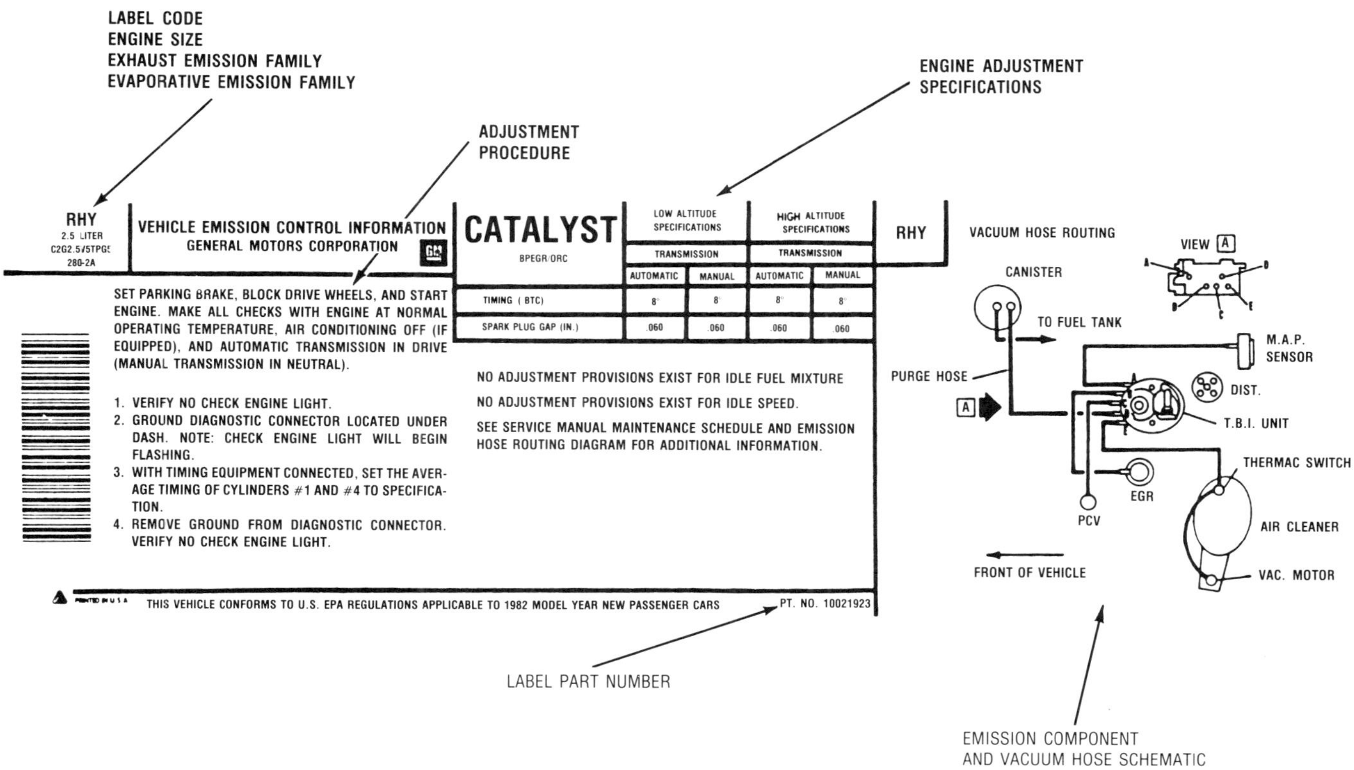

FIGURE 2-18 Vehicle emission control information label *(Courtesy of General Motors Product Service Training).*

ation enleanment mode and is based on MAP sensor values, engine rpm, and TPS signals. These factors are used to guide the ECM calculations and are stored in the PROM discussed earlier. The ECM can actually momentarily shut off the injector.

Battery Voltage Correction Mode

The battery voltage correction mode enables the ECM to try to compensate for changes in system voltage and to try to stabilize voltage to the electric fuel pump and fuel injector. Battery voltage correction can occur in any of the computer system operating modes.

Fuel Cutoff Mode

The fuel cutoff mode is activated whenever the ECM does not receive reference pulses from the EST module in the distributor. This includes whenever the ignition switch is turned off.

SYSTEM SERVICE

Fuel injection system service is not difficult, but procedures must be carefully followed to retain the sensitivity and accuracy of the system.

Setting Ignition Timing

The vehicle emissions sticker in the engine compartment describes the procedure for setting ignition timing (Fig. 2-18). Generally, two procedures are used. The first procedure requires the technician to jumper the correct terminals of the assembly line communication link (ALCL) connector or the assembly line diagnostic link (ALDL) connector (Fig. 2-19). The correct terminals are identified in the manufacturer's service manual. The second procedure requires the technician to disconnect or "open" the set timing connector in the engine compartment (Fig. 2-20). The timing should be set preferably with a magnetic induction direct-reading meter. The use of a timing light is a second choice.

All timing checks should be done with the engine at operating temperature, computer system is closed loop, and engine idle speed set to or below the specified rpm. Ad-

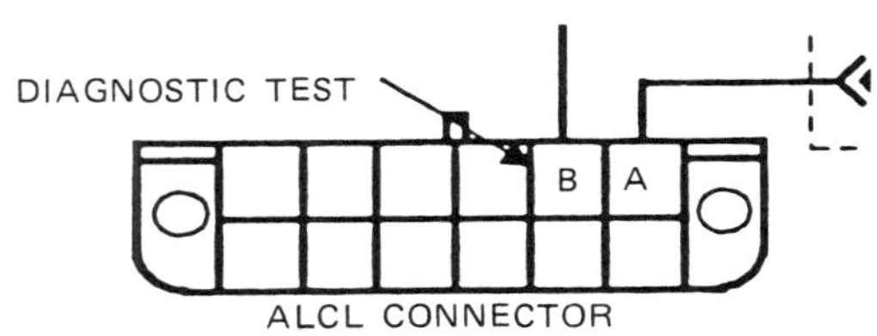

FIGURE 2-19 Assembly line communication link connector *(Courtesy of Buick Motor Division).*

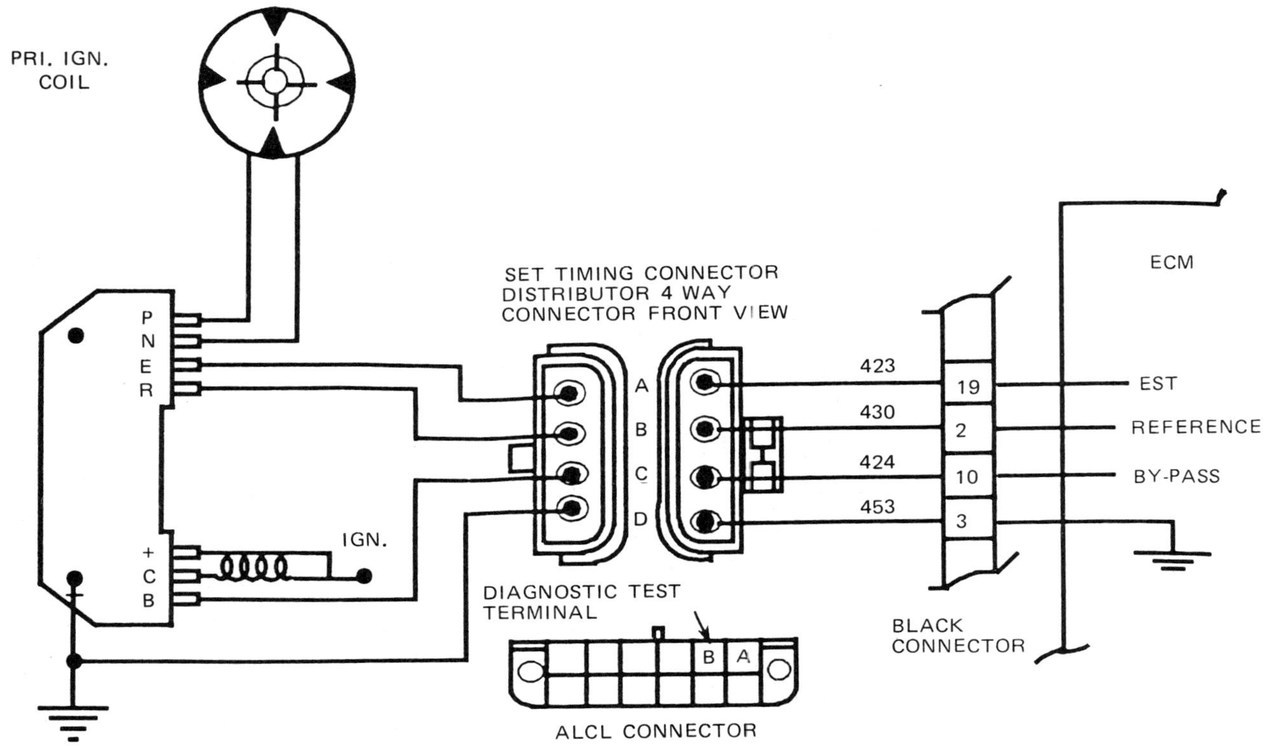

FIGURE 2-20 Set timing connector *(Courtesy of Chevrolet Motor Division).*

ditional instructions vary by model. Refer to the emissions sticker or service manual and follow the steps closely.

Idle Speed Adjustment

This adjustment should be made only if the TPS requires adjustment or if the throttle body parts have been replaced. The engine should be at operating temperature when this adjustment is made. The idle speed adjustment is sometimes referred to as the minimum idle speed or minimum air rate adjustment.

All models require the use of a special tool called an idle air plug (Figs. 2-21 and 2-22). Others require the use of a water manometer. Instructions for actually making adjustments are very specific for each model and should be followed closely. These instructions are described best in the service manual.

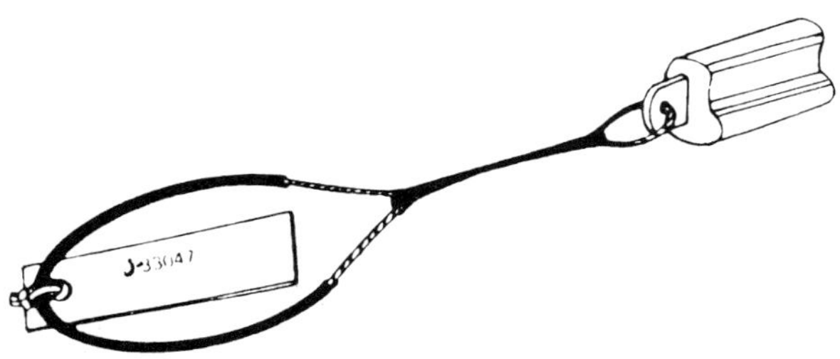

FIGURE 2-21 Tool J-33047 *(Courtesy of Oldsmobile Division).*

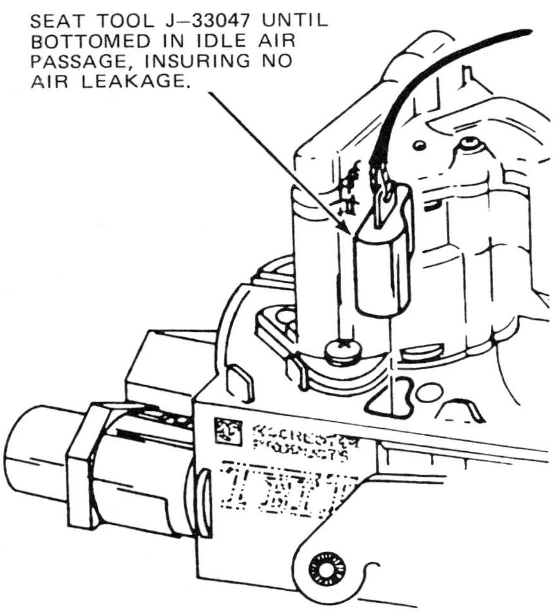

FIGURE 2-22 Installing tool J-33047 *(Courtesy of Chevrolet Motor Division).*

ECM Service

Replacement of the PROM is the only service that may be required for the ECM (Fig. 2-23). The ECM lights a CHECK ENGINE light or a SERVICE ENGINE SOON

ELECTRONIC CONTROL MODULE (ECM) REPLACEMENT 2 BOARD EFI — SERVICE NUMBER 1225610

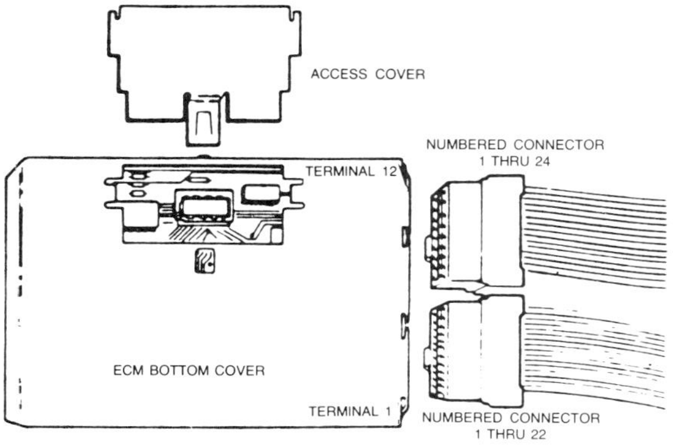

ELECTRONIC CONTROL MODULE (ECM) MOUNTING HARDWARE NOT ILLUSTRATED. HARDWARE CONFIGURATION WILL VARY WITH CAR DIVISION

A. 1. Disconnect the two connectors from the ECM.
   2. Remove the ECM mounting hardware.
   3. Remove the ECM from the passenger compartment.
   4. Turn the ECM so the bottom cover is facing up.
   5. Remove the slide off PROM access cover by depressing locking tab.

REPLACEMENT ELECTRONIC CONTROL MODULE (ECM) IS SUPPLIED WITHOUT AN ENGINE CALIBRATION UNIT (PROM) SO CARE SHOULD BE TAKEN WHEN REMOVING THE PROM FROM THE DEFECTIVE ECM AS IT WILL BE REUSED IN THE NEW ECM.

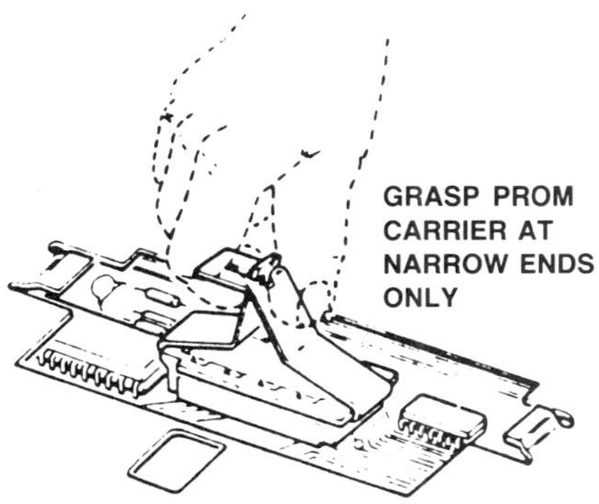

B. 1. Using the PROM removal tool, grasp the engine calibration unit (PROM) carrier at the narrow ends. Gently rock the carrier from end to end while applying a firm upward force.
   2. Note the reference end of the PROM carrier and carefully set aside.

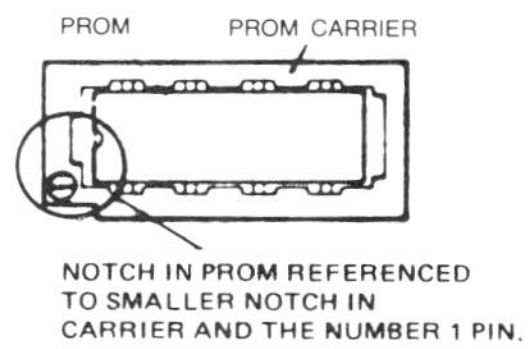

C. 1. Take the new electronic control module out of its packaging and check the service number to make sure it is the same as the defective ECM.

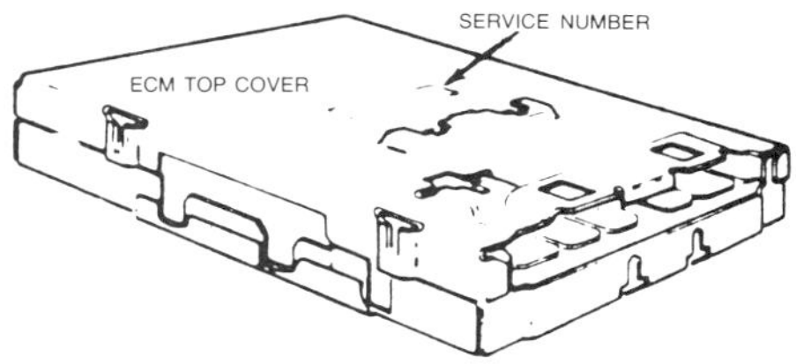

DO NOT press on PROM - ONLY CARRIER.

D. 1. Take the PROM mounted in the PROM carrier (which you had previously set aside) and position the carrier squarely over the PROM socket with the small notched end of the carrier aligned with the small notch in the socket at the pin 1 end. Press on PROM carrier until it is firmly seated in socket.

DO NOT press on PROM - ONLY CARRIER.

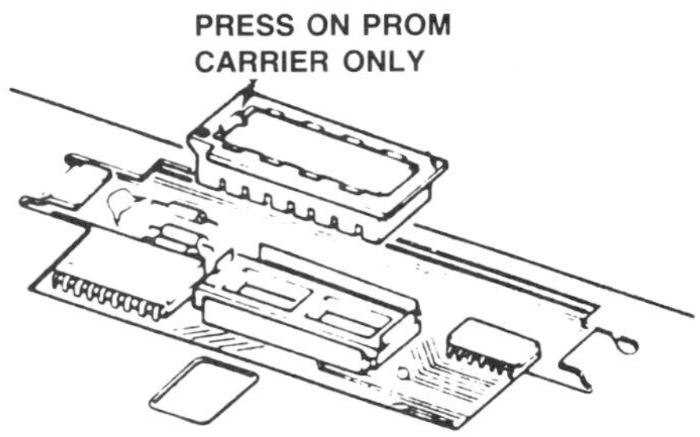

FIGURE 2-23 Electronic control module service *(Courtesy of General Motors Products Service Training).*

light on the instrument panel if there is a problem with the PROM. After activating the ECM diagnostic mode, this same telltale light blinks a code 51, indicating a faulty PROM, bent pins, corroded pins, or incorrect installation. The self-diagnostic feature of the ECM will be discussed more after the servicing section.

Fuel Filter Service and Pressure Relief

Prior to servicing the fuel filter, pressure in the lines should be relieved. To do this, the fuel pump fuse should be removed from the fuse panel in the passenger compartment or from the fuse holder under the hood. The engine should

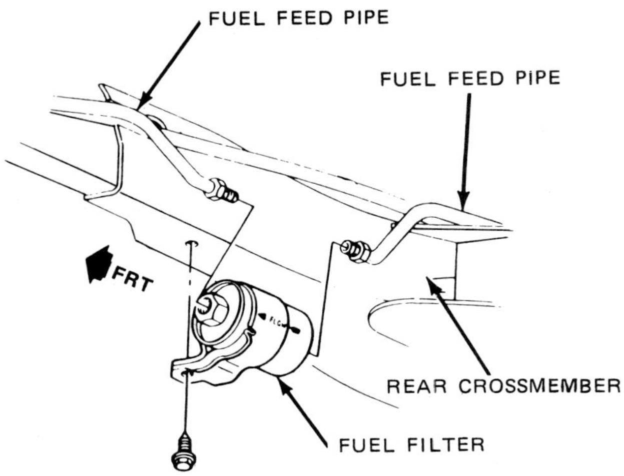

FIGURE 2-24 Fuel filter *(Courtesy of Pontiac Motor Division).*

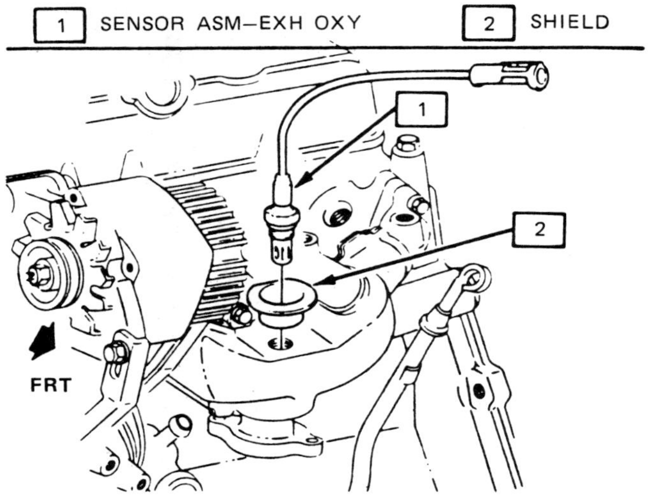

FIGURE 2-25 Oxygen sensor mounting, 2.0 liter *(Courtesy of Oldsmobile Division).*

be cranked; it will start and run until fuel in the fuel lines is used. Then crank the engine an additional 3 to 5 seconds to be sure all pressure is relieved. With the ignition turned off, replace the fuel pump fuse. This procedure should be followed whenever any fuel line needs to be disconnected or when servicing the throttle body. A dry chemical, class B fire extinguisher should also be kept near the work area.

The fuel filter is an in-line type and is replaced as an assembly. It is located either under the hood in the fuel supply line to the throttle body (Fig. 2-3) or it is mounted under the car, to the rear of the fuel tank in the fuel supply line coming from the electric fuel pump (Fig. 2-24). The inlet and outlet fuel lines are sealed into the filter fittings with rubber O-rings. These seals are made of a special rubber and should be replaced only with an equivalent material and whenever the filter is serviced. The fuel line nuts should be tightened to about 18 pound-feet. Always check for leaks after pressurizing the system.

Engine Sensor Service

The oxygen sensor should be kept clean and free of dirt, grease, and other contaminants (Fig. 2-25). Its operation is dependent on a comparison of exhaust oxygen to outside air oxygen. If air exchange passages are clogged, its voltage generation and accuracy will be affected. Do not use solvents to clean the louvered end. Generated voltage generally varies from a minimum of 100 millivolts to about 900 millivolts maximum. The ECM monitors this voltage and then issues commands to the fuel injector. For example, when in closed loop, a lean exhaust stream produces low voltage at the oxygen sensor, and the ECM will issue a wider injector pulse for a richer intake mixture. If an open occurs in the oxygen sensor, the ECM issues a wider injector pulse for a richer intake mixture. An open in the oxygen sensor or circuit should set a code 13 in the ECM memory. A shorted sensor or circuit should set a code 44, and high voltage in the sensor circuit should set a code 45 in the ECM memory. If any of these codes is set, the engine should run in the open-loop mode. Do not attempt to test the oxygen sensor with a multimeter as damage to the sensor may result.

The MAP sensor measures changes in intake manifold pressure and reads the opposite of what a vacuum gauge measures (Fig. 2-13). The ECM sends a 5-volt reference to the MAP sensor, whose electrical resistance changes with manifold pressure. A higher pressure (lower vacuum) means more engine load and the ECM issues a longer injector pulse width. The ECM uses MAP sensor data to regulate fuel delivery, torque converter clutch, and ignition timing. If the MAP sensor or circuit should fail, a code 33 or code 34 should be set in the ECM memory. The MAP sensor cannot be tested using ordinary equipment. The MAP voltage will be either high or low. When the ECM receives a low MAP voltage, it provides more spark advance through the EST module. If the MAP voltage is high, the ECM provides less spark advance. The coolant temperature sensor works in conjunction with the MAP sensor to influence the ECM control of the spark advance. When coolant temperature is low, more spark advance is provided. When coolant temperature is high, less spark advance is provided.

When these factors are considered, detonation or pinging could be caused by low MAP output or by high resistance in the coolant sensor or circuit. Poor performance could be caused by high MAP output or low resistance in the coolant sensor or circuit.

The engine coolant temperature sensor resistance changes with coolant temperature (Fig. 2-26). This type of variable resistor is called a *thermistor.* Low temperature (−40°F) provides a resistance of about 100,000 ohms and

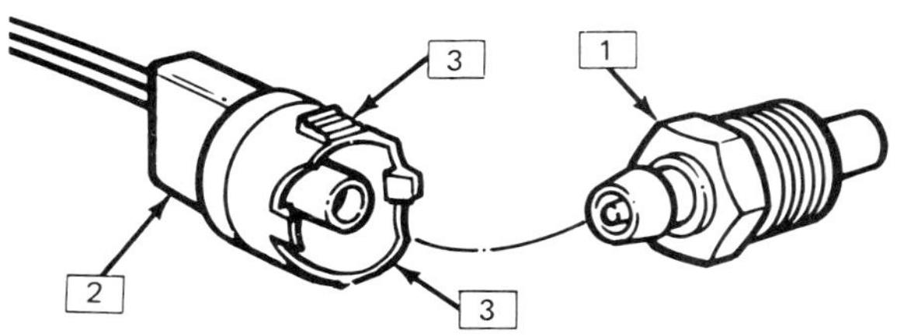

1 COOLANT TEMP. SENSOR
2 ECM HARNESS--CONNECTOR TO COOLANT TEMP. SENSOR
3 DEPRESS TABS TO RELEASE

FIGURE 2-26 Engine coolant temperature sensor *(Courtesy of Oldsmobile Division).*

high temperature (266°F) provides about 70 ohms. The ECM sends a 5-volt reference signal to the CTS and measures what returns as an indication of coolant temperature. When the engine is cold, voltage is high, and when the engine is hot, voltage is low. Engine coolant temperature affects most of the systems controlled by the ECM. A code 14 or code 15 should be set in the ECM memory if a problem occurs in the coolant temperature sensor or circuit. The coolant temperature sensor resistance can be checked with a digital multimeter and should be checked at more than one temperature to be sure it is within the intended range of operation.

The TPS is a potentiometer or variable resistor that is fed a 5-volt reference from the ECM at one terminal and is grounded at another terminal (Fig. 2-27). A third terminal moves along the resistor as throttle blade angle changes. The ECM monitors the voltage at this third terminal. At closed throttle, the TPS output is low, (approximately 500 millivolts). At wide-open throttle, TPS output should be approximately 5 volts. In this manner, the ECM knows what driver demands are in terms of throttle blade angle. An open TPS or circuit should set a code 22. A shorted TPS or circuit should set a code 21. If the TPS is loose or broken, the ECM will think the throttle is moving and intermittent pulses will be sent to the injector, resulting in an unstable idle or surge at constant speed. If a code 21 or 22 is set in memory, the ECM uses a programmed value as a substitute for the TPS signal in order to restore vehicle performance.

The TPS is mounted to the throttle body with screws. Sometimes these screws are welded. If replacement of the sensor is necessary, the screws must be drilled out and replaced also. Follow instructions in the service manual closely. Adjustment of the TPS is done by using special jumper leads and a 10-megohm impedance digital multimeter (Fig. 2-28). Use of an analog meter or the wrong digital meter may damage circuits. Operation of the TPS can be checked by disconnecting it, connecting special jumper leads between the connector and sensor A and C terminals, and turning to ON the ignition switch; then with a 10-megohm impedance digital multimeter connected to the B terminal and ground on the dc volts scale, operate the throttle slowly to monitor a gradual change in the voltage reading. As the throttle is opened, the change in reading should be consistent. A skip in the reading may indicate a faulty sensor.

The PARK/NEUTRAL (P/N) switch is serviced and tested similar to past models. It is located at the base of the steering column in the passenger compartment or in the console in the passenger compartment. The ECM uses the signal from the P/N switch to control idle speed and the vehicle speed sensor diagnostics.

Some engine distributors use a Hall-effect switch to provide distributor reference pulses to the ECM. It is located in the EST distributor and can be tested by using a 12-volt battery and a voltmeter. Follow the service manual instructions for each model.

The vehicle speed sensor (VSS) can be serviced by removing the instrument cluster (Fig. 2-11). The VSS is mounted on the back of it. The vehicle should not be driven without the VSS as poor idle quality may result. A code 24 is set in the ECM memory if a problem exists in the VSS or circuit.

Wiring Harness and Connectors

Wire used for fuel injection and ECM harnesses is standard copper wire. The connectors, however, are environmentally protected and have positive locks, and more critical sensors or components have secondary locks to prevent terminals from backing out during installation. New terminals are used to provide better circuit connections. These circuits are lower voltage than in prior years (a minimum of 500 millivolts and a maximum of 5 volts). They are also low-current-carrying circuits, almost always less than 250 milliamperes. For these reasons, the connectors must be sealed to prevent corrosion. The voltage is not enough to overcome small resistances in the form of oxidized terminals. When replacing terminals or when splicing, the terminal or splice clip should always be soldered to reduce resistance due to oxidation. Never probe through the rubber seals in the connectors. The environmentally protected connectors are known as Weather-Pak connectors and require the use of a special tool for terminal removal. Another tool known as a *pick* is used for general terminal removal. The shielded cable used for the transmission of the distributor reference pulses from the EST module to the ECM should be repaired only as shown in the service manual.

Throttle Body Assembly Service

A throttle body identification number is stamped on the TPS side of the assembly (Fig. 2-29). Vacuum tubes are identified with letters for correct vacuum hose installation.

1 FUEL METER ASSEMBLY

2 GASKET – FUEL METER BODY

3 SCREW & WASHER ASSY– ATTACH. (3)

4 FUEL INJECTOR KIT

5 FILTER – FUEL INJECTOR NOZZLE

6 SEAL – SMALL "O" RING

7 SEAL – LARGE "O" RING

8 BACK-UP WASHER – FUEL INJECTOR

9 GASKET – FUEL METER COVER

10 DUST SEAL – PRESS, REGULATOR

11 GASKET – FUEL METER OUTLET

12 SCREW & WASHER ASSY – LONG (3)

13 SCREW & WASHER ASSY – SHORT (2)

14 NUT – FUEL INLET

15 GASKET – FUEL INLET NUT

16 NUT – FUEL OUTLET

17 GASKET – FUEL OUTLET NUT

18 FUEL METER BODY ASSEMBLY

19 THROTTLE BODY ASSEMBLY

20 SCREW – IDLE STOP

21 SPRING – IDLE STOP SCREW

22 LEVER – TPS

23 SCREW – TPS LEVER ATTACHING

24 SENSOR – THROTTLE POSITION KIT

25 RETAINER – TPS (2)

26 SCREW – TPS ATTACHING (2)

27 WASHER – TPS SCREW (2)

28 IDLE AIR CONTROL VALVE

29 GASKET – CONTROL VALVE TO T.B.

30 GASKET – FLANGE MOUNTING

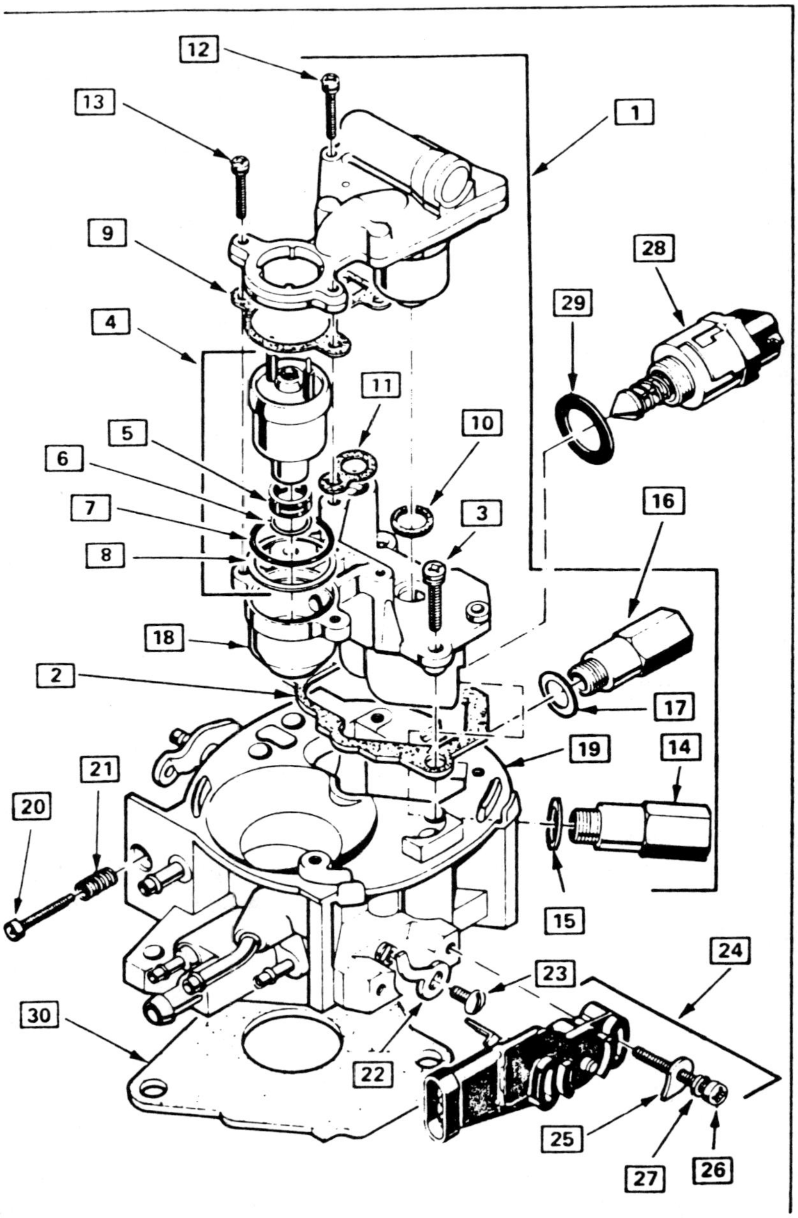

FIGURE 2-27 Throttle position sensor (typical) *(Courtesy of Chevrolet Motor Division).*

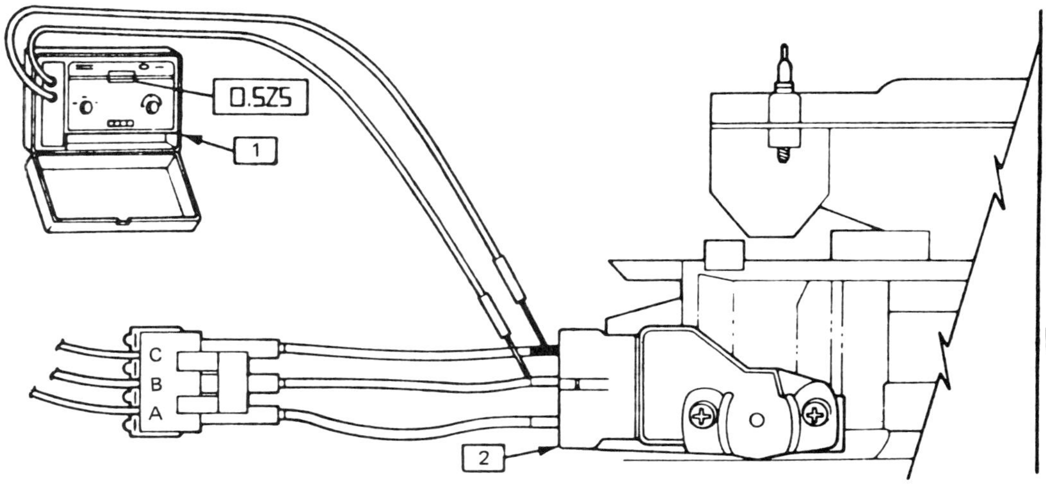

FIGURE 2-28 Throttle position sensor adjustment *(Courtesy of Oldsmobile Division).*

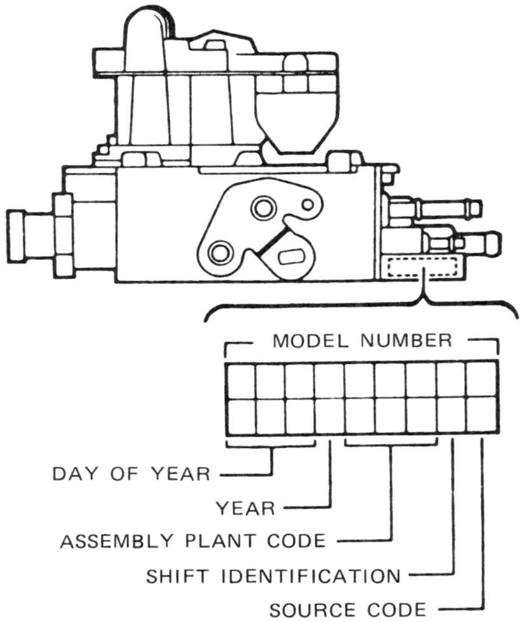

THROTTLE BODY INJECTION – IDENTIFICATION

(TYPICAL SINGLE INJECTOR TYPE)

FIGURE 2-29 Throttle body identification *(Courtesy of Chevrolet Motor Division).*

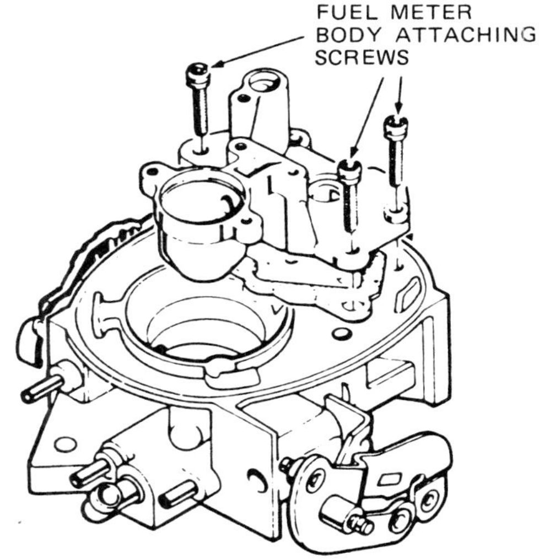

FIGURE 2-30 Fuel meter body and gasket *(Courtesy of Oldsmobile Division).*

Before servicing the fuel metering assembly and pressure regulator or fuel injector, the system fuel pressure should be relieved as discussed earlier. The fuel pressure regulator is serviced as part of the fuel metering assembly cover (Figs. 2-5 and 2-30). Do not remove the four screws securing the regulator to the cover. The regulator pressure is factory preset and should not be field adjusted or driveability symptoms will result. The fuel metering assembly cover should not be cleaned or immersed in any solvent, as damage to the regulator diaphragm may result. Always reassemble with new gaskets and dust seal.

The fuel injector should be removed carefully to avoid damaging the electrical connectors, the injector fuel filter (screen), and its nozzle (Fig. 2-31). The injector cannot be disassembled, nor should it be cleaned or immersed in solvents of any kind. Reinstall the injector using new O-ring seals; make sure it is seated and that the electrical terminals are parallel to the throttle shaft.

The throttle body casting, with TPS, IAC, fuel metering assembly cover, fuel injector, and all seals and gaskets removed, can be cleaned with a cold immersion-type

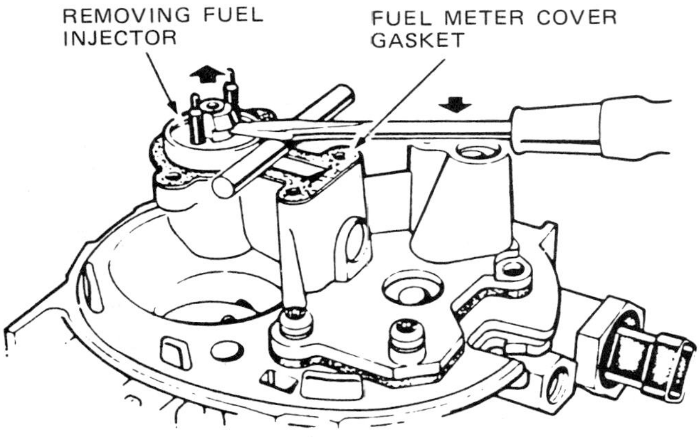

FIGURE 2-31 Removing fuel injector *(Courtesy of Oldsmobile Division).*

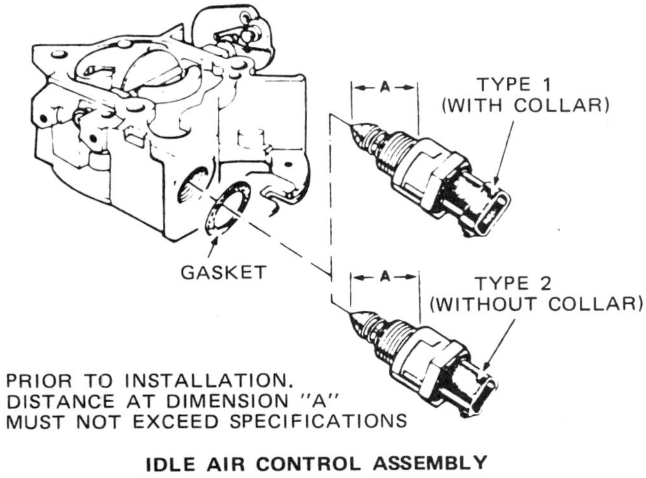

FIGURE 2-33 Types 1 and 2 air control assembly *(Courtesy of Oldsmobile Division).*

solvent such as carburetor cleaner, Carbon X, or an equivalent. Blow dry and make sure all passages are clear. Allow any thread-locking compound in the IAC bore to remain.

If a new IAC is needed, identify the original as being a type 1 or 2 (Figs. 2-32 and 2-33). A type 1 has a rectangular collar around the electrical connector end. Type 2 does not have the collar. Before installing the IAC, the pintle extension position must be measured. If it is extended too far, damage may be done to the pintle as it is screwed into the throttle body. Follow service manual specifications for correct adjustment. After installation, the IAC may need to be reset. If, after installation, the idle speed is high (1800 to 2000 rpm) and it will not slow down to specification after a few moments (engine at operating temperature), the IAC resetting procedure should be followed. Operate the vehicle at 40 mph and the engine at normal temperature. The ECM will extend the IAC until fully seated and will therefore establish a reference point for IAC pintle position. The engine should idle normally after completing this procedure.

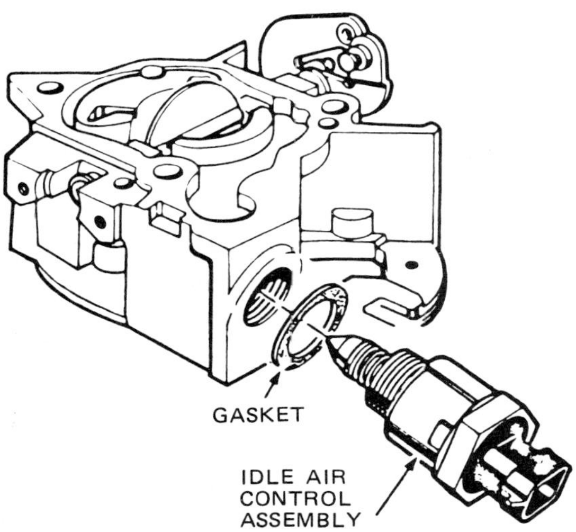

FIGURE 2-32 Idle air control assembly *(Courtesy of Oldsmobile Division).*

SYSTEM DIAGNOSIS

Code Types

The throttle body fuel injection system in combination with the computer system has a self-diagnostic capability. If the CHECK ENGINE or SERVICE ENGINE SOON light on the instrument panel lights and stays on, the system problem is present. This is known as a hard or current problem and will set a hard or current code in the ECM memory. If the CHECK ENGINE or SERVICE ENGINE SOON light comes on only for a short time and then goes out, a problem occurred and went away by itself. This is known as an intermittent or history problem and will set a code in the ECM memory. In either case, a code will be stored in the ECM memory, which should be checked first. This may indicate the cause of the customer complaint.

Service manual diagnostic procedures are based on symptoms that are described at the end of this chapter. The technician should become familiar with these definitions to be sure that everyone is describing the same problem. The diagnostic procedures also are written to help diagnose problems on vehicles that "ran well at one time." The conclusion is that time and/or possibly wear have caused the symptom to appear. It is impossible to cover all causes of a symptom, so when repair work is done and the symptom is still present, either the wrong symptom was used to diagnose or a more detailed analysis of the problem is warranted.

If the complaint involves a CHECK ENGINE or

SERVICE ENGINE SOON light, but no codes are stored, refer to the driver complaint chart in the service manual.

Self-Diagnostic Mode

The ECM can be operated in the self-diagnostic mode by grounding a test terminal in the ALCL connector under the instrument panel with the ignition turned to ON and the engine not running (Fig. 2-19). When this is done, the CHECK ENGINE light (CEL) blinks once, pauses, and then blinks twice. It repeats this same sequence two more times. This is a code 12, which means "no distributor reference signal." With the ignition on and the engine not running, this should be anticipated. The code 12 also means the self-diagnostic mode is working. After the third 12, if any codes are stored in the ECM memory, they will be shown in numerical order from lowest to highest. For example, if the light blinks once, pauses, and then blinks five times, and this is repeated two more times, a code 15 for "Coolant Sensor Voltage High" has been stored in memory. Then the CEL may blink four times, pause, and then blink two times, and this is repeated two more times; this is code 42 for "Electronic Spark Timing" is stored in memory. When all codes have been given, the CEL begins again by giving a code 12. This sequence of code display keeps repeating as long as the mode is energized.

The trouble codes indicate a problem in the circuit that corresponds with the code number. The circuit includes the ECM, edgeboard connector at the ECM, the wire and any connectors within it, the coolant sensor (code 14 or 15), and the wire returning to the ECM.

While the ECM is in the self-diagnostic mode as just described, it will also energize all ECM-controlled relays.

Field Service Mode

Grounding the test terminal at the ALCL connector while the engine is running puts the computer system into the FIELD SERVICE mode. While operating in this mode, the CEL blinks at either a slow rate, once per second, or at a faster rate, twice per second. When the CEL is blinking at the slower rate, it is an indication that the engine is in closed-loop operation. When the CEL is blinking at the faster rate, the engine is in an open-loop operation.

Erasing Codes

Once repairs have been accomplished, the next step involves erasing the ECM memory to eliminate previously stored codes. This can be done in two ways depending on the vehicle model. The positive battery cable can be disconnected for 10 seconds or the positive battery cable pigtail can be disconnected for 10 seconds (Fig. 2-34). In any case, make sure the ignition is off when this is done; otherwise, damage may be done to the ECM.

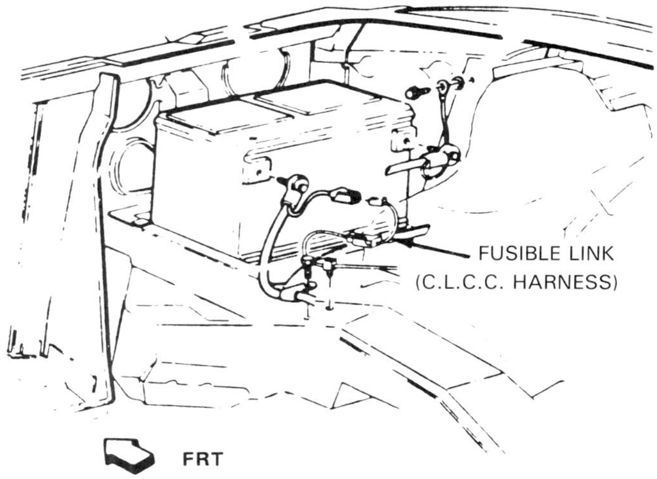

FIGURE 2-34 Battery pigtail *(Courtesy of General Motors Product Service Training).*

PROM Replacement

To make the ECM more versatile, a PROM or programmable read only memory is used. The PROM matches the ECM computer circuitry to the vehicle into which it is installed. The PROM is also serviceable, which means the ECM functions can be modified or updated as needed even after the vehicle has accumulated mileage. It is located inside the ECM and is accessible by removing a cover (Fig. 2-35). It contains information concerning vehicle weight, engine type, transmission/transaxle type, final drive/rear axle ratio, and several other bits of information. The PROM, then, is very specific and is programmed for a particular vehicle only. Always be sure to replace it with the correct part number.

A replacement ECM is supplied without a PROM, so the technician will be required to install the PROM from the defective ECM. It must be removed and installed carefully as its pins are fragile. A PROM problem sets a code 51 in the ECM memory.

CALPAK

Some model engines use a CALPAK (calibration pack). The CALPAK contains artificial, fixed engine sensor values for operation of the fuel injection system should the normal ECM circuitry fail. This keeps the engine running, although performance and economy will suffer. The CALPAK is also serviceable by removing a cover plate on the ECM. Replacement procedures are the same as for the PROM. A replacement ECM is supplied without a CAL-

ELECTRONIC CONTROL MODULE (ECM) REPLACEMENT 2 BOARD EFI — SERVICE NUMBER 1225610

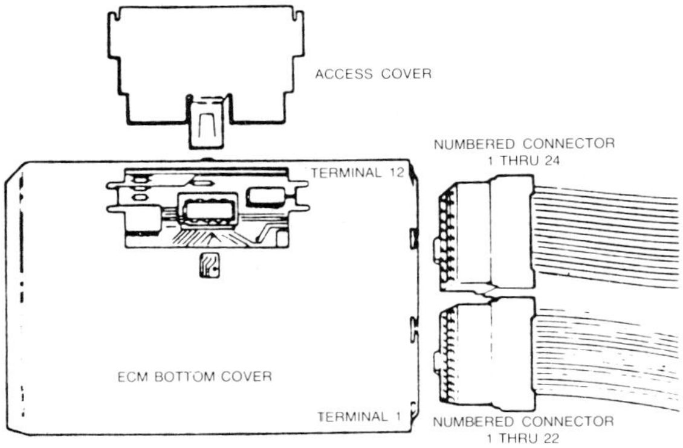

ECM MOUNTING IS DESCRIBED IN SECTION 8. WHILE PERFORMING SERVICE ON THE ECM, IT IS IMPERATIVE THAT THE IGNITION SWITCH BE IN THE "OFF" POSITION.

A. 1. Disconnect the two connectors from the ECM.
   2. Remove the ECM mourting hardware.
   3. Remove the ECM from the passenger compartment.
   4. Turn the ECM so the bottom cover is facing up.
   5. Remove the slide off PROM access cover by depressing locking tab.

REPLACEMENT ELECTRONIC CONTROL MODULE (ECM) IS SUPPLIED WITHOUT AN ENGINE CALIBRATION UNIT (PROM) SO CARE SHOULD BE TAKEN WHEN REMOVING THE PROM FROM THE DEFECTIVE ECM AS IT WILL BE REUSED IN THE NEW ECM.

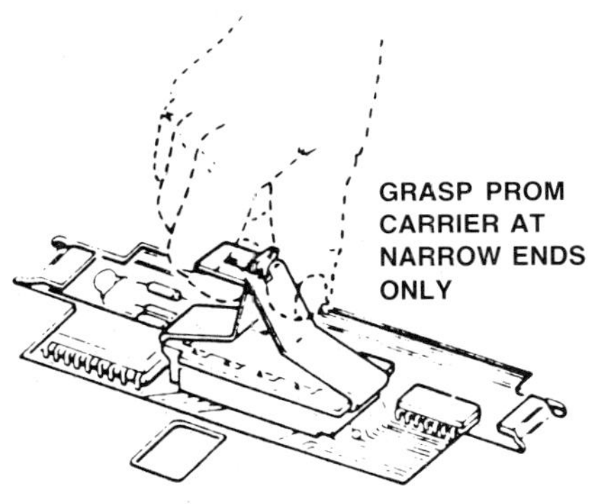

B. 1. Using the PROM removal tool, grasp the engine calibration unit (PROM) carrier at the narrow ends. Gently rock the carrier from end to end while applying a firm upward force.
   2. Note the reference end of the PROM carrier and carefully set aside.

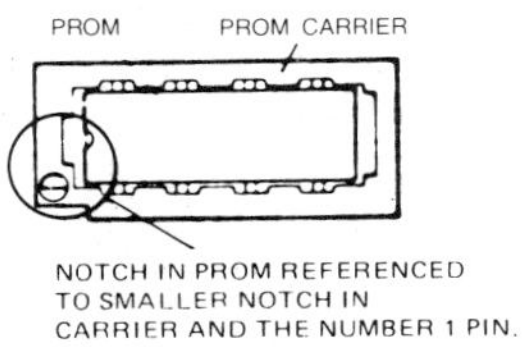

C. 1. Take the new electronic control module out of its packaging and check the service number to make sure it is the same as the defective ECM.

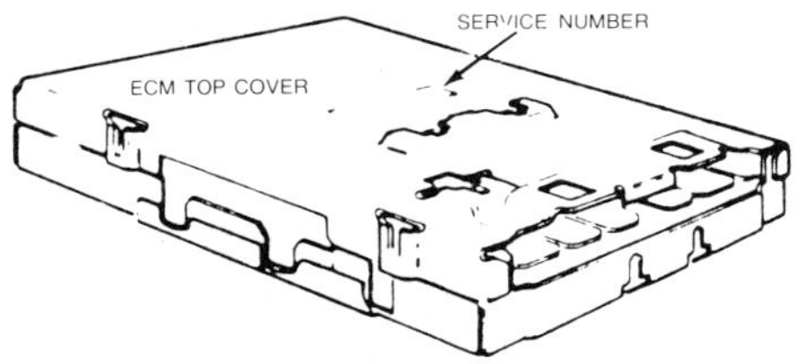

DO NOT press on PROM - ONLY CARRIER.

D. 1. Take the PROM mounted in the PROM carrier (which you had previously set aside) and position the carrier squarely over the PROM socket with the small notched end of the carrier aligned with the small notch in the socket at the pin 1 end. Press on PROM carrier until it is firmly seated in socket.

DO NOT press on PROM - ONLY CARRIER.

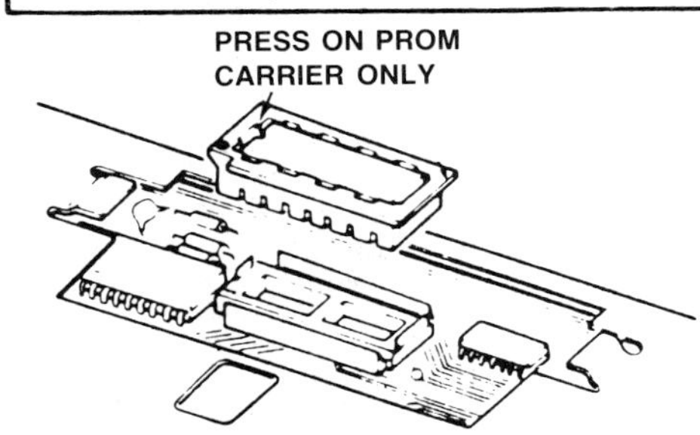

FIGURE 2-35 Electronic control module service *(Courtesy of Oldsmobile Division).*

PAK and, therefore, the technician must transfer it from the defective ECM.

Fuel System Pressure Test

Before making a fuel system pressure test, it is necessary to relieve fuel pressure as discussed earlier. A special fuel pressure gauge should be installed in the fuel line between the fuel filter and the throttle body. On some models it may be easier to install the gauge on the throttle body side of the fuel filter at the rear of the car near the fuel tank. The gauge should be capable of reading 9 to 15 psi.

Start the car and make a reading. The gauge should show between 9 and 13 psi. If it does not, refer to the diagnostic charts or decision trees in the service manual.

Relieve fuel pressure again before removing the gauge. Always check for leaks after reinstalling the fuel line.

SPECIAL TOOLS

Special diagnosis tools are shown in Fig. 2-36 and described in the following list:

1. *Digital volt/ohmmeter:* A digital volt/ohmmeter with a 10-megohm impedance is required to perform checks on any General Motors computer system. Do not use a 12-volt test light as damage to the computer or its circuits will result.
2. *Idle air plugs:* Idle air passage plugs are needed to set minimum idle speed.
3. *Fuel line wrenches:* Fuel line wrenches are needed for servicing the throttle body, fuel lines, and fuel filter.
4. *Wire repair tools:* To service the harness connectors and terminals, a Weather-Pak terminal removal tool is needed, as well as a terminal pick.

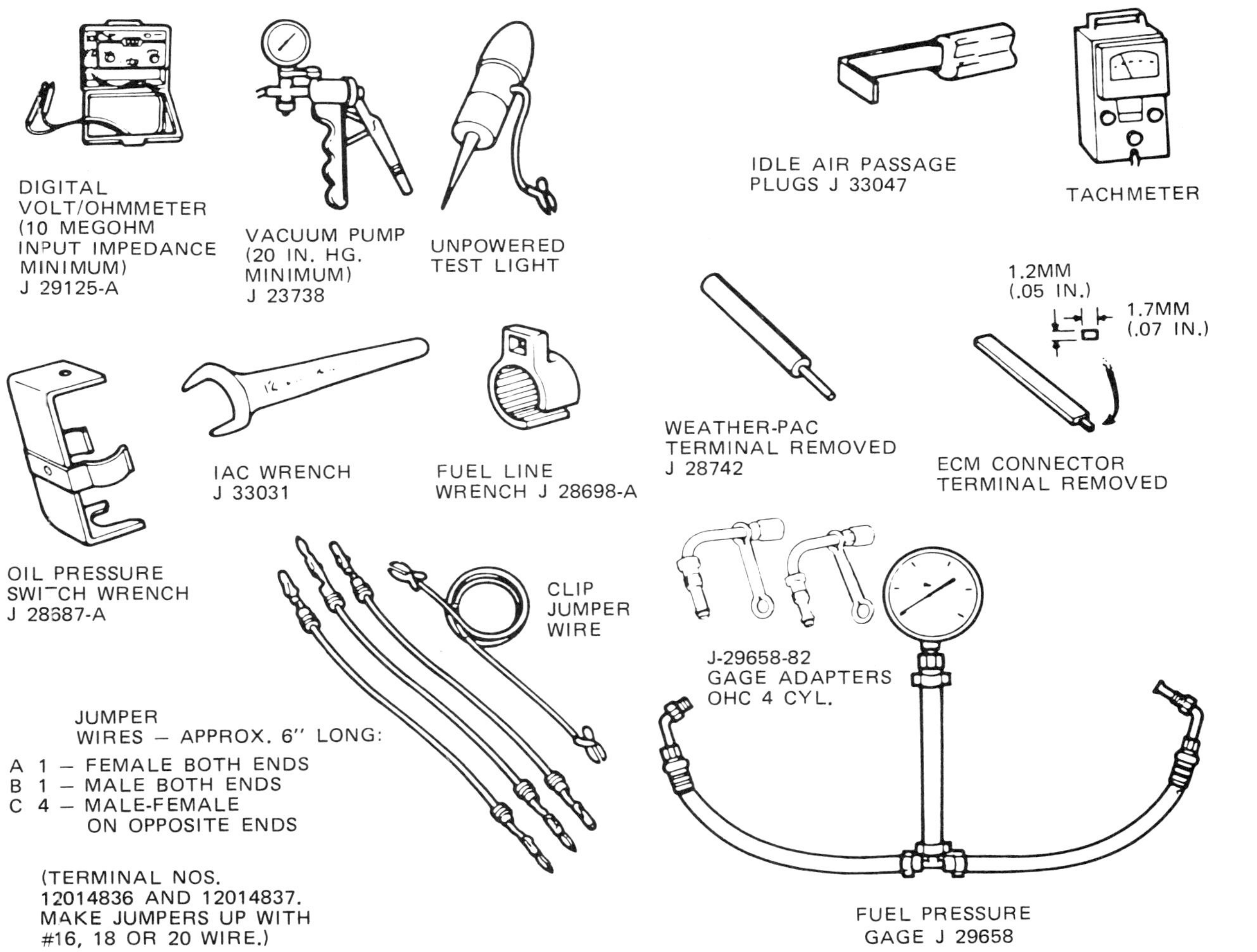

FIGURE 2-36 Diagnosis Tools *(Courtesy of Oldsmobile Division).*

5. *Jumper leads:* Jumper leads equipped with Weather-Pak terminals are needed for TPS adjustments. One jumper should have female terminals on both ends. A second jumper should have male terminals on both ends. A third jumper should have a male terminal at one end and a female terminal at the other end. These jumpers should be about 6 inches long. Four of each should be enough for most testing.
6. *Fuel injection pressure gauge:* A fuel injection pressure gauge and adapters are needed for pressure tests and system diagnosis.

Other tools needed are usually part of the technician's normal tool inventory.

DIAGNOSTIC CHARTS AND DECISION TREES

The diagnostic charts in the service manual provide a step-by-step guideline for system testing and analysis. Where applicable, an electrical diagram or vacuum diagram is included on the chart to help with the diagnostic procedure. Specifications are included, and in some cases a pictorial view of the component or electronic circuitboard may also be shown.

Charts are provided for diagnosing the data sensors, ECM, PROM, CALPAK, fuel system, ECM self-diagnostic function, and an "engine cranks but no start" condition.

The throttle body fuel injection system is an improvement over the older EFI system discussed in Chapter 1, but it is far more complex. As we continue, further improvements will be made to the injection systems. Complexity will remain about the same, so it is very important that a basic understanding of this system be achieved.

Diagnostic Charts

Diagnostic charts aid the technician in effecting repairs. Those available for TBI and CFI systems are listed here.

ECM Connector Identification
Diagnostic Circuit Check
Field Service Mode Check
Driver's Complaint
No Check Engine Light
Won't Flash Code 12, Check Engine Light On Steady
Engine Cranks—Will Not Start, 1.8 L, 2.0 L, 2.5 L
Engine Cranks—Will Not Start, 5.0 L, 5.7 L
Fuel System Diagnosis, 1.8 L, 2.0 L, 2.5 L
Fuel System Diagnosis, 5.0 L, 5.7 L
HEI Diagnosis—Integral Coil
HEI Diagnosis—Remote Coil
Driveability Symptoms
EGR Check—1983 2.0 L, 5.0 L, 5.7 L
EGR Check—1982 5.0 L, 5.7 L
Coolant Fan Control—1.8 L, 2.5 L
Air Management Check—2.0 L, 5.0 L, 5.7 L
Pulse Air Solenoid Check—1983 2.0 L
Canister Purge Valve—1983 2.0 L, 5.0 L, 5.7 L
Canister Purge Valve—1982
Transmission Converter Clutch
Hood Louver Solenoid—5.0 L, 5.7 L
A/C Clutch Control—1983 2.0 L
A/C Clutch Control—2.5 L
Park/Neutral Switch
Crank Signal
Electronic Spark Control—1983
Electronic Spark Control—1982
Idle Air Control—1983
Idle Air Control—1982
Idle Air Control, Power Steering Pressure Signal—2.0 L
Power Steering Pressure Signal—2.5 L X Series 1983 1/2 Automatic except 2.84 axle

ECM Diagnostic Trouble Code Charts

Code 13—0_2 Sensor Circuit
Code 14—Coolant Sensor Voltage Low
Code 15—Coolant Sensor Voltage High
Code 21—Throttle Position Sensor Voltage High
Code 22—Throttle Position Sensor Voltage Low
Code 24—Vehicle Speed Sensor—Remote, 1.8 L, 2.5 L
Code 24—Vehicle Speed Sensor—Remote, 2.0 L, 5.7 L
Code 24—Vehicle Speed Sensor—In Cluster, 1.8 L, 2.5 L
Code 24—Vehicle Speed Sensor—In Cluster, 2.0 L, 5.0 L
Chart 24A—Vehicle Speed Sensor—Cluster Connectors
Code 33—Manifold Absolute Pressure Sensor Voltage High
Code 34—Manifold Absolute Pressure Sensor Voltage Low
Code 42—Electronic Spark Timing
Code 43—Electronic Spark Timing—1983

Code 43—Electronic Spark Timing—1982
Code 44—Lean Exhaust Indication
Code 45—Rich Exhaust Indication
Code 51/55—PROM/ECM

ABBREVIATIONS

ALCL Assembly line communication link

BARO Barometric pressure

CCP Canister purge

Conv. Converter

Diagnostic "test" term Lead or ALCL connector terminal, which is grounded to obtain a trouble code

ECM Electronic control module

EFI Electronic fuel injection

EGR Exhaust gas recirculation

ESC Electronic spark control

EST Electronic spark timing

HALL Hall-effect module

HEI High-energy ignition

IAC Idle air control

MAP Manifold absolute pressure

PCV Positive crankcase ventilation

P/N Park/Neutral switch

Port Exhaust ports

PROM Programmable read-only memory (engine calibration unit)

TBI Throttle body injector

TCC Torque converter clutch

TPS Throttle position sensor

Vac. Vacuum

VIN Vehicle identification number

VSS Vehicle speed sensor

UNIT I
General Motors Corporation Systems

3 Cadillac Digital Electronic Fuel Injection, 1980

OVERVIEW

The 1980 Cadillac Eldorado and Seville are equipped with a 6.0-liter engine that is provided fuel by a digital electronic fuel injection (DEFI) system (Fig. 3-1). This system is designed as an improvement over the earlier EFI system discussed in Chapter 1.

The DEFI system controls exhaust emissions while it maintains or improves fuel economy. It is similar in operation to TBI discussed in Chapter 2. It has a self-diagnostic capability, stores trouble codes that can be "called out" by the service technician, and has a CHECK ENGINE light to warn the driver of circuit problems, whether intermittent or hard. This system is a speed/density fuel injection system where the MAP, BARO, and MAT data are used by the ECM to calculate the density and volume of the air entering the engine. After also receiving rpm data from the EST module, the ECM can make high-speed digital calculations to determine fuel injector pulse width. This system does not use an oxygen sensor; therefore, there is no feedback feature by which the ECM can compensate for its adjustment of the air/fuel mixture. The DEFI system is known as an open-loop system and in this regard is similar to the Cadillac EFI system discussed in Chapter 1. The Cadillac EFI system is an analog system, whereas the DEFI system is a digital system. Both were produced by Bendix Corporation for the Cadillac Motor Car Division of General Motors Corporation.

Instead of a carburetor, the DEFI system uses a throttle body with two electronically controlled fuel injectors providing the fuel spray. Fuel flow from the injectors is dependent on engine sensor input data to the ECM, which then calculates the fuel injector pulse width. The two injectors are pulsed alternately.

Fuel pressure supplied to the injectors is sufficient to maintain a consistent pressure across the injector inlets during operation. When the injectors are pulsed, the ball valve inside moves to the full-open position. Since the pressure to the injectors remains constant, the amount of fuel injected is changed by varying the amount of time the injector is open or the pulse width. The manifold absolute pressure sensor (MAP), the manifold air temperature sensor (MAT), and the engine rpm are used by the ECM to calculate pulse width for the injectors. Under normal conditions, each injector is pulsed twice for every revolution of the crankshaft.

The ignition system consists of an EST distributor with a seven-terminal module. The ECM calculates spark advance, so the vacuum and centrifugal advance mechanisms are not needed.

Engine idle speed is controlled at closed throttle by the ECM. The ECM uses an idle speed control (ISC) motor, which is a small reversible electric motor mounted on the throttle body, to control the idle speed. Idle speed is

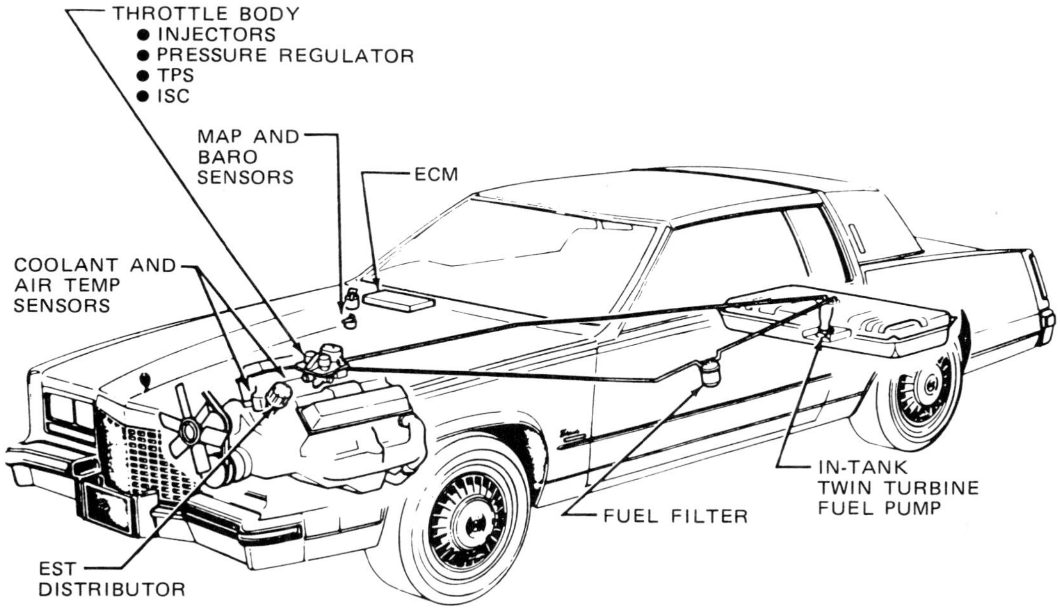

FIGURE 3-1 Digital electronic fuel injection system components *(Courtesy of Cadillac Motor Car Division).*

controlled in response to changes in engine load or input from engine data sensors.

SYSTEM SUBSYSTEMS

The DEFI system consists of several smaller systems, including the fuel supply system and fuel injectors, the air induction system, engine data sensors, the electronic control module, the electronic spark timing system, the idle speed control system, and the EGR valve control system. Each subsystem will be discussed in detail.

Fuel Supply System

The fuel supply system consists of an in-tank electric fuel pump, the fuel tank and cap, an in-line fuel filter, the fuel lines, and the throttle body, which consists of a two-piece casting, two fuel injectors, and a pressure regulator (Fig. 3-2).

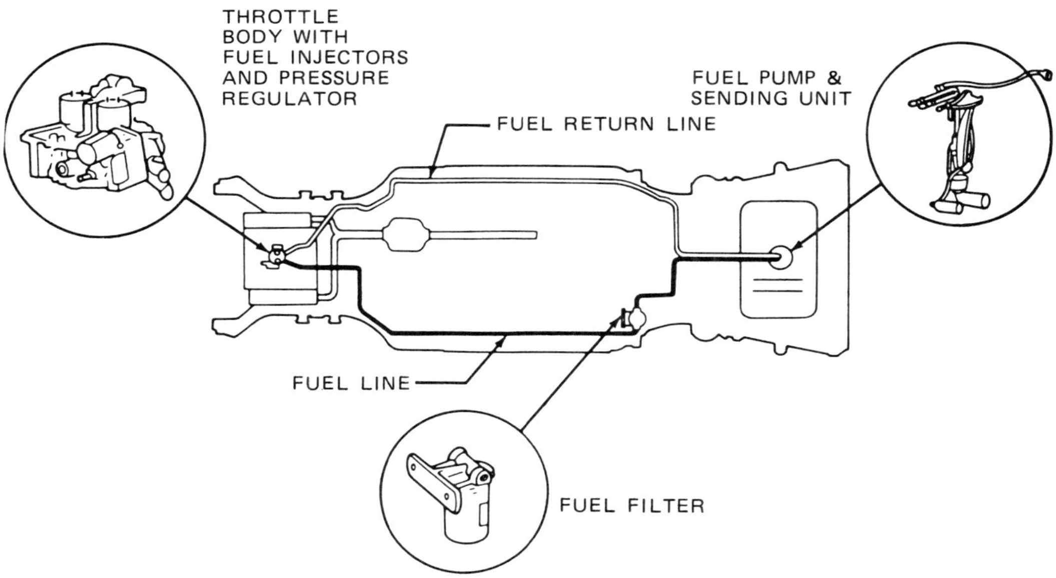

FIGURE 3-2 Fuel supply system *(Courtesy of Cadillac Motor Car Division).*

In-tank Electric Fuel Pump

The in-tank electric fuel pump is a twin-turbine type of pump that is mounted on the fuel tank float unit (Fig. 3-2). It is serviceable separate from the sending unit. The pump itself is not repairable. The ECM controls the fuel pump relay, located in the relay center under the instrument panel, and the relay operates the fuel pump. This circuit is protected by a 10-ampere fuse located in the mini fuse block under the instrument panel. The ECM energizes the relay when the ignition switch is turned to ON or to START. When the switch is turned to ON, if the engine is not cranked within 1 second, the ECM de-energizes the relay and pump. Fuel is pumped from the tank through the fuel filter and lines to the throttle body to supply the injectors. The pressure regulator returns excess fuel to the tank.

Fuel Tank

The fuel tank contains a reservoir beneath the fuel pump pickup screen. This reservoir provides a sure supply of fuel when the level is low or during severe maneuvering conditions. The return fuel from the pressure regulator empties into this reservoir. The fuel tank cap is a pressure-vacuum release emission control type.

Fuel Filters

The system has three fuel filters (Fig. 3-2). The first is a woven plastic screen or filter on the fuel pump pickup tube in the fuel tank reservoir. It can be replaced separately from the rest of the pump/sending unit assembly. The second is an in-line filter mounted on the frame rail near the left rear wheel. This filter can remove contaminants as small as 10 microns (0.0004 inch) in size. The filter element is contained in a screw-on can and can be replaced along with a new square-cut rubber seal at the recommended interval. The last filter is a fine-mesh screen surrounding the inlet of the fuel injector.

Fuel Pressure Regulator

The fuel pressure regulator is an integral part of the fuel metering body of the throttle body assembly (Fig. 3-3). It is a diaphragm-operated relief valve that has fuel pressure on one side and atmospheric pressure on the other. Fuel operating pressure is set by a calibrated spring. Pressure is held constant across the injectors to about 10.5 psi. The pressure regulator and fuel metering body cover are serviced as an assembly.

Fuel Injectors

The two fuel injectors are pulsed alternately by the ECM. They meter and direct fuel spray into the throttle body bores. The injectors are, basically, solenoids whose plungers have a ball valve at the bottom end (Fig. 3-4). When the injector is pulsed, the ball valve is raised off its seat against spring tension. At this point, fuel is sprayed into the throttle body bore. When the injector is de-energized, the spring returns the ball valve to its seat, sealing off the flow of fuel. The fuel injectors may be replaced individually.

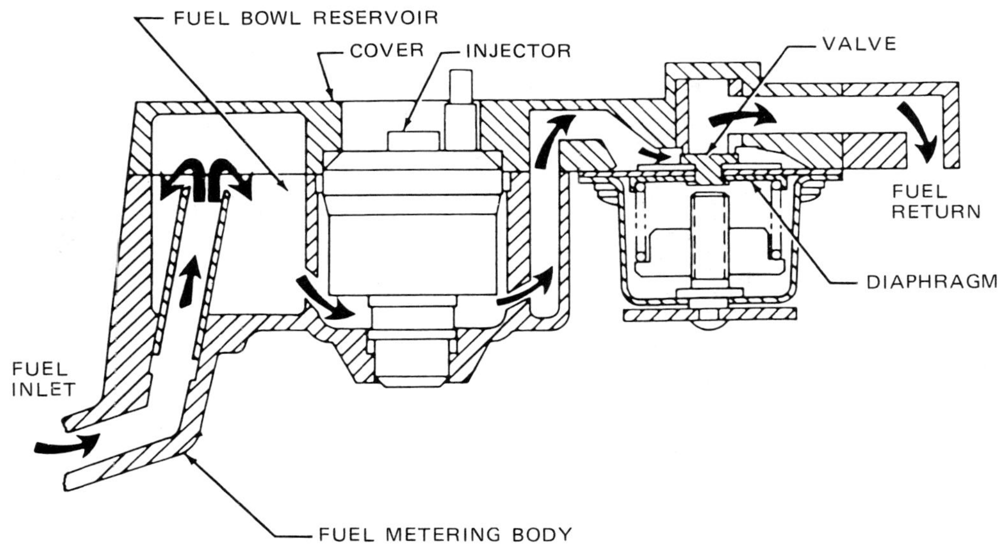

FIGURE 3-3 Fuel pressure regulator *(Courtesy of Cadillac Motor Car Division).*

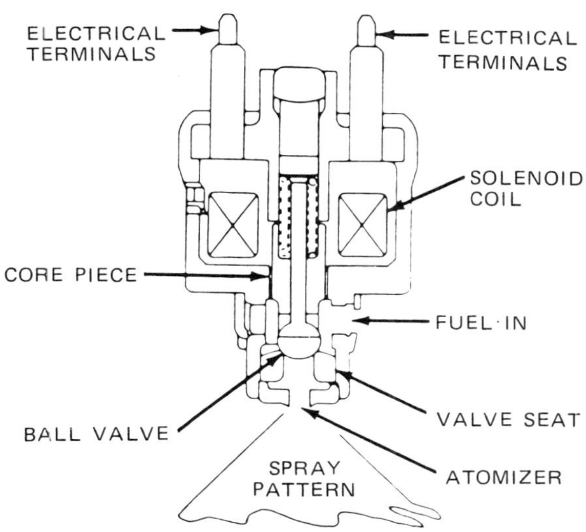

FIGURE 3-4 Fuel injector *(Courtesy of Cadillac Motor Car Division).*

Fuel Lines

The fuel supply line is 3/8-inch inside diameter and runs along the left-side frame rail (Fig. 3-2). A braided stainless steel mesh-covered Teflon hose connects this fuel line to the steel line on the engine. This allows flexibility and yet protection against chafing and heat. The return line is 5/16-inch inside diameter and is mounted along the right-side frame rail. A braided stainless steel mesh-covered Teflon hose connects the engine line to this line also. Replace these lines with the same-sized hose or tubing or pressure drops and driveability symptoms will result.

AIR INDUCTION SYSTEM

The air induction system consists of the model 100 throttle body, the idle speed control motor, and the intake manifold (Fig. 3-5). Air enters the throttle body and is distributed to the cylinders by the intake manifold. The throttle blades control the volume of incoming air. To raise idle speed for cold starts, the idle speed control motor will hold the throttle blades open farther than normal.

Throttle Body

The throttle body assembly, model 100, consists of two main castings, the fuel metering body and the throttle body (Fig. 3-6). An idle speed control motor (ISC) is mounted on the left side, and a throttle position sensor is mounted on the right side. The sensor has a pigtail connector instead of the connector being an integral part of the sensor. The ISC function will be described later as one of the ECM-controlled parameters.

Intake Manifold

The intake manifold is a dual-plane design and is made of aluminum. Dual plane refers to the two levels of the runners that carry the air/fuel mixture to the cylinders. The

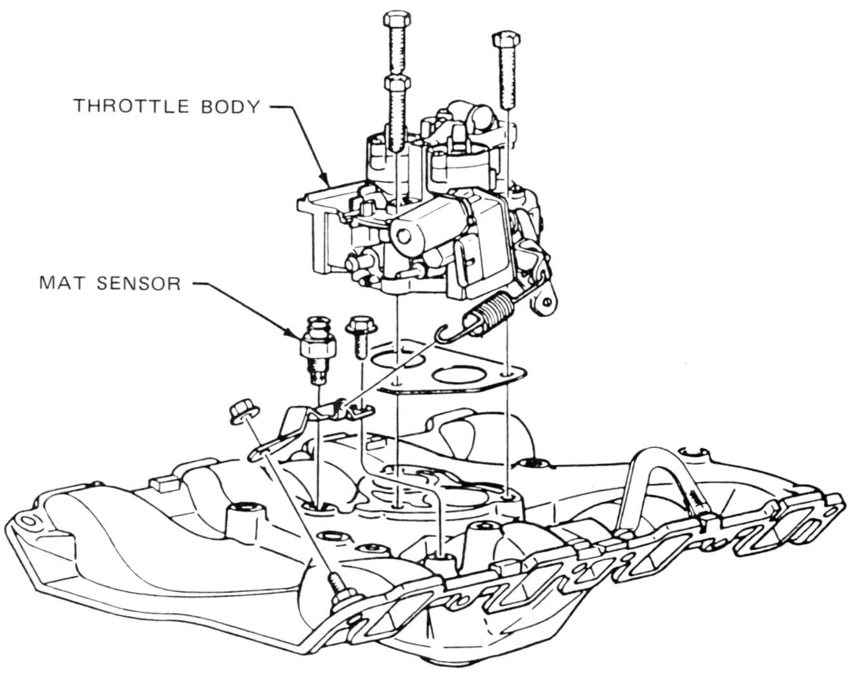

FIGURE 3-5 Throttle body installation *(Courtesy of Cadillac Motor Car Division).*

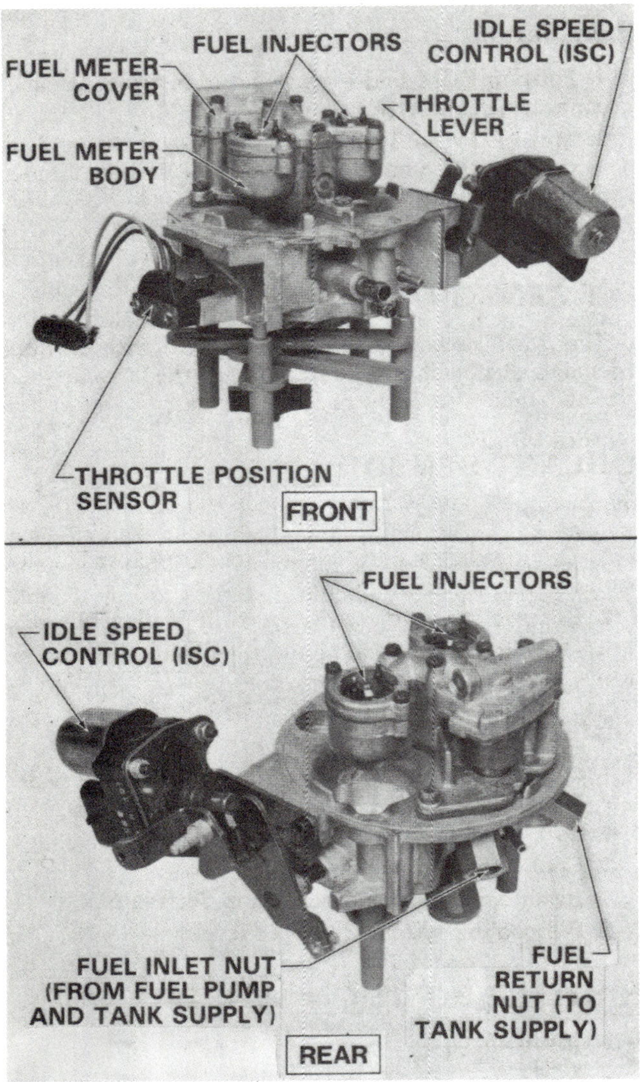

FIGURE 3-6 Throttle body assembly (DEFI) *(Courtesy of Cadillac Motor Car Division).*

manifold also has a heat crossover passage for the early fuel evaporation emission control system to be described later in this chapter.

ENGINE DATA SENSORS

The engine data sensors include a manifold air temperature sensor (MAT), an engine coolant temperature sensor (CTS), a manifold absolute pressure sensor (MAP), a barometric pressure sensor (BARO), a throttle position sensor (TPS), and an rpm sensor which is part of the electronic spark timing (EST) module.

Manifold Air Temperature Sensor

The manifold air temperature sensor (MAT) is screwed into the intake manifold in front of the throttle body (Fig. 3-1). It measures the temperature of the air/fuel mixture and sends these data to the electronic control module (ECM). This sensor is a thermistor whose resistance value changes with temperature. When the temperature is low, its resistance is high. As the temperature rises, the sensor resistance decreases.

Engine Coolant Temperature Sensor

The engine coolant temperature sensor (CTS) is similar in operation to the MAT. It is located at the right front corner of the engine near the thermostat housing (Fig. 3-1). It senses engine coolant temperature, changes its resistance in accordance with changes in temperature, and sends these data to the ECM. If coolant temperature is low, the sensor resistance will be high. As temperature increases, the CTS resistance decreases. The ECM uses CTS data for fuel enrichment during cold starts, idle speed control, ignition timing, and exhaust gas recirculation (EGR) operation.

Manifold Absolute Pressure Sensor

The manifold absolute pressure sensor (MAP) uses an internal transducer to monitor changes in intake manifold pressure (or engine vacuum). These changes are a result of engine load variations, speed changes, and gradual normal wear. As engine load increases, intake manifold pressure increases, and the MAP sensor transducer converts the pressure changes into electrical signals. Under heavier loads, more fuel is needed. The MAP sensor indicates this need to the ECM, which, in turn, lengthens the fuel injector pulse width. As engine load decreases, the data sent to the ECM by the MAP sensor cause a decrease in injector pulse width. The MAP sensor is located under the instrument panel near the right-side air-conditioning outlet, is connected electrically to the ECM, and is connected to a port on the throttle body where it gets its pressure (or vacuum) signal (Fig. 3-7).

Barometric Pressure Sensor

The barometric pressure sensor (BARO) provides data representing ambient or atmospheric pressure to the ECM. Ambient pressure changes as weather or altitude changes. The BARO data, in conjunction with MAT and MAP data, are used to determine air volume and speed density. From these calculations, the ECM can better regulate injector

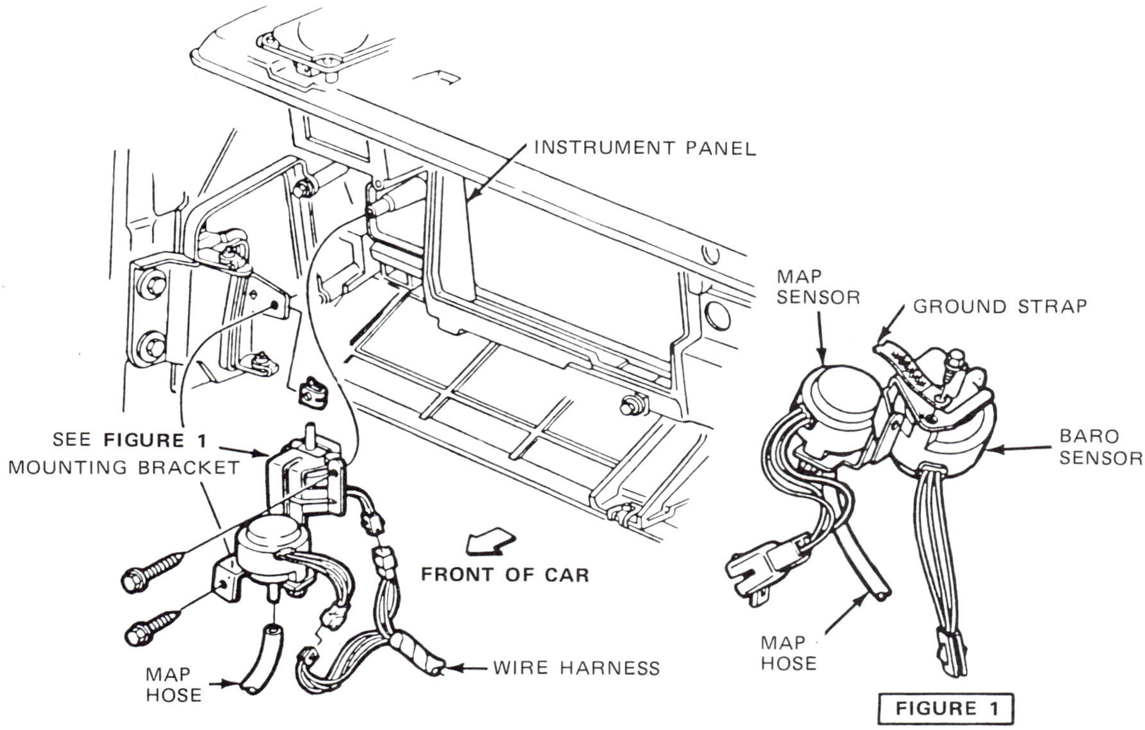

FIGURE 3-7 Barometric and manifold absolute pressure sensor installation *(Courtesy of Cadillac Motor Car Division).*

pulse width. The BARO is located under the instrument panel near the right-side air-conditioning outlet (Fig. 3-7). Its atmospheric port is covered by a foam filter.

Throttle Position Sensor

The throttle position sensor (TPS) is mounted on the right side of the throttle body and senses movements of the throttle shaft (Fig. 3-8). As the throttle blades open and close, the sensor, which is a variable resistor, varies a voltage sent to it by the ECM. The ECM interprets this change in voltage as accelerator pedal demand from the driver.

Engine Speed Sensor

The engine speed or rpm sensor is part of the electronic spark timing module in the distributor. Pulses are sent to the ECM by the EST module. The time lapse between pulses is used by the ECM to calculate engine speed and spark advance.

DEFI sensor signals are fed into the ECM where all calculations and commands for control of the system are processed. The ECM is mounted under the instrument panel

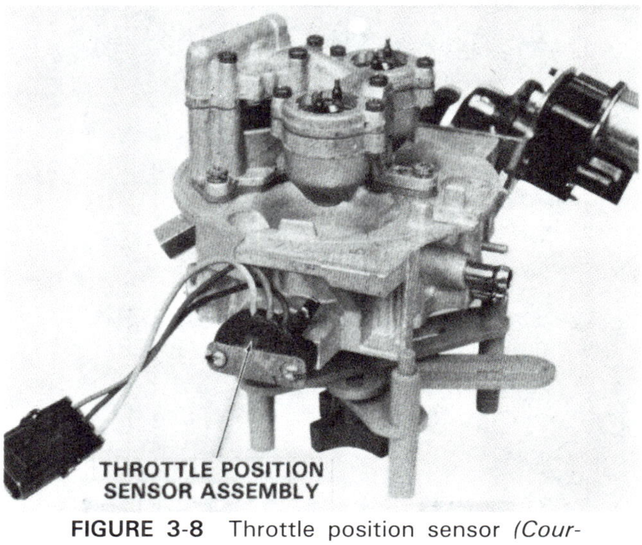

FIGURE 3-8 Throttle position sensor *(Courtesy of Cadillac Motor Car Division).*

(Fig. 3-9). ECM input data include signals from the engine coolant temperature sensor, throttle position sensor, barometric pressure sensor, manifold absolute pressure sensor, manifold air temperature sensor, and the engine speed of

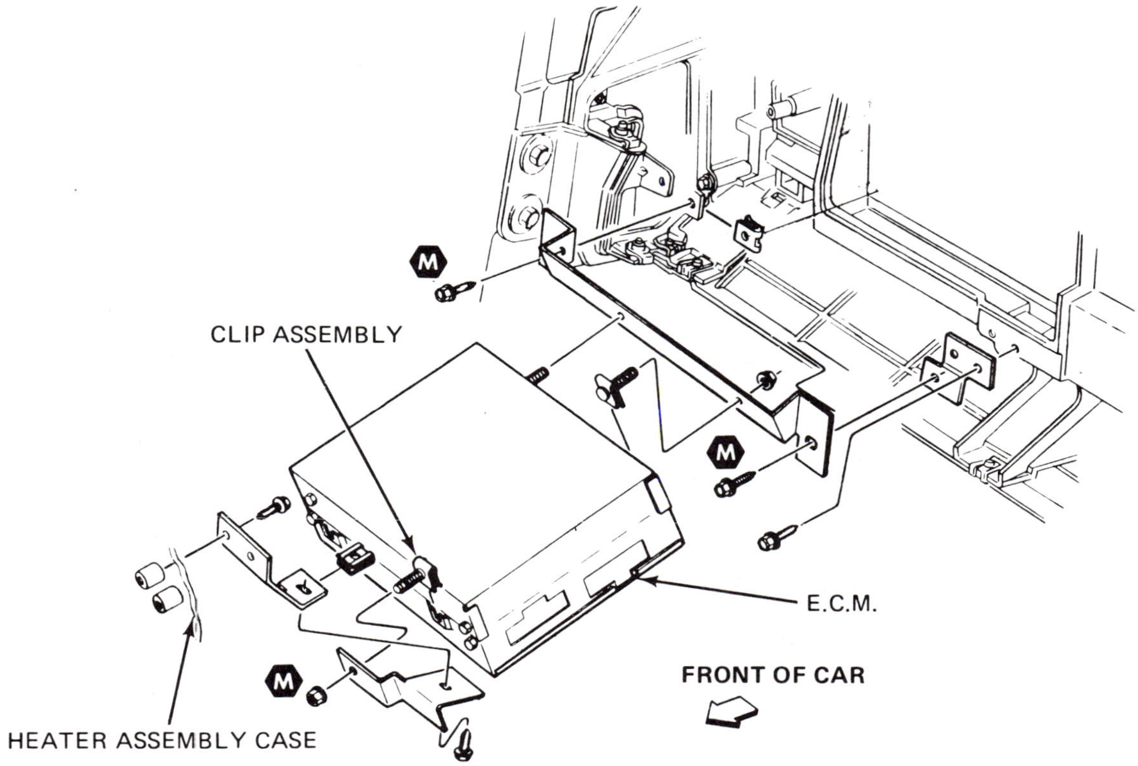

FIGURE 3-9 Electronic control module removal *(Courtesy of General Motors Product Service Training).*

the EST module. The ECM makes high-speed calculations based on these signals, and its microprocessor can issue commands to the electric fuel pump, the EGR solenoid, the fuel injectors, the idle speed control motor, and the spark advance control of the EST module. When the engine is running, the input signals and the ECM program are used to calculate a pulse width for the injectors to provide a desirable air/fuel ratio.

FIGURE 3-10 Electronic control module *(Courtesy of Cadillac Motor Car Division).*

ELECTRONIC CONTROL MODULE SYSTEM CONTROL

The electronic control module controls the fuel delivery system, the electronic spark timing system, and the idle speed control system (Fig. 3-10).

Fuel Control

The ECM has three modes of controlling the fuel delivery system: cranking, part throttle, and wide-open throttle. A voltage signal is sent to the ECM through a wire from the ignition switch when the engine is cranked. The ECM, realizing that the engine is cranking, issues one long *prime* pulse to the injectors, followed by several *starting* pulses until the cranking voltage signal stops. There is also a *clear-flood* operation circuit available where, if the throttle is held wide open and the engine is cranked for longer than 5 seconds, smaller alternating injector pulses are issued. This allows for clearing the intake manifold of excess fuel and is similar in function to the "unloader" operation of the carburetor.

Once the engine starts, the injector pulse width is varied to match the engine operation, such as idle, part throttle, acceleration, deceleration, altitude, air and coolant temperature, and engine load. By comparing the MAP and BARO sensor readings, the ECM can widen the pulse width for additional enrichment. As the difference between the two readings becomes less and less, the ECM realizes that the throttle blades are approaching a wide-open position. The TPS reinforces this conclusion.

Electronic Spark Timing

The electronic spark timing system is controlled by the ECM for spark advance. The sensor input data enable the ECM to calculate the optimum spark advance for best fuel economy, performance, and lowest emissions. The ECM sends a signal to the HEI module in the distributor, which, in turn, fires the spark plugs. The ECM can modify, delay, or advance this signal issued to the HEI module, thereby regulating the spark advance.

The ECM control of the spark advance operates in one of two operating modes: cranking/bypass mode and normal mode. During the cranking/bypass mode, the ignition timing is at a *reference* setting. This means the spark plugs fire at a point determined by setting the distributor. In years past, this was known as *initial* or *base* timing. The ignition operates in this mode during cranking in spite of other sensor data input. During normal operation, the ECM can add to or change the base timing in accordance with the sensor data and the ECM program.

Idle Speed Control

The idle speed control system is controlled by the ECM to provide normal idle, fast idle, and an antistall condition when decelerating or closing the throttle quickly. No adjustments for idle speed are possible as it is programmed into the ECM. The computer monitors engine rpm and then makes adjustments in its signal to the idle speed control motor (Fig. 3-11). For starting, the ISC motor holds the throttle blades open for a period of time determined by the air and coolant temperature sensor data. This provides a fast idle during warmup conditions.

Park/Neutral Signal

A voltage signal that indicates when the transmission/transaxle shift lever is moving from PARK or NEUTRAL to any other gear is provided by the automatic door lock circuit. When the shift lever is moved from PARK or NEUTRAL, the door lock circuit is energized, automatically locking the doors. This circuit sends a voltage signal to the ECM, which, in anticipation of the load about to be placed on the engine by the transmission, maintains idle speed to compensate for the load.

FIGURE 3-11 Idle speed control *(Courtesy of Cadillac Motor Car Division).*

Air Conditioning Signal

When the air-conditioning compressor clutch circuit is energized, the ECM, which monitors the voltage in this circuit, compensates for the load of the compressor by maintaining idle speed.

Throttle Switch

A throttle switch, sometimes referred to as a nose switch, is incorporated into the ISC motor to indicate to the ECM whether or not the driver's foot is on the accelerator pedal. The signal to the ECM from this switch tells the ECM to control the idle speed (switch closed) or not to control the idle speed (switch open because the driver has moved the throttle linkage off of the ISC motor plunger). The ECM fully extends the plunger and then stops issuing idle speed commands.

The ISC motor acts as an antistall dashpot during quick deceleration. It holds the throttle blades off idle for a brief period to prevent stalling and to improve emission levels.

ELECTRONIC SPARK TIMING SYSTEM

The electronic spark timing/HEI module in the distributor is controlled by the ECM when the engine is running. The circuitry for controlling spark advance is in the ECM, and the circuitry for monitoring engine rpm and for triggering the ignition circuit is in the HEI module. The EST distributor is the same as its HEI predecessor except it has no centrifugal or vacuum advance mechanisms. The EST/HEI module has seven terminals, whereas the older HEI module had only four terminals. The three new terminals are for engine rpm reference, EST, and bypass. The reference terminal provides an rpm signal from the pickup coil to the ECM. The ECM sends a signal to the bypass terminal to indicate that a reference signal was received. This bypass signal occurs only when the engine rpm is 200 or more. Sensor input data help the ECM calculate spark advance, and that advance signal is sent to the EST terminal for control of the ignition primary circuit.

If the HEI module does not receive a bypass signal or if its reference signal does not reach the ECM, it will continue to operate the distributor at base timing in accordance with a program in the module. This timing will be slightly retarded from normal operation and will vary somewhat with engine speed. In this condition, sensor data are not used.

When the distributor reference pulses reach 200 rpm and the ECM receives this signal, the ECM sends out a bypass signal to the HEI module. When this signal reaches the module, the ignition primary circuit control is transferred from the HEI module circuitry to the ECM EST circuitry. The signal from the ECM is transmitted over the EST wire and module terminal. If this wire is open, the HEI module will not receive the ECM EST signal and the engine will stop. A crank, start and run, then stop condition will exist.

IDLE SPEED CONTROL SYSTEM

The idle speed is controlled by the ECM, which uses a small reversing electric motor to move the throttle linkage (Fig. 3-11). Idle speed is maintained in accordance with the ECM program and sensor input data, as discussed earlier. The throttle switch is built into the idle speed control motor and tells the ECM whether or not to control the idle speed. After the engine is shut off, the ECM commands the ISC motor plunger to extend slightly in preparation for the next cranking cycle.

EXHAUST GAS RECIRCULATION VALVE CONTROL

The exhaust gas recirculation valve is controlled by the ECM through the use of an EGR solenoid. If engine coolant temperature is below 120°F, the ECM issues a voltage signal to the EGR solenoid. The solenoid is then energized and blocks vacuum signals to the EGR valve, rendering it inoperative. The EGR solenoid is also energized while cranking and at wide-open throttle.

When coolant temperature is above 120°F, the ECM de-energizes the solenoid and the EGR valve operates in response to its ported vacuum and exhaust back-pressure signals.

SYSTEM OPERATION

When the ignition switch is turned to ON or to CRANK, the ECM receives power from the battery. At this point, the ECM receives the following input data: throttle position, engine coolant temperature, engine rpm, manifold air temperature, manifold absolute pressure, and barometric pressure. As a result, the following commands are issued by the ECM: the electric fuel pump is turned on, the EGR solenoid may or may not be energized depending on coolant temperature, the ISC motor extends slightly, the spark advance system is ready for activation, and the fuel injectors are activated during cranking. In conclusion, the DEFI system is energized when the ignition switch is ON. In addition, the CHECK ENGINE light on the instrument panel is lit as a bulb check (Fig. 3-12).

IGNITION	ENGINE	BULB
OFF	OFF	OFF
ON	OFF	ON
ON	CRANK	ON
ON	ON	OFF

FIGURE 3-12 CHECK ENGINE light functions *(Courtesy of General Motors Product Service Training).*

When the engine is started, the fuel pump is activated for continual operation, the fuel pressure regulator maintains 10.5 psi across the injectors, the ISC motor begins controlling idle speed if the throttle or nose switch is closed, the spark advance system shifts from bypass or base timing to the EST mode, and the ECM continually monitors the engine coolant temperature, barometric pressure, intake manifold air temperature, manifold absolute pressure, engine rpm, and throttle position sensor.

To operate the two fuel injectors, the ECM provides grounds for them alternately. The power feed wire provides 12 volts to the injectors, and the ECM completes the circuit by providing the grounds alternately through its circuitry.

SYSTEM SERVICE

Idle Speed

With the engine at operating temperature, check the idle speed specification on the emission decal under the hood and compare it to a tachometer reading. There are no adjustments possible. If the idle speed is not correct, refer to the appropriate chart in the service manual.

TPS Adjustment

A tachometer and a 10-megohm impedance digital multimeter are necessary for setting the TPS. Follow the steps listed in the service manual for meter connections and the adjustment procedures (Fig. 3-13). If these procedures are not followed, driveability symptoms may result.

Minimum Air Rate Adjustment

The minimum air rate adjustment is necessary only when throttle body parts have been replaced or when the TPS adjustment requires it. A tachometer is the only meter needed to make this adjustment. Refer to the service manual for correct procedures (Fig. 3-14).

Idle Speed Control Motor Adjustment

This adjustment is required only as a result of ISC motor replacement. A tachometer is required. The service manual provides steps to be followed to make this adjustment correctly.

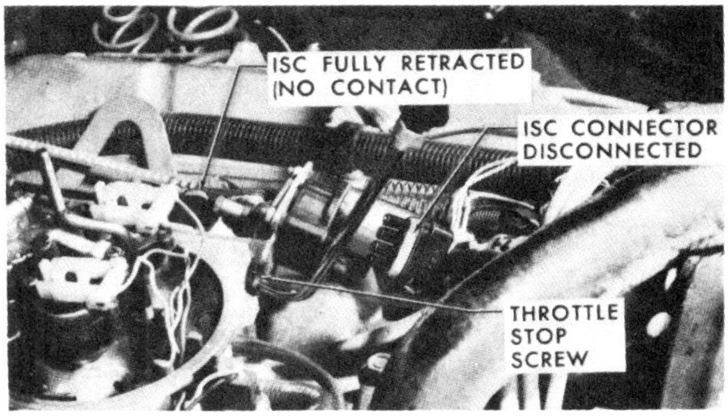

FIGURE 3-14 Adjusting throttle stop *(Courtesy of Cadillac Motor Car Division).*

Throttle Body Service

Prior to servicing the throttle body, it is necessary to relieve the fuel system pressure by following the prescribed procedure in the service manual.

The ISC motor should not be immersed in any cleaning solvent. It cannot be repaired and should be replaced as an assembly (Fig. 3-15).

The fuel metering body cover and pressure regulator are serviced as an assembly and also should not be immersed in cleaning solvent (Fig. 3-16).

The fuel injectors can be replaced individually and should not be immersed in cleaning solvent. When installing the injector, be sure the steel backup washer is under the large O-ring seal used at the top of the injector (Fig. 3-17). Incorrect installation will result in a fuel leak.

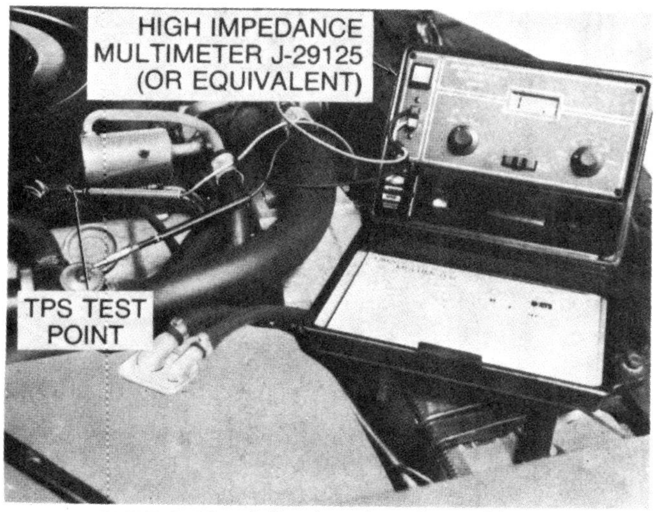

FIGURE 3-13 Adjusting throttle position sensor *(Courtesy of Cadillac Motor Car Division).*

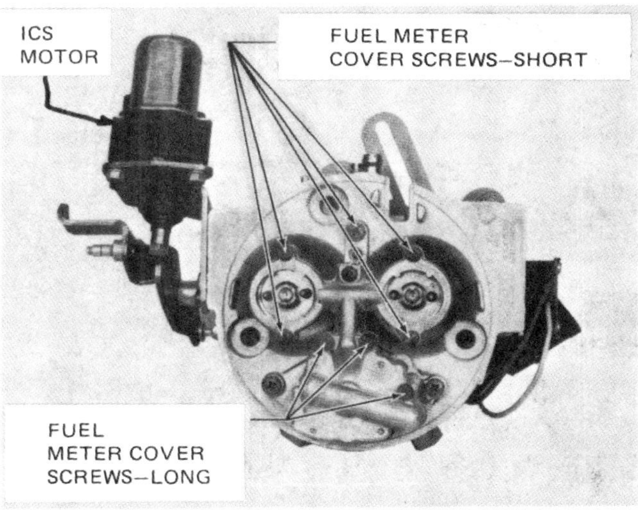

FIGURE 3-15 Idle speed control motor *(Courtesy of Cadillac Motor Car Division).*

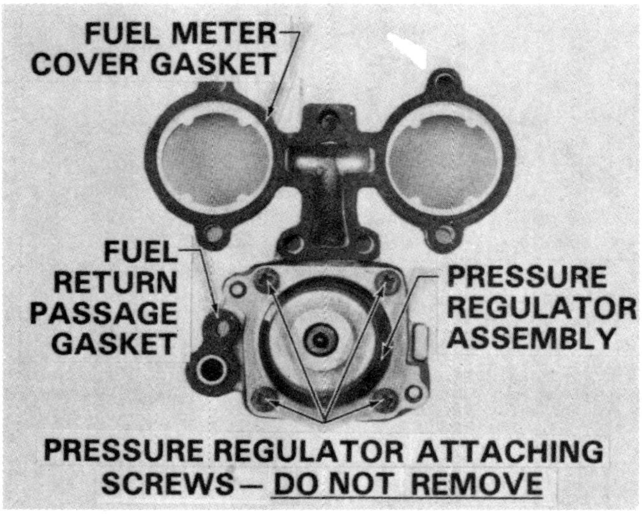

FIGURE 3-16 Fuel meter cover, bottom view *(Courtesy of Cadillac Motor Car Division).*

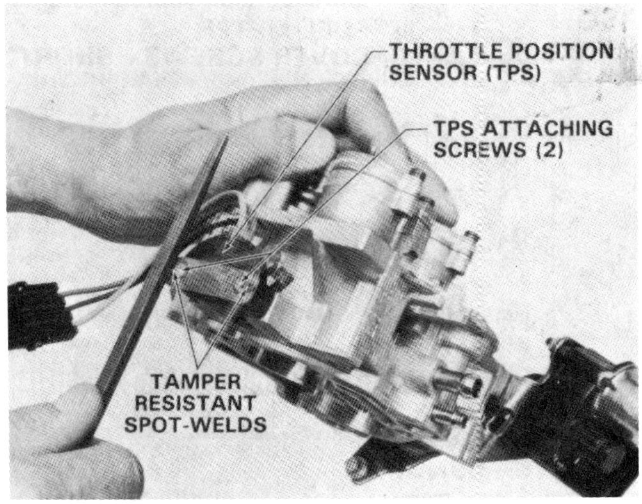

FIGURE 3-18 Removing throttle position sensor *(Courtesy of Cadillac Motor Car Division).*

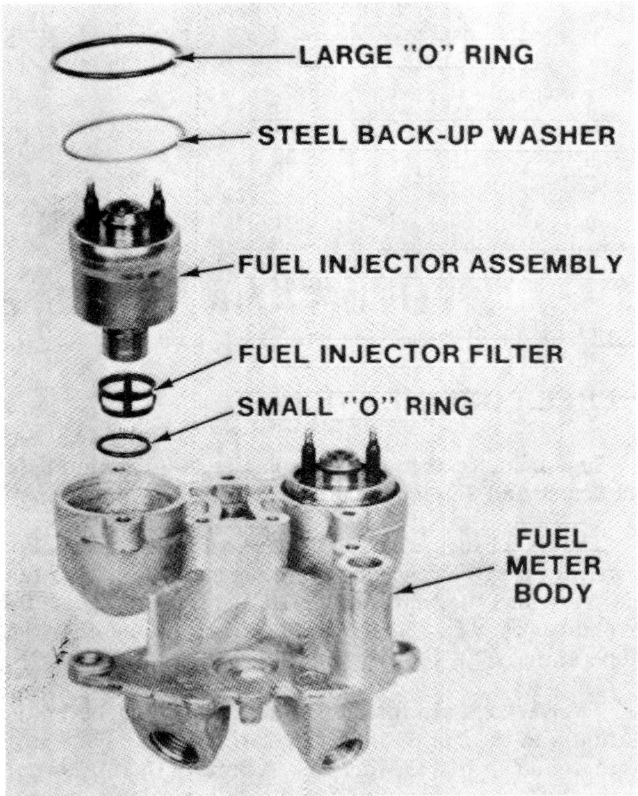

FIGURE 3-17 Fuel injector installation *(Courtesy of Cadillac Motor Car Division).*

The TPS is mounted on the throttle body with two screws (Fig. 3-18). These screws are spot welded in place. To remove them, the spot welds must be filed off. The TPS should not be immersed in cleaning solvents.

The throttle body casting can be cleaned using a cold immersion-type solvent. New gaskets and seals should be used for reassembly. Rubber O-ring seals should be installed carefully to prevent nicking and fuel leaks. Before starting the engine, pressurize the fuel system by turning the ignition switch to ON and check for fuel leaks.

SENSOR SERVICE

Manifold Absolute Pressure Sensor

The manifold absolute pressure sensor (MAP) is located under the extreme right end of the instrument panel. The barometric pressure sensor (BARO) is mounted on the same bracket that secures the MAP sensor to the underside of the instrument panel (Fig. 3-19). The MAP and BARO sensors are pressure-sensing potentiometers. They can be tested using a 10-megohm impedance multimeter and some jumper wires. Follow the procedure in the service manual diagnostic charts. A malfunction in the MAP circuit should set a code 31, 32, 33, or 34 in the ECM memory. Recalling these codes will be discussed in the next section of this chapter. The BARO sensor circuit should set a code 33, 35, or 36 if a malfunction occurs.

Manifold Air Temperature Sensor

The manifold air temperature sensor is located in the intake manifold in front of the throttle body (Fig. 3-20). Whenever it is removed, the threads should be cleaned and coated with a nonhardening sealer. A problem in the MAT sensor circuit should set a code 37 or 38 in the ECM memory.

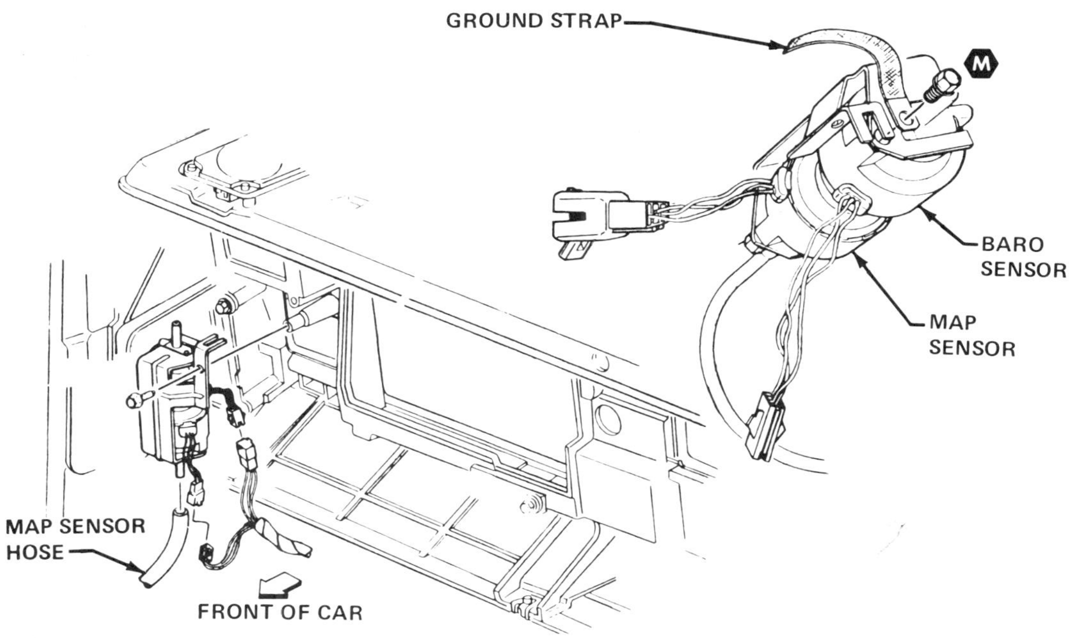

FIGURE 3-19 Manifold absolute pressure sensor removal *(Courtesy of General Motors Product Service Training).*

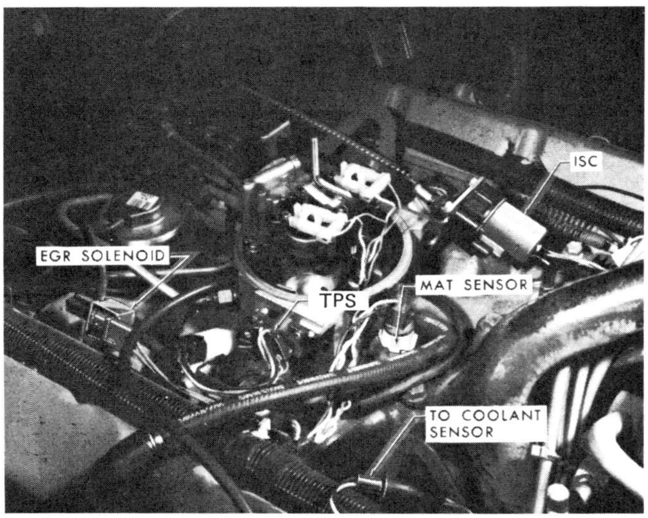

FIGURE 3-20 Engine controls *(Courtesy of Cadillac Motor Car Division).*

The MAT sensor is a thermistor whose resistance lowers as the temperature rises.

Coolant Temperature Sensor

The coolant temperature sensor is located at the right front corner of the engine near the thermostat housing (Fig. 3-20). Whenever it is removed, the threads should be coated with a nonhardening sealer. A malfunction in the CTS circuit should set a code 14 or 15 in the ECM memory. The CTS is a thermistor whose resistance gets less as the temperature increases.

Both the MAT and CTS can be tested using a 10-megohm impedance digital multimeter and some jumper leads. Follow the diagnostic procedure provided in the service manual.

ECM Service

The ECM is located under the instrument panel at the right end. A malfunction in the ECM circuitry should set a code 55 or 56. This necessitates replacement of it as it is not serviceable nor can it be tested using normal procedures.

In addition to the sensor codes referred to above, there are other codes that can be set to indicate problems in other areas. These will be discussed in the next section.

SYSTEM DIAGNOSIS

The DEFI system has a self-diagnostic capability that can be operated from the driver's seat.

Circuit and component problems may cause a code to be set in the ECM memory. Two kinds of codes may be set: intermittent or history codes and current or hard codes. Most intermittent or history codes are caused by electrical connections. The circuit problem occurs and then goes away; a temporary malfunction causes an intermittent code to be set. When a malfunction occurs and remains, a

FIGURE 3-21 CHECK ENGINE light *(Courtesy of Cadillac Motor Car Division).*

current or hard code is set. In either case, the codes can be recalled by the technician servicing the vehicle.

As the vehicle is being driven, if the CHECK ENGINE light (CEL) comes on and then goes out, an intermittent or history code was set (Fig. 3-21). If the CEL comes on and stays on, the code is set and the malfunction remains and is a hard or current code. The ECM keeps codes in memory until erased by the technician or until 20 cycles of the ignition switch have occurred.

Self-diagnostic Mode

To place the DEFI system in its self-diagnostic mode, the technician should use the electronic climate control (ECC) panel in the instrument panel (Fig. 3-22). The ignition switch must be turned to ON for 5 seconds and then the OFF and WARMER buttons on the ECC should be pressed simultaneously and held until a 00 is displayed on the ECC digital display. When this appears, the buttons can be released and the system will automatically begin displaying digital codes stored in memory. The codes are displayed in numerical order beginning with the lowest code.

Trouble codes can be cleared or erased from the ECM memory by pressing the OFF and HI buttons simultaneously while in the diagnostic mode (Fig. 3-22). The buttons should be held until 00 is displayed.

To get the system to return to normal operation, any ECC function button, such as ECON or AUTO, can be pushed (Fig. 3-22) or turn the ignition switch off for at least 10 seconds. Trouble codes will not be erased by using this procedure. Use this procedure for exiting diagnostics.

As the vehicle is being driven, if a malfunction in one of the DEFI circuits occurs, the CHECK ENGINE light will come on. The malfunction may or may not affect engine performance depending on the circuit affected. As the technician, to determine which circuit has a problem, you should follow the diagnostic procedure.

DIAGNOSTIC PROCEDURE

The following procedure will not only enable the technician to cause the DEFI system to enter its self-diagnostic mode, but it will also prepare the system for another self-diagnostic feature known as *switch tests.*

By turning the ignition switch to ON for 5 seconds, pushing the OFF and WARMER buttons on the ECC simultaneously and holding them until 00 is displayed, two digits, 88, should appear. This display ensures accurate

FIGURE 3-22 Diagnostic functions *(Courtesy of Cadillac Motor Car Division).*

TROUBLE CODES THE FOLLOWING TROUBLE CODES HAVE BEEN PROGRAMMED INTO THE ECM.		
TROUBLE CODE	CIRCUIT AFFECTED	SEE PAGE
00	ALL DIAGNOSTICS COMPLETE	11
12	NO TACH SIGNAL	14
14 15	SHORTED COOLANT SENSOR CIRCUIT OPEN COOLANT SENSOR CIRCUIT	16 18
21 22	SHORTED THROTTLE POSITION SENSOR CIRCUIT OPEN THROTTLE POSITION SENSOR CIRCUIT	20 22
28 29 30	SHORTED IDLE SPEED CONTROL CIRCUIT IDLE SPEED CONTROL CIRCUIT IDLE SPEED CONTROL CIRCUIT	24 26 26
31 32	SHORTED MAP SENSOR CIRCUIT OPEN MAP SENSOR CIRCUIT	28 30
33	MAP/BARO SENSOR CORRELATION	33
34	MAP HOSE	34
35 36	SHORTED BARO SENSOR CIRCUIT OPEN BARO SENSOR CIRCUIT	36 38
37 38	SHORTED MAT SENSOR CIRCUIT OPEN MAT SENSOR CIRCUIT	40 42
55 56	ECM	43

FIGURE 3-23 Trouble codes *(Courtesy of Cadillac Motor Car Division).*

display of codes. The double-eight display is digital and consists of seven segments for each digit. If one of the segments is not functional and the technician is unaware of it, the code display will be inaccurate and diagnosis will not be correct.

Once the 88 segment check is finished, the ECM automatically begins displaying the stored codes as described earlier (Fig. 3-23). The lowest-numbered code is displayed first for about 3 seconds. The remaining, if there are any, higher-numbered codes are then displayed in sequence for about 3 seconds each. When the end of the sequence is reached, 88 is again shown and the code display sequence is repeated. When this second code display sequence is finished, the system repeats the process a third time. At the end of the third code display sequence, a digital 70 is shown to indicate the system is ready to perform the switch tests.

SWITCH TESTS

The DEFI system has an additional self-diagnostic function in that it can indicate problems in the brake light circuit, throttle or nose switch circuit, automatic door lock, drive circuit, reverse circuit, the MPG miles per gallon, "average" circuit, and the MPG reset circuit (Fig. 3-24). Activating the switch tests diagnostic procedure will display codes indicating problem areas.

TROUBLE CODE	SWITCH
71	Brake Light Switch
72	ISC Throttle Switch
73	Drive (ADL) Switch
74	Backup Lamp Switch
78	MPG Panel "AVG" Switch
79	MPG Panel "RESET" Switch

FIGURE 3-24 Switch tests codes *(Courtesy of Cadillac Motor Car Division).*

Brake Switch

A code 70 or 71 indicates a malfunction in the brake light switch circuit. This circuit is related to the electronic cruise control system. The ECM monitors this function in order to maintain cruise control speeds.

Throttle Switch

The throttle switch circuit is monitored by the ECM for correct idle speed control, as discussed earlier. A malfunction in this circuit should set a code 72.

Automatic Door Lock Circuit

The automatic door lock or DRIVE circuit is monitored by the ECM so that it can anticipate the load of the transmission/transaxle when shifting from PARK or NEUTRAL, as discussed earlier. A code 73 is set in memory if a problem occurs.

Backup Light Switch

A code 74 indicates a problem in the backup light switch circuit, which is energized when the transmission/transaxle is shifted to reverse.

MPG Sentinel

A code 78 helps to diagnose a problem with the MPG sentinel panel in the instrument panel. This panel provides information related to fuel remaining in the tank and a display of overall or instantaneous fuel economy. The code 78 applies to the miles per gallon "average" circuit.

The MPG sentinel also has a push button that enables the driver to reset the system. A problem in this circuit sets a code 79.

SWITCH TESTS DIAGNOSTIC PROCEDURE

After recalling codes as discussed earlier, the ECM displayed 70 on the ECC panel. At this point, the technician can enter the switch tests diagnostic mode of the computer.

When ready to begin the test sequence, the service brake pedal should be depressed. A code 71 should be displayed indicating the beginning of the ECM testing of the switch circuits. The service brake pedal should be depressed a second time. If the brake light switch passes the test, the display will advance to code 72. If there is a problem, the display will advance to code 72 after 10 seconds, and a code 71 will be stored in memory as a "failed" indication. If the brake pedal is not depressed the second time, the display will advance after 10 seconds to code 72 and a failed code 71 will be stored in memory. This time the failed code 71 means the technician did not operate the system correctly, which means that a problem does not necessarily exist in the circuit.

When code 72 is first displayed, the throttle should be opened to wide-open throttle and then released. If the nose switch passes the test, the display will advance to code 73. As before, if the switch fails, the display will advance to code 73 after 10 seconds, a code 72 will be stored in memory as a failed code. If the accelerator pedal is not depressed, a failed code 72 will be stored and the display will advance to code 73 after 10 seconds. Again, this could be misleading if allowed to happen.

When code 73 is displayed, the transmission/transaxle shift lever should be shifted from PARK to DRIVE and then to NEUTRAL. The display advances to code 74 when this test is completed.

A code 74 means the transmission/transaxle shift lever should be shifted to REVERSE and then to PARK. The display advances to code 78.

With code 78 displayed, the AVERAGE button on the MPG sentinel should be pushed. If the vehicle is not equipped with the MPG sentinel, a code 78 will be stored as a failed code. Under these conditions, this would be normal. The display advances to code 79.

When code 79 is displayed, the RESET button should be pushed. This completes the switch tests sequence.

The ECM now displays, in order, the failed codes. As they are displayed, they will not disappear until the problem has been corrected or tested again for correct operation. Vehicles not equipped with the MPG sentinel will be stuck in code 78. This is the end of the corrections needed for these vehicles.

Upon completion of all the switch tests problem corrections, the ECM remains in the diagnostic mode and displays 00 until the mode is exited.

Diagnostic charts in the service manual should be followed when switch tests problems are encountered.

FUEL SYSTEM DIAGNOSIS

The condition in the fuel supply system of low or no pressure should be diagnosed by first determining if the fuel pump is operating. Turn the ignition switch off, on, off, on, and off and listen for the distinct whine of the pump. The clicking of the fuel pump relay may also be heard. The fuel pump relay is in the relay center under the instrument panel (Fig. 3-25). If the pump does not operate, the circuitry should be checked. If it does work, a pressure check should be made.

A special fuel pressure gauge should be installed between the fuel supply line and the throttle body (Fig. 3-26). If the fuel return line is disconnected and the throttle body return line fitting is plugged, a pressure of at least 9 psi should be seen. If so, the cause of low pressure is the pressure regulator and it should be replaced. If the pressure is below 9 psi, the fuel supply line is restricted. Kinks in the line, a collapsed rubber hose, or a restricted fuel filter could cause this symptom.

Excessively high fuel pressure is caused by either a malfunctioning pressure regulator or a restricted fuel return

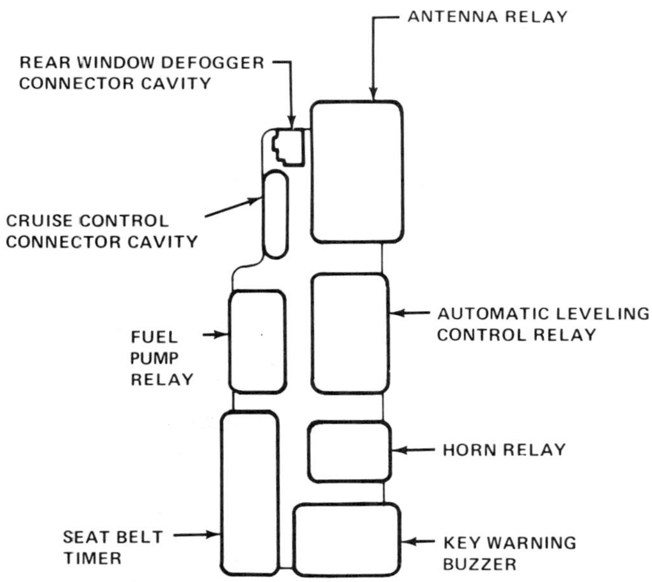

FIGURE 3-25 Relay center *(Courtesy of General Motors Product Service Training).*

FIGURE 3-26 Fuel pressure check *(Courtesy of Cadillac Motor Car Division).*

line. To diagnose this problem, disconnect the fuel return line and connect a hose to the fitting on the throttle body. With the pressure gauge installed on the supply line, activate the fuel pump and check the pressure gauge reading. Make sure the hose is inserted into an unbreakable container. If the pressure remains above 12 psi, the pressure regulator is malfunctioning and should be replaced. If the pressure drops to normal, the return line is restricted and should be inspected for kinks or collapsed rubber hoses.

DIAGNOSTIC CHARTS

The diagnostic charts available in the service manual to aid the technician are given next:

ECM Trouble Code Charts

Code 12, Distributor Pulses
Code 14, Coolant Sensor Circuit, Signal Low
Code 15, Coolant Sensor Circuit, Signal High
Code 21, Throttle Position Sensor, Signal High
Code 22, Throttle Position Sensor, Signal Low
Code 28, Idle Speed Control, Shorted
Code 29 or 30, Idle Speed Control, No Control
Code 31, Manifold Absolute Pressure Sensor, Signal High
Code 32, Manifold Absolute Pressure Sensor, Signal Low
Code 33, Manifold Absolute Pressure Sensor or Barometric Pressure Sensor, Signal Comparison
Code 34, Manifold Absolute Pressure Sensor, Signal Same as BARO
Code 35, Barometric Pressure Sensor, Signal High
Code 36, Barometric Pressure Sensor, Signal Low
Code 37, Manifold Air Temperature Sensor, Signal Low
Code 38, Manifold Air Temperature Sensor, Signal High
Code 55 or 56, ECM Malfunction

Switch Tests Codes Charts

Code 70 or 71, Brake Light Circuit
Code 72, Throttle Switch Circuit
Code 73, Drive Switch Circuit
Code 74, Reverse Circuit
Code 78, MPG Average Circuit
Code 79, MPG Reset Circuit

Engine Performance Charts

No Start or Stall after Start
Hard Start—Cold, No Check Engine Light
Check Engine Light on, No Codes Stored
Check Engine Light Inoperative
Fuel System Diagnosis
Poor Performance
Trouble Codes Cannot Display or Diagnostic Display Problems
MPG Sentinel Display Problems
Improper Idle Speed

ABBREVIATIONS

BARO (barometric) sensor Same device as MAP sensor, except senses surrounding (ambient) atmospheric pressure.

DEFI Digital electronic fuel injection system.

ECM Electronic control module (on-board computer).

EFE Early fuel evaporation system (heat trap).

ISC Idle speed control (motor). Electric motor, controlled by the ECM, that controls and maintains idle speed settings.

MAP (sensor) Manifold absolute pressure sensor. Solid-state electronic device that senses intake manifold pressures.

MAT (sensor) Manifold air temperature sensor. Device that measures degree of hotness of air/fuel mixture in intake manifold.

MICRON Unit of length equal to 1 thousandth (0.001) of a millimeter. A 10-micron filter will remove particles of 0.0004-inch particle size.

VOM Volt/ohm meter

UNIT I
General Motors Corporation Systems

4
Cadillac Digital Fuel Injection, 1981

OVERVIEW

The digital fuel injection (DFI) system used by Cadillac in 1981 is based on the DEFI system design but is different from it in the way it operates. The DFI system can accept input signals in analog form, which means variable voltage signals, or in digital form, which means on and off voltage signals. The analog input signals are converted to digital signals, which is what the system computer or electronic control module (ECM) understands. The ECM analyzes the incoming information, makes high-speed calculations in accordance with its programs, and then issues commands in digital voltage form. If the component being controlled is an analog device, the ECM will convert its digital signal to an analog signal before it leaves the ECM (Fig. 4-1).

The high-speed calculations, in part, are used to control fuel delivery by the fuel injectors. Fuel delivery is based on engine speed and air density. Engine speed information is provided by the electronic spark timing module (EST). The air density data must be derived from manifold air temperature (MAT) and intake manifold absolute pressure (MAP). At this point, the ECM can calculate a fuel injector pulse width, or ON time, to match the air/fuel ratio to engine operation. Another engine sensor monitors the overall results of adjustments in pulse width to enable the ECM to make corrections in the air/fuel ratio due to changes in engine or vehicle operation. This is referred to as *closed-loop* operation.

SUBSYSTEMS

The DFI system consists of a fuel delivery subsystem, an air induction subsystem, fuel management sensors, other sensors and switches, electronic spark timing, electronic control module, idle speed control, emission controls, cruise control, and the miles per gallon (MPG) sentinel.

FUEL DELIVERY SYSTEM

The fuel delivery system includes the fuel tank and cap, in-tank electric fuel pump, fuel filter, fuel supply line, throttle body, and the fuel return line (Fig. 4-2).

Fuel Tank

The fuel tank has a sump area into which the in-tank electric fuel pump is installed. The sump area prevents fuel from running away from the pump when fuel level is low and during severe maneuvering. The fuel return line empties into this area to aid in avoiding a run-dry condition. The fuel tank cap is a pressure/vacuum relief design emission control cap.

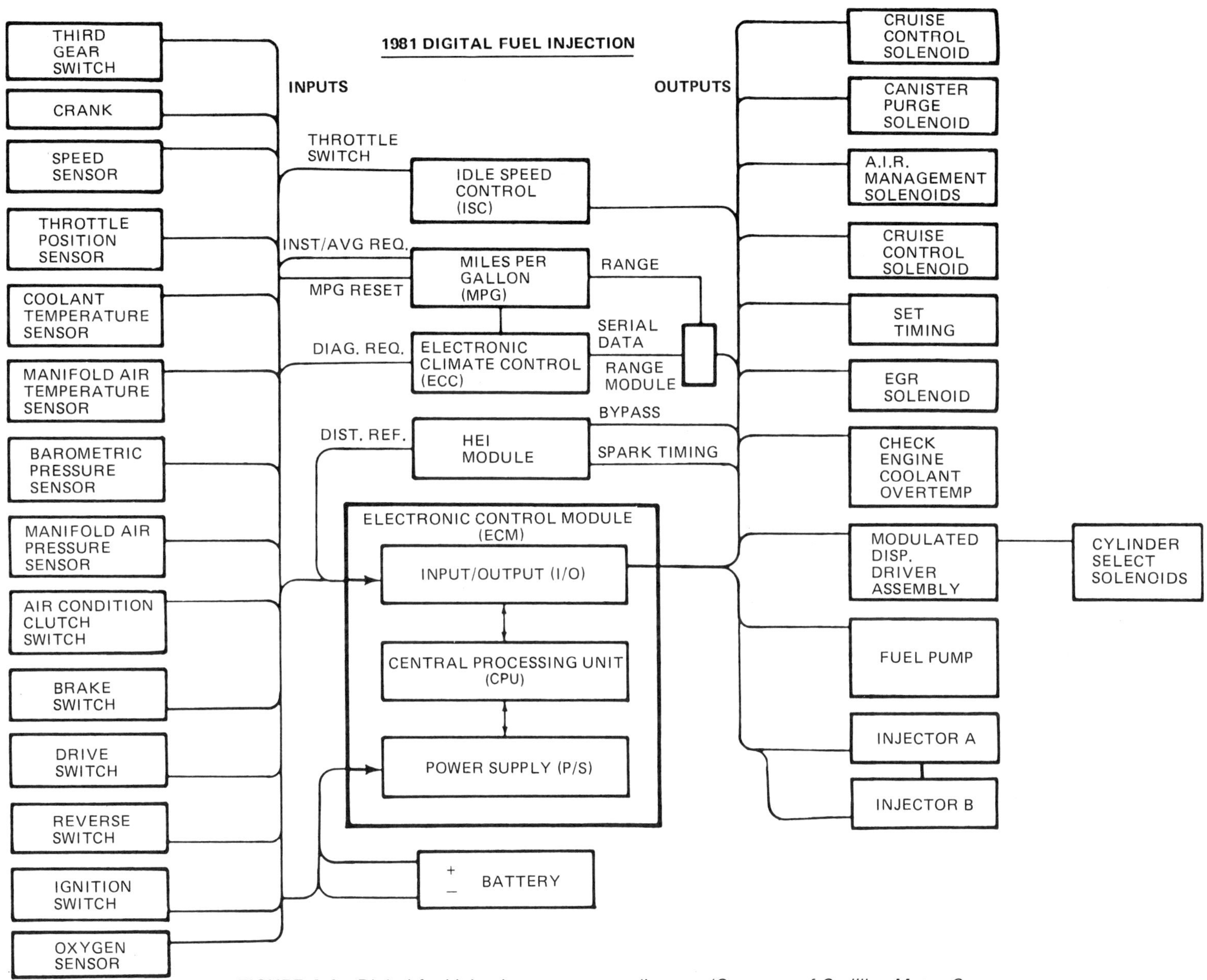

FIGURE 4-1 Digital fuel injection component diagram *(Courtesy of Cadillac Motor Car Division).*

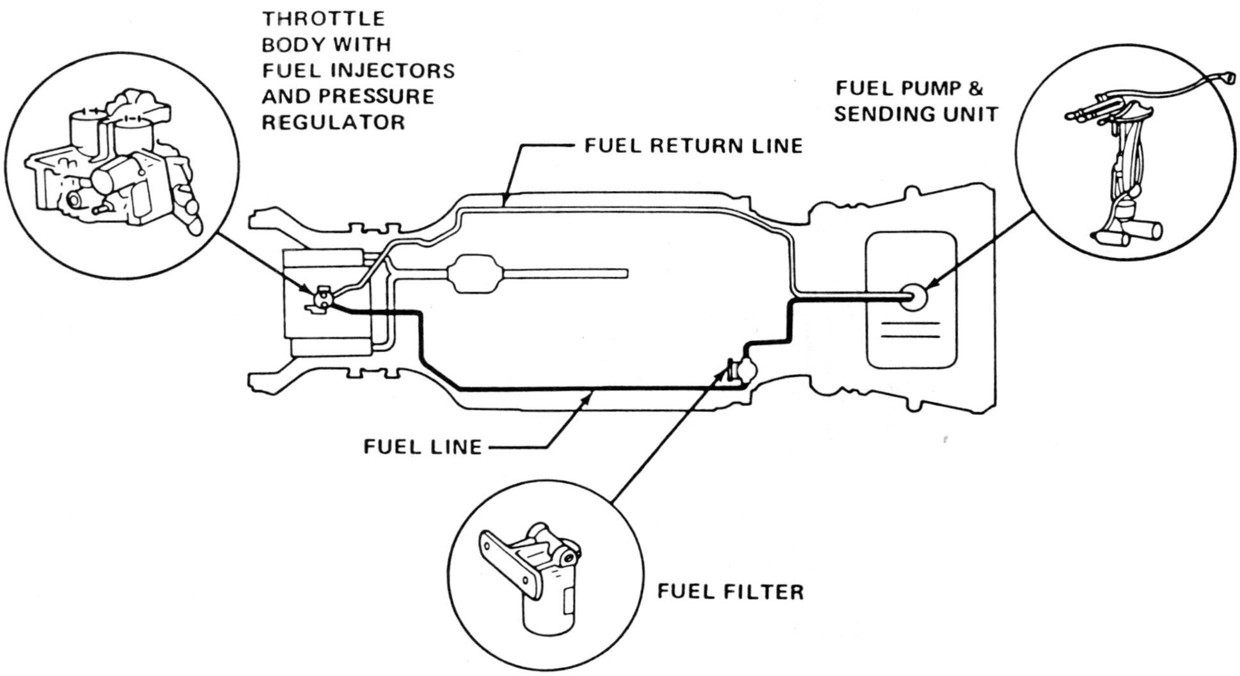

FIGURE 4-2 Fuel supply system *(Courtesy of Cadillac Motor Car Division).*

In-tank Electric Fuel Pump

The in-tank electric fuel pump is of the twin turbine design and is attached to the fuel tank sending unit (Fig. 4-3). The pump can be replaced separately from the float sending unit and the pickup tube screen. Pump operation is dependent on a signal from the ECM to the fuel pump relay. The relay completes the circuit to the fuel pump. This circuit is protected by a 10-ampere fuse in the mini fuse block. When the ignition switch is turned to ON, the ECM energizes the fuel pump relay, which, in turn, energizes the fuel pump. If the ECM does not get an engine cranking signal within 1 second, it de-energizes the fuel pump relay and the fuel pump shuts off.

Fuel Filter

The fuel filter is a replaceable paper element type that can remove particles as small as 10 microns or 0.0004 inch. The filter housing is a metal screw-on can and square-cut rubber seal and is located near the left rear wheel on the vehicle frame (Fig. 4-4).

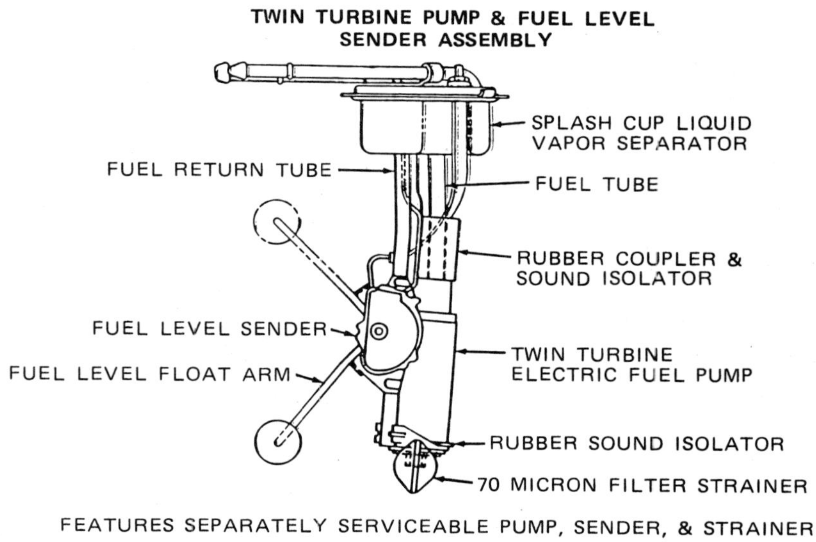

FIGURE 4-3 Twin-turbine pump and fuel level sender assembly *(Courtesy of General Motors Product Service Training).*

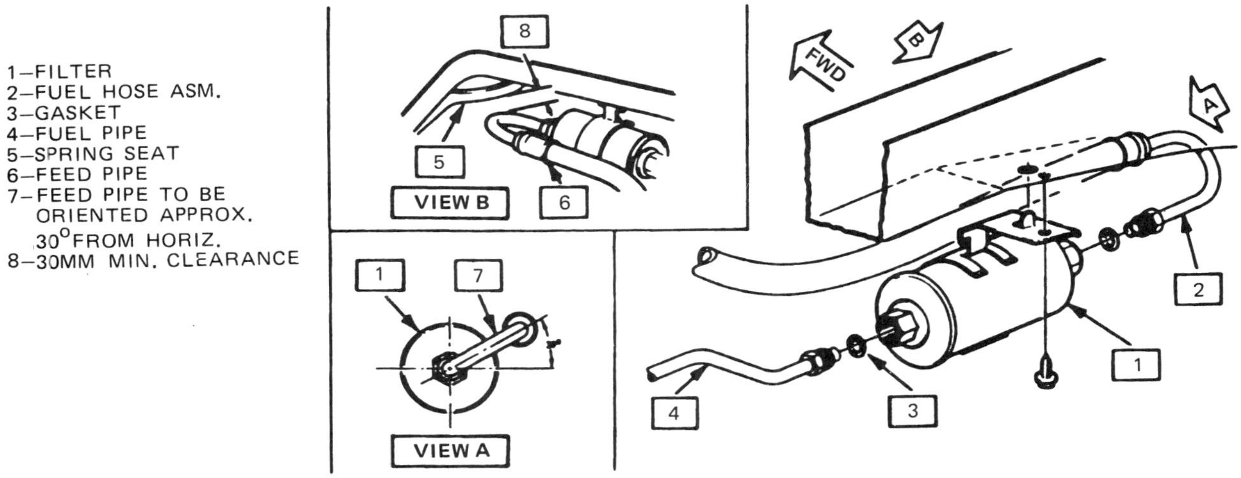

FIGURE 4-4 Fuel filter *(Courtesy of Cadillac Motor Car Division).*

Fuel Supply Line

The fuel supply line is 3/8-inch in size and runs along the left-side frame rail to the throttle body (Fig. 4-2). In the engine compartment, this steel line connects to a Teflon hose protected by a braided stainless steel covering. This provides the flexibility necessary due to engine vibration and also prevents wear-through due to rubbing.

Throttle Body

The throttle body fuel metering assembly contains the fuel pressure regulator and two fuel injectors (Fig. 4-5). The pressure regulator maintains 10.5-psi pressure across the injectors. The regulator is a diaphragm that is subjected to fuel pressure on one side and atmospheric pressure on the other side. A calibrated spring provides a means of obtaining the 10.5-psi pressure needed for system operation. Excess pressure unseats a valve integral with the diaphragm. This excess pressure is relieved through the fuel return line to the fuel tank.

FIGURE 4-5 Throttle body assembly (DFI) *(Courtesy of Cadillac Motor Car Division).*

Fuel Return Line

The fuel return line is a 5/16-inch diameter steel line beginning at the throttle body connection (Fig. 4-1). Another Teflon hose with a braided stainless steel cover is used to connect this line to the steel line that runs along the right-side frame rail to the fuel tank.

Fuel Injectors

The two fuel injectors are activated by the ECM once the fuel pump relay contacts have closed. Each injector has its own 3-ampere fuse in the mini fuse panel. The amount of time the injector is open or energized is known as its pulse width.

AIR INDUCTION SYSTEM

The throttle body, intake manifold, and idle speed control motor are components of the air induction system (Fig. 4-6). The throttle body is a model 200 unit. It uses a new-design throttle position sensor that does not have a pigtail connector (Fig. 4-7). The connection for the wiring harness is integral with the sensor body. To improve fuel mixing and distribution, two inverted cones were added to the throttle bores above the throttle plates. The throttle plates control airflow into the engine.

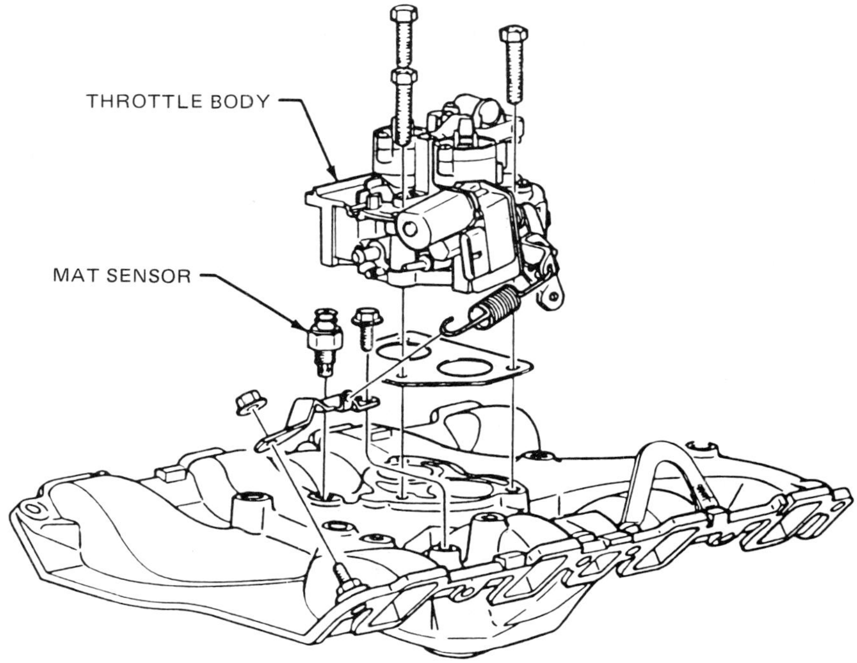

FIGURE 4-6 Throttle body installation *(Courtesy of Cadillac Motor Car Division).*

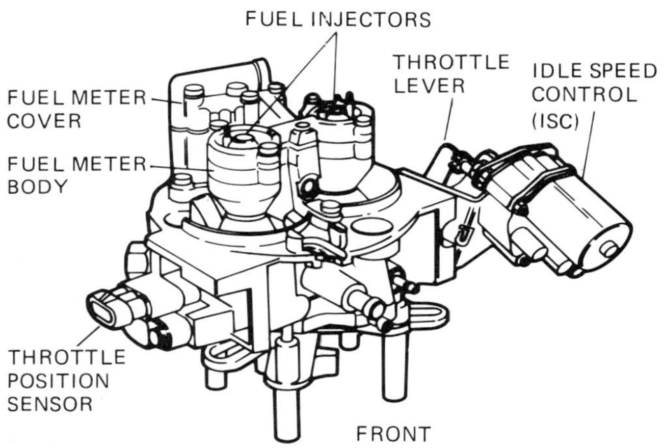

FIGURE 4-7 Throttle position sensor *(Courtesy of Cadillac Motor Division).*

Idle Speed Control Motor

The idle speed control motor is used by the ECM to regulate idle speed. It is a reversible electric motor-driven plunger that can extend or retract upon command from the ECM. A detailed discussion of the ISC circuitry is included later in this chapter.

Intake Manifold

The intake manifold is aluminum and is of the single-plane design where all runners are on the same level in the manifold. These features improve air/fuel distribution. It also has an exhaust heat crossover passage to accommodate the vacuum-operated early fuel evaporation (EFE) emission control.

FUEL MANAGEMENT SENSORS

The fuel management sensors provide the data to the ECM necessary for correct delivery of fuel. The engine speed sensor is integral with the electronic spark timing module in the distributor. It sends voltage signals or pulses to the ECM where the time interval between the pulses is interpreted to mean engine rpm.

Manifold Air Temperature Sensor

The manifold air temperature (MAT) sensor is located in the intake manifold in front of the throttle body (Fig. 4-6). Its voltage signal to the ECM is an indication of the temperature of the air/fuel mixture in the intake manifold. The MAT is a thermistor whose resistance value decreases as the temperature increases.

Manifold Absolute Pressure Sensor

The manifold absolute pressure (MAP) sensor is sensitive to the changes in intake manifold pressure as a result of changes in engine operation, such as load and speed (Fig. 4-8). The MAP sensor transmits the changes in pressure to the ECM in the form of changing voltage signals. If engine load increases, intake manifold pressure increases; the re-

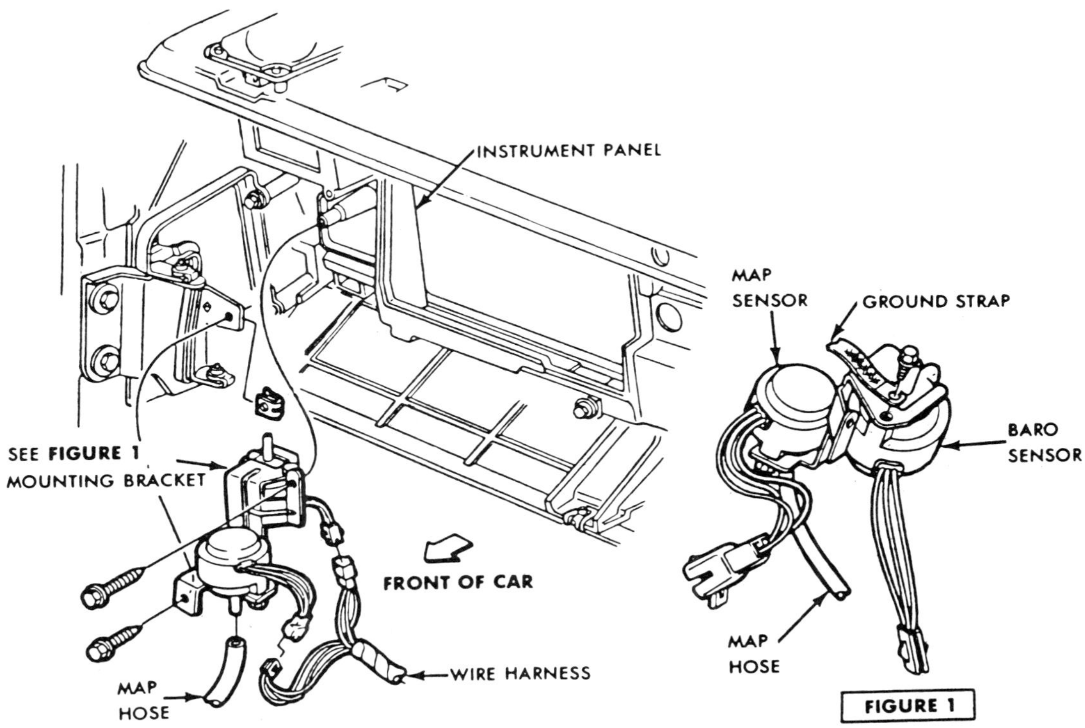

FIGURE 4-8 Barometric and manifold absolute pressure sensor installation *(Courtesy of Cadillac Motor Car Division).*

sulting signal from the MAP to the ECM causes the ECM to respond by commanding a wider pulse width.

OTHER SENSORS AND SWITCHES

To control other parts of the DFI system, the ECM monitors additional sensors and switches.

Coolant Temperature Sensor

The coolant temperature sensor (CTS) is located at the right front corner of the engine near the thermostat housing. It operates in a similar manner to the manifold air temperature sensor discussed earlier. The CTS senses coolant temperature and sends a voltage signal to the ECM, where it is converted into a temperature reading. As the coolant temperature rises, the resistance of the CTS decreases. If coolant temperature falls, the CTS resistance increases. For these reasons, the CTS is called a *thermistor.* In the actual circuit, the ECM feeds a 5-volt signal to the MAT or CTS. The resistance of the sensor causes a voltage drop. A return wire from the sensor back to the ECM carries a signal of less than 5 volts depending on the resistance of the sensor. The ECM compares its 5-volt output to what it receives from the sensor and calculates an equivalent temperature reading.

Barometric Pressure Sensor

A barometric pressure sensor (BARO) is used to keep the ECM informed of changes in atmospheric pressure. The BARO calibrates the fuel system for operation in any part of the world regardless of altitude. It is mounted on the same bracket as the MAP sensor under the instrument panel at the right end (Fig. 4-8). The ECM provides a 5-volt signal to each of these sensors. The sensor resistance changes with pressure changes, and the voltage returning to the ECM varies with the changing resistance. The returning voltage is generally between 0 and 5 volts, depending on the sensor resistance. When atmospheric pressure is 14.7 psi, or at sea level, the BARO voltage signal to the ECM is about 5 volts. As this sea-level pressure drops, the BARO signal also drops. The ECM monitors signals from both the MAP and BARO sensors, and, depending on the difference between the signals, calculates an accurate fuel injector pulse width.

Throttle Position Sensor

The throttle position sensor (TPS) is mounted on the throttle shaft on the right side of the throttle body (Fig. 4-5). It is a variable resistor whose resistance changes with changes in throttle opening. The ECM sends a 5-volt signal to the TPS. The throttle blade position determines the TPS re-

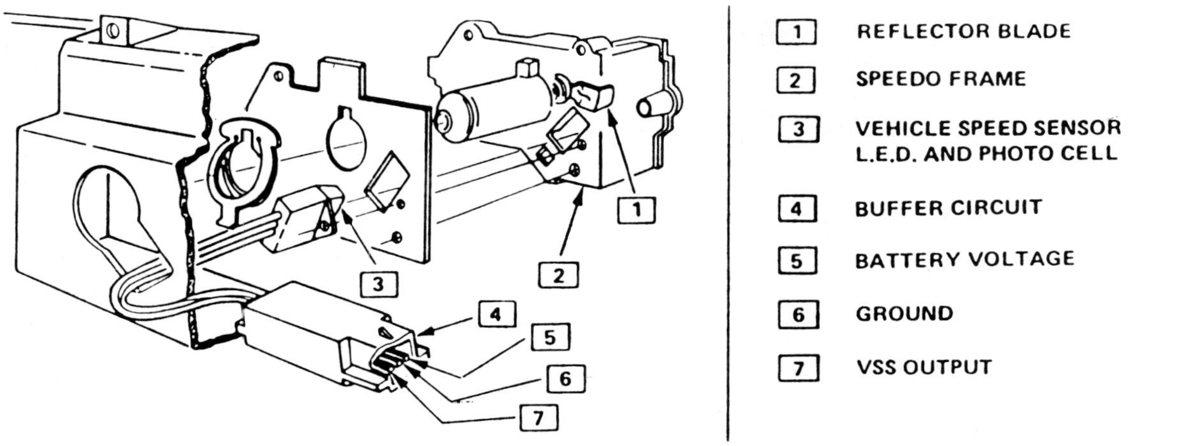

FIGURE 4-9 Vehicle speed sensor *(Courtesy of General Motors Product Service Training).*

sistance, which, in turn, affects the ECM's 5-volt signal. The return signal from the TPS to the ECM is an indication of throttle position. The ECM can control idle speed and fuel enrichment by knowing throttle position.

Vehicle Speed Sensor

A vehicle speed sensor (VSS) is located behind the instrument cluster in the instrument panel (Fig. 4-9). It consists of a light-emitting diode and photo-transistor and a metal blade fastened to the speedometer head. As the speedometer cable turns, the metal blade reaches a position where it reflects light from the light-emitting diode onto the photo-transistor. The photo-transistor produces a minute on-off voltage signal when the light strikes it. This signal is sent to a buffer amplifier module also located under the instrument panel. The buffer amplifier increases the VSS weak signal and ''conditions'' it to be sent to the ECM as an indication of vehicle road speed. To operate the VSS, the ECM provides an 8-volt source to the buffer amplifier, which then powers the sensor components. The sensor and buffer amplifier signal to the ECM varies with road speed.

Oxygen Sensor

Exhaust gas oxygen content is constantly measured by using an oxygen sensor (O_2) (Fig. 4-10). This sensor is mounted in the exhaust manifold so that its end extends into the exhaust gas stream. The end of the sensor is made of zirconia oxide which, when it comes into contact with oxygen, generates a voltage. The exhaust and sensor temperature must be high for this to happen. The amount of oxygen in the exhaust stream determines how much voltage is generated. The voltage generated varies from about 200 to about 900 millivolts. Higher voltages are produced when the exhaust is rich (lower oxygen content), and lower voltages are produced when the exhaust is lean (higher oxygen content). When the ECM receives this varying voltage signal, it can make compensations in its signal to the fuel injectors in an attempt to maintain a 14.7 to 1 air/fuel ratio.

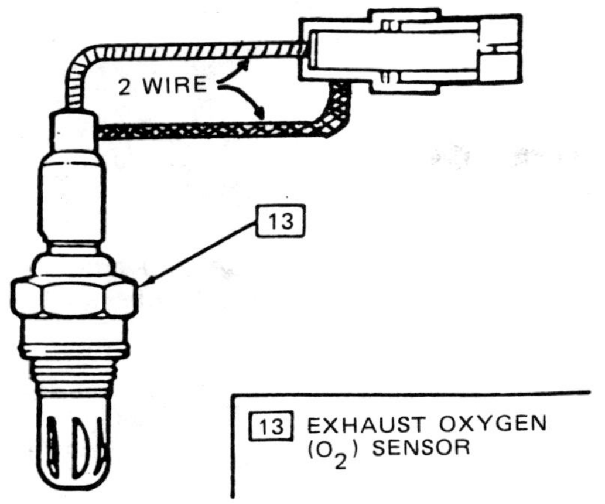

FIGURE 4-10 Exhaust oxygen (O_2) sensor *(Courtesy of Cadillac Motor Car Division).*

ECM Power

Power for the ECM is provided by two sources. Battery voltage is available through a 20-ampere ECS battery fuse in the mini fuse panel when the ignition switch is off. When the ignition switch is turned to ON, the ECM receives battery voltage through a 15-ampere ECS ignition fuse in the mini fuse panel.

Crank Signal

The ECM also receives a cranking signal from the ignition switch through a 3-ampere crank fuse in the mini fuse panel. This signal is provided by the ignition switch when it is turned to the START position. The ECM keeps the fuel pump circuit energized and the ignition spark retarded for quicker starting when the cranking signal is available.

Brake Switch

The brake switch is not only for brake light operation but also disengages the cruise control system. The ECM self-diagnostic feature tests this function of the brake switch.

Drive and Reverse Switches

A drive switch and a reverse switch are built into the park/neutral switch at the base of the steering column. As the gear selector lever is moved from park or neutral into drive or reverse, the drive or the reverse switch sends a voltage signal to the ECM. The ECM can maintain the idle speed when it receives these signals. Without the drive or reverse switch signals, as the transmission/transaxle shifts into gear, the idle speed will lower, quality may suffer, or at worst, the engine may stall.

Air-conditioning Switch

An air-conditioning switch provides a voltage signal to the ECM when the air-conditioning system is operating. The ECM can maintain idle speed and quality even though the air-conditioning compressor clutch places extra load on the engine.

Cruise Control Switches

The cruise control system provides three switch input signals to the ECM for correct operation. The cruise control enable switch provides a 12-volt signal to the ECM as an indication that the driver wants to use the cruise control system. The set/coast switch, when depressed, sends a voltage signal to the ECM. The ECM then energizes the cruise control system if certain operating conditions are met. The resume/accelerate switch sends a voltage signal to the ECM, which either enables the cruise control system to resume a speed to which it was previously set or to increase vehicle speed to a new setting.

Throttle Switch

A throttle switch is built into the idle speed control motor and provides a voltage signal to the ECM to indicate driver control of the throttle. When the throttle linkage is resting against the idle speed control motor plunger, the throttle switch is closed. The closed throttle switch provides a signal to the ECM that enables the ECM to control the idle speed. If the throttle linkage moves off the ISC plunger, the throttle switch opens and the ECM no longer tries to control idle speed.

ELECTRONIC CONTROL MODULE

The electronic control module (ECM) monitors the sensors and switches just discussed. It makes high-speed calculations and decisions based on data from those sensors and switches and according to its program information. Once calculations concerning a particular function, such as pulse width, are complete, a command or control signal is issued by the ECM to the fuel injectors.

To arrive at a pulse-width calculation, the ECM uses data from its program and from the manifold air temperature sensor, exhaust gas recirculation valve control circuit, closed-loop operating circuit, fuel pump power supply circuit, manifold absolute pressure sensor, barometric pressure sensor, throttle position sensor, and the commanded air/fuel ratio.

ECM Modes of Operation

The ECM can issue fuel injector pulse-width commands in different modes of operation. Cold engine enrichment, cranking enrichment, clear flood, and power enrichment are the four modes that cover the complete range of engine operation.

The cold engine enrichment mode provides wider injector pulses according to the coolant temperature sensor for quicker cold starting conditions.

The cranking enrichment mode aids better starting by supplying additional fuel injector pulses.

The clear flood mode operates whenever the driver holds the throttle wide open for more than 3 seconds while cranking the engine. The ECM remains in this mode until the engine starts or the driver stops cranking the engine or the throttle is allowed to close slightly.

A power enrichment mode is activated when the throttle blades are wide open or nearly wide open. Intake manifold pressure rises, and the ECM response is to widen the injector pulse.

The ECM can operate the fuel injectors either simultaneously or alternately depending on the operating needs of the engine. Alternate pulsing of the injectors occurs for most engine operating conditions, and one injector energizes each time a spark plug fires. Alternate pulsing of the fuel injectors is known as *synchronous injection.*

When both fuel injectors are energized simultaneously, it is known as *asynchronous injection.* Under this condition, the injectors energize every 12.5 milliseconds. Tip-in enrichment pulses, prime pulses, or acceleration enrichment pulses fall into this category, known as *transient fuel delivery.* During coastdown and deceleration enleanment, both injectors still energize at the same time, but the ECM determines the intervals between pulses. This is called *overrun fuel delivery.*

Tip-in enrichment pulses provide fuel for the transition from closed-throttle to part-throttle operation. This is similar to the idle-to-off-idle circuits in a carburetor.

Prime pulses are needed for richer mixtures for startup. They are about 10 milliseconds in duration. Coolant temperature determines how many prime pulses are necessary.

Acceleration enrichment pulses provide additional fuel for increasing speed. The MAP sensor pressure increases when the throttle blades are opened. The TPS indicates the change in throttle blade angle. Coolant temperature and the rate of throttle blade angle change are used by the ECM to determine pulse width for the injectors.

Coastdown or deceleration enleanment occurs when input data require a command fuel injector pulse that is too small for accurate injector operation. The ECM will issue a pulse width that is wide but will issue fewer of them.

OPEN- AND CLOSED-LOOP OPERATION

The 1981 DFI system has a feedback feature that enables the ECM to measure the results of its control of the fuel injectors. The oxygen (O_2) sensor is the component that indicates to the ECM the results of its attempts to control emissions. As discussed earlier, the oxygen sensor temperature must be high, approximately 600°F, before it is capable of feeding information back to the ECM. Until it reaches its operating temperature, the oxygen sensor is ignored by the ECM and the engine runs with the DFI system in an open-loop mode. The ECM determines fuel injector pulse width based solely on the engine sensor data. The DFI system operates in the open-loop mode during startup, warmup, wide-open throttle, and when specified diagnostic codes have been set in memory.

Once the oxygen sensor reaches its operating temperature, the engine sensors have stabilized (usually after about 70 seconds), coolant temperature is above 130°F, and the engine rpm is above 300, the DFI system operates in a closed-loop mode. At this time, the ECM uses the oxygen sensor signal to help calculate the fuel injector pulse width. Based on engine sensor data, the ECM can calculate a fuel injector pulse width, issue that command to the injector, and then measure the effectiveness of its efforts through the oxygen sensor. A compensation in injector pulse width can be made if necessary. This cycle of operation happens at extremely high speed, which contributes to the accuracy and effectiveness of the system.

If engine operation changes in such a way that one of the conditions for closed-loop operation is no longer being met, the DFI system reverts back to its open-loop mode. To determine if the DFI system is in open or closed loop, the computer can be put into its self-diagnostic mode. This will be discussed later in this chapter.

IDLE SPEED CONTROL

The engine idle speed is controlled by the ECM whenever the throttle linkage closes the throttle switch. The ECM uses an idle speed control motor, which is a 12-volt dc reversing motor, to regulate idle speed (Fig. 4-5). The throttle switch is an integral part of the ISC and indicates to the ECM that the driver has allowed the throttle plates to close. The ISC is mounted on the throttle body and has an extendible and retractable plunger that positions the throttle linkage for correct idle speed settings.

The ECM uses several inputs to determine idle speed. An air-conditioning clutch engagement signal indicates additional engine load and, therefore, a need to extend the ISC plunger to maintain idle speed. The electronic spark timing module provides an indication of engine speed. The coolant temperature sensor indicates to the ECM a need for increased idle speed during cold engine operation. The manifold absolute pressure sensor informs the ECM of deceleration and the need to control idle speed to prevent stalling. The park/neutral switch provides a signal that enables the ECM to anticipate transmission/transaxle load on the engine and to respond by extending the ISC plunger. The throttle switch tells the ECM when to control idle speed. The throttle position sensor indicates throttle angle and whether or not it corresponds to other input data.

Idle speed control is necessary for the following engine operating conditions: idle, deceleration, cranking, and startup.

ELECTRONIC SPARK TIMING

The electronic spark timing system (EST) is a more accurate way of controlling spark advance as compared to its vacuum and centrifugal counterpart. This sensitive and precise control of spark timing complements the DFI system. The older high-energy ignition system module is replaced by an EST module in the distributor. This new module has seven terminals that allow it to integrate with the existing DFI system. Engine sensors, the ECM, and the EST module work together to calculate and provide spark timing for optimum performance and fuel economy. This system is functional in all engine operating conditions except cranking.

The MAP sensor provides engine load information to the ECM. This is comparable to the previously used vacuum advance unit. Coolant temperature data are obtained by the ECM from the CTS. This information, which was not considered in older ignition systems, enables the ECM to change ignition timing as engine temperature changes. The BARO sensor provides input data concerning atmospheric pressure changes due to altitude or weather. The high-gear switch informs the ECM of the transmission/transaxle operating range. Neither the BARO nor the high-gear switch information was available on older ignition systems. The EST module provides data related to engine speed. This is comparable to the centrifugal weight mechanism on older vehicles. The result of all the data inputs to the ECM for spark advance is a much more precisely controlled timing of the spark for any particular set of engine operating conditions. When coupled with the DFI system, the overall conclusion is that the total system provides a highly calibrated, very accurately controlled delivery of fuel and ignition.

EMISSION CONTROL SYSTEM

In addition to the accuracy of the DFI system, other emission controls are used to further reduce emission levels and to improve the effectiveness of the catalytic converter. The emission control systems used in conjunction with the DFI system include an early fuel evaporation (EFE) system, an air injection reaction (AIR) system, an evaporative emission control (EEC) system, and an exhaust gas recirculation (EGR) system. The ECM must also be considered as it controls the operation of some of these systems.

Early Fuel Evaporation

The early fuel evaporation system consists of a vacuum-diaphragm-operated valve in the exhaust manifold. When vacuum is applied to the diaphragm, the valve closes and redirects warm exhaust to the base of the throttle body for improved fuel vaporization and distribution during the warmup period. If vacuum to the diaphragm is denied, the valve opens and exhaust leaves the engine normally. A thermal vacuum switch (TVS) is used to direct or deny vacuum to the EFE diaphragm in accordance with the engine coolant temperature. When coolant temperature is low, vacuum can pass the TVS to the diaphragm. When coolant temperature reaches about 120°F, the TVS blocks vacuum allowing the EFE valve to open. Full manifold vacuum is supplied to the TVS.

Air Management

An air injection reaction system is used to continue the combustion process in the exhaust manifold. This does not mean that there is an explosion or open burning in the exhaust system. An AIR pump, driven by an engine V-belt, provides airflow through a diverter and air management valve. The diverter valve is operated by an ECM-controlled solenoid. When energized, the solenoid causes the diverter valve to allow airflow to pass on to the air management valve. When de-energized, the solenoid causes the diverter valve to direct the air flow into the air cleaner housing.

The air management valve directs airflow to the exhaust manifold ports when its ECM-controlled solenoid is energized. When the ECM de-energizes the air management valve solenoid, airflow is directed to the catalytic converter. Check valves are used in the hoses between the exhaust manifold and the air management valve to prevent hot exhaust gases from backing up into the AIR system.

Catalytic Converter

The catalytic converter has two beds of pellets instead of the single bed used on previous models. The front bed contains pellets coated with rhodium and helps reduce oxides of nitrogen (NO_x). The rear bed contains pellets coated with platinum and palladium and helps reduce carbon monoxide (CO) and hydrocarbons (HC). Between the two beds is a small chamber where the AIR system can pump fresh air into the catalytic converter to help reduce emissions. The combustion process referred to earlier is actually a chemical reaction between the exhaust and the fresh air where heat is produced, but not fire or explosion.

The ECM controls the AIR system with the diverter and air management valve solenoids. When engine coolant temperature is below 120°F and when the engine is not under load, the diverter and air management solenoids are energized. This allows airflow to enter the exhaust manifold. As the engine coolant temperature rises to about 120°F and the engine is not under load, the diverter valve solenoid is energized, but the air management valve is de-energized. Airflow is directed to the catalytic converter. When the engine is under load or is decelerating, both solenoids are de-energized. Air flow is directed to the air cleaner.

Exhaust Gas Recirculation

An exhaust gas recirculation (EGR) system is used to help reduce oxides of nitrogen. Vacuum and exhaust pressure must be present to open the EGR valve. The valve has an integral back-pressure transducer. Vacuum is directed to the EGR valve diaphragm by an EGR solenoid, which is controlled by the ECM. When the EGR solenoid is energized, it blocks vacuum to the EGR valve. Solenoid operation is dependent on engine coolant temperature. When coolant temperature is below 150°F, the ECM energizes the solenoid, which prevents operation of the EGR valve. When coolant temperature rises above 150°F, the solenoid is de-energized and vacuum is allowed to act on the EGR valve diaphragm. However, the valve still remains closed unless exhaust back pressure is sufficient to cause the valve to open. The vacuum source for the EGR system is a ported vacuum tap on the throttle body. The ported vacuum tap is above the throttle plate, which means there will be no vacuum at idle. Thus, the EGR valve will not operate under any normal conditions while the engine is idling. In addition, the EGR system will not operate at wide-open throttle because there is not enough vacuum to operate the valve.

Evaporative Emission Control

The evaporative emission control system consists of a vapor canister, a purge valve and solenoid, and a vacuum source. The vapor canister stores unburned hydrocarbon vapors collected from the fuel tank. The ECM controls the purge valve solenoid, which in turn controls the vacuum for purge valve operation. When engine coolant temperature is below 178°F, the ECM energizes the purge valve solenoid, which blocks vacuum to the purge valve. When coolant temperature rises above 178°F, the ECM de-energizes the solenoid, which allows vacuum to the purge valve. Stored vapors are drawn from the canister into the throttle body and are burned with the air/fuel mixture. Ported vacuum is used to purge the canister.

ELECTRONIC CLIMATE CONTROL AND MILES PER GALLON SENTINEL PANELS

The electronic climate control and the miles per gallon sentinel utilize light-emitting diodes to provide a digital readout of information for the driver. The MPG sentinel provides instantaneous fuel economy figures, average fuel economy, driving range for fuel remaining, or the number of active cylinders in the case of a V8-6-4 engine (Fig. 4-11).

The ECC panel can read out temperature comfort settings, windshield defogger operation and rear window defogger operation and contains mode push buttons for the electronic climate control system operation (Fig. 4-12). The

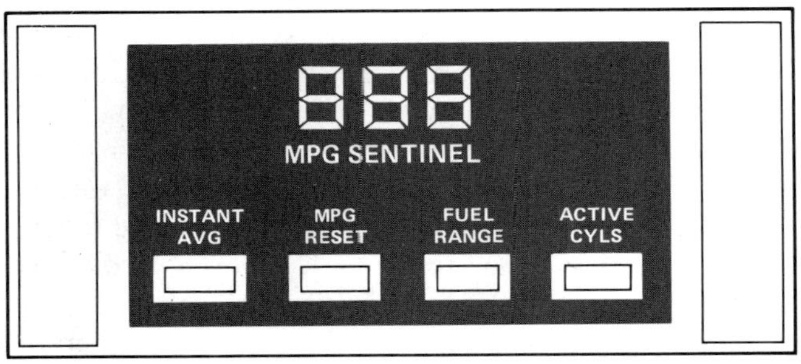

FIGURE 4-11 Miles per gallon sentinel *(Courtesy of Cadillac Motor Car Division).*

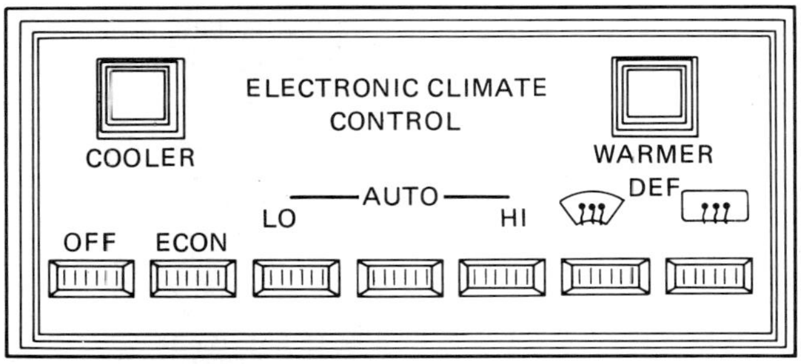

FIGURE 4-12 Electronic climate control panel *(Courtesy of Cadillac Motor Car Division).*

ECC panel is also interfaced or linked to the DFI system for purposes of self-diagnosis. The DFI system is capable of self-diagnosing circuit problems and, if a problem exists, stores a trouble code in its memory. The ECC panel provides a display where, when requested, the DFI system can reveal stored trouble codes in digital form. This self-diagnostic feature and the necessary procedures for operating it will be discussed later in this chapter.

CRUISE CONTROL SYSTEM

The electronic cruise control system is also interrelated with the DFI system. Because of this, the DFI self-diagnostic capability can be extended to include the sensors, solenoids, and switches of the cruise control system. When in the diagnostic mode, the MPG sentinel is also interfaced with the DFI system. Although its display face is blank, the push buttons can be used to check system switches and sensors. Engine solenoids can also be activated and then touched or heard for correct operation. The ECM operates the cruise control system according to input data from various sensors previously discussed.

SYSTEM OPERATION

With the ignition switch off, the ECM is inactive except for the storage of codes in its memory. The ISC plunger is fully extended, which will slightly open the throttle plates for startup.

If the engine switch is turned on, the ECM activates all its sensors and the electric fuel pump. The fuel system pressurizes to 10.5 psi, and then the pump shuts off if a cranking signal is not received by the ECM within 1 second.

Crank Mode

When the ignition switch is turned to crank the engine, the ECM issues pulses, as described earlier, to the fuel injectors. These asynchronous prime pulses change to synchronous pulses when the engine starts, and the distributor reference or engine rpm has been established.

Clear Flood Mode

Should the engine become flooded, the accelerator pedal can be held to the floor and the engine cranked for 3 seconds; the ECM will then enter its clear flood mode of operation. Under these conditions, the pulse width to the injectors is very short. The ECM continues to issue a very short pulse width until the accelerator pedal returns from wide-open throttle or until the engine starts.

Run Mode

With the engine started, the ECM issues injector pulse-width signals that reflect sensor data inputs. Idle speed is controlled by the ISC motor, and spark timing is controlled by the ECM. The engine operates with the DFI system in the open-loop mode until conditions for closed-loop operation are met, as discussed earlier.

Acceleration Enrichment Mode

As the driver steps down on the accelerator pedal, the ECM enters its off idle or tip-in enrichment mode. This happens as a result of the TPS signal changing with throttle blade angle. Asynchronous fuel injector pulses are issued by the ECM. The ISC motor plunger retracts because the throttle switch opens.

Part-throttle Mode

Part-throttle operation can occur in one of two different DFI modes, open or closed loop. In the open-loop mode, the injector pulse width is based primarily on engine sensor data except for the oxygen sensor. The closed-loop mode not only involves engine sensor data but also utilizes the oxygen sensor data. Fuel injector pulses in either mode are synchronous.

When acceleration is required, the TPS and MAP signals cause the ECM to issue asynchronous pulses to the injectors. The increase in frequency of pulses is determined by the rate of change in the TPS signal.

At wide-open throttle or power enrichment, the ECM increases pulse width to the injectors as much as it is allowed based on coolant temperature. This continues to happen until the throttle stops moving.

Deceleration Enleanment Mode

When decelerating, the ECM shortens the injector pulse width according to data received from the TPS, throttle switch, vehicle speed sensor, and MAP. These pulses are asynchronous. The ISC motor plunger extends to prevent stalling.

SYSTEM SERVICE

Idle Speed

Check idle speed with the engine at operating temperature. No adjustments are possible. If the idle speed is out of specification, check the ISC motor adjustment and also check to see if trouble codes have been stored. Follow prescribed procedures in the service manual.

Throttle Position Sensor Adjustment

The TPS can be adjusted by following prescribed procedures in the service manual. Some TPS mounting screws are spot welded. Be sure to break these welds before removing the screws. Tools needed include a tachometer, a 10-megohm impedance digital multimeter, an ISC plunger adjusting tool, and special jumper leads. After completing the TPS adjustment, if the CHECK ENGINE light is lit on the instrument panel, enter diagnostics and clear any codes that may have been stored as a result of the adjustment procedures (Fig. 4-13).

Minimum Air Rate Adjustment

This adjustment is necessary only when throttle body parts have been replaced or when required by the TPS adjustment. Detailed instructions in the service manual should be followed closely. Tools needed for this adjustment include

FIGURE 4-13 CHECK ENGINE light *(Courtesy of Cadillac Motor Car Division).*

a tachometer, special jumper leads, and an ISC plunger adjusting tool.

Idle Speed Control Motor Adjustment

The ISC motor adjustment should be done only as a result of replacement. It establishes the initial position of the motor and affects idle speed. Incorrect adjustment may cause the ECM to continue to extend the plunger in an attempt to maintain idle speed. When the plunger reaches the end of its travel, the ISC motor causes the plunger to "ratchet." This produces a varying idle speed and may also cause eventual ISC burnout. Follow adjustment procedures in the service manual closely.

Throttle Body Service

Before disconnecting fuel lines, the system pressure should be relieved. This is done by removing the fuel pump fuse and starting the engine. It will start and run and then stall as it runs out of fuel. Crank the engine an additional 3 seconds to relieve any residual pressure.

The throttle body consists of the fuel metering body and the throttle body casting. The fuel metering body includes the two fuel injectors and the fuel pressure regulator, which is integral with the fuel metering body cover (Fig. 4-14). The throttle body casting includes the TPS, ISC motor, the throttle plates, the vacuum hose connections, and the fuel inlet and return fittings (Fig. 4-15).

The fuel metering body is serviceable and should be disassembled and cleaned as required. The pressure regulator is integral with the fuel metering body cover and should not be disassembled. It is serviced as an assembly. The fuel injectors can be replaced individually, and care should

FIGURE 4-14 Fuel meter body *(Courtesy of Cadillac Motor Car Division).*

FIGURE 4-15 Fuel inlet and return nut removal *(Courtesy of Cadillac Motor Car Division).*

be taken when installing them through the rubber O-ring seals. The upper seal has a backup washer that should be installed under the O-ring (Fig. 4-16). No attempt should be made to clean the injectors in an immersion-type solvent.

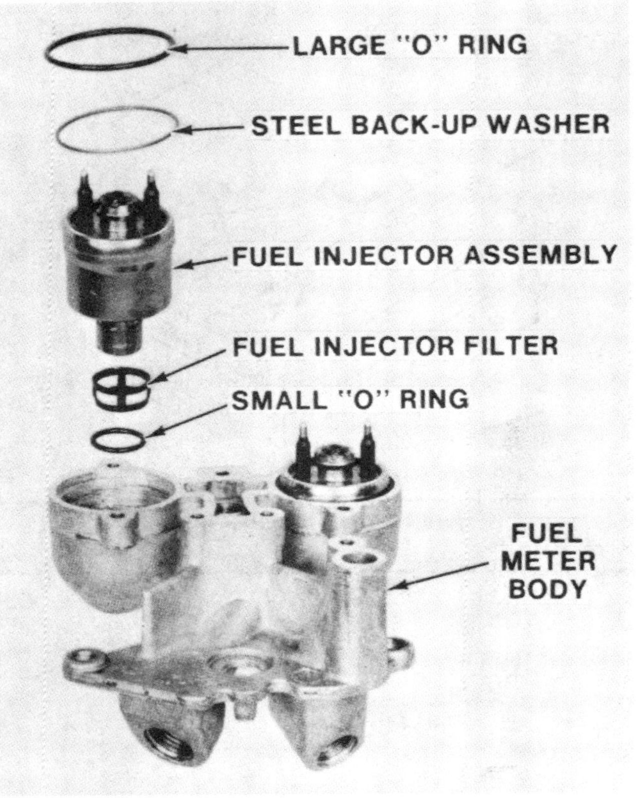

FIGURE 4-16 Fuel injector installation *(Courtesy of Cadillac Motor Car Division).*

The throttle body casting can be disassembled and cleaned. Before removing the TPS mounting screws, check to see if they are spot welded in place. Spot welds must be removed first; otherwise, damage to the screw thread hole will occur. The TPS should not be cleaned in an immersion-type solvent. The ISC motor can be removed. It is not a serviceable unit and should be replaced as an assembly. The remaining throttle body casting can be placed in an immersion-type solvent to be cleaned as a carburetor would. Follow service manual procedures for more detailed disassembly, cleaning and inspection, and reassembly information.

Sensor Service

The MAP and BARO sensors are serviced by replacement only. They can be tested through use of the system's self-diagnostic mode of operation. They are located on a common mounting bracket under the right side of the instrument panel (Fig. 4-8).

The CTS and MAT are also serviced by replacement. Their threads should be coated with a nonhardening sealer before installation. These sensors can be tested by entering the DFI self-diagnostic mode. The MAT is located in the intake manifold in front of the throttle body (Fig. 4-6). The CTS is located in the engine block near the thermostat housing.

The ECM is located under the right side of the instrument panel. It is serviced as an assembly except for two PROMs, or programmable read-only memory (Fig. 4-17). The PROM contains information that pertains to a particular vehicle. The ECM uses this information to calculate signals for the components it controls. The PROMs are located under a cover on the ECM. Each PROM is surrounded by an installation and removal guide or carrier. One carrier is clear plastic and the other is green plastic.

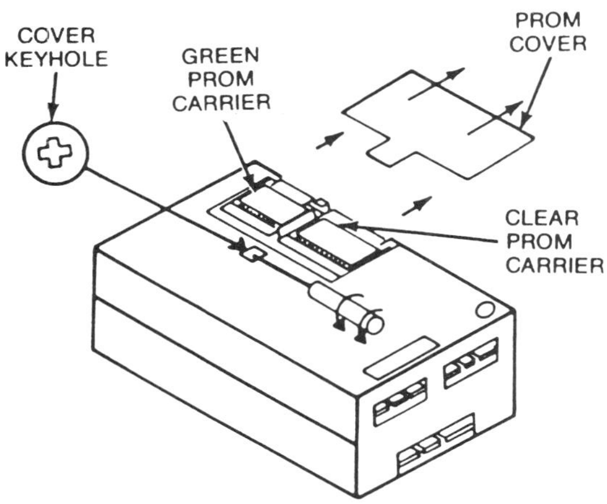

FIGURE 4-17 Programmable read-only memory removal *(Courtesy of Cadillac Motor Car Division).*

When servicing, remove one at a time so that correct installation will be easier. Make sure the reference dimple or boss on the PROM is aligned with the dimple or boss in the PROM socket as it is installed.

Fuel Filter Service

The fuel filter contains a replaceable paper element that should be changed at the recommended interval specified in the maintenance schedule. The filter container is sealed to the filter body by a square-cut rubber seal. This seal should be replaced whenever the filter is disassembled. The fuel filter is located on the frame rail ahead of the left rear wheel.

SYSTEM DIAGNOSIS

The DFI self-diagnostic mode functions to detect and reveal problems in the input and output circuits of the ECM. The system has four kinds of tests to help the technician diagnose the system: engine sensor tests, switch tests, engine data tests, and output cycling tests (Fig. 4-18).

Trouble Codes

Two kinds of trouble codes can be stored in the ECM memory if a problem in the input or output circuitry appears. Intermittent codes are a result of a problem that occurs and then corrects itself. An intermittent code causes the CHECK ENGINE telltale light to show on the instrument panel. When the problem corrects itself, the light goes out, but a code will remain in the ECM memory until the technician clears it or until the ignition switch is cycled off and on 20 times. Hard codes are stored in the ECM memory when a problem occurs in the circuitry. A hard code is the result of a problem that does not correct itself. This causes the CHECK ENGINE light to show, but this time it remains on until the problem is corrected and the code is cleared by the technician.

There are 35 codes that can be set in the ECM memory that are related to the DFI system and 16 codes that can be set that are related to the switch circuits and the cruise control circuits. A list of these codes can be found at the end of this chapter (Fig. 4-24).

Self-diagnostic Mode

To enter the diagnostic mode, the ignition switch should be on. The technician then should push the OFF and WARMER buttons simultaneously on the electronic climate control panel (Fig. 4-19). The display on the ECC will show an 88 or segment check. If any segments of the digital numbers are missing, the ECC display should be

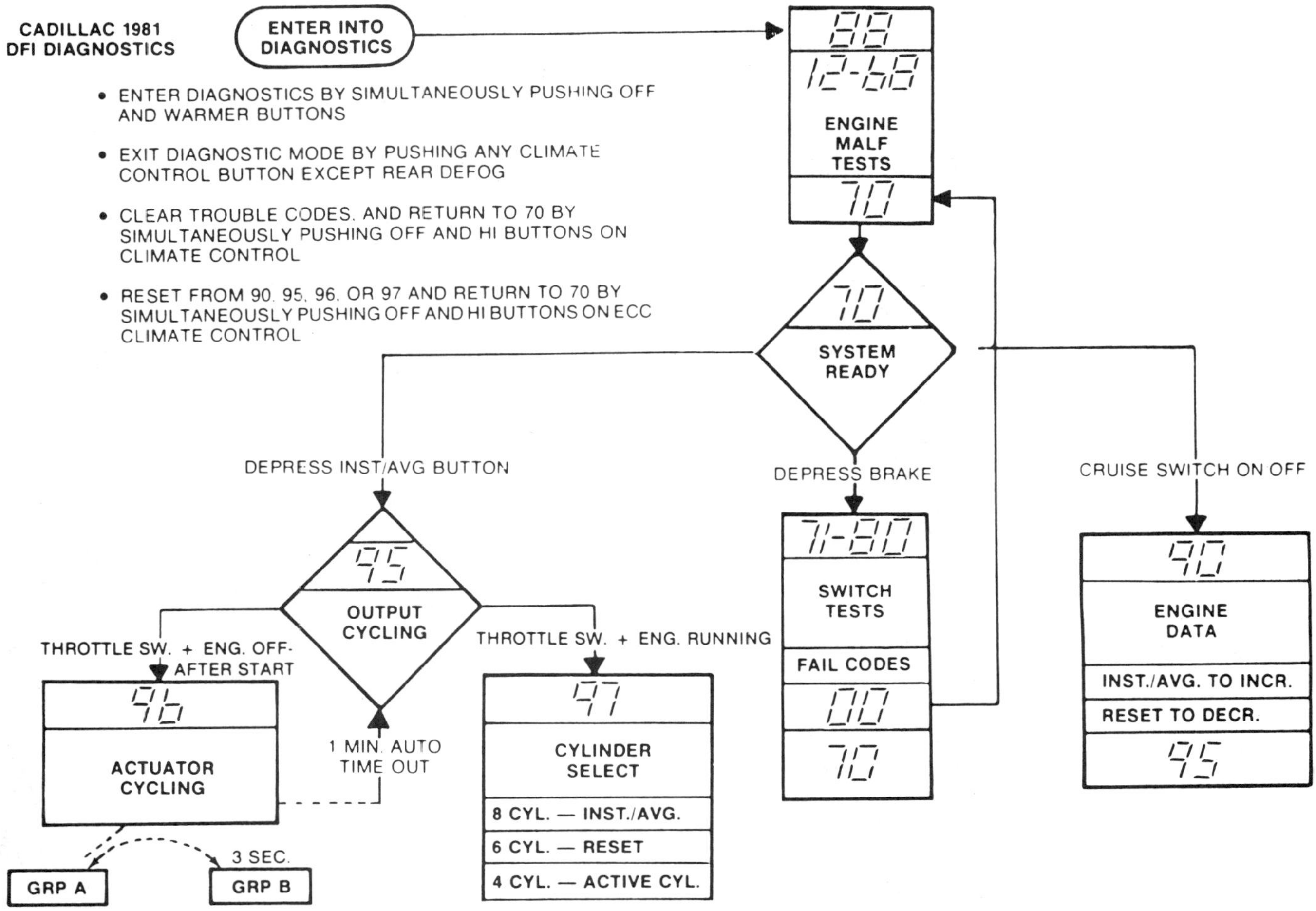

FIGURE 4-18 Digital fuel injection diagnostics *(Courtesy of Cadillac Motor Car Division).*

FIGURE 4-19 Electronic climate control head *(Courtesy of Cadillac Motor Car Division).*

repaired first. Any stored trouble codes are displayed in numerical ascending order after the segment check. After all codes have been displayed, a digital 70 appears. This indicates that the system is ready for the technician to select the next testing sequence.

Engine Data Tests

By turning the cruise control switch to ON and then OFF, the system will enter the engine data tests (Fig. 4-20). This test sequence consists of ten parameters or checks on various sensors. A chart of specifications is provided in the service manual so that the technician can compare the ECC display reading with a specified range of operation or parameters (Fig. 4-21). The tenth parameter indicates whether or not the DFI system is operating in the open- or closed-loop mode.

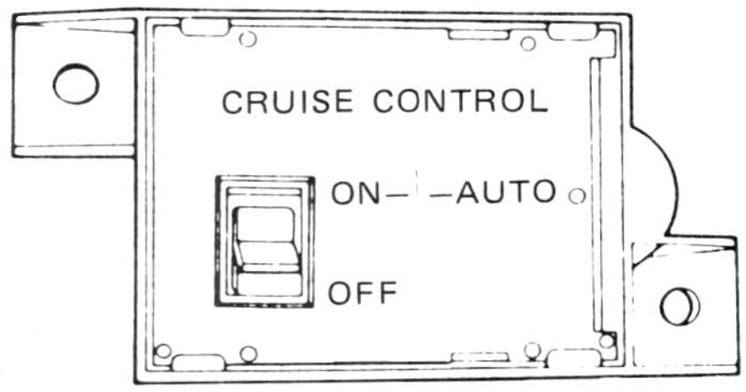

FIGURE 4-20 Cruise control switch *(Courtesy of Cadillac Motor Car Division).*

Switch Tests

If, instead, the technician depresses the brake pedal when the system is displaying 70, the system will enter the switch tests mode. There are ten switch tests that are performed from the driver's seat (Fig. 4-22). Any switch that fails the

1981 DIAGNOSTIC ROUTINE
ENGINE DATA SERIES

VARIOUS ENGINE SENSOR AND ENGINE CONTROL VALUES CAN BE DISPLAYED AT THE TECHNICIAN'S OPTION

ENGINE DATA DISPLAY

PARAMETER NO.	PARAMETER	VALUE	UNITS
01	TPS–(THROTTLE ANGLE)	XX	DEGREES
02	MAP	XX	kPa
03	BARO	XX	kPa
04	COOLANT	XX	0 = −40°C 99 = 150°C
05	MAT	XX	0 = −40°C 99 = 150°C
06	INJECTOR PULSE WIDTH	X.X	MSEC
07	O_2 SENSOR	.XX	VOLTS
08	SPARK ADVANCE	XX	DEGREES
09	IGNITION CYCLES	XX	KEY CYCLES
10	OPEN/CLOSED-LOOP INDICATOR	X	0 = OPEN LOOP 1 = CLOSED LOOP
11	BATTERY VOLTAGE	X.X	VOLTS — WILL NOT DISPLAY 10'S DIGIT

FIGURE 4-21 Engine data display *(Courtesy of Cadillac Motor Car Division).*

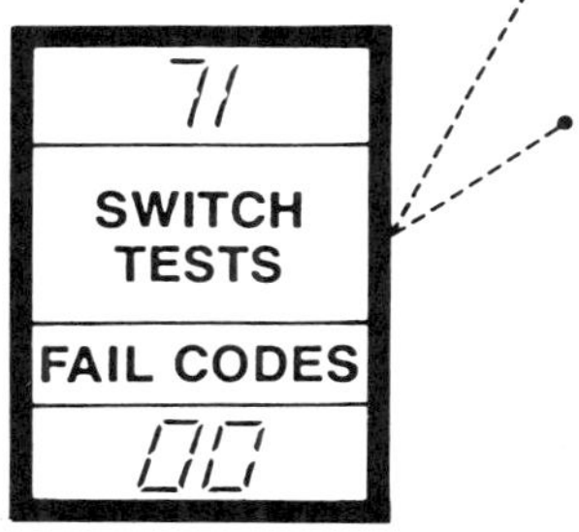

1981 DFI DIAGNOSTIC ROUTINE
SWITCH TEST SERIES

- DURING THIS TEST SERIES THE TECHNICIAN CYCLES CERTAIN SWITCHES AND THE ECM ANALYZES THIS ACTION TO DETERMINE IF THE SWITCHES ARE OPERATING PROPERLY.
- SWITCH TEST CODES:

70 SYSTEM READY
71 CRUISE CONTROL BRAKE SWITCH
72 ISC THROTTLE SWITCH
73 DRIVE (ADL) SWITCH
74 BACKUP LAMP SWITCH
★75 CRUISE OFF/ON
★76 CRUISE SET/COAST
★77 CRUISE RESUM/ACCEL.
78 AVG. MPG SWITCH
79 RESET MPG SWITCH
★80 A/C CLUTCH (PRESS REAR DEFOG SWITCH)

★NEW FOR 1981

FIGURE 4-22 Switch test series *(Courtesy of Cadillac Motor Car Division).*

test causes a code to be set. At the completion of the test sequence, these codes are displayed one at a time until each is repaired.

Output Cycling Test

If the technician pushes the INSTANT/AVG button on the MPG sentinel when the system displays 70, the system enters the output cycling mode (Fig. 4-23). The ECC displays 95 when this is accomplished. Partially depressing the accelerator pedal when the engine is off causes the system to activate solenoids and motors controlled by the ECM. The technician can go into the engine compartment and check the operation of these components by either touch or sound. Partially depressing the accelerator pedal while the engine is running causes the system to enter a cylinder select mode. This is applicable to the modulated displacement engine or V8-6-4 only. While in this mode, the technician can select 8-, 6-, or 4-cylinder engine operation.

If any of the codes on the list in Fig. 4-24 are displayed during the diagnostic mode of operation, it is necessary to refer to the appropriate trouble chart of decision tree procedures to determine the cause of the problem. These charts provide a step-by-step procedure for checking the malfunctioning circuit or component. The service manual contains a chart for each code.

Once the problems have been repaired, it is necessary to clear stored codes from the memory. This can be accomplished if the system is in the diagnostic mode and in the 70, system ready, condition. By depressing OFF and HI simultaneously, the codes will be cleared and the ECC panel will display 00 to indicate the completion of the clearing process (Fig. 4-19). The system then returns to the 70 condition.

Exiting Diagnostic Mode

To exit the diagnostic mode, or get the system to function normally again, the technician can push any one (other than REAR DEFOG) of the ECC function buttons, such as AUTO or ECON, or the ignition switch can be turned off for at least 10 seconds. This procedure will not clear trouble codes. The ECC panel should then display the temperature setting.

Fuel System Diagnosis

Fuel system diagnosis is aided through the use of diagnostic charts in the service manual. Pressure testing requires the use of a special pressure gauge and adapters. Generally, the pressure should be between 9 and 12 psi. If it is out of range, the chart will guide the technician as the pressure regulator, fuel filter, fuel lines, and fuel pump are checked. This service manual chart is easy to follow, but none of the steps should be bypassed.

The 1981 DFI system is an improvement over its

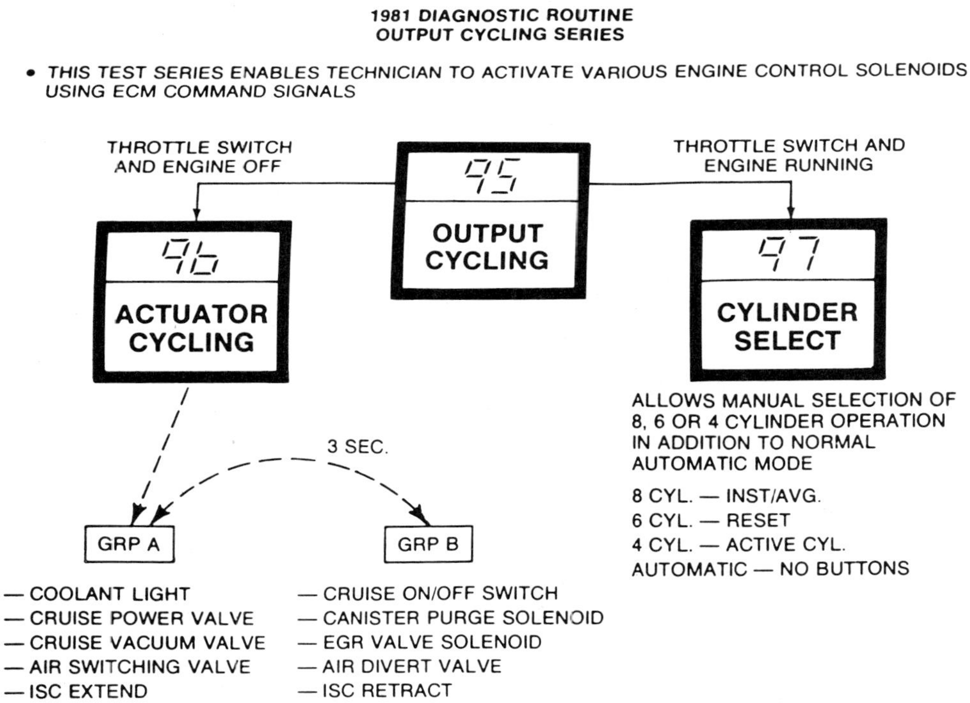

FIGURE 4-23 Output cycling tests *(Courtesy of Cadillac Motor Car Division).*

THE FOLLOWING CODES ARE PROGRAMMED INTO THE ECM.

CODE	CIRCUIT AFFECTED
12	NO TACH SIGNAL
13	O2 SENSOR NOT READY
14	SHORTED COOLANT SENSOR
15	OPEN COOLANT SENSOR CIRCUIT
16	GENERATOR VOLTAGE OUT OF RANGE
17	CRANK SIGNAL CIRCUIT HIGH
18	OPEN CRANK SIGNAL CIRCUIT
19	FUEL PUMP CIRCUIT HIGH
20	OPEN FUEL PUMP CIRCUIT
21	SHORTED THROTTLE POSITION SENSOR CIRCUIT
22	OPEN THROTTLE POSITION SENSOR CIRCUIT
23	EST/BYPASS CIRCUIT SHORTED OR OPEN
24	SPEED SENSOR FAILURE
25	MODULATED DISPLACEMENT FAILURE
26	SHORTED THROTTLE SWITCH CIRCUIT
27	OPEN THROTTLE SWITCH CIRCUIT
30	IDLE SPEED CONTROL CIRCUIT
31	SHORT MAP SENSOR CIRCUIT
32	OPEN MAP SENSOR CIRCUIT
33	MAP/BARO SENSOR CORRELATION
34	MAP HOSE
35	SHORTED BARO SENSOR CIRCUIT
36	OPEN BARO SENSOR CIRCUIT
37	SHORTED MAT SENSOR CIRCUIT
38	OPEN MAT SENSOR CIRCUIT
44	O2 SENSOR LEAN
45	O2 SENSOR RICH
51	PROM INSERTION FAULTY
60	DRIVE (ADL) SWITCH CIRCUIT
61	SET AND RESUME SWITCH CIRCUIT
62	CAR SPEED EXCEEDS MAXIMUM LIMIT
63	CAR AND SET SPEED TOLERANCE EXCEEDED
64	CAR ACCELERATION EXCEEDS MAXIMUM LIMIT
65	COOLANT TEMPERATURE EXCEEDS MAXIMUM LIMIT
66	ENGINE RPM EXCEEDS MAXIMUM LIMIT
68	SET AND RESUME SWITCH CIRCUIT
70	SYSTEM READY — SWITCH TESTS
71	BRAKE LIGHT SWITCH
72	ISC THROTTLE SWITCH
73	DRIVE (ADL) SWITCH
74	BACK-UP LAMP SWITCH
75	CRUISE ON/OFF CIRCUIT
76	SET/COAST CIRCUIT
77	RESUME/ACCELERATION CIRCUIT
78	INSTANT/AVERAGE MPG BUTTON
79	RESET MPG BUTTON
80	A/C CLUTCH CIRCUIT
88	DISPLAY CHECK
90	SYSTEM READY TO DISPLAY ENGINE DATA
95	SYSTEM READY FOR ACTUATOR CYCLING
96	ACTUATOR CYCLING
97	MD CYLINDER SOLENOID CYCLING
00	ALL DIAGNOSTIC COMPLETE

FIGURE 4-24 Digital fuel injection diagnostic codes *(Courtesy of Cadillac Motor Car Division).*

predecessors in that the system aids the service technician in locating and making accurate repairs. The system also responds to changes in engine operation and to outside influencing factors in a more accurate way.

ABBREVIATIONS

A/F Air/fuel (A/F Ratio)

AIR Air injection reaction

ALDL Assembly line data link

BARO Barometric pressure

CCP Climate control panel

Conv. Converter (catalytic)

DFI Digital fuel injection

DVM (10 Meg.) Digital voltmeter with 10-million ohms resistance

ECC Electronic climate control

ECM Electronic control module (controller)

EECS Evaporative emissions control system

EFE Early fuel evaporation

EGR Exhaust gas recirculation

EST Electronic spark timing

FDP Fuel data panel

HEI High-energy ignition

ISC Idle speed control

MAP Manifold absolute pressure

O_2 Oxygen (sensor)

PCV Positive crankcase ventilation

P/N Park/neutral

PROM Programmable read only memory (engine calibrator)

TBI Throttle body injection (unit)

THERMAC Thermostatic air cleaner

TPS Throttle position sensor

VSS Vehicle speed sensor (Speedo buffer sensor)

WOT Wide-open throttle

UNIT I
General Motors Corporation Systems

5
Cadillac Digital Fuel Injection, 1982–1985

OVERVIEW

The Cadillac digital fuel injection (DFI) system used on 1982 through 1985 vehicles is an expanded version of the 1981 system that also has some major differences. It is a speed/density fuel system that accurately provides an air/fuel mixture to obtain desirable performance and emission levels. Fuel is metered by two electronically controlled injectors that are different from those used in previous models. The system utilizes engine sensors, various switches, and other sensors to provide data to a computer where high-speed calculations are made. These calculations enable the computer to accurately control the fuel injectors. The DFI system can be broken down into subsystems for more thorough discussion. It consists of fuel delivery, air induction, switch signals and data sensors, electronic controls and interface, electronic spark timing, idle speed control, emissions, torque converter clutch, cruise control, fuel data panel, and self-diagnostic subsystems.

FUEL DELIVERY SYSTEM

The fuel delivery system consists of the fuel tank and cap, an electric in-tank fuel pump, fuel filter, fuel pressure regulator, fuel injector, fuel supply, and fuel return lines.

Fuel Tank

The fuel tank contains a sump area where fuel is drawn from the tank by the electric fuel pump. The sump area prevents fuel from running away from the pump during severe maneuvering or turns. The tank is the same as the one used on 1981 DFI systems. The fuel tank cap is a pressure/vacuum relief type emission control cap.

In-Tank Electric Fuel·Pump

The in-tank electric fuel pump is a twin-turbine type and is integral with the fuel gauge sending unit. It is, however, serviceable separate from the sending unit. It is the same as the pump used on the 1981 DFI systems.

Fuel Filter

The fuel filter is new for this system (Fig. 5-1). It consists of a metal casing with a paper element enclosed. It is serviced by replacement of the entire assembly. The paper element inside can remove particles as small as 10 microns or 0.0004 inch.

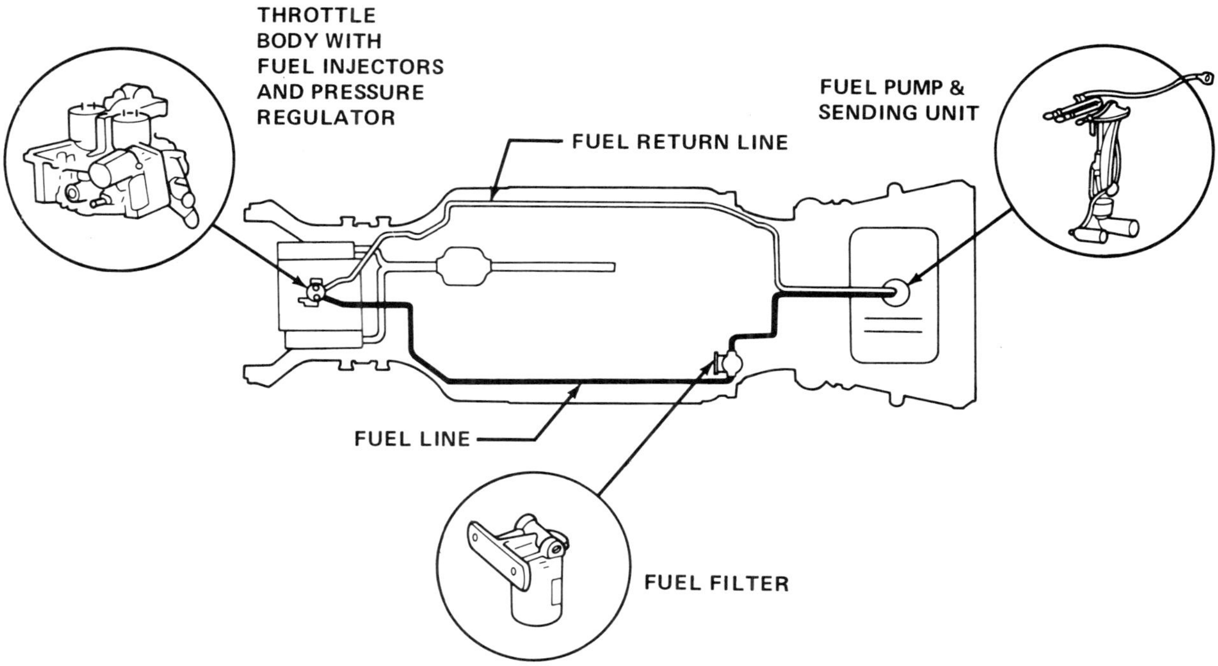

FIGURE 5-1 Fuel supply system *(Courtesy of Cadillac Motor Car Division).*

Fuel Pressure Regulator

A fuel pressure regulator is an integral part of the fuel metering body cover on the throttle body (Fig. 5-2). It is calibrated to maintain a constant pressure of 10 psi across the two fuel injectors. This regulator is a diaphragm-operated fuel pressure relief valve. Fuel pressure is applied to one side of the diaphragm and atmospheric pressure plus a calibrated spring is applied to the other side. Pressure in excess of 10 psi is allowed to return to the fuel tank be-

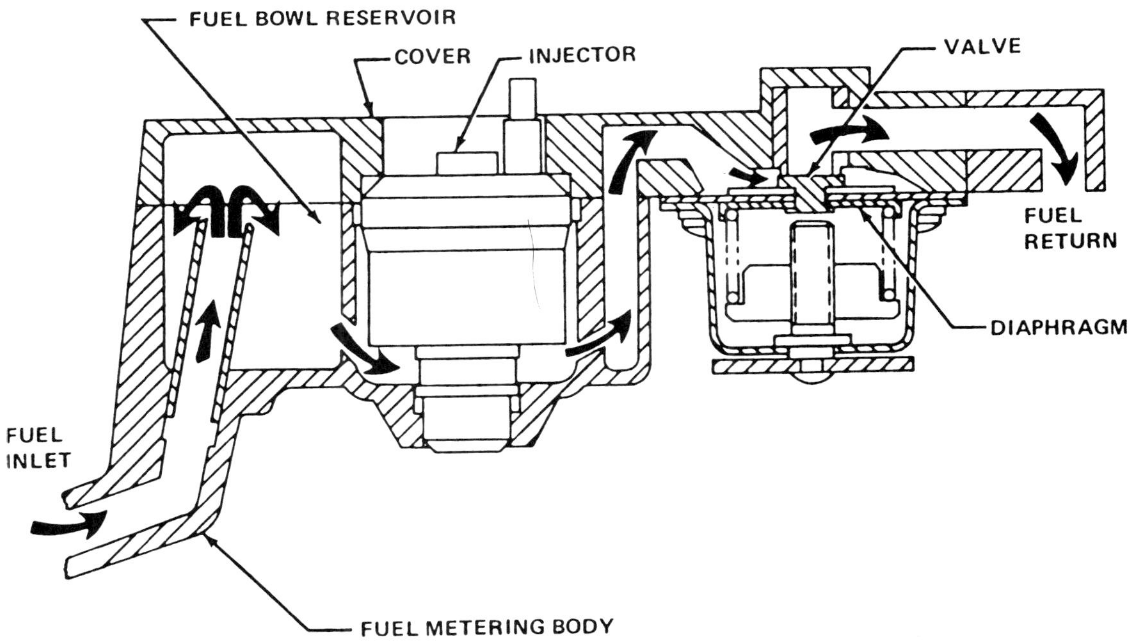

FIGURE 5-2 Fuel pressure regulator *(Courtesy of Cadillac Motor Car Division).*

cause it will unseat the relief valve. Adjustment of the regulator is done during factory assembly and is not field serviceable.

Fuel Injectors

The fuel injectors are mounted in the fuel metering body. They are electronically activated and provide atomized fuel directed into the throttle body bores above the throttle plates. Each injector is a basic solenoid whose plunger or pintle valve is raised off its seat when the injector winding is energized or pulsed. A return spring closes the pintle valve when the injector winding is de-energized. The amount of fuel injected is determined by the amount of time the injector is energized or its pulse width. This is so because the supply pressure is held constant at 10 psi by the pressure regulator. The fuel injectors for the 1982 system are not the same as earlier models even though their outward appearances are similar (Fig. 5-3). The internal structure is completely redesigned to reduce the possibility of leaking or sticking. These new injectors also have a different flow rate and, therefore, are not interchangeable with earlier models. The computer program was also modified to match the new flow rate.

Fuel Supply Line

The fuel supply line is a 3/8-inch diameter steel line that runs along the left-side frame rail (Fig. 5-1). A rubber fuel line hose connects the electric fuel pump to this steel line. The fuel filter is fastened into the steel line with flared fittings. At the engine, a Teflon rubber hose connects the steel line to the steel line that fastens to the throttle body. This Teflon rubber hose is covered with braided stainless steel to protect it against chafing and heat. The fuel return line is 5/16 inch in diameter. Another braided stainless steel covered Teflon rubber hose connects the steel line from the throttle body to the steel line that runs along the right-side frame rail (Fig. 5-1). A rubber fuel line hose connects this steel line to the return line entering the fuel tank.

AIR INDUCTION SYSTEM

Intake Manifold

An aluminum, single-plane intake manifold and a model 200 throttle body make up the air induction system (Fig. 5-4). Earlier model years used a dual-plane cast-iron or

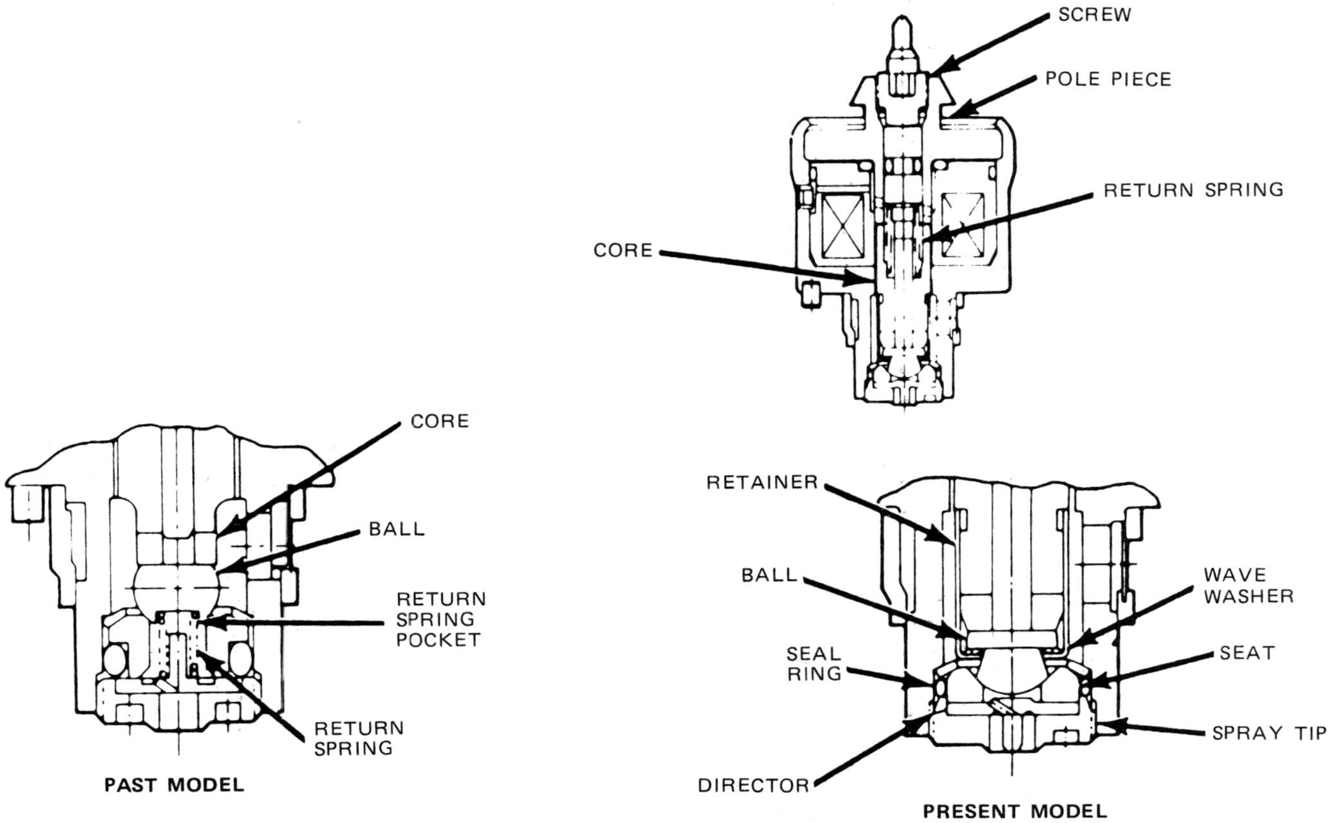

FIGURE 5-3 Injector comparison *(Courtesy of General Motors Product Service Training).*

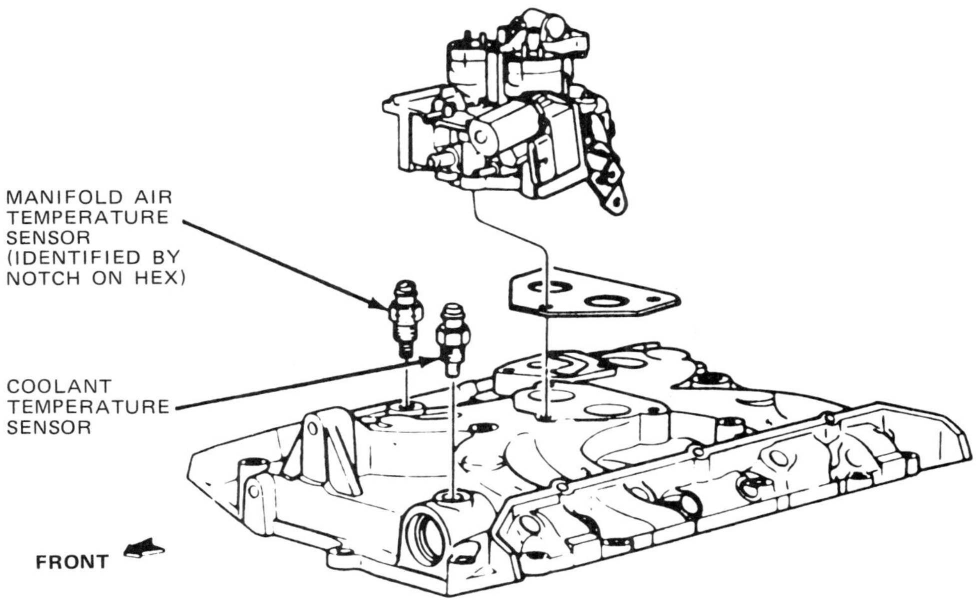

FIGURE 5-4 Air induction *(Courtesy of General Motors Product Service Training).*

aluminum intake manifold. Single-plane designs refer to the fact that the runners or passageways in the manifold are at the same level. Dual-plane manifold runners are at two different levels. The use of either design is determined by the distribution characteristics of the engine.

Throttle Body

The throttle body is different from earlier model years in that its bore size is smaller. Even though it is a model 200 unit, which is the same model as the 1981 unit, it is not interchangeable. The computer program is also changed to match the new bore size. A second difference in the throttle body for 1984 is a new-design throttle position sensor. This sensor connector is in a vertical position, whereas the 1981–1983 unit was horizontal (Figs. 5-5 and 5-6).

DATA SENSORS AND SWITCH SIGNALS

Data sensors for the 1982–1984 DFI system include the manifold air temperature sensor (MAT), coolant temperature sensor (CTS), barometric pressure sensor (BARO), manifold absolute pressure sensor (MAP), engine speed sensor (ESS), vehicle speed sensor (VSS), throttle position sensor (TPS), and oxygen sensor (0_2).

Manifold Air Temperature Sensor

The manifold air temperature sensor (MAT) is located in the intake manifold in front of the throttle body (Fig. 5-4). The hex shape of the sensor has a notch cut around it to prevent confusion in identifying it and the coolant temperature sensor. The MAT sensor measures the air/fuel mixture temperature and sends these data to the electronic control module. Mixture temperature is used to compute speed/density, which ultimately will affect fuel injector pulse width or ON time. The MAT sensor is a thermistor whose resistance rises as the temperature falls.

Engine Coolant Temperature Sensor

Another thermistor is used to measure engine coolant temperature. The coolant temperature sensor (CTS) is located in the intake manifold at the front (Fig. 5-4). Its hex does not have a notch cut around it. Information from the CTS is sent to the electronic control module. This information is used to control or determine closed-loop operation, ignition timing, exhaust gas recirculation valve operation, air management valve operation, torque converter clutch, idle speed control motor, and fuel enrichment for cold engine operation.

Barometric Pressure Sensor

A barometric pressure sensor (BARO) and a manifold absolute pressure sensor (MAP) are located on a common bracket under the right side of the instrument panel (Fig. 5-7). The BARO sensor measures changes in ambient pressure and sends these data to the electronic control module. Ambient pressure or atmospheric pressure changes as the result of changes in altitude and/or weather. These data are used to compute speed/density and will affect pulse width for the fuel injectors. The MAP sensor measures changes

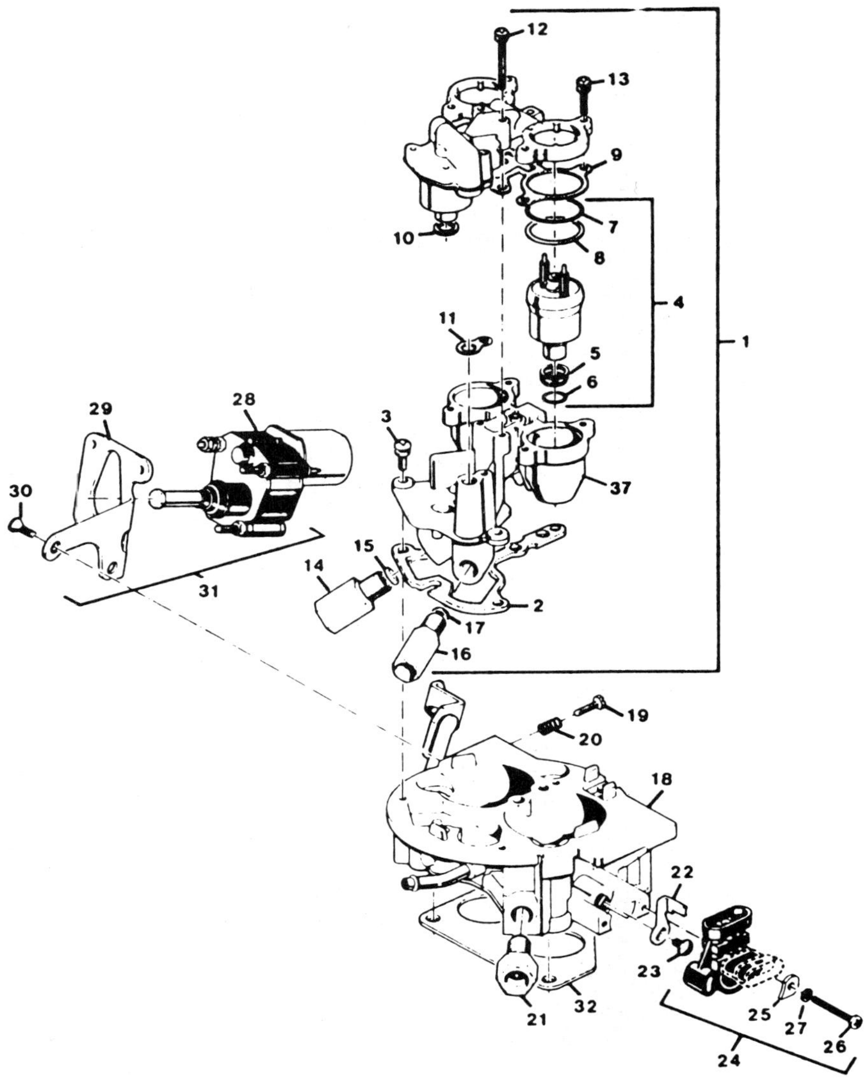

FUEL METERING PARTS		THROTTLE BODY PARTS	
1.	FUEL METER ASSEMBLY	18.	THROTTLE BODY ASSEMBLY
2.	GASKET - FUEL METER BODY	19.	SCREW - IDLE STOP
3.	SCREW - FUEL METER BODY (3)	20.	SPRING - IDLE STOP SCREW
4.	FUEL INJECTOR ASSEMBLY	21.	FITTING - POWER BREAK LINE
5.	FILTER - INJECTOR NOZZLE	22.	LEVER - T.P.S.
6.	SEAL - SMALL "O" RING (2)	23.	SCREW - T.P.S. LEVER
7.	SEAL - LARGE "O" RING (2)	24.	THROTTLE POSITION SENSOR ASSEMBLY (T.P.S.)
8.	BACK UP WASHER - INJECTOR (2)		
9.	GASKET FUEL METER COVER	25.	PLATE - T.P.S. RETAINER (2)
10.	DUST SEAL - REGULATOR	26.	SCREW - T.P.S. ATTACHING (2)
11.	GASKET - FUEL METER OUTLET	27.	WASHER - T.P.S. SCREW
12.	SCREW - LONG (3)	28.	IDLE SPEED CONTROL ASSEMBLY (ISC)
13.	SCREW - SHORT (5)	29.	BRACKET - ISC
14.	FUEL INLET NUT	30.	SCREW - ISC BRACKET ATTACHING
15.	GASKET - FUEL INLET NUT	31.	IDLE SPEED CONTROL/BRACKET ASSEMBLY
16.	FUEL OUTLET NUT	32.	GASKET - FLANGE MOUNTING
17.	GASKET - FUEL OUTLET NUT		

FIGURE 5-5 Throttle body assembly (disassembled view), 1984 *(Courtesy of Cadillac Motor Car Division).*

FIGURE 5-6 Throttle body assembly (DFI), 1982 *(Courtesy of Cadillac Motor Car Division).*

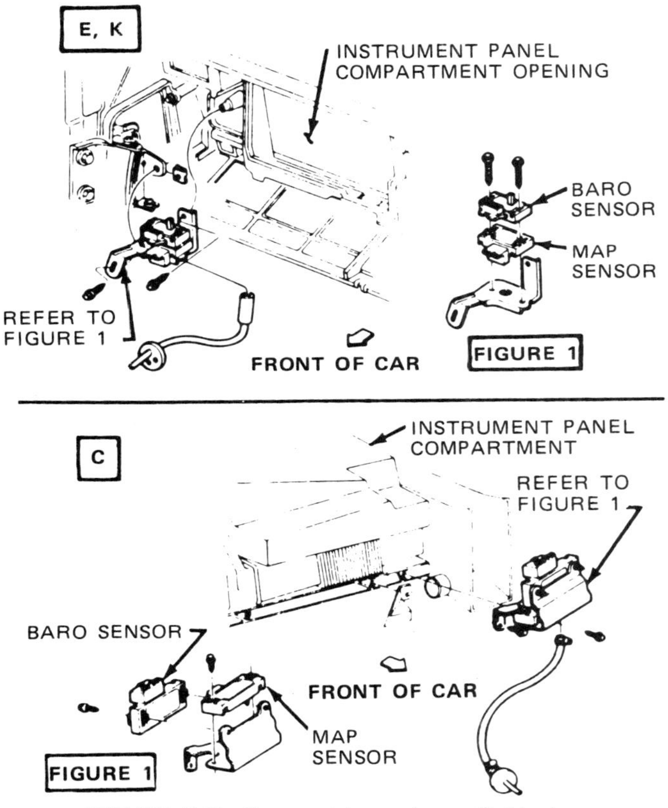

FIGURE 5-7 Barometric and manifold absolute pressure sensor installation *(Courtesy of Cadillac Motor Car Division).*

in intake manifold pressure and sends this information to the electronic control module. Intake manifold pressure changes with throttle blade position and also with altitude and/or weather. Intake manifold pressure is also affected by the mechanical condition of the engine, such as valve timing, and piston rings. An increase in intake manifold pressure means the engine requires more fuel; therefore, the fuel injector pulse width will be increased. The vacuum hose connecting the MAP to the throttle body tap should be suspended from brackets or braces in the engine compartment. This prevents the collection of water vapor and, in cold weather, the formation of ice crystals in the hose.

Engine Speed Sensor

The engine speed sensor (ESS) is integral with the electronic module in the distributor (Fig. 5-8a). This module is a combination high-energy ignition and electronic spark timing module. It provides data in the form of electrical pulses, representing engine rpm, to the electronic control module. These pulses are used to compute idle speed, ignition timing, exhaust gas recirculation valve operation, fuel delivery, and air management valve operation. The distributor electronic module is not interchangeable with earlier model years.

Vehicle Speed Sensor

The vehicle speed sensor (VSS) is located at the back of the instrument cluster on the speedometer head (Fig. 5-8b). It consists of a light-emitting diode (LED), a reflective blade that spins with the speedometer cable, and a photocell. As the reflective blade reaches the right position, the light beam from the LED is reflected onto the photo-transistor. The photo-transistor produces an on-off voltage signal which is sent to a buffer/amplifier module under the instrument panel. The amplified signal is sent to the electronic control module as an indication of vehicle road speed. The lower the signal frequency is, the lower the road speed. This information is used to compute torque converter clutch application, ignition timing, idle speed, and air management valve operation. Previous model years used an 8-volt source to operate the VSS. The 1982–1984 DFI VSS uses a 12-volt source. This improves circuit reliability and reduces sensor circuit problems due to VSS circuit shorts.

Throttle Position Sensor

The throttle position sensor (TPS) provides to the electronic control module an indication of the throttle plate angle (Fig. 5-6). The 1984 DFI system, as mentioned earlier, uses a new TPS that has its connector mounted in a vertical position (Fig. 5-5). This TPS is also different from earlier versions in that it provides a direct linear change in voltage as the throttle plates are opened. The 1981 TPS had a stepped output by comparison. This new TPS is located on the right side of the throttle body and is a rotary potentiometer that senses movement of the throttle shaft.

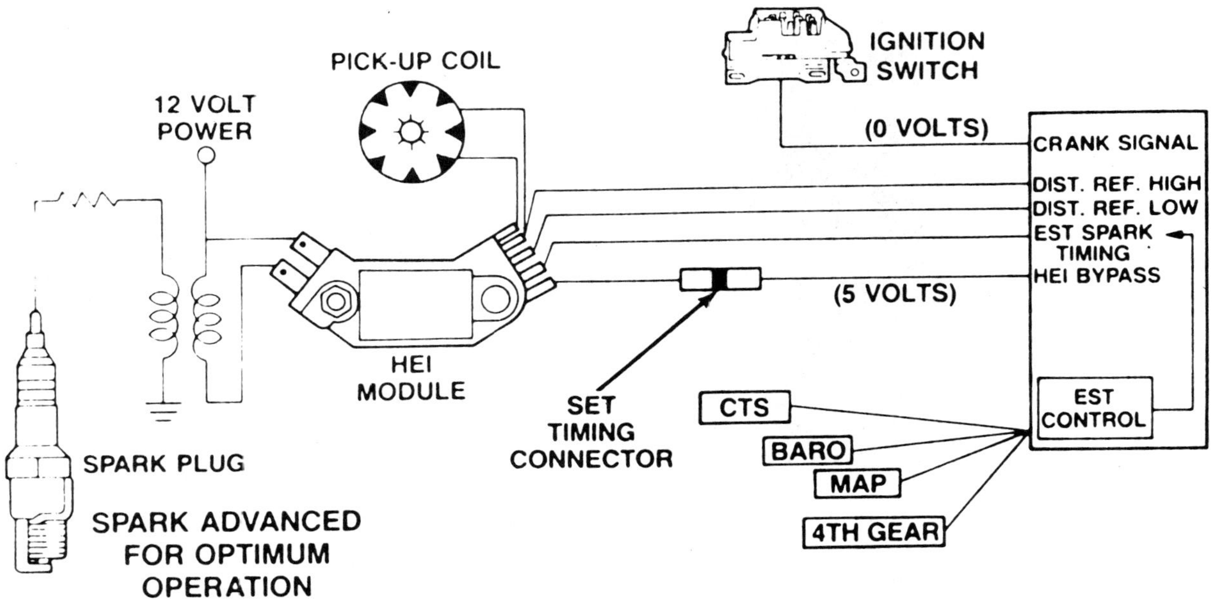

ELECTRONIC SPARK TIMING
ENGINE RUNNING

(a)

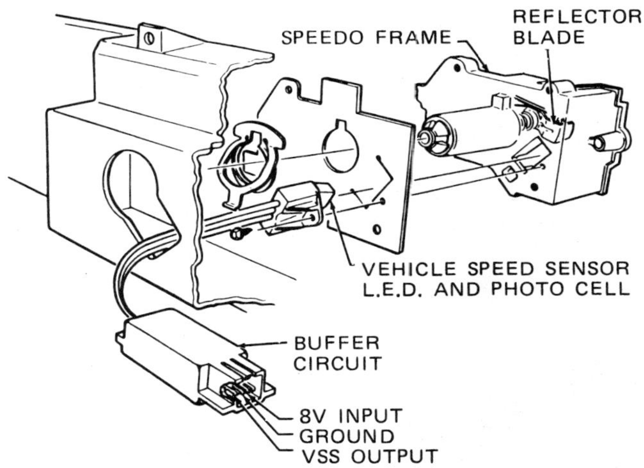

(b)

FIGURE 5-8 (a) Electronic spark timing, engine running *(Courtesy of General Motors Product Service Training).* (b) Vehicle speed sensor *(Courtesy of Cadillac Motor Car Division).*

Oxygen Sensor

The oxygen sensor (0_2) is mounted in the exhaust manifold and extends into the exhaust gas stream (Fig. 5-9). It consists of platinum plates separated by a zirconia electrolyte. One platinum plate is exposed to the hot exhaust gases and the other is exposed to outside air. When the zirconia electrolyte is exposed to oxygen gas, it becomes an electrical conductor. The voltage produced between the two platinum plates is determined by the oxygen content of the exhaust gas. This voltage varies during engine operation between 100 and 900 millivolts. This varying voltage signal is sent to the electronic control module representing exhaust gas oxygen content. The electronic control module uses this information to help compute fuel injector pulse width. High exhaust gas oxygen content, produced by lean air/fuel mix-

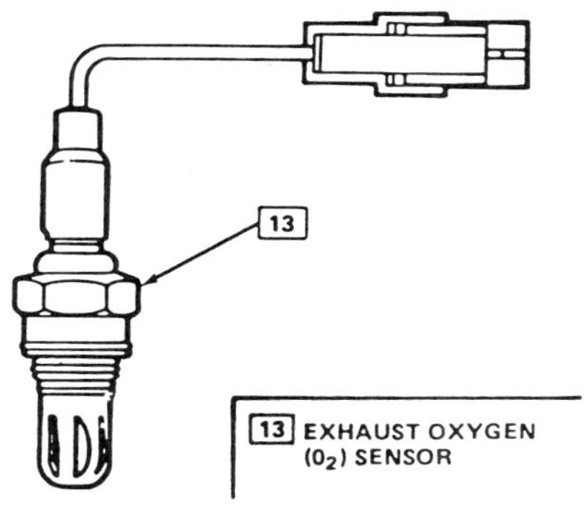

FIGURE 5-9 Exhaust oxygen (O_2) sensor *(Courtesy of Cadillac Motor Car Division).*

tures, causes the oxygen sensor to produce lower voltages. Low exhaust gas oxygen content, produced by rich air/fuel mixtures, causes the oxygen sensor to produce higher voltages. The oxygen sensor signal is used by the electronic control module only during closed-loop operation.

SWITCH SIGNALS

The switch signals for the 1982–1984 DFI system include the ignition switch, crank signal, brake switch, drive switch, reverse switch, fourth-gear switch, air-conditioning switch, cruise control switch, and throttle switch.

ECM Power

The electronic control module receives its power from two sources. A 20-ampere ECS battery fuse in the mini fuse panel provides power when the ignition switch is off. This keeps the ECM memory active. When the ignition switch is turned on, power is provided through a 15 ampere ECS ignition fuse in the mini fuse panel.

Crank Signal

A cranking signal is provided to the electronic control module through a 3-ampere crank fuse in the mini fuse panel. When the ignition switch is turned to START, this signal is sent to the electronic control module to activate the fuel pump circuit and to retard the ignition timing for easier starting.

Brake Switch

The brake switch has a dual purpose. It operates the brake lights and also disengages the cruise control system. Since the cruise control system is interfaced with the DFI system, the brake switch also sends a signal to the electronic control module.

Drive and Reverse Switches

A drive switch and a reverse switch are incorporated into the park/neutral switch at the base of the steering column. As the driver moves the gear selector lever into drive or reverse, a signal is sent to the electronic control module. Since the ECM controls idle speed, this signal enables it to anticipate the load of the transmission/transaxle. It can then extend the idle speed control motor plunger and maintain idle speed and quality.

Fourth-gear Switch

The torque converter clutch is operated by the electronic control module. A fourth-gear switch indicates to the ECM when the transmission/transaxle shifts into fourth gear. If other conditions are met, such as vehicle speed, the ECM can engage the torque converter clutch.

Air-conditioning Switch

The method of monitoring the air-conditioning clutch circuit is different from previous years. A signal from the electronic climate control panel requesting activation of the air-conditioning system first goes to the ECM. The ECM delays this signal by 0.7 second while it extends the plunger of the idle speed control motor. Thus the engine idle speed will not vary as the air-conditioning system cycles on and off. This feature was not as important in earlier models because a 6.0-liter engine was used, which was not affected by the load of the compressor clutch. The 1982–1984 DFI system is used on a 4.1-liter engine that is more sensitive to such load changes.

Cruise Control Switches

The cruise control system is interfaced with the DFI system. There are three input signals to the ECM provided by the electronic cruise control system. The first comes from the cruise enable switch on the instrument panel. This switch sends a 12-volt signal to the ECM when the driver wants to activate the cruise control system. The second signal comes from the set/coast switch. This signal, when received by the ECM, allows the ECM to activate the cruise control system but only if certain operating conditions are met. The third signal is provided by the resume/accelerate switch. This voltage signal causes the ECM to either resume a speed to which the system was previously set or accelerate to a new vehicle speed setting.

Throttle Switch

The throttle switch is an integral part of the idle speed control (ISC) motor. Sometimes referred to as a nose switch, it provides a signal to the ECM when the throttle is closed. As a result, the ECM controls the throttle position by energizing the ISC motor. If the driver steps on the accelerator pedal, the nose switch will open and the signal to the ECM is interrupted. The ECM will not attempt to control idle speed.

ELECTRONIC CONTROL MODULE

The electronic control module (ECM) receives sensor and switch data, calculates at very high speed, and issues command signals or pulses to control the torque converter

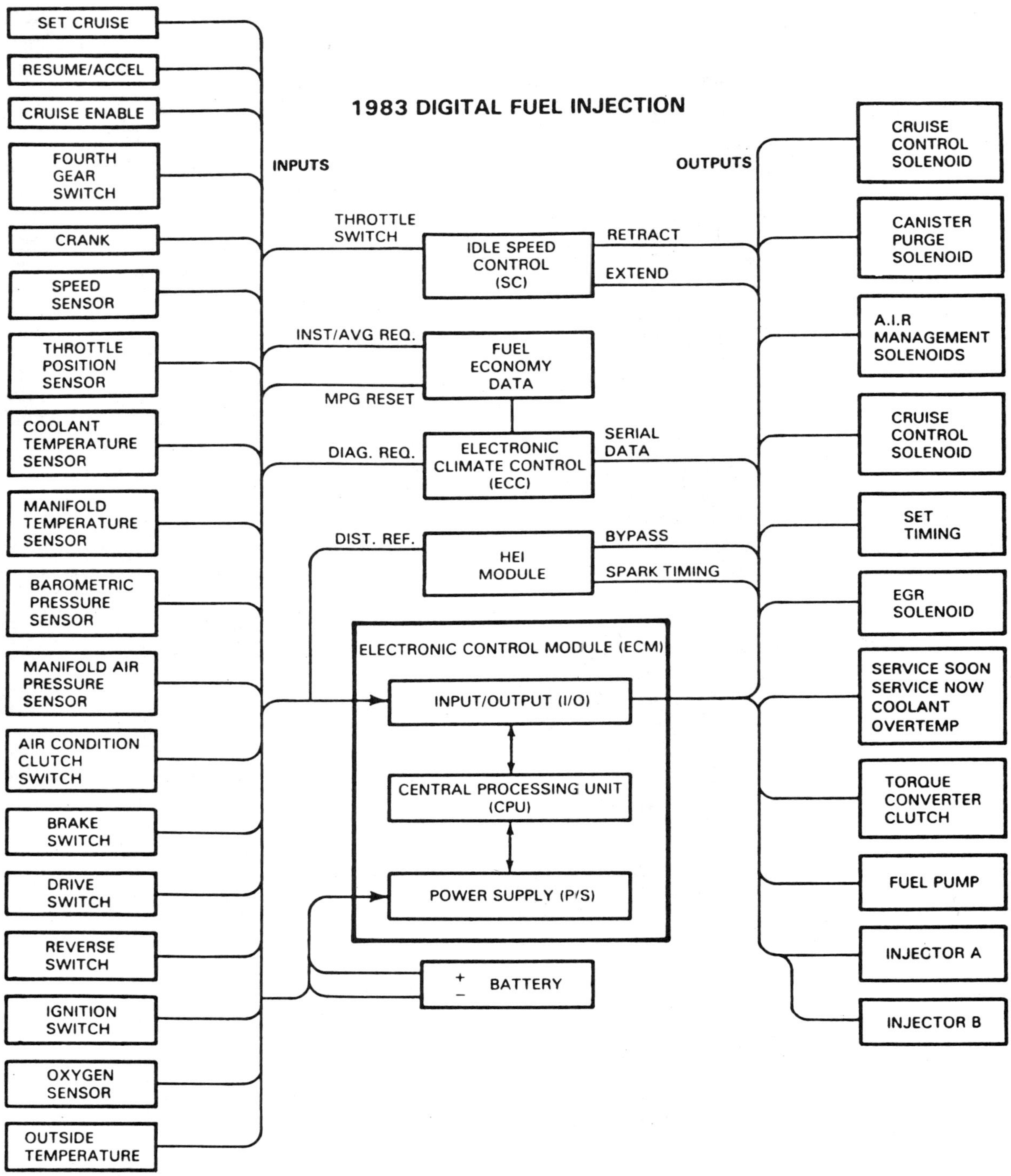

FIGURE 5-10 Digital fuel injection component diagram *(Courtesy of Cadillac Motor Car Division).*

clutch, fuel delivery system, ignition timing, canister purge, air management valve, exhaust gas recirculation valve, idle speed control motor, cruise control system, the self-diagnostic mode of operation, and the electronic climate control display panel (Fig. 5-10). The ECM has three kinds of memories to assist in the calculations. A read-only memory (ROM) is a permanent memory. A programmable read-only memory (PROM) is also a permanent memory but is replaceable by the service technician. The ECM has two PROMs, one in a clear plastic installation device called a carrier and one in a green plastic carrier. Their positions must not be exchanged. The third memory is a random-access memory (RAM). The RAM is where diagnostic codes and temporary information are stored. The RAM is not permanent and all information in it is lost if its power is removed.

When the ignition switch is turned to ON, the ECM receives that signal and activates the electric fuel pump. If a cranking signal is not received in 2 seconds, the ECM turns the fuel pump off. All sensors are energized and begin sending input data to the ECM. The exhaust gas recirculation valve solenoid is energized to block vacuum to the valve diaphragm as long as coolant temperature is below 175°F. The CHECK ENGINE telltale light will light as a bulb check (Fig. 5-11). The high-energy ignition module is operated in the HEI mode where it provides a programmed base ignition timing.

When the engine is cranked, the ECM receives that signal and either keeps the fuel pump running or reactivates it, depending on the time lapse from when the ignition switch was turned on. The ECM also issues a series of short fuel injector pulses, or prime pulses, and then pulses the injectors alternately at each distributor reference pulse from the ignition module. The engine sensors continue to deliver input data.

FIGURE 5-11 CHECK ENGINE light *(Courtesy of Cadillac Motor Car Division).*

After the engine starts, the crank signal is no longer sent to the ECM. However, the fuel pump runs continuously because distributor reference pulses are supplied by the ignition module. The fuel injectors are pulsed alternately for each distributor reference pulse. The high-energy ignition module control of the timing is bypassed, and the ECM controls timing through its electronic spark timing circuitry. The ISC begins to control the idle speed if the throttle switch is closed. The fuel system pressure is maintained by the pressure regulator at about 10 psi.

The ECM can control fuel delivery in one of three modes of operation. The cranking mode was just discussed, as was the normal delivery mode. If the engine should flood, a third mode of operation, wide-open throttle, can be entered by the ECM. The throttle should be held wide open and the engine should crank for at least 3 seconds. By doing this, the ECM will deliver smaller alternating pulses to the injectors, enabling the engine to "clear flood" and start.

ELECTRONIC SPARK TIMING

Components of the electronic spark timing (EST) system include the electronic control module, the HEI distributor and module, the coolant temperature sensor, the manifold absolute pressure sensor, the barometric pressure sensor, and the ignition switch.

The EST system has two modes of operation, cranking and running. During the cranking mode, the ECM receives a 12-volt crank signal and then allows the HEI/EST module to control timing (Fig. 5-12). When the cranking signal drops to 0 volts, the ECM issues a 5-volt signal to the module. This causes the module to bypass the HEI control of the timing and allows the ECM to gain control instead. (Fig. 5-8). This is known as the running mode of operation.

IDLE SPEED CONTROL

The idle speed control (ISC) system consists of the ECM, vehicle speed sensor, throttle position sensor, throttle switch, engine speed sensor, air-conditioning clutch signal, ignition switch, park/neutral switch, coolant temperature sensor, and manifold absolute pressure sensor. The ISC motor controls throttle plate angle for cranking, idling, and decelerating (Fig. 5-6).

While cranking, the throttle may be held open more than normal for cold starting or may be held in a normal position for hot starting. The ECM uses input data from the sensors of the system to determine correct throttle plate angle for idle. This idle is maintained even when additional loads are placed on the engine.

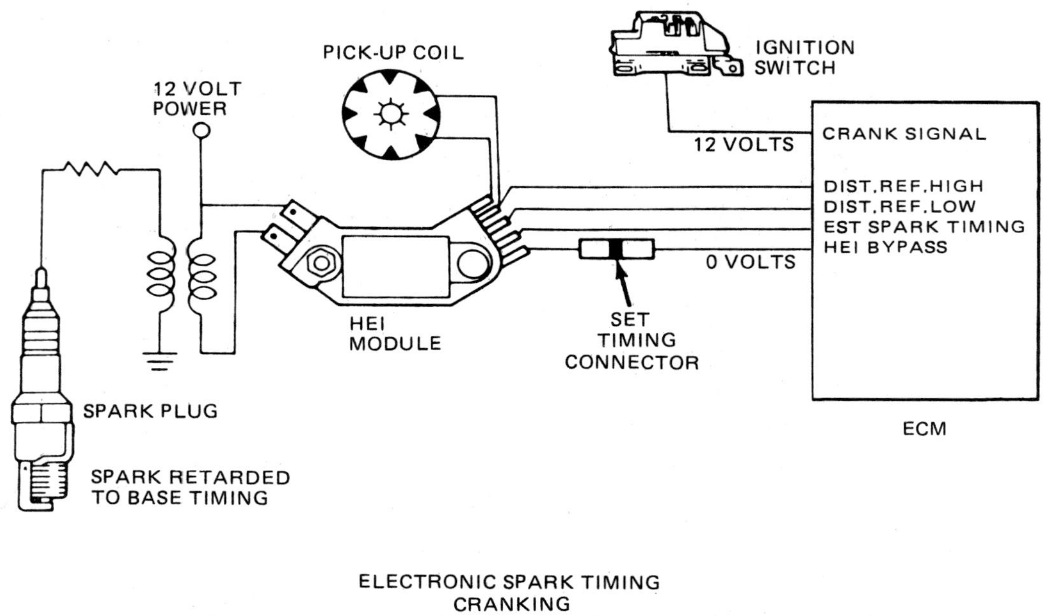

FIGURE 5-12 Electronic spark timing, cranking *(Courtesy of General Motors Product Service Training).*

Throttle plate angle when decelerating is important to control not only to prevent stalling, but also to reduce hydrocarbon emissions. When the throttle plates close suddenly, intake manifold pressure drops and causes condensed droplets of fuel or puddling to vaporize. The resulting rich mixture produces an increase in hydrocarbon emissions. To offset this rise in emissions, the ECM extends the ISC plunger and opens the throttle plates. This adds air to the intake manifold mixture and reduces emissions.

EMISSIONS SYSTEMS

The emission control systems controlled by the DFI system include an evaporative emission control (EEC) system, an exhaust gas recirculation (EGR) system, and an air management (AIR) system.

Evaporative Emission Control

The evaporative emission control system operates the same as earlier versions except that the ECM has control over the purging of the canister vapors. A canister purge control solenoid either blocks or allows vacuum to the purge valve. The ECM provides a ground for the solenoid, energizing it, and purging of the canister is stopped. If the ECM does not provide a ground, the solenoid de-energizes, and vacuum can cause the purge valve to allow vapors to be drawn from the canister.

Exhaust Gas Recirculation

The exhaust gas recirculation valve is vacuum operated. The vacuum source can be interrupted by an EGR solenoid that is controlled by the ECM. When the solenoid is energized, it blocks vacuum to the valve, making it inoperable. When the ECM de-energizes the solenoid, a ported vacuum signal can act on the valve diaphragm. The valve, however, opens only if exhaust system pressure is sufficient because the EGR valve is a back-pressure transducer design.

The 1983–1984 DFI system controls the EGR valve in a different way. The ECM can pulse the EGR solenoid on and off at a varying frequency. The frequency of pulsing determines how much vacuum is allowed to act on the EGR valve diaphragm. This is known as *pulse-width modulation.* The pulse-width signal to the solenoid is varying or modulating. This method provides a more accurate means of controlling the EGR valve. When the ECM does not provide a ground for the solenoid, it is de-energized and allows vacuum to pass. The amount of time the solenoid allows vacuum to pass is known as the *duty cycle.* A 90% duty cycle means the EGR valve has vacuum applied to its diaphragm 90% of the time. The duty cycle for different operating conditions is programmed into the PROM.

Air Management

The air management system is an ECM-controlled air injection reaction system. An engine-driven air pump delivers fresh air to a divert valve. The divert valve directs the air to either the air cleaner or the switching valve. The switching valve directs the air to either the center chamber of the catalytic converter or to the exhaust manifolds.

The ECM controls solenoids that operate the divert and switching valves. When the engine is cold, both solenoids are energized and fresh air is sent into the exhaust manifolds. When the engine warms up, the ECM keeps the divert valve solenoid energized and de-energizes the switching valve solenoid. This allows fresh air to enter the catalytic converter. If the engine decelerates, the ECM de-energizes the divert valve solenoid and the air is sent to the air cleaner.

TRANSMISSION CONVERTER CLUTCH

To reduce the power loss at the torque converter at normal, constant road speeds, a torque converter clutch is used (Fig. 5-13). The torque converter operates normally to multiply engine torque and provide a hydraulic coupling between the engine and the transmission/transaxle. When the converter clutch is applied, the clutch facing provides a direct mechanical connection between the engine and the transmission. By doing this, the slippage and power loss due to the hydraulic connection are eliminated; however, engine torque can no longer be multiplied. For this reason, the converter clutch can be applied only during certain driving conditions.

The ECM controls a torque converter clutch (TCC) solenoid located inside the automatic four-speed transmission. When the solenoid is energized, the TCC is applied. When the solenoid is shut off, the TCC disengages and the torque converter operates like any other standard torque converter. The ECM energizes the TCC solenoid when engine coolant temperatures are between 185° and 285°F, BARO pressure is above 12.5 psi, vehicle speed is above 45 mph, and the transmission is in fourth gear. When these conditions are met, 13 seconds will pass and then the ECM will energize the TCC solenoid to apply the clutch.

The brake switch de-energizes the solenoid when the brakes are applied; the ECM de-energizes the solenoid when the transmission shifts out of fourth gear, vehicle speed is less than 45 mph while in fourth gear, or BARO pressure is less than about 12 psi (high altitude).

CRUISE CONTROL

The electronic cruise control system is interfaced with the DFI system. Signals from the cruise control engagement switch, instrument panel switch, brake switch, drive switch, and vehicle speed sensor are sent to the ECM to be processed along with input data from the DFI system. The ECM then issues commands to the vacuum control solenoid valve and the servo unit solenoid valve to control vehicle speed.

FUEL DATA PANEL

As an improvement over the MPG panel display used in 1981 and to fully utilize the capabilities of the ECM, a fuel data display panel has been incorporated into the instrument panel (Fig. 5-14). Information that can be displayed

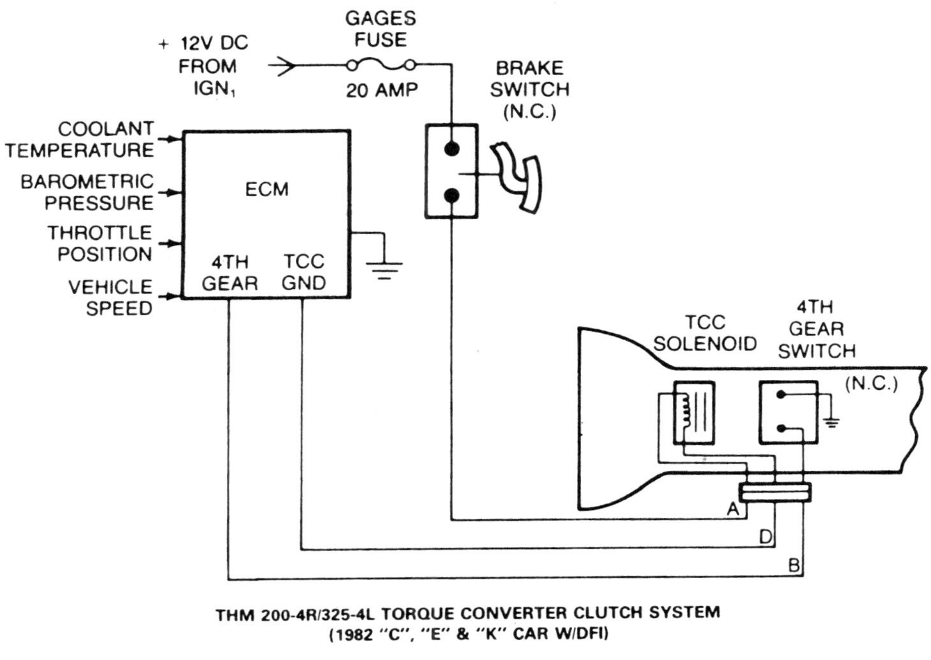

FIGURE 5-13 THM 200-4R/325-4L torque converter clutch system (1982 C, E, and K car with digital fuel injection) *(Courtesy of General Motors Product Service Training).*

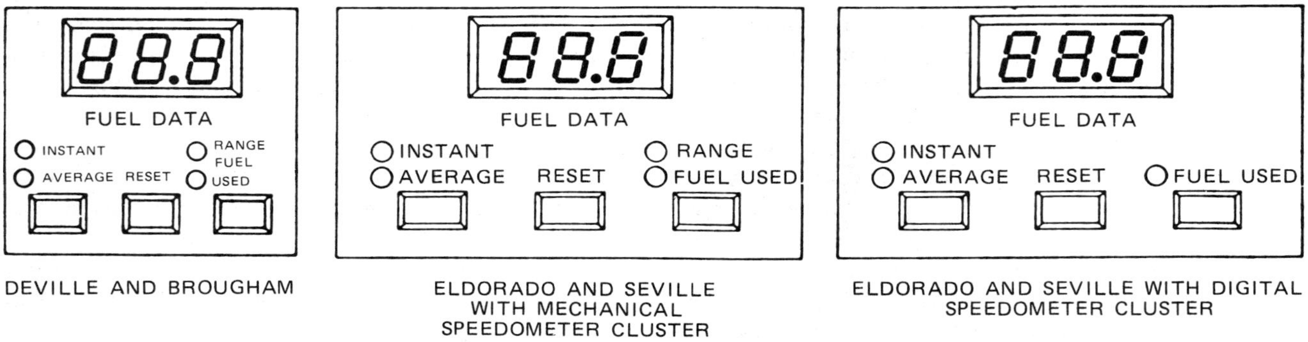

FIGURE 5-14 Fuel data panel (all) *(Courtesy of Cadillac Motor Car Division).*

include fuel used, range of driving for remaining fuel, fuel economy average for complete trip or for instantaneous readouts, and a reset feature that enables the driver to set the system to start over for a new trip.

SYSTEM DIAGNOSTICS

The electronic climate control panel is the control for the self-diagnostic function of the DFI system (Fig. 5-15). The panel is different from earlier model years and is operated with new procedures. Operation of the diagnostic mode will be discussed later in this chapter. There are 31 diagnostic trouble codes for the DFI system, 6 cruise control codes, and 10 switch tests. The system can also display operating ranges or parameters for the DFI sensors. An output cycling function assists the service technician in checking operation of components controlled by the ECM.

SYSTEM SERVICE

Idle Speed

Engine idle speed is controlled by the ECM and ISC motor. If idle speed is not correct, the ignition timing, TPS, throttle linkage, ISC motor adjustment, and minimum air rate should be checked. Follow detailed instructions in the service manual.

Minimum Air Rate Adjustment

The throttle stop screw is factory adjusted. It can be adjusted if necessary by following the service manual procedure. If an adjustment is made, it is necessary to adjust the TPS.

Throttle Position Sensor Adjustment

The minimum air rate adjustment must be correct before adjustments to the TPS are made. A 10-megohm impedance digital multimeter and special jumper leads are required to make this adjustment. It may be necessary to remove the throttle body to remove TPS mounting screw spot welds before making adjustments (Fig. 5-16). Follow service manual procedures and also check the ISC adjustment when finished.

ISC Motor Adjustment

The ISC motor plunger should be fully extended before adjustments are made. Also, the minimum air rate and TPS adjustments should have been done. Follow the procedure in the service manual for accurate results.

Throttle Body Assembly

Complete disassembly, cleaning, inspection, and reassembly including new gaskets and seals are required for an

FIGURE 5-15 Electronic climate control panel *(Courtesy of General Motors Product Service Training).*

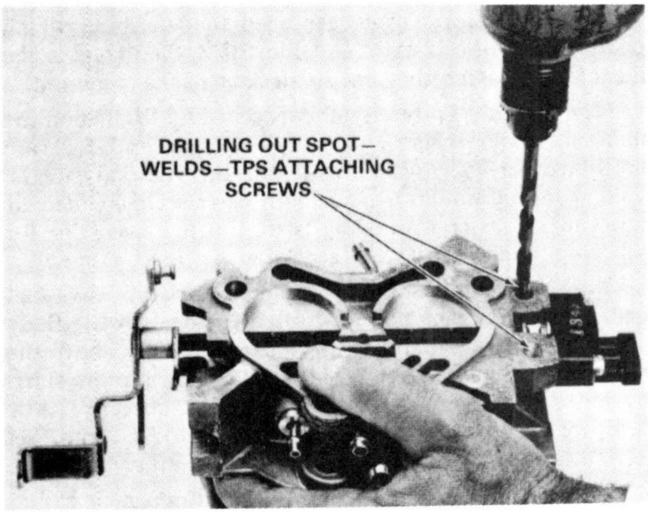

FIGURE 5-16 Throttle position sensor spot weld removal *(Courtesy of Cadillac Motor Car Division).*

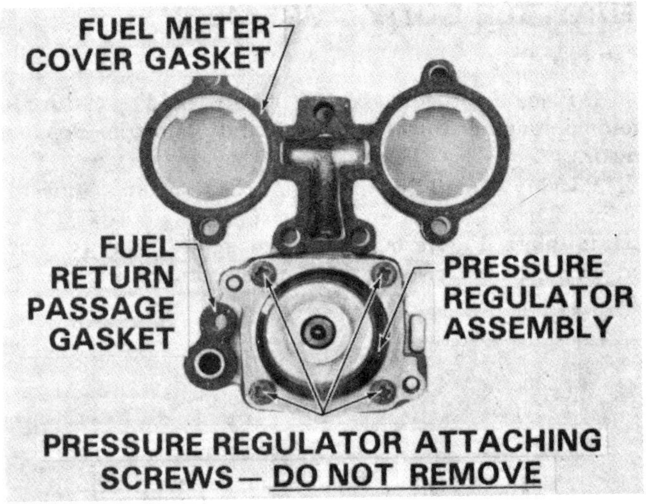

FIGURE 5-17 Fuel meter cover, bottom view *(Courtesy of Cadillac Motor Division).*

effective servicing of the throttle body. The fuel pressure regulator is integral with the fuel metering body cover and is replaced as an assembly (Fig. 5-17). The fuel injectors can be replaced separately, and new seals should be used whenever an injector is removed (Fig. 5-18). The ISC motor and TPS can be removed, and the remaining throttle body casting can be cleaned in an immersion-type solvent. Do not clean injectors, TPS, ISC, or the pressure regulator in an immersion-type solvent.

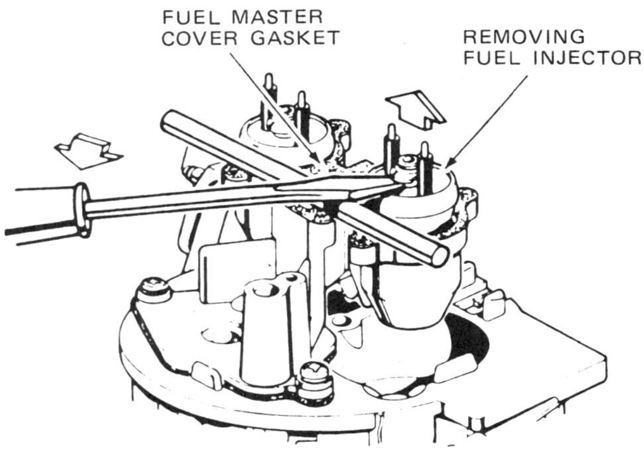

FIGURE 5-18 Injector removal *(Courtesy of Cadillac Motor Car Division).*

Sensor Service

The MAP and BARO sensors are not serviceable and can only be replaced. On some models they are located on a common bracket under the right end of the instrument panel (Fig. 5-19). Testing procedures are outlined in the service manual and should be followed to determine if replacement

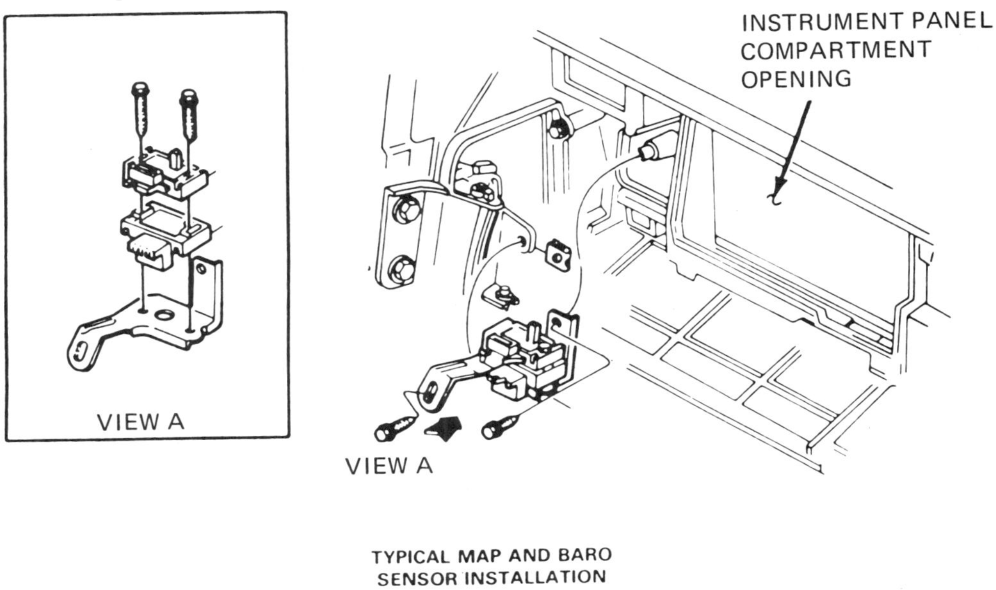

FIGURE 5-19 Typical barometric and manifold absolute pressure sensor installation *(Courtesy of General Motors Product Service Training).*

is required. A 10-megohm impedance digital multimeter and jumpers are needed to perform these tests.

The MAT and CTS sensors should be coated with a nonhardening sealer whenever they are removed. This prevents leakage past the threads. Test procedures are provided in the service manual to determine if replacement is needed. A 10-megohm impedance digital multimeter is required for these tests.

ECM Service

The ECM is located under the right end of the instrument panel (Fig. 5-20). The ECM contains two PROMs located under a removable cover (Fig. 5-21). These PROMs are serviceable and can be replaced individually. They should not be interchanged, and installation position is very important. Each PROM is mounted in a removal and installation tool called a carrier. To make identification easier, one carrier is clear plastic and the other is green plastic. Be sure to install them correctly. If a PROM is installed backwards, it will be destroyed when the ignition switch is turned on.

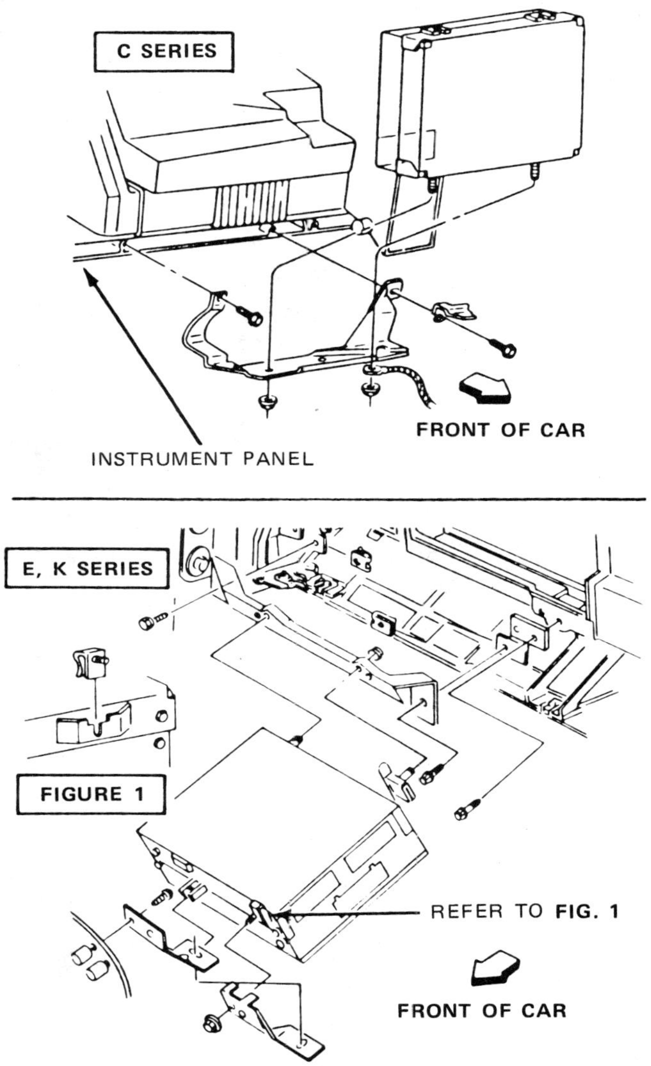

FIGURE 5-20 Electronic control module mounting *(Courtesy of Cadillac Motor Car Division).*

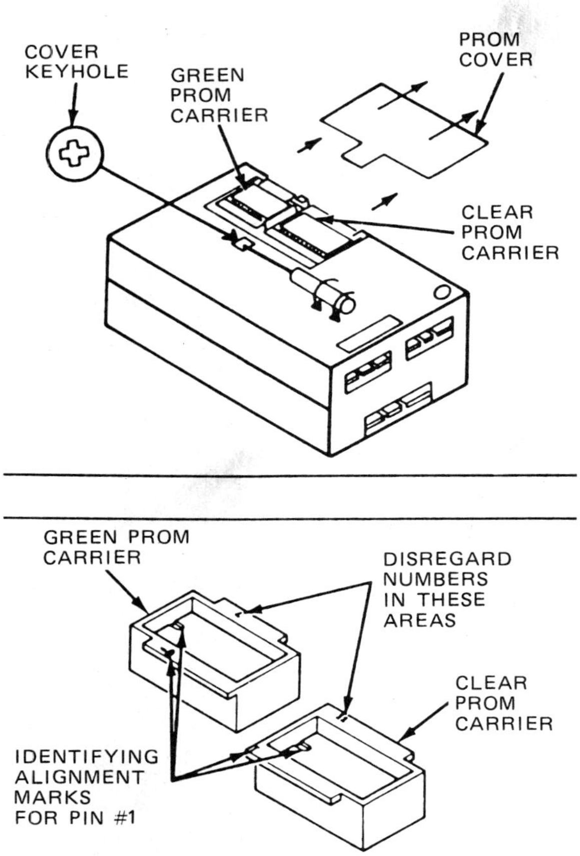

FIGURE 5-21 PROM-to-carrier orientation *(Courtesy of Cadillac Motor Car Division).*

Filter Service

The fuel filter assembly should be replaced at the specified maintenance interval. The paper element and metal casing are one piece in construction and are replaced as an assembly (Fig. 5-1). Always check for leaks after replacement.

SYSTEM DIAGNOSIS

Trouble codes are set to indicate problems in many different DFI system circuits. A list of trouble codes is provided at the end of this chapter (Figs. 5-27 to 5-30).

When a problem occurs in a circuit, a trouble code may be stored or set in the ECM memory. An intermittent code is set if the problem appears and then corrects itself. This type of code may be caused by a loose connector in the circuit. A hard code is set if the problem appears and remains. This type of code indicates a permanent problem.

Diagnostic charts are provided in the service manual to assist the service technician in diagnosing circuits. These charts cannot be used for intermittent codes. As was just discussed, intermittent codes can be caused by deformed connectors (melted) or bent terminals, loose connections, corrosion or sealers on terminal connections, terminals that have "backed out" of the connector, poor connection between the wire and the terminal, or connector halves not fully seated.

Self-diagnostic Mode

To get the system to enter its diagnostic mode, the service technician should turn the ignition switch on and push the OFF and WARMER buttons simultaneously on the ECC panel (Fig. 5-15). These buttons should be depressed until the ECC panel displays two decimal points . . . The panel will then display −1.8.8, which serves as a segment check. If any segment of this digital readout is missing, diagnostic trouble codes may be misinterpreted. Once the segment check is complete, the diagnostic codes are displayed in ascending order. At the same time, small light-emitting diodes (LEDs) above each push button and alongside the outside temperature button, begin to provide additional diagnostic information (Fig. 5-22).

The LED above the OFF button indicates the ECM mode of operation. If it is lit, the ECM is in the closed-loop mode. If it is not lit, the ECM is in the open-loop mode.

The LED above the ECON button indicates the oxygen sensor operation. When it is lit, the oxygen sensor is rich; when it is off, the oxygen sensor is lean.

The LO button LED shows the throttle switch position. If it is lit, the throttle switch is closed, and if it is off, the throttle switch is open.

The AUTO button LED indicates TCC operation. If it is lit, the ECM has enabled the TCC solenoid. If the LED is not lit, the ECM has disabled the TCC solenoid. This LED only indicates ECM control of the solenoid, not operation of the converter clutch itself.

The LED above the HI button is an indication of fourth-gear operation in the transmission. When the LED lights, the transmission is in fourth gear. If it is off, the transmission is not in fourth gear.

The OUTSIDE TEMP button LED indicates operation of the air-conditioning compressor clutch. When the LED lights, the compressor clutch is engaged. When the LED is off, the compressor clutch is disengaged.

After all trouble codes have been displayed, the display sequence is repeated. When the second sequence is complete, a third sequence begins, but only hard or current codes will be shown. Following this third sequence is another segment check −1.8.8, and then .7.0, which means the system is ready for further diagnostic instructions.

Engine Data Tests

At this point, there are three areas that the system can enter for additional diagnostic information (Fig. 5-23). The engine data display can be entered by pushing the MPG reset button on the fuel economy data panel (Fig. 5-14). The ECC panel displays .9.0 and, when the instant button on the fuel economy data panel is pushed, the system begins to display parameters or ranges of operation for engine sensors and other related data. A list of sensed parameters is included at the end of this chapter (Fig. 5-26). When en-

STATUS LIGHT DISPLAY

	ECM OPERATING MODE	O_2 SENSOR INPUT	THROTTLE SWITCH INPUT	TCC OUTPUT	4TH GEAR INPUT	A/C CLUTCH OUTPUT
LIGHT ON	CLOSED LOOP	RICH	CLOSED THROTTLE	TCC ENABLED	IN 4TH GEAR	CLUTCH ENABLED
LIGHT OFF	OPEN LOOP	LEAN	OPEN THROTTLE	TCC DISABLED	NOT IN 4TH GEAR	CLUTCH DISABLED

FIGURE 5-22 Status light display *(Courtesy of Cadillac Motor Car Division).*

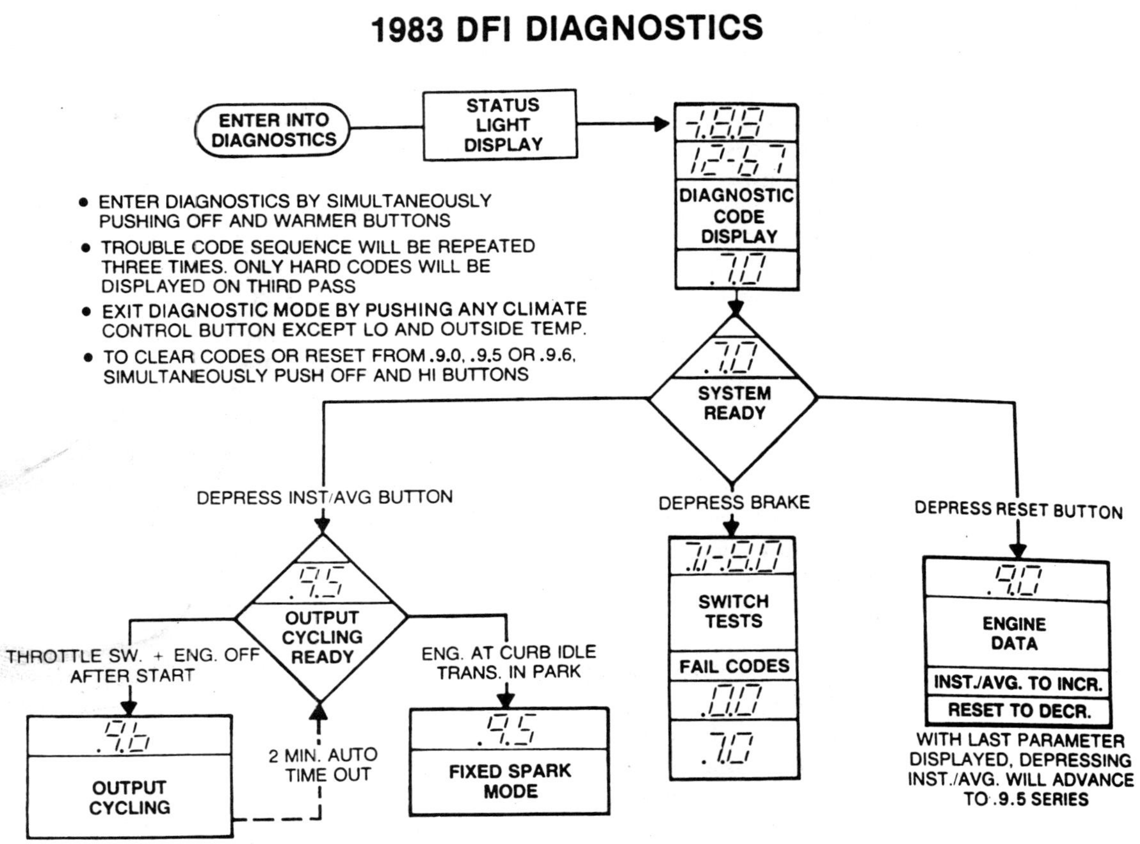

FIGURE 5-23 Digital fuel injection diagnostics *(Courtesy of Cadillac Motor Car Division).*

gine data display begins, the ECC panel shows .0.1 as the parameter number and then the range value is shown. The system repeats this cycle until the instant-average button on the fuel economy data panel is pushed. This causes the system to advance to the next parameter number. The service technician can repeat this procedure for all 13 parameters.

Switch Tests

When the ECC panel displays .7.0, the system performs the switch tests by stepping on the brake pedal (Fig. 5-24). There are ten switch tests. The first is for the cruise control brake switch. By stepping on the brake pedal a second time, the system checks the operation of this switch and

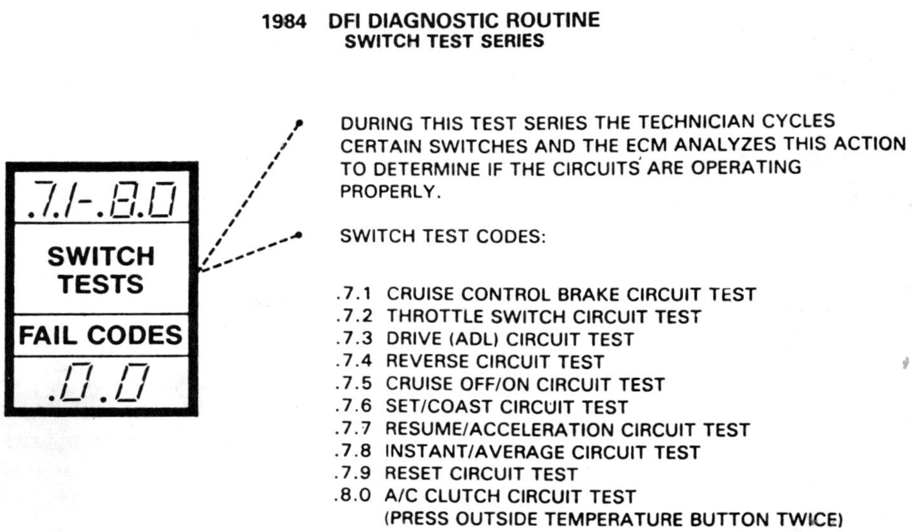

FIGURE 5-24 Switch test series *(Courtesy of Cadillac Motor Car Division).*

advances to the next switch, the throttle switch. The technician should step on the accelerator pedal for this test and the system will advance again. The ECC display indicates by number which switch is being tested. The brake switch is .7.1, and .7.2 is the throttle switch. In each case, the technician is required to operate the switch in question in order to get the system to advance. All these tests can be done from the driver's seat. When all the switch tests have been completed, the ECC panel displays the first switch that failed the test. This display remains until the switch is tested again and passes or until repairs are made and it is retested.

Output Cycling Tests

If, when the ECC panel displays .7.0, the instant-average button is depressed, the system enters its output cycling diagnostic mode (Fig. 5-25). The ECC panel displays .9.5 and, if the accelerator pedal is depressed and released, .9.6 will appear. The system now cycles the ISC motor retract, cruise vacuum light, coolant light, air divert solenoid, air switching solenoid, ISC extend, canister purge solenoid, TCC solenoid, and EGR solenoid. To get reliable results from this test, the engine should have been running just prior to beginning the test and then shut off. The service technician can determine if these components are operating either by touch or by sound. The system continues cycling these components for about 2 minutes and then returns to .9.5.

Fixed Spark Mode

While the ECC panel is displaying .9.5, the ECM is in a fixed spark mode. In this mode, the ECM provides a fixed timing of 20°BTDC. At this point, the service technician can check ignition timing control without disconnecting connectors and possibly causing poor connections.

Clearing Codes

Stored codes can be erased or cleared from the ECM memory by entering the diagnostic mode and then pushing the OFF and HI buttons simultaneously. Hold the buttons in until .0.0 appears on the ECC control panel. The ECM then displays .7.0 and the ignition switch can be turned off.

Exiting the Self-diagnostic Mode

To exit or get out of the diagnostic mode, any of the ECC panel function buttons can be pushed. This includes ECON, AUTO, or HI. The diagnostic mode can also be exited by turning the ignition switch off for 10 seconds.

While in the diagnostic mode, the ECC operates in the "low" condition. Air temperature will be regulated to the setting at the time the diagnostic mode was entered.

This DFI system is more sophisticated than its earlier version and has more capabilities. There is room for improvement and for expanded control and diagnosis.

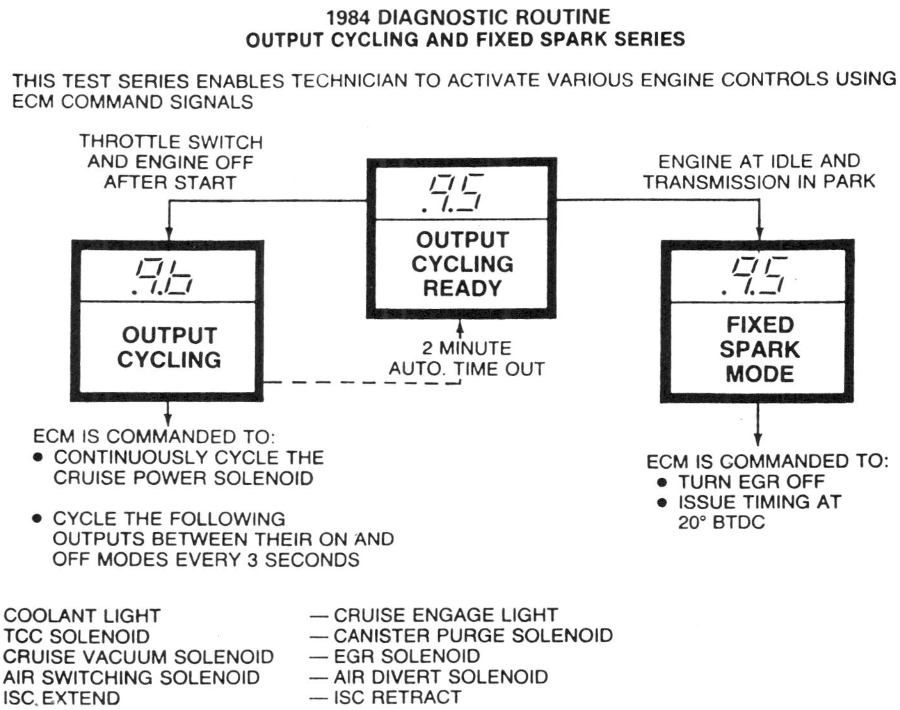

FIGURE 5-25 Output cycling test and fixed spark mode *(Courtesy of Cadillac Motor Car Division).*

1983 DIAGNOSTIC ROUTINE
ENGINE DATA SERIES

VARIOUS ENGINE SENSOR AND ENGINE CONTROL VALUES CAN BE DISPLAYED AT THE TECHNICIAN'S OPTION

ENGINE DATA DISPLAY

PARAMETER NO.	PARAMETER	RANGE	DISPLAY UNITS
.0.1	THROTTLE POSITION	−10 to 90	DEGREES
.0.2	MAP	14 to 108	kPa
.0.3	BARO	14 to 108	kPa
.0.4	COOLANT	−40 to 151	°C
.0.5	MAT	−40 to 151	°C
.0.6	INJECTOR PULSE WIDTH	0 to 19.9	MSEC
.0.7	O_2 SENSOR	0 to 1.99	VOLTS
.0.8	SPARK ADVANCE	0 to 52	DEGREES
.0.9	IGNITION CYCLES	0 to 50	KEY CYCLES
.1.0	BATTERY VOLTAGE	0 to 19.9	VOLTS
.1.1	ENGINE RPM	0 to 1990	RPM ÷ 10
.1.2	VEHICLE SPEED	0 to 199	MPH
.1.3	PROM I.D.	0 to 199	CODE

FIGURE 5-26 Engine data display *(Courtesy of Cadillac Motor Car Division).*

1982 DIAGNOSTIC CODES

CODE	CIRCUIT AFFECTED
12	NO DISTRIBUTOR (TACH) SIGNAL
13	O_2 SENSOR NOT READY
14	SHORTED COOLANT SENSOR CIRCUIT
15	OPEN COOLANT SENSOR CIRCUIT
16	GENERATOR VOLTAGE OUT OF RANGE
18	OPEN CRANK SIGNAL CIRCUIT
19	SHORTED FUEL PUMP CIRCUIT
20	OPEN FUEL PUMP CIRCUIT
21	SHORTED THROTTLE POSITION SENSOR CIRCUIT
22	OPEN THROTTLE POSITION SENSOR CIRCUIT
23	EST CIRCUIT PROBLEM IN RUN MODE
24	SPEED SENSOR CIRCUIT PROBLEM
25	EST CIRCUIT PROBLEM IN BYPASS MODE
26	SHORTED THROTTLE SWITCH CIRCUIT
27	OPEN THROTTLE SWITCH CIRCUIT
28	OPEN FOURTH GEAR CIRCUIT
29	SHORTED FOURTH GEAR CIRCUIT
30	ISC CIRCUIT PROBLEM
31	SHORTED MAP SENSOR CIRCUIT
32	OPEN MAP SENSOR CIRCUIT
33	MAP/BARO SENSOR CORRELATION
34	MAP SIGNAL TOO HIGH
35	SHORTED BARO SENSOR CIRCUIT
36	OPEN BARO SENSOR CIRCUIT
37	SHORTED MAT SENSOR CIRCUIT
38	OPEN MAT SENSOR CIRCUIT
39	TCC ENGAGEMENT PROBLEM
44	LEAN EXHAUST SIGNAL

CODE	CIRCUIT AFFECTED
45	RICH EXHAUST SIGNAL
51	PROM ERROR INDICATOR
52	ECM MEMORY RESET INDICATOR
60	TRANSMISSION NOT IN DRIVE
63	CAR AND SET SPEED TOLERANCE EXCEEDED
64	CAR ACCELERATION EXCEEDS MAX. LIMIT
65	COOLANT TEMPERATURE EXCEEDS MAXIMUM LIMIT
66	ENGINE RPM EXCEEDS MAXIMUM LIMIT
67	SHORTED SET OR RESUME CIRCUIT
.7.0	SYSTEM READY FOR FURTHER TESTS
.7.1	CRUISE CONTROL BRAKE CIRCUIT TEST
.7.2	THROTTLE SWITCH CIRCUIT TEST
.7.3	DRIVE (ADL) CIRCUIT TEST
.7.4	REVERSE CIRCUIT TEST
.7.5	CRUISE ON/OFF CIRCUIT TEST
.7.6	"SET/COAST" CIRCUIT TEST
.7.7	"RESUME/ACCELERATION" CIRCUIT TEST
.7.8	"INSTANT/AVERAGE" CIRCUIT TEST
.7.9	"RESET" CIRCUIT TEST
.8.0	A/C CLUTCH CIRCUIT TEST
−1.8.8	DISPLAY CHECK
.9.0	SYSTEM READY TO DISPLAY ENGINE DATA
.9.5	SYSTEM READY FOR OUTPUT CYCLING OR IN FIXED SPARK MODE
.9.6	OUTPUT CYCLING
.0.0	ALL DIAGNOSTICS COMPLETE

FIGURE 5-27 Digital fuel injection diagnostic codes, 1982 *(Courtesy of Cadillac Motor Car Division).*

1983 DIAGNOSTIC CODES

CODE	CIRCUIT AFFECTED
■ ■ 12	NO DISTRIBUTOR (TACH) SIGNAL
□ 13	O_2 SENSOR NOT READY
□ 14	SHORTED COOLANT SENSOR CIRCUIT
□ 15	OPEN COOLANT SENSOR CIRCUIT
■ ■ 16	GENERATOR VOLTAGE OUT OF RANGE
□ 18	OPEN CRANK SIGNAL CIRCUIT
□ 19	SHORTED FUEL PUMP CIRCUIT
■ ■ 20	OPEN FUEL PUMP CIRCUIT
□ 21	SHORTED THROTTLE POSITION SENSOR CIRCUIT
□ 22	OPEN THROTTLE POSITION SENSOR CIRCUIT
□ 23	EST/BYPASS CIRCUIT PROBLEM
□ 24	SPEED SENSOR CIRCUIT PROBLEM
□ 26	SHORTED THROTTLE SWITCH CIRCUIT
□ 27	OPEN THROTTLE SWITCH CIRCUIT
□ 28	OPEN FOURTH GEAR CIRCUIT
□ 29	SHORTED FOURTH GEAR CIRCUIT
□ 30	ISC CIRCUIT PROBLEM
■ ■ 31	SHORTED MAP SENSOR CIRCUIT
■ ■ 32	OPEN MAP SENSOR CIRCUIT
■ ■ 33	MAP/BARO SENSOR CORRELATION
□ 34	MAP SIGNAL TOO HIGH
□ 35	SHORTED BARO SENSOR CIRCUIT
□ 36	OPEN BARO SENSOR CIRCUIT
□ 37	SHORTED MAT SENSOR CIRCUIT
□ 38	OPEN MAT SENSOR CIRCUIT
□ 39	TCC ENGAGEMENT PROBLEM
■ ■ 44	LEAN EXHAUST SIGNAL
■ ■ 45	RICH EXHAUST SIGNAL
■ ■ 51	PROM ERROR INDICATOR
▼ 52	ECM MEMORY RESET INDICATOR
▼ 53	DISTRIBUTOR SIGNAL INTERRUPT
▼ 60	TRANSMISSION NOT IN DRIVE
▼ 63	CAR AND SET SPEED TOLERANCE EXCEEDED
▼ 64	CAR ACCELERATION EXCEEDS MAX. LIMIT
▼ 65	COOLANT TEMPERATURE EXCEEDS MAX. LIMIT
▼ 66	ENGINE RPM EXCEEDS MAXIMUM LIMIT
▼ 67	SHORTED SET OR RESUME CIRCUIT
.7.0	SYSTEM READY FOR FURTHER TESTS
.7.1	CRUISE CONTROL BRAKE CIRCUIT TEST
.7.2	THROTTLE SWITCH CIRCUIT TEST
.7.3	DRIVE (ADL) CIRCUIT TEST
.7.4	REVERSE CIRCUIT TEST
.7.5	CRUISE ON/OFF CIRCUIT TEST
.7.6	"SET/COAST" CIRCUIT TEST
.7.7	"RESUME/ACCELERATION" CIRCUIT TEST
.7.8	"INSTANT/AVERAGE" CIRCUIT TEST
.7.9	"RESET" CIRCUIT TEST
.8.0	A/C CLUTCH CIRCUIT TEST
−1.8.8	DISPLAY CHECK
.9.0	SYSTEM READY TO DISPLAY ENGINE DATA
.9.5	SYSTEM READY FOR OUTPUT CYCLING OR IN FIXED SPARK MODE
.9.6	OUTPUT CYCLING
.0.0	ALL DIAGNOSTICS COMPLETE
■ ■	TURNS ON "SERVICE NOW" LIGHT
□	TURNS ON "SERVICE SOON" LIGHT
▼	DOES NOT TURN ON ANY TELLTALE LIGHT

NOTE: CRUISE IS DISENGAGED WITH ANY "SERVICE NOW" LIGHT OR WITH CODES 60-67.

FIGURE 5-28 Digital fuel injection diagnostic codes, 1983 *(Courtesy of Cadillac Motor Car Division).*

1984 DIAGNOSTIC CODES

CODE	CIRCUIT AFFECTED
■ ■ 12	NO DISTRIBUTOR (TACH) SIGNAL
□ 13	O_2 SENSOR NOT READY
□ 14	SHORTED COOLANT SENSOR CIRCUIT
□ 15	OPEN COOLANT SENSOR CIRCUIT
■ ■ 16	GENERATOR VOLTAGE OUT OF RANGE
□ 18	OPEN CRANK SIGNAL CIRCUIT
□ 19	SHORTED FUEL PUMP CIRCUIT
■ ■ 20	OPEN FUEL PUMP CIRCUIT
□ 21	SHORTED THROTTLE POSITION SENSOR CIRCUIT
□ 22	OPEN THROTTLE POSITION SENSOR CIRCUIT
□ 23	EST/BYPASS CIRCUIT PROBLEM
□ 24	SPEED SENSOR CIRCUIT PROBLEM
□ 26	SHORTED THROTTLE SWITCH CIRCUIT
□ 27	OPEN THROTTLE SWITCH CIRCUIT
□ 28	OPEN FOURTH GEAR CIRCUIT
□ 29	SHORTED FOURTH GEAR CIRCUIT
□ 30	ISC CIRCUIT PROBLEM
■ ■ 31	SHORTED MAP SENSOR CIRCUIT
■ ■ 32	OPEN MAP SENSOR CIRCUIT
■ ■ 33	MAP/BARO SENSOR CORRELATION
■ ■ 34	MAP SIGNAL TOO HIGH
□ 35	SHORTED BARO SENSOR CIRCUIT
□ 36	OPEN BARO SENSOR CIRCUIT
□ 37	SHORTED MAT SENSOR CIRCUIT
□ 38	OPEN MAT SENSOR CIRCUIT
□ 39	TCC ENGAGEMENT PROBLEM
■ ■ 44	LEAN EXHAUST SIGNAL
■ ■ 45	RICH EXHAUST SIGNAL
■ ■ 51	PROM ERROR INDICATOR
▼ 52	ECM MEMORY RESET INDICATOR
▼ 53	DISTRIBUTOR SIGNAL INTERRUPT
▼ 60	TRANSMISSION NOT IN DRIVE
▼ 63	CAR AND SET SPEED TOLERANCE EXCEEDED
▼ 64	CAR ACCELERATION EXCEEDS MAX. LIMIT
▼ 65	COOLANT TEMPERATURE EXCEEDS MAX. LIMIT
▼ 66	ENGINE RPM EXCEEDS MAXIMUM LIMIT
▼ 67	SHORTED SET OR RESUME CIRCUIT
.7.0	SYSTEM READY FOR FURTHER TESTS
.7.1	CRUISE CONTROL BRAKE CIRCUIT TEST
.7.2	THROTTLE SWITCH CIRCUIT TEST
.7.3	DRIVE (ADL) CIRCUIT TEST
.7.4	REVERSE CIRCUIT TEST
.7.5	CRUISE ON/OFF CIRCUIT TEST
.7.6	"SET/COAST" CIRCUIT TEST
.7.7	"RESUME/ACCELERATION" CIRCUIT TEST
.7.8	"INSTANT/AVERAGE" CIRCUIT TEST
.7.9	"RESET" CIRCUIT TEST
.8.0	A/C CLUTCH CIRCUIT TEST
−1.8.8	DISPLAY CHECK
.9.0	SYSTEM READY TO DISPLAY ENGINE DATA
.9.5	SYSTEM READY FOR OUTPUT CYCLING OR IN FIXED SPARK MODE
.9.6	OUTPUT CYCLING
.0.0	ALL DIAGNOSTICS COMPLETE
■ ■	TURNS ON "SERVICE NOW" LIGHT
□	TURNS ON "SERVICE SOON" LIGHT
▼	DOES NOT TURN ON ANY TELLTALE LIGHT

NOTE: CRUISE IS DISENGAGED WITH ANY "SERVICE NOW" LIGHT OR WITH CODES 60-67.

FIGURE 5-29 Digital fuel injection diagnostic codes, 1984 *(Courtesy of Cadillac Motor Car Division).*

1985 DIAGNOSTIC CODES

CODE	CIRCUIT AFFECTED
■■ 12	NO DISTRIBUTOR (TACH) SIGNAL
□ 13	O_2 SENSOR NOT READY
□ 14	SHORTED COOLANT SENSOR CIRCUIT
□ 15	OPEN COOLANT SENSOR CIRCUIT
■■ 16	GENERATOR VOLTAGE OUT OF RANGE
□ 18	OPEN CRANK SIGNAL CIRCUIT
□ 19	SHORTED FUEL PUMP CIRCUIT
■■ 20	OPEN FUEL PUMP CIRCUIT
□ 21	SHORTED THROTTLE POSITION SENSOR CIRCUIT
□ 22	OPEN THROTTLE POSITION SENSOR CIRCUIT
□ 23	EST/BYPASS CIRCUIT PROBLEM
□ 24	SPEED SENSOR CIRCUIT PROBLEM
□ 26	SHORTED THROTTLE SWITCH CIRCUIT
□ 27	OPEN THROTTLE SWITCH CIRCUIT
□ 28	OPEN FOURTH GEAR CIRCUIT
□ 29	SHORTED FOURTH GEAR CIRCUIT
□ 30	ISC CIRCUIT PROBLEM
■■ 31	SHORTED MAP SENSOR CIRCUIT
■■ 32	OPEN MAP SENSOR CIRCUIT
■■ 33	MAP/BARO SENSOR CORRELATION
■■ 34	MAP SIGNAL TOO HIGH
□ 35	SHORTED BARO SENSOR CIRCUIT
□ 36	OPEN BARO SENSOR CIRCUIT
□ 37	SHORTED MAT SENSOR CIRCUIT
□ 38	OPEN MAT SENSOR CIRCUIT
□ 39	TCC ENGAGEMENT PROBLEM
■■ 44	LEAN EXHAUST SIGNAL
■■ 45	RICH EXHAUST SIGNAL
■■ 51	PROM ERROR INDICATOR
▼ 52	ECM MEMORY RESET INDICATOR
▼ 53	DISTRIBUTOR SIGNAL INTERRUPT
▼ 60	TRANSMISSION NOT IN DRIVE
▼ 63	CAR AND SET SPEED TOLERANCE EXCEEDED
▼ 64	CAR ACCELERATION EXCEEDS MAX. LIMIT
▼ 65	COOLANT TEMPERATURE EXCEEDS MAX. LIMIT
▼ 66	ENGINE RPM EXCEEDS MAXIMUM LIMIT
▼ 67	SHORTED SET OR RESUME CIRCUIT
.7.0	SYSTEM READY FOR FURTHER TESTS
.7.1	CRUISE CONTROL BRAKE CIRCUIT TEST
.7.2	THROTTLE SWITCH CIRCUIT TEST
.7.3	DRIVE (ADL) CIRCUIT TEST
.7.4	REVERSE CIRCUIT TEST
.7.5	CRUISE ON/OFF CIRCUIT TEST
.7.6	"SET/COAST" CIRCUIT TEST
.7.7	"RESUME/ACCELERATION" CIRCUIT TEST
.7.8	"INSTANT/AVERAGE" CIRCUIT TEST
.7.9	"RESET" CIRCUIT TEST
.8.0	A/C CLUTCH CIRCUIT TEST
−1.8.8	DISPLAY CHECK
.9.0	SYSTEM READY TO DISPLAY ENGINE DATA
.9.5	SYSTEM READY FOR OUTPUT CYCLING OR IN FIXED SPARK MODE
.9.6	OUTPUT CYCLING
.0.0	ALL DIAGNOSTICS COMPLETE
■■	TURNS ON "SERVICE NOW" LIGHT
□	TURNS ON "SERVICE SOON" LIGHT
▼	DOES NOT TURN ON ANY TELLTALE LIGHT
NOTE:	CRUISE IS DISENGAGED WITH ANY "SERVICE NOW" LIGHT OR WITH CODES 60-67.

FIGURE 5-30 Digital fuel injection diagnostic codes, 1985 *(Courtesy of Cadillac Motor Car Division).*

ABBREVIATIONS

A/F Air/fuel (A/F ratio)

AIR Air injection reaction

ALDL Assembly line data link

BARO Barometric pressure

CCP Climate control panel

Conv. Converter (catalytic)

DFI Digital fuel injection

DVM (10 meg.) Digital voltmeter with 10-million ohms resistance

ECC Electronic climate control

ECM Electronic control module (controller)

EECS Evaporative emissions control system

EFE Early fuel evaporation

EGR Exhaust gas recirculation

ESS Engine speed sensor

EST Electronic spark timing

FDP Fuel data panel

HEI High-energy ignition

ISC Idle speed control

MAP Manifold absolute pressure

O_2 Oxygen (sensor)

PCV Positive crankcase ventilation

P/N Park/neutral

PROM Programmable read-only memory (engine calibrator)

TBI Throttle body injection (unit)

THERMAC Thermostatic air cleaner

TPS Throttle position sensor

VSS Vehicle speed sensor

WOT Wide-open throttle

UNIT I
General Motors Corporation Systems

6
Cadillac 1985 FWD, ECM-BCM Digital Fuel Injection

OVERVIEW

The 1985 front wheel drive Cadillac DeVille and Fleetwood sedan (C-body) use a new digital fuel injection system. Some of the operation is the same as 1984, as discussed in Chapter 5. This information will not be repeated. New information will supercede or be added to the DFI information in Chapter 5.

The 1985 FWD DFI System uses a second computer, a body control module (BCM). This DFI system has a new fuel system component layout, monolith catalytic converter, single-wire oxygen sensor, thermac air cleaner, electric EFE, park/neutral switch, vehicle speed sensor and buffer, power steering switch input to the ECM, set timing procedure and ignition distributor module, viscous converter clutch, EGR solenoid, vacuum pump that runs continuously, service electrical system telltale light, ECM to BCM communication line, procedure for diagnostic mode operation, and list of diagnostic codes.

FUEL SYSTEM COMPONENT LAYOUT

Basic operation of the fuel system is the same as that described in Chapter 5. The fuel pump relay is located under the instrument panel to the right of the radio in the relay panel. The fuel pump circuit is protected by a 10-ampere fuse in the mini fuse panel under the instrument panel, to the right of the ash tray. Fuel supply and return lines are both located on the frame rail on the left side of the car (Fig. 6-1 and 6-2).

MONOLITH CATALYTIC CONVERTER

The exhaust system is completely different from earlier model years. The heat stove is an integral part of the exhaust crossover pipe (Fig. 6-3). This preheats intake air with improved efficiency.

The catalytic converter is a monolith design that is only half as heavy as the previously used pellet-type converter (Fig. 6-4). The exhaust pipes are stainless steel.

OXYGEN SENSOR

A single-wire oxygen (O_2) sensor is used (Fig. 6-5). This sensor has a metal boot to reduce silicon contamination and has a stainless steel wave washer inside instead of a mica washer, as was used in earlier years. This helps to prevent ceramic breakage and sensor failure.

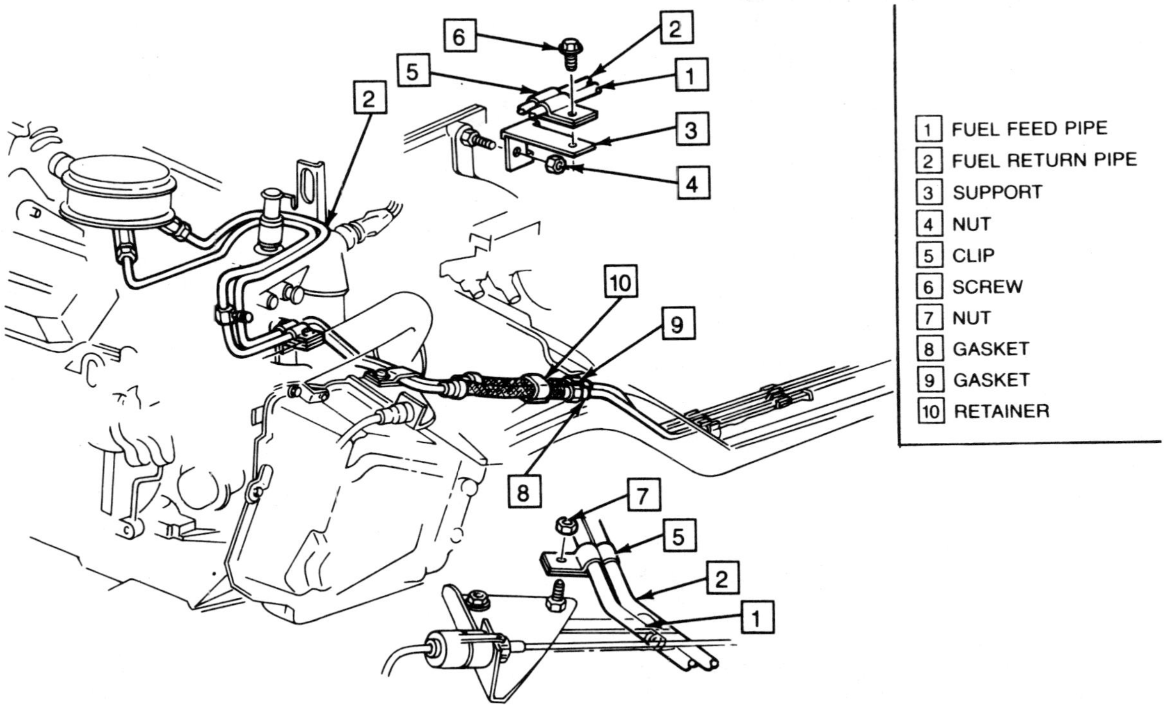

FIGURE 6-1 Fuel line routing *(Courtesy of General Motors Product Service Training).*

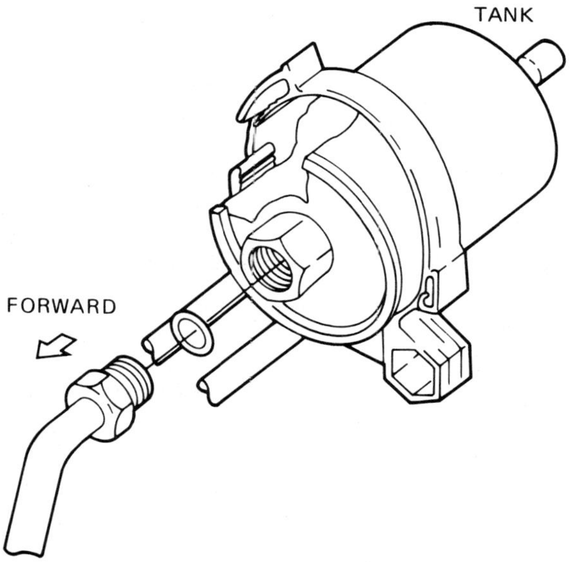

FIGURE 6-2 Detail of fuel filter *(Courtesy of General Motors Product Service Training).*

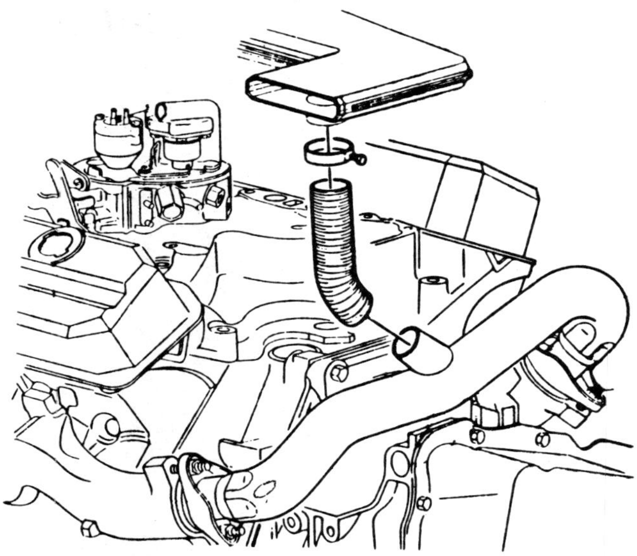

FIGURE 6-3 Engine air heat stove and tube *(Courtesy of General Motors Product Service Training).*

THERMAC AIR CLEANER

The thermac air cleaner inlet faces forward for better engine performance (the engine is transversely mounted) (Fig. 6-6). During cold engine operation, inlet air is preheated by the heat stove surrounding the crossover exhaust pipe. Manifold vacuum is used to operate the thermac air cleaner, where earlier models used a vacuum pump as a source. Airflow restrictions in the system have been reduced. This produces better cold, wide-open throttle performance. Pre-

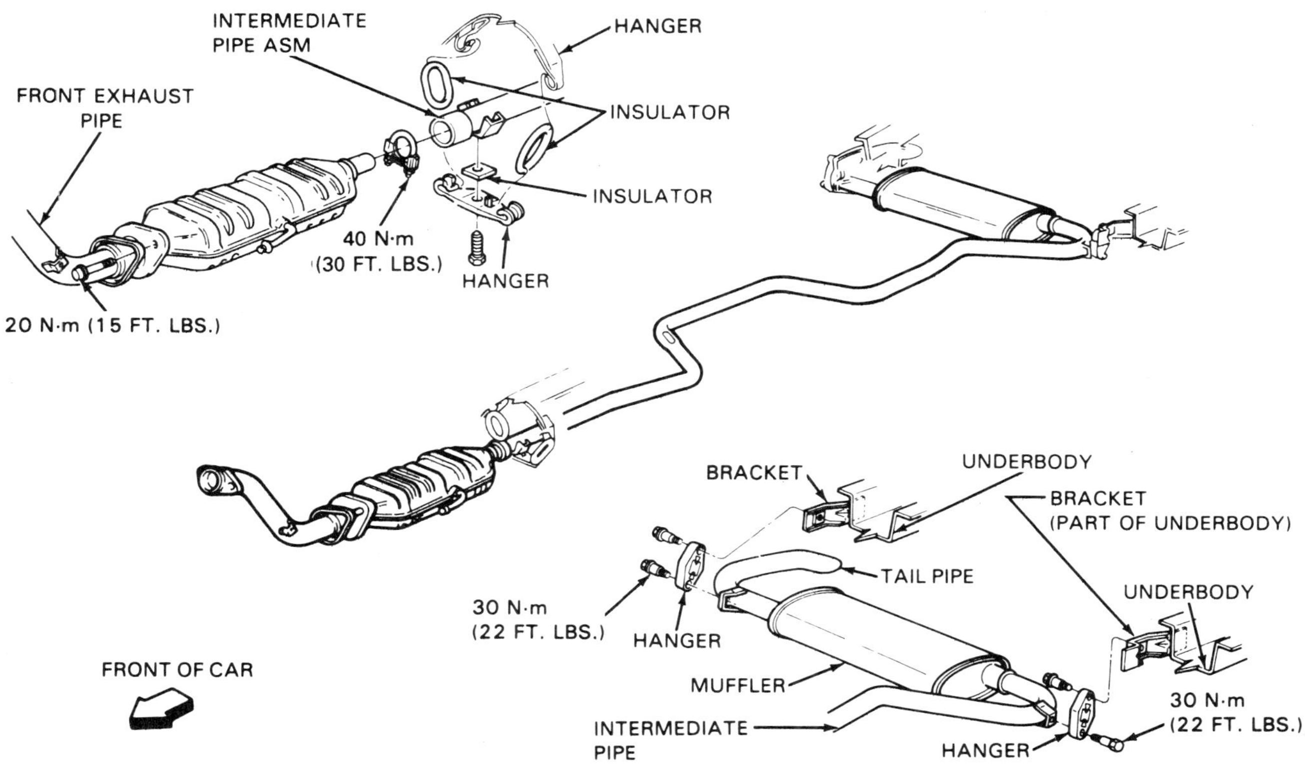

FIGURE 6-4 Exhaust system, C *(Courtesy of Cadillac Motor Car Division).*

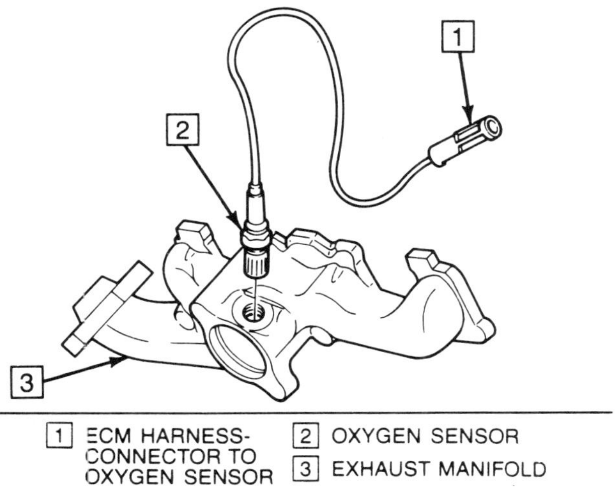

FIGURE 6-5 Oxygen sensor *(Courtesy of Cadillac Motor Car Division).*

heated air is provided if the air cleaner air temperature is below about 85°F. Outside air is provided if air cleaner temperature is above about 135°F. A blend of preheated and outside air is provided between 85° and 135°F air cleaner air temperature.

ELECTRIC EARLY FUEL EVAPORATION

Previous models used a vacuum-operated early fuel evaporation (EFE) valve. A new electric EFE grid is now used under the throttle body (Fig. 6-7). The ECM controls an EFE relay, which controls the EFE heated grid. As the air/fuel mixture passes through the grid, when energized, it is heated. This provides better control of the air/fuel ratio and improved efficiency during cold engine operation. The ECM energizes the relay and grid when the MAT sensor indicates an air temperature of lower than about 65°F, the CTS indicates a coolant temperature of lower than about 220°F, and battery voltage is higher than 10 volts. All three of these conditions must be met; otherwise, the ECM will de-energize the grid.

If the ECM shuts the EFE circuit off while the engine is operating, the grid will come on again if the MAT is less than 100°F, the CTS is less than 170°F, and battery voltage is greater than 12 volts.

The ECM turns the circuit on if the TPS indicates the throttle is about halfway open and the MAT is less than about 140°F. The circuit is energized for a minimum of 15 seconds to prevent icing of the grid.

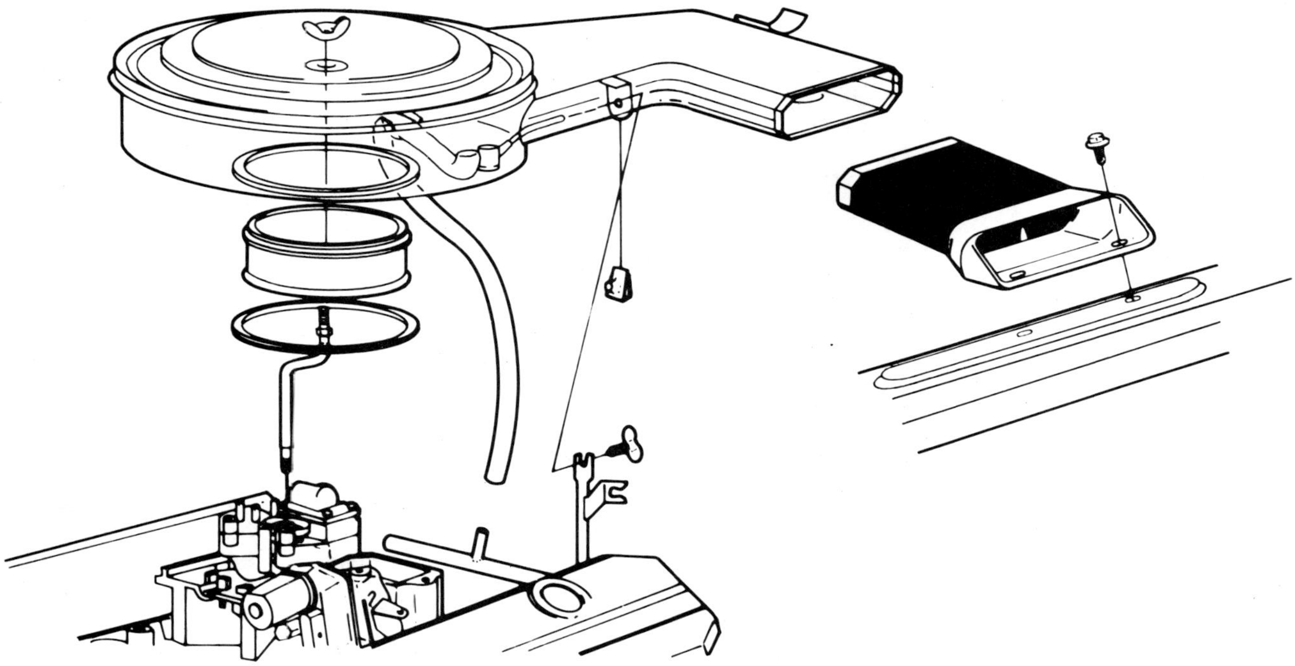

FIGURE 6-6 New thermac air cleaner *(Courtesy of General Motors Product Service Training).*

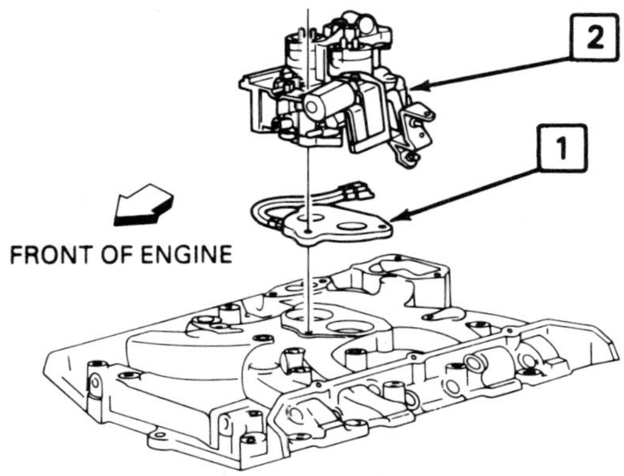

[1] EFE HEATER ASSEMBLY [2] THROTTLE BODY ASSEMBLY

FIGURE 6-7 Early fuel evaporation grid *(Courtesy of Cadillac Motor Car Division).*

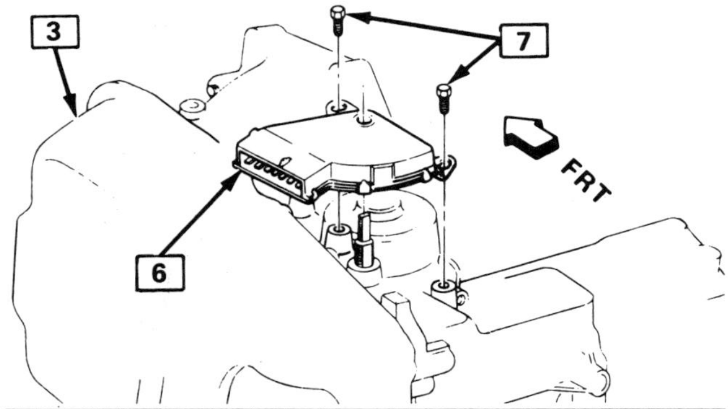

[3] TRANSAXLE ASSM.
[6] NEUTRAL START & BACK-UP LAMP SWITCH
[7] BOLT (27 N·m/20 FT. LBS.)

FIGURE 6-8 Neutral start and backup lamp switch *(Courtesy of General Motors Product Service Training).*

PARK/NEUTRAL SWITCH

This new park/neutral switch is located on top of the transaxle under the hood (Fig. 6-8). It replaces the older separate drive and reverse switches. Correct adjustment of this switch is important.

VEHICLE SPEED SENSOR

The vehicle speed sensor (VSS) is a permanent magnet generator located in the transaxle (Fig. 6-9). A speedometer cable is no longer used. The permanent-magnet (PM) generator is driven by the transaxle speedometer gear and

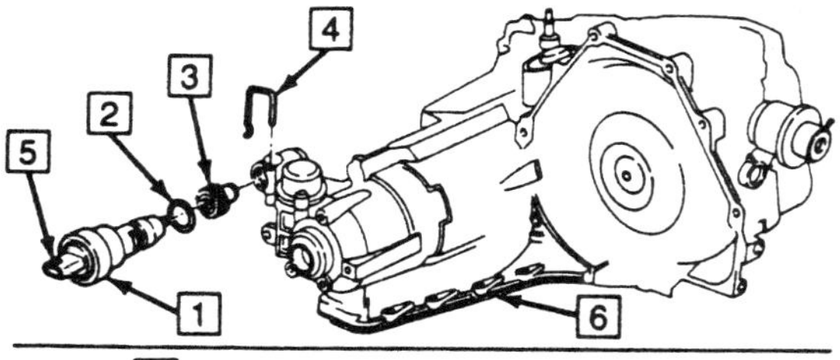

1 SENSOR ASM. — SPEEDO
2 GASKET
3 GEAR — 28 100TH (YELLOW)
4 RETAINER
5 ELECTRICAL CONNECTOR
6 TRANSMISSION ASM.

FIGURE 6-9 Speed sensor and gear assembly *(Courtesy of General Motors Product Service Training).*

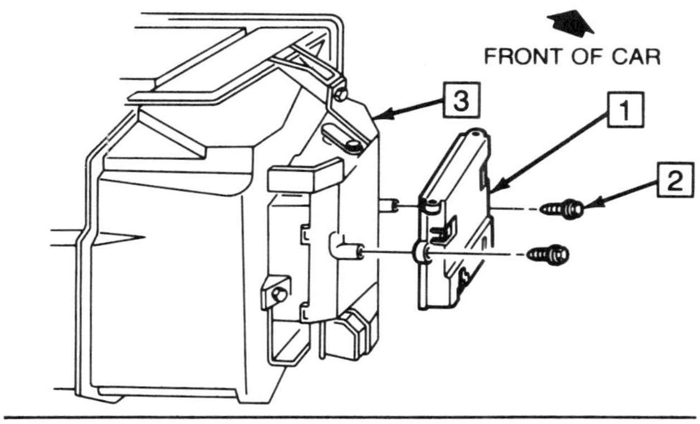

1 SPEEDO BUFFER ASM.
2 SCREW
3 HEATER A/C PROGRAMMER

FIGURE 6-10 Speed sensor buffer *(Courtesy of General Motors Product Service Training).*

generates an alternating-current signal. This signal is sent to a new buffer/amplifier, where it is "conditioned" and translated into a digital signal (Fig. 6-10). This signal is sent to the ECM and the speedometer. The buffer/amplifier is located on the air-conditioning system programmer under the right side of the instrument panel (Fig. 6-11).

POWER STEERING SWITCH

The power steering switch is located in the high-pressure line near where it enters the steering gear. When power steering pressure rises above about 300 psi, the switch opens and the ECM compensates for this additional load by adjusting the idle speed. This improves idle quality and prevents stalling during tight maneuvering situations.

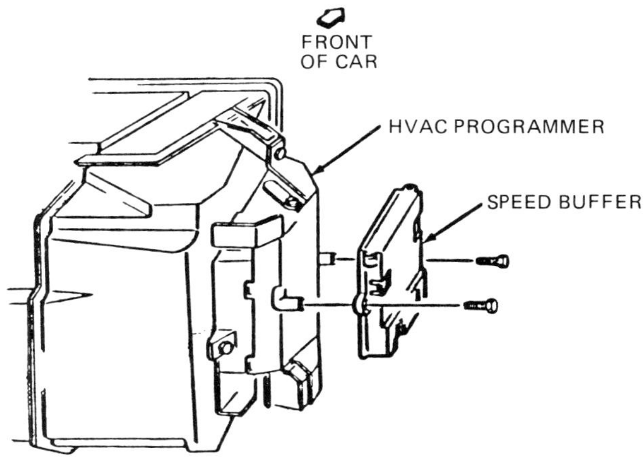

FIGURE 6-11 Speed buffer*(Courtesy of General Motors Product Service Training).*

SET TIMING PROCEDURE

Earlier models required the service technician to open the set timing connector or to enter the diagnostic mode .9.5 in order to check and adjust ignition timing. On this vehicle, connecting a jumper between the A and B terminals on the assembly line diagnostic link (ALDL) connector places the system in a fixed spark mode of operation (Fig. 6-12). The timing should be adjusted to 10° BTDC at any rpm less than 800. The new ignition distributor module allows this procedure to be used.

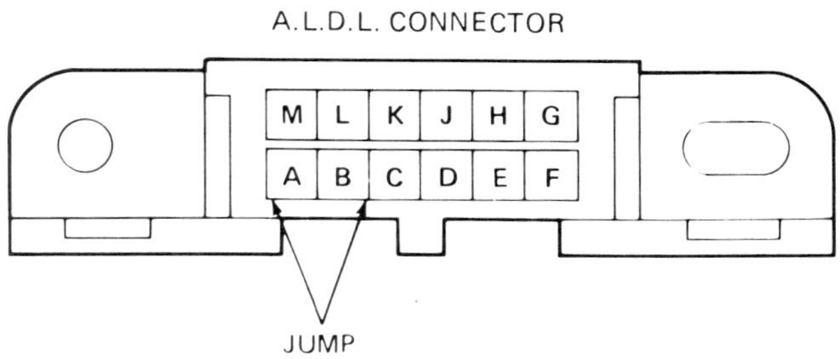

FIGURE 6-12 Set base timing *(Courtesy of General Motors Product Service Training).*

VISCOUS CONVERTER CLUTCH

Prior models used a torque converter clutch. This model uses a viscous converter clutch (Fig. 6-13). It can be engaged at lower road speeds and, as a result, improved fuel economy can be obtained over a wider range of road speeds. This clutch engages at about 25 to 45 mph, whereas the TCC engaged at about 45 mph. The viscous converter

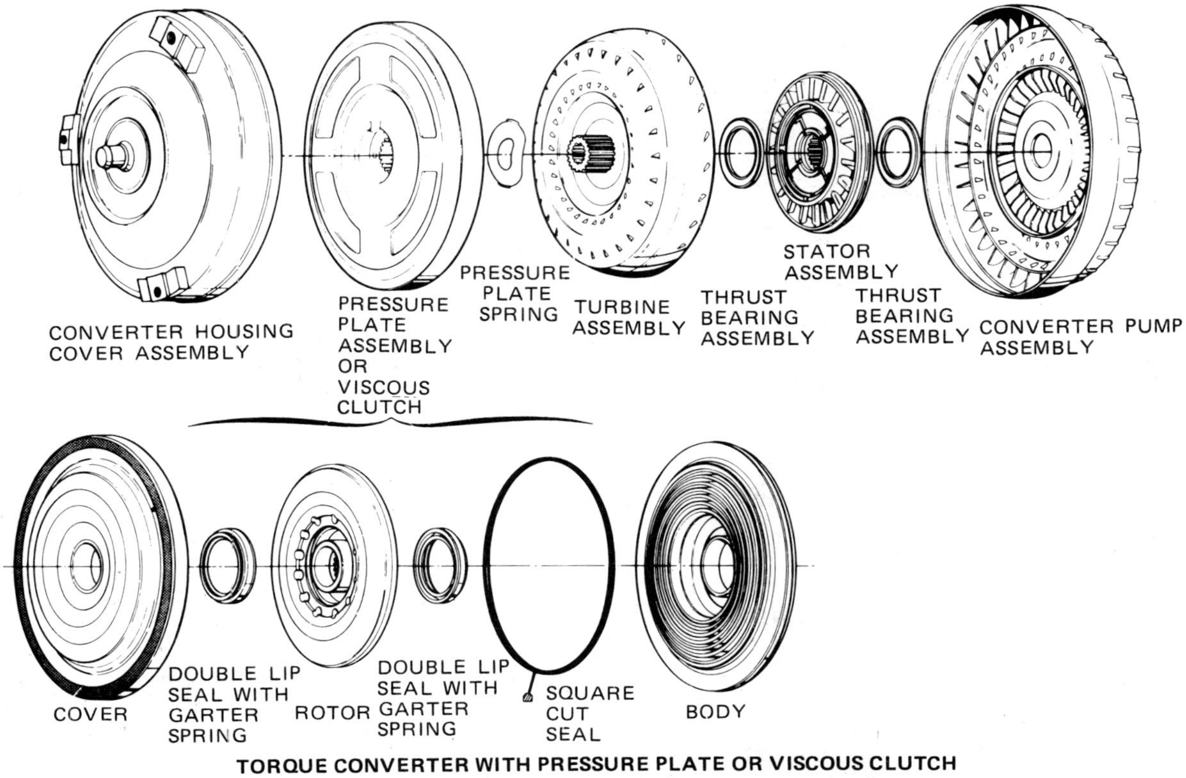

FIGURE 6-13 Torque converter with pressure plate or viscous clutch *(Courtesy of General Motors Product Service Training).*

clutch provides about 95% improvement in city-driving fuel economy, whereas the TCC provided about 10% improvement.

The ECM controls the VCC solenoid in the transaxle by monitoring engine coolant temperature, transaxle gear range, throttle angle, vehicle speed, barometric pressure, transaxle oil temperature, the brake switch and a VCC oil temperature switch.

If initial engine coolant temperature is above 60°F, the ECM allows the VCC to apply when engine coolant temperature reaches 140°F. If initial coolant temperature is below 60°F, the ECM uses a timer circuit, 5 minutes at 5 mph, to determine when to apply the VCC. When the engine overheats, engine coolant temperature rises above 250°F; the VCC will be applied in second gear to assist in engine cooling.

Transaxle gear range is indicated to the ECM by a third-gear switch and a fourth-gear switch. The ECM will not apply the VCC unless these switches indicate third- or fourth-gear operation, with the exception of overheating as discussed earlier.

Throttle plate angle is used along with vehicle speed to determine VCC application. The vehicle speed determines when the VCC is disengaged as well. The VCC remains engaged during most coastdown conditions.

Barometric pressure is calculated from an initial MAP sensor reading by the ECM. If the vehicle is above 4500 feet, the VCC will not apply until vehicle speed reaches 45 mph.

Transaxle oil temperature is sensed by a thermistor in the transaxle. When oil temperature is below 200°F, the VCC can apply at 25 mph, and when oil temperature rises above 200°F, the VCC cannot apply until vehicle speed reaches 40 mph.

The brake switch is normally closed. When the brake pedal is depressed, the switch opens and causes the VCC to disengage.

The VCC oil temperature switch or overtemperature protection switch is also normally closed. When VCC oil temperature rises above 315°F, the switch opens and causes the VCC to disengage.

During VCC operation, the ECM is monitoring engine speed and vehicle speed. The difference between them is interpreted by the ECM as being slippage. If the slippage becomes excessive, the ECM lights a SERVICE SOON telltale light and a code 39 is set in its memory.

EXHAUST GAS RECIRCULATION VALVE SOLENOID

The EGR solenoid has a different calibration from earlier models and is not interchangeable with them. To make identification of this solenoid easier, it has a red dot on it. This EGR system is not pulse width modulated, as the earlier models discussed in Chapter 5 were.

VACUUM PUMP

An electric vacuum pump, designed to help maintain vacuum levels for operation of controls and accessories, runs full time. It is located under the battery tray at the left front corner of the vehicle.

SERVICE ELECTRICAL SYSTEM TELLTALE

A new SERVICE ELECTRICAL SYSTEM telltale is used on the instrument panel (Fig. 6-14). The ECM can light this telltale if generator output either exceeds or falls below a specified value. The ECM also lights the telltale if excessive demand causes a decrease in generator voltage output. The ECM monitors the system voltage through a new third terminal on the generator marked F. This circuit does not replace the generator telltale that has been in use for several years; however, the 40-ohm resistor used in earlier generators to calibrate the generator telltale has now been eliminated. The remainder of the generator operates the same as in the past.

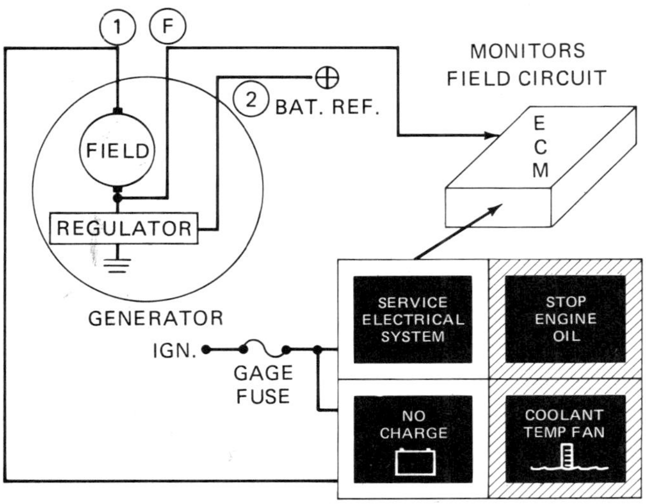

FIGURE 6-14 Service electrical system *(Courtesy of General Motors Product Service Training).*

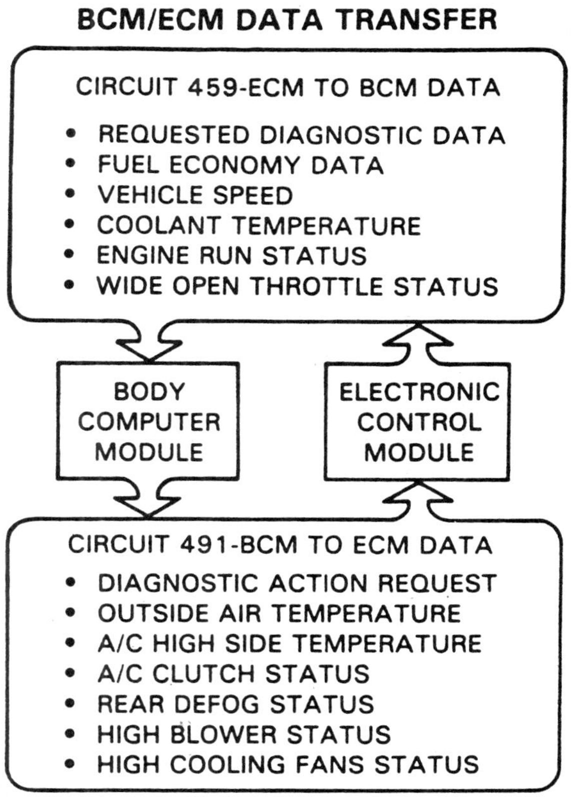

FIGURE 6-15 Body control module to electronic control module data transfer *(Courtesy of Cadillac Motor Car Division).*

ECM–BCM COMMUNICATIONS

Some of the components formerly controlled by the ECM are now controlled by the BCM. For the BCM to control them, it must be aware of engine data and operating conditions. Since the ECM receives engine data from the sensors, the BCM can get this information from the ECM by means of a communications link. The ECM also needs to be aware of some of the BCM functions. This information can also be obtained through a communications link. This is known as interfacing the computers (Fig. 6-15).

The BCM, for example, controls the engine cooling fans. It must know what the engine coolant temperature is in order to turn on the fans. This information is provided to the BCM through the ECM. Also, the BCM controls the air-conditioning compressor clutch. The ECM needs to know when the clutch engages in order to maintain the idle speed. The ECM gets this information from the BCM.

There are several other examples of this interfaced system communication. The electronic climate control panel is also interfaced with the BCM for transfer of information and for display purposes while in the system diagnostic mode.

DIAGNOSTIC MODE OPERATION

The system diagnostic mode can be entered if the ignition switch is on and the OFF and WARMER buttons on the ECC are pushed simultaneously and held until −1.8.8 appears (Fig. 6-16). This is the segment check and signals the beginning of the status light operation. Diagnostic codes are displayed next. This DFI system can display 42 malfunction codes (Fig. 6-17), and it has 11 sensor failsoft or compensating values, 5 system operation displays, 7 switch tests, and 13 engine data displays; it can activate 11 components for testing. In addition, the BCM has 15 malfunction codes (Fig. 6-18), 4 sensor failsoft values, 6 system operation displays, and 12 body data displays.

Code Display

The ECM codes are shown by first displaying ..E on the ECC panel. This indicates that any numbers following are ECM history or intermittent codes (Fig. 6-19). Next E.E is displayed, which indicates that any numbers following are ECM current or hard codes (Fig. 6-20). Then . . F is displayed. Any numbers following are BCM history or intermittent codes (Fig. 6-21). When .F.F is displayed, any numbers following are BCM current or hard codes (Fig. 6-22). At any point in this sequence, .7.0 can be displayed if no further codes have been set (Fig. 6-23). When the code display cycle is complete, .7.0 is displayed to indicate that the system is ready for additional instructions.

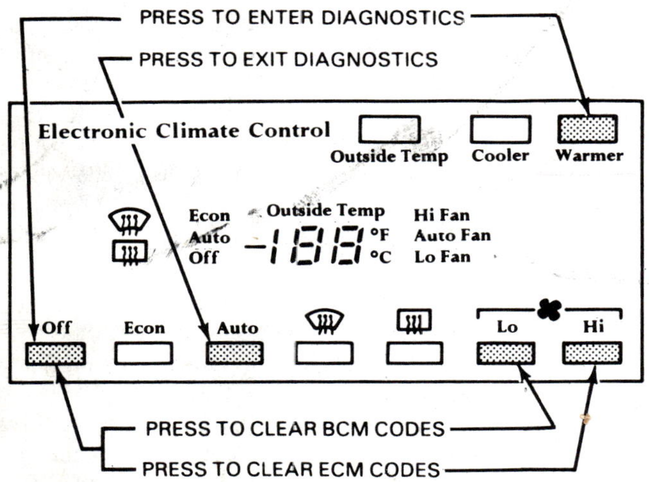

FIGURE 6-16 Climate control panel *(Courtesy of Cadillac Motor Car Division).*

ECM DIAGNOSTIC CODES

CODE		MALFUNCTION
■■	E12	NO DISTRIBUTOR SIGNAL
□	E13	OXYGEN SENSOR NOT READY (CANISTER PURGE)
□	E14	SHORTED COOLANT SENSOR CIRCUIT
□	E15	OPEN COOLANT SENSOR CIRCUIT
■■	E16	GENERATOR VOLTAGE OUT OF RANGE (ALL SOLENOIDS)
□	E18	OPEN CRANK SIGNAL CIRCUIT
□	E19	SHORTED FUEL PUMP CIRCUIT
■■	E20	OPEN FUEL PUMP CIRCUIT
□	E21	SHORTED THROTTLE POSITION SENSOR CIRCUIT
□	E22	OPEN THROTTLE POSITION SENSOR CIRCUIT
□	E23	EST/BYPASS CIRCUIT PROBLEM (AIR)
□	E24	SPEED SENSOR CIRCUIT PROBLEM (VCC)
□	E26	SHORTED THROTTLE SWITCH CIRCUIT
□	E27	OPEN THROTTLE SWITCH CIRCUIT
□	E28	OPEN THIRD OR FOURTH GEAR CIRCUIT
□	E30	ISC CIRCUIT PROBLEM
■■	E31	SHORTED MAP SENSOR CIRCUIT (AIR)
■■	E32	OPEN MAP SENSOR CIRCUIT (AIR)
■■	E34	MAP SENSOR SIGNAL TOO HIGH (AIR)
□	E37	SHORTED MAT SENSOR CIRCUIT
□	E38	OPEN MAT SENSOR CIRCUIT
□	E39	VCC ENGAGEMENT PROBLEM
□	E40	OPEN POWER STEERING PRESSURE CIRCUIT
■■	E44	LEAN EXHAUST SIGNAL (AIR & CL & CANISTER PURGE)
■■	E45	RICH EXHAUST SIGNAL (AIR & CL & CANISTER PURGE)
□	E47	BCM - ECM DATA PROBLEM
■■	E51	ECM PROM ERROR
▼	E52	ECM MEMORY RESET INDICATOR
▼	E53	DISTRIBUTOR SIGNAL INTERRUPT
▼	E59	VCC TEMPERATURE SENSOR CIRCUIT
▼	E60	TRANSMISSION NOT IN DRIVE
▼	E63	CAR SPEED AND SET SPEED DIFFERENCE TOO HIGH
▼	E64	CAR ACCELERATION TOO HIGH
▼	E65	COOLANT TEMPERATURE TOO HIGH
▼	E66	ENGINE RPM TOO HIGH
▼	E67	CRUISE SWITCH SHORTED DURING ENABLE

ECM AND CRUISE CONTROL COMMENTS:

■■	TURNS ON "SERVICE NOW" LIGHT
□	TURNS ON "SERVICE SOON" LIGHT
▼	DOES NOT TURN ON ANY TELLTALE LIGHT
( )	FUNCTIONS WITHIN BRACKETS ARE DISENGAGED WHILE SPECIFIED MALFUNCTION REMAINS CURRENT (HARD)

E16 & E24 DISABLE VCC FOR ENTIRE IGNITION CYCLE

E24 & E67 DISABLE CRUISE FOR ENTIRE IGNITION CYCLE

CRUISE IS DISENGAGED WITH CODE(S) E16, E51 OR E60 - E67

FIGURE 6-17 Digital fuel injection diagnostic codes *(Courtesy of Cadillac Motor Car Division).*

BCM DIAGNOSTIC CODES	
CODE	CIRCUIT AFFECTED
▼F10	OUTSIDE TEMP SENSOR CKT
▼F11	A/C HIGH SIDE TEMP SENSOR CKT
▼F12	A/C LOW SIDE TEMP SENSOR CKT
▼F13	IN-CAR TEMP SENSOR CKT
▼F14	DIESEL COOLANT SENSOR CKT
▼F30	CCP TO BCM DATA CKT
▼F31	FDC/DDC TO BCM DATA CKT
▼F32	ECM-BCM DATA CKT'S
▼F40	AIR MIX DOOR PROBLEM
▼F41	COOLING FANS PROBLEM
☑F46	LOW REFRIGERANT WARNING
☑F47	LOW REFRIGERANT CONDITION
☑F48	LOW REFRIGERANT PRESSURE
▼F49	HIGH TEMP CLUTCH DISENGAGE
▼F51	BCM PROM ERROR

☑ TURNS ON "SERVICE AIR COND" LIGHT
▼ DOES NOT TURN ON ANY LIGHT

COMMENTS:
F11 TURNS ON COOLING FANS WHEN A/C CLUTCH IS ENGAGED
F12 DISENGAGES A/C CLUTCH
F14 & F32 TURN ON COOLING FANS
F30 TURNS ON FT. DEFOG AT 75° F
F41 TURNS ON "COOLANT TEMP/FANS" LIGHT WHEN FANS SHOULD BE ON
F47 & F48 SWITCHES FROM "AUTO" TO "ECON"

FIGURE 6-18 Body control module diagnostic codes *(Courtesy of Cadillac Motor Car Division).*

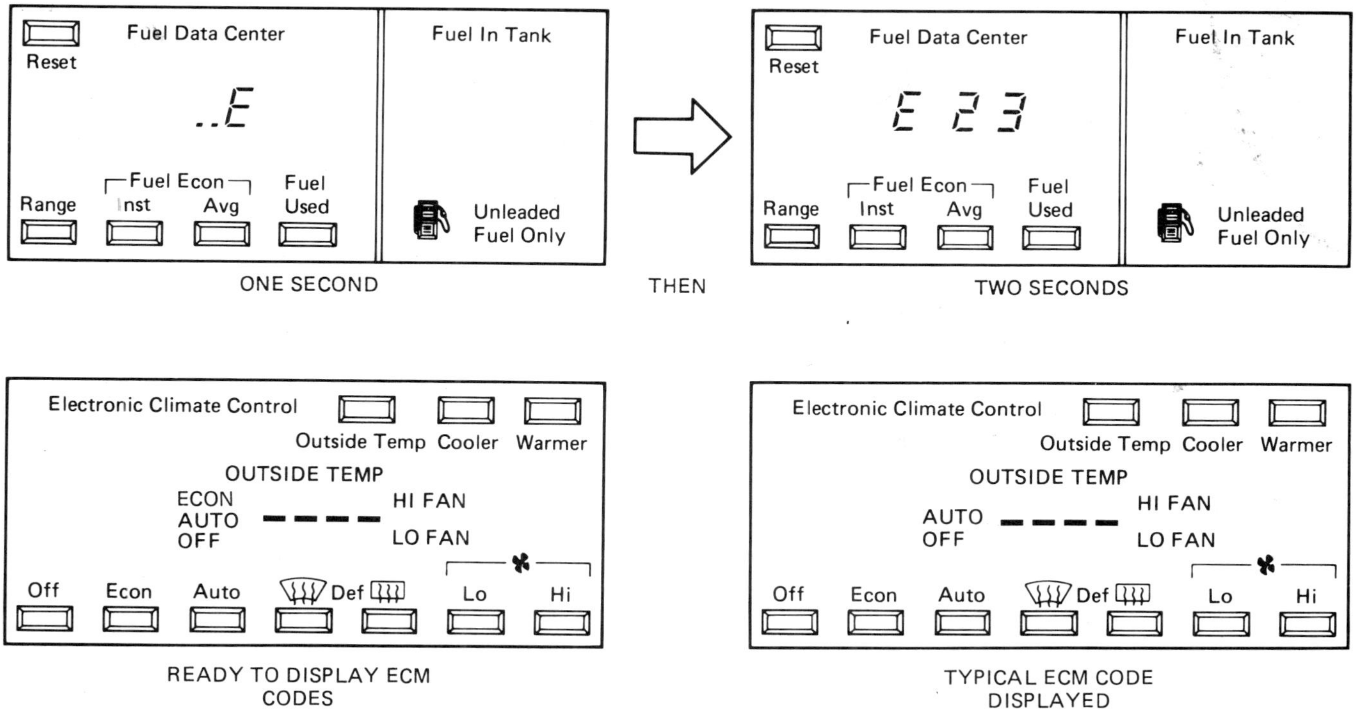

FIGURE 6-19 Display of electronic control module history codes *(Courtesy of General Motors Product Service Training).*

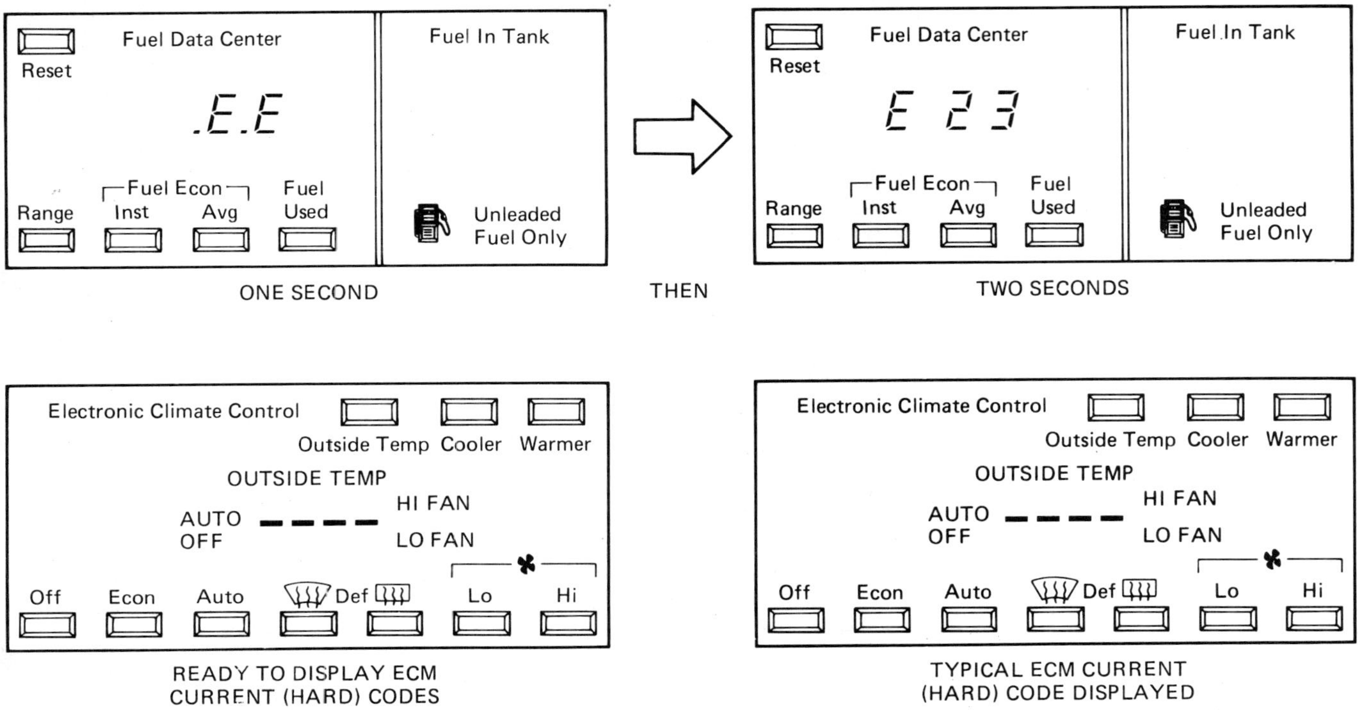

FIGURE 6-20 Display of electronic control module current (hard) codes *(Courtesy of General Motors Product Service Training).*

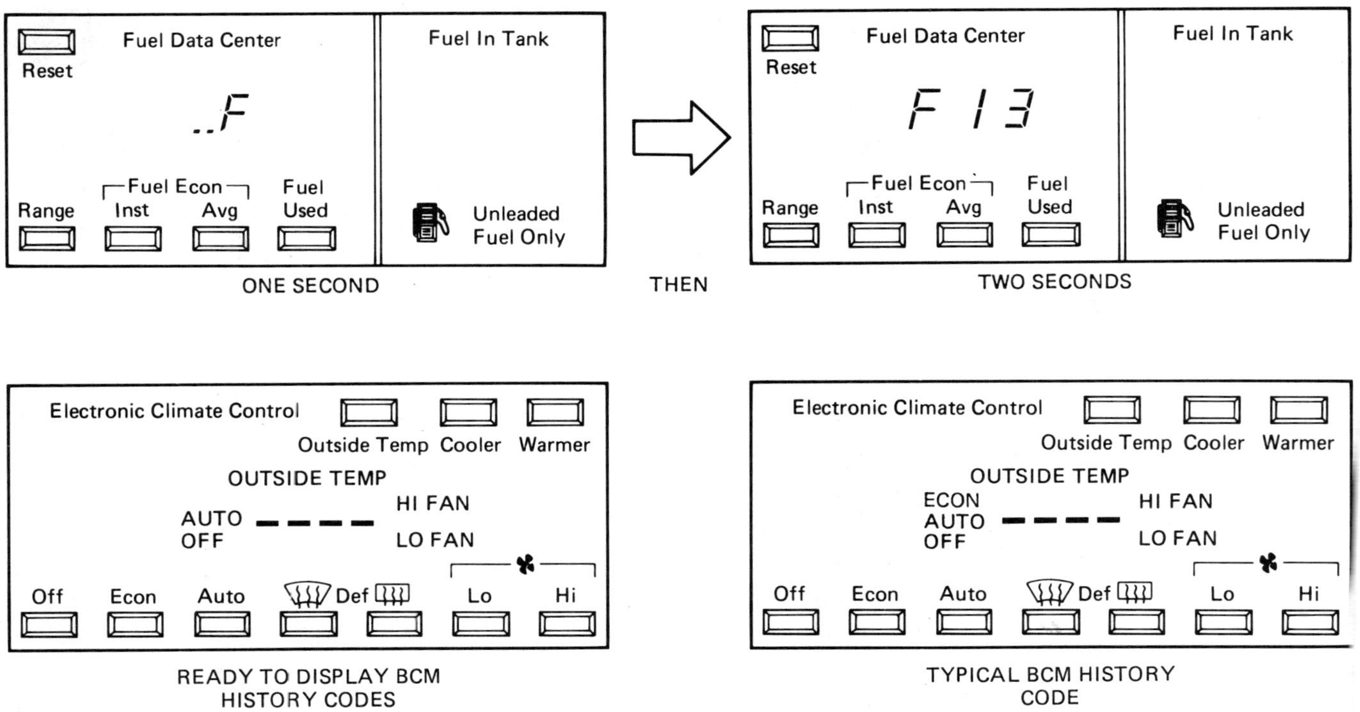

FIGURE 6-21 Display of body control module history codes *(Courtesy of General Motors Product Service Training).*

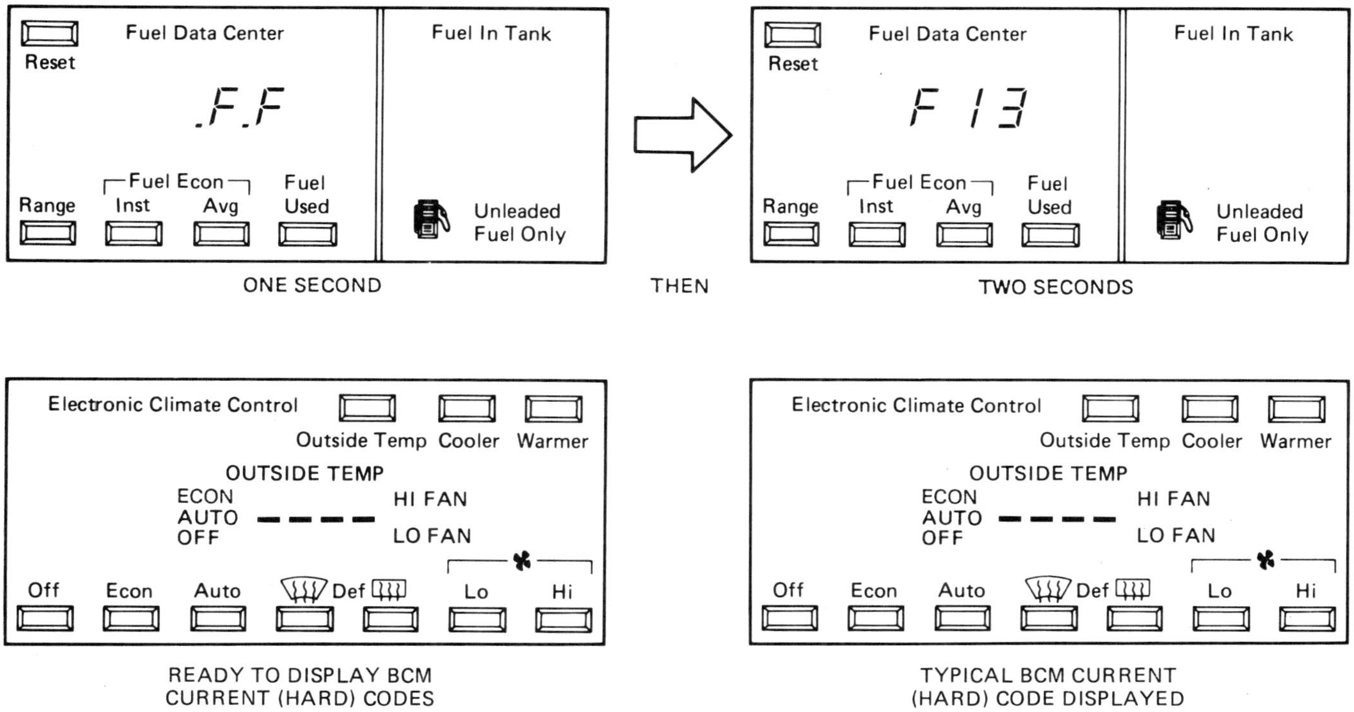

FIGURE 6-22 Display of body control module current (hard) codes *(Courtesy of General Motors Product Service Training).*

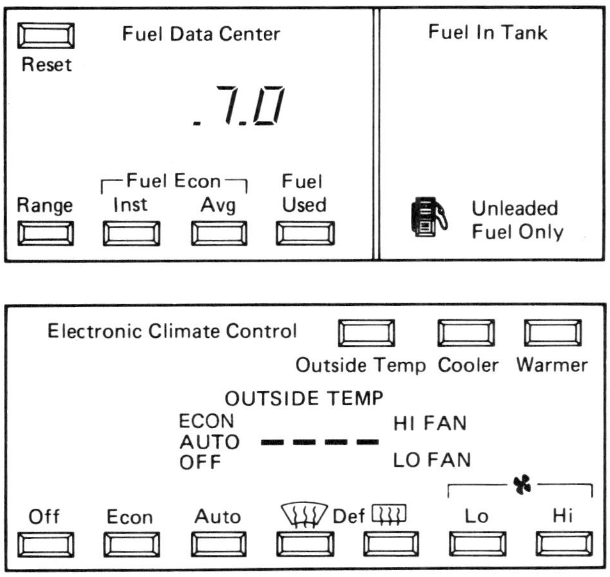

FIGURE 6-23 "System ready" *(Courtesy of General Motors Product Service Training).*

System Tests

By depressing the brake pedal, the service technician can perform the ECM switch tests (Fig. 6-24). If the OUTSIDE TEMP button is pushed instead, the BCM data parameters will be shown. Pushing the ECON button allows the service technician to manually operate and test the engine cooling fans. Pressing the LO button displays the ECM data parameters (Fig. 6-25). Pushing the HI button with the engine off after having started it causes the system to cycle controlled components such as motors and solenoids (Fig. 6-26). All these operations are done from the driver's seat.

The status light display on the ECC panel indicates operation of the fourth-gear input signal, VCC output signal, throttle switch input signal, oxygen sensor input signal, ECM operating mode, A/C clutch output, compressor low pressure switch input, heater control valve output, A/C-defroster mode door output, cooling fans status, and the up/down mode door output (Fig. 6-27). The status lights on the ECC panel are not LEDs as used in earlier models, but are words or symbols that will illuminate to indicate operation or inactivity of both ECM and BCM components.

Clearing Codes

After repairs have been made, the service technician can erase or clear codes from the ECM and BCM memories (Fig. 6-28). By pressing and holding the OFF and HI buttons simultaneously while in the diagnostic mode, E.O.O will be displayed. Then .7.0 will be displayed. This indicates that any ECM codes that were in memory are now cleared. By pressing and holding the OFF and LO buttons

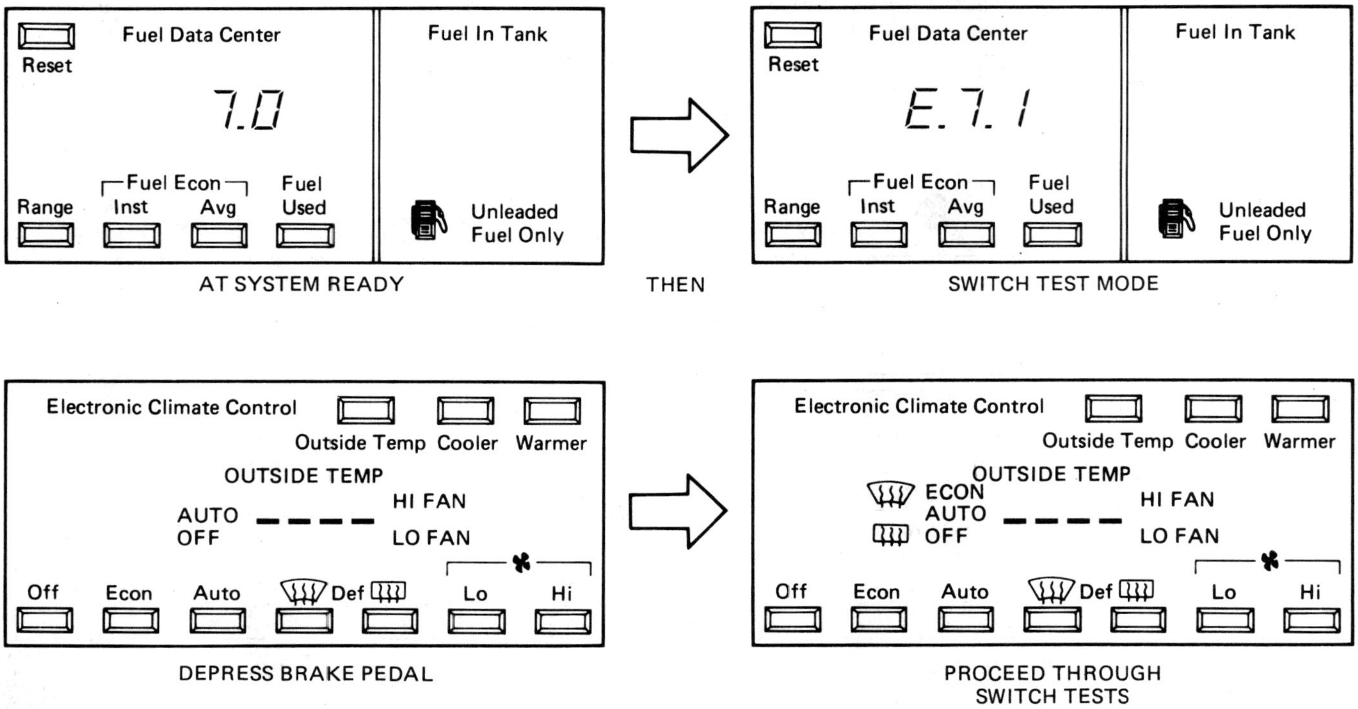

FIGURE 6-24 Electronic control module switch tests *(Courtesy of General Motors Product Service Training).*

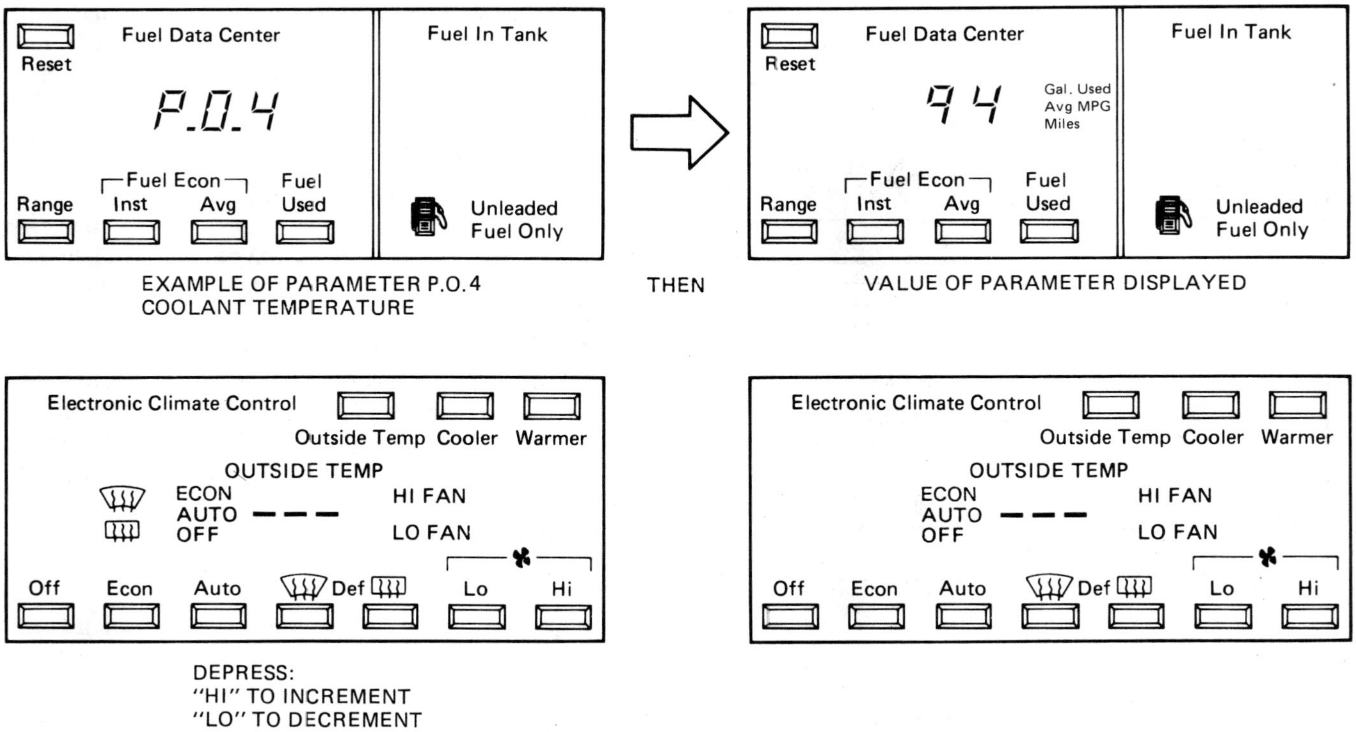

FIGURE 6-25 Engine data parameters *(Courtesy of General Motors Product Service Training).*

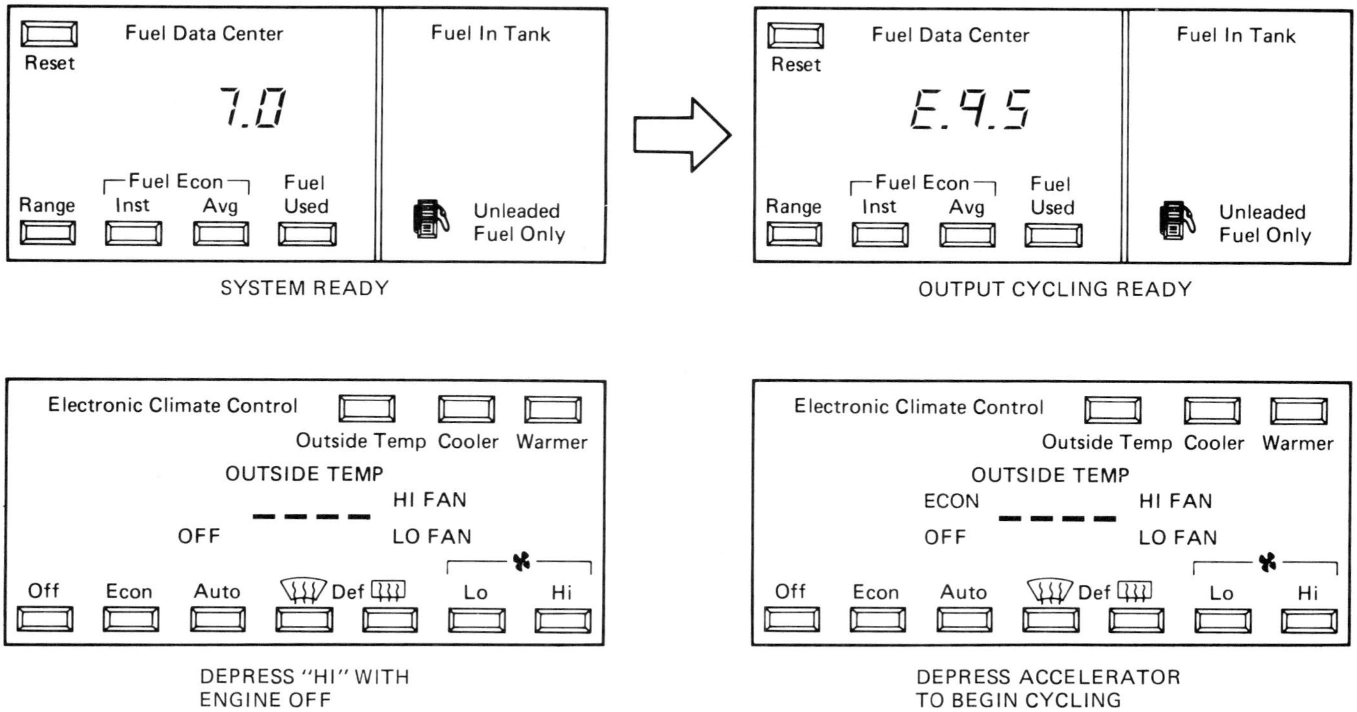

FIGURE 6-26 Entering output cycling mode *(Courtesy of General Motors Product Service Training).*

simultaneously while in the diagnostic mode, F.O.O will be displayed followed by .7.0. This indicates that any BCM codes that were in memory are now erased.

Exiting Diagnostic Mode

When the AUTO, FRONT DEF, or REAR DEF buttons are pushed (any one), the system returns to normal operation and displays temperature (Fig. 6-29). The system also returns to normal if the ignition switch is turned off or if the engine is cranked.

SYSTEM SERVICE

The 1985 FWD DFI system is serviced in a similar manner as the 1984 DFI system. Follow service manual procedures. A summary of system component relationships is shown in Fig. 6-30.

As you can see, the electronics on the Cadillac models has progressed over the years. The 1986 Seville and Eldorado also use the BCM and add an IPC or instrument panel computer. The ECM continues to monitor the DFI circuits, while the BCM responsibilities are expanded to include twilight sentinel circuits, instrument panel lamp illumination control circuits, door ajar warning circuits, and cruise control circuits. The IPC is responsible for headlamp circuits, dimming sentinel circuits, and twilight sentinel circuits. These computers are interfaced and capable of self-diagnosis.

ABBREVIATIONS

A/F Air/fuel (A/F ratio)

AIR Air injection reaction

ALDL Assembly line data link

BARO Barometric pressure

BCM Body control module

CCP Climate control panel

Conv. Converter (catalytic)

DFI Digital fuel injection

DVM (10 Meg.) Digital voltmeter with 10-million ohms resistance

ECC Electronic climate control

ECM Electronic control module (controller)

EECS Evaporative emissions control system

EFE Early fuel evaporation

EGR Exhaust gas recirculation

EST Electronic spark timing

FDP Fuel data panel

HEI High-energy ignition

ISC Idle speed control

MAP Manifold absolute pressure

N.C. Normally closed

E.9.0 ENGINE DATA DISPLAY

PARAMETER NUMBER	PARAMETER	PARAMETER RANGE	DISPLAY UNITS
P.0.1	Throttle Position	-10 - 90	Degrees
P.0.2	MAP	14 - 109	kPa
P.0.3	Computed BARO	61 - 103	kPa
P.0.4	Coolant Temperature	-40 - 151	°C
P.0.5	MAT	-40 - 151	°C
P.0.6	Injector Pulse Width	0 - 99.9	ms
P.0.7	Oxygen Sensor Voltage	0 - 1.14	Volts
P.0.8	Spark Advance	0 - 52	Degrees
P.0.9	Ignition Cycle Counter	0 - 50	Key Cycles
P.1.0	Battery Voltage	0 - 25.5	Volts
P.1.1	Engine RPM	0 - 6370	RPM ÷ 10
P.1.2	Car Speed	0 - 255	MPH
P.1.3	ECM PROM I.D.	0 - 255	Code

F.8.0 BCM DATA DISPLAY

PARAMETER NUMBER	PARAMETER	PARAMETER RANGE	DISPLAY UNITS
P.2.0	Commanded Blower Voltage	-3.3 - 18.0	Volts
P.2.1	Coolant Temperature	-40 - 215	°C
P.2.2	Commanded Air Mix Door Position	0 - 100	%
P.2.3	Actual Air Mix Door Position	0 - 100	%
P.2.4	Air Delivery Mode 0 = Max A/C 4 = Off 1 = A/C 5 = Normal Purge 2 = Intermediate 6 = Cold Purge 3 = Heater 7 = Front Defog	0 - 7	Code
P.2.5	In-Car Temperature	-40 - 102	°C
P.2.6	Actual Outside Temperature	-40 - 93	°C
P.2.7	High Side Temperature (Condenser Out)	-40 - 215	°C
P.2.8	Low Side Temperature (Evaporator In)	-40 - 93	°C
P.2.9	Actual Fuel Level	0 - 19.0	Gallons
P.3.0	Ignition Cycle Counter	0 - 99	Key Cycles
P.3.1	BCM PROM I.D.	0 - 255	Code

ECM PROM I.D.

ECM PROM I.D. is Parameter .1.3 of Engine Data and is displayed as a numerical code as follows:

X X X

FINAL DRIVE RATIO
2 = 3.33:1
(2.97:1 Effective Ratio)

EMISSIONS SYSTEM
1 = Federal
2 = California
3 = Export
4 = Altitude

ECM PROM CALIBRATION
Number varies with individual calibration.

BCM PROM I.D.

BCM PROM I.D. is Parameter .3.1 of BCM Data and is displayed as a numerical code as follows:

X X X

ENGINE SYSTEM
Blank = Gas
1 = Diesel

BCM PROM CALIBRATION
Numbers vary with individual calibration.

ECM STATUS LIGHT DISPLAY

LIGHT ON	IN 4th GEAR	VCC ENABLED	CLOSED THROTTLE	RICH	CLOSED LOOP
LIGHT OFF	NOT IN 4TH GEAR	VCC DISABLED	OPEN THROTTLE	LEAN	OPEN LOOP
INDICATOR			Off	Econ	Auto
FUNCTION	**4TH GEAR INPUT**	**VCC OUTPUT**	**THROTTLE SWITCH INPUT**	**OXYGEN SENSOR INPUT**	**ECM OPERATING MODE**

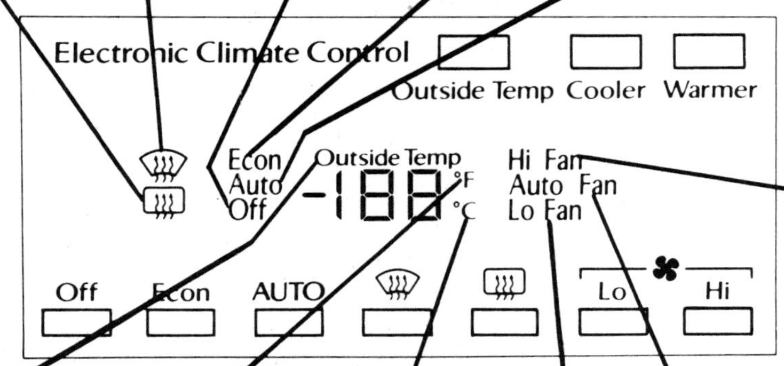

BCM STATUS LIGHT DISPLAY

FUNCTION	**A/C CLUTCH OUTPUT**	**COMPRESSOR LOW PRESSURE SWITCH INPUT**	**HEATER WATER VALVE OUTPUT**	**A/C-DEF MODE DOOR OUTPUT**	**COOLING FANS STATUS**	**UP/DOWN MODE DOOR OUTPUT**
INDICATOR	Outside Temp	°F	°C	Lo Fan	Auto Fan	Hi Fan
LIGHT ON	ENERGIZED	OPEN (LOW PRESSURE)	CLOSED (NO WATER FLOW)	A/C	FANS RUNNING	UP
LIGHT OFF	DE-ENERGIZED	CLOSED	OPEN	DEF	FANS OFF	DOWN

FIGURE 6-27 Diagnostic flow diagram *(Courtesy of Cadillac Motor Car Division).*

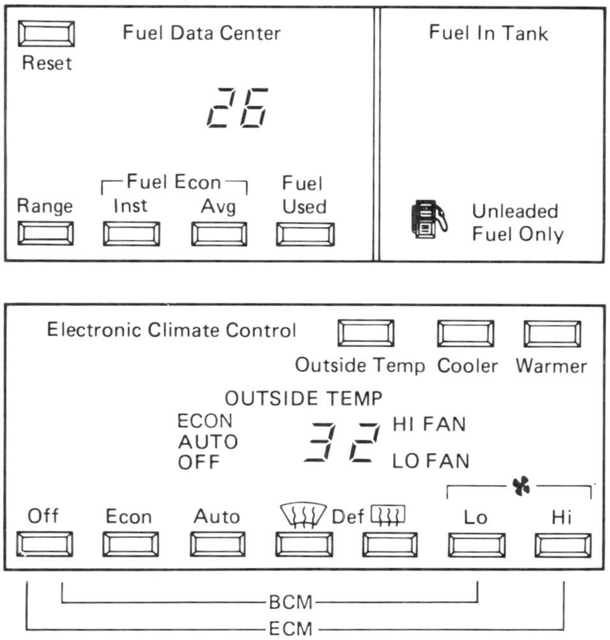

FIGURE 6-28 Clearing codes. Each system is cleared separately *(Courtesy of General Motors Product Service Training).*

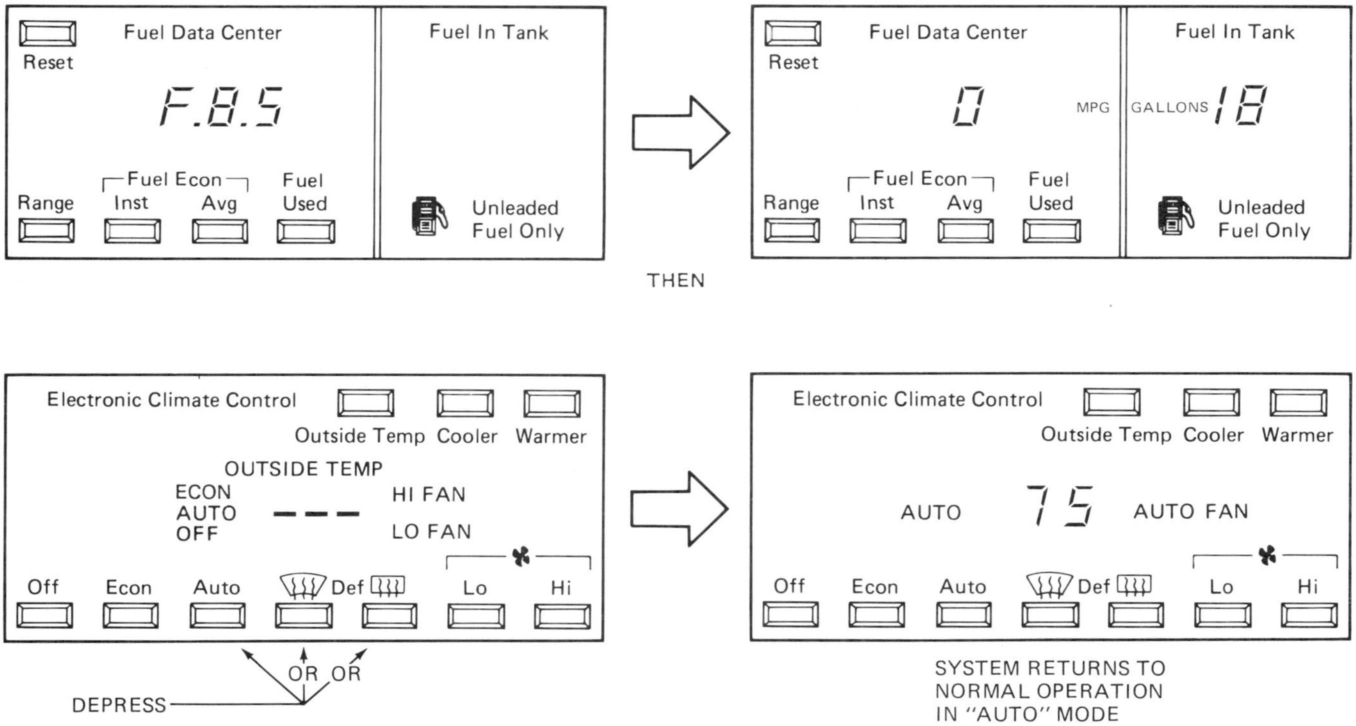

FIGURE 6-29 Exiting diagnostics *(Courtesy of General Motors Product Service Training).*

N.O. Normally open
O_2 Oxygen (sensor)
PCV Positive crankcase ventilation
P/N Park/neutral
Port Exhaust port
PROM Programmable read-only memory (engine calibrator)
TBI Throttle body injection (unit)
THERMAC Thermostatic air cleaner
TPS Throttle position sensor
TVS Thermal vacuum switch
Vac. Vacuum
VIN Vehicle identification number
VSS Vehicle speed sensor (speedo buffer sensor)
WOT Wide-open throttle

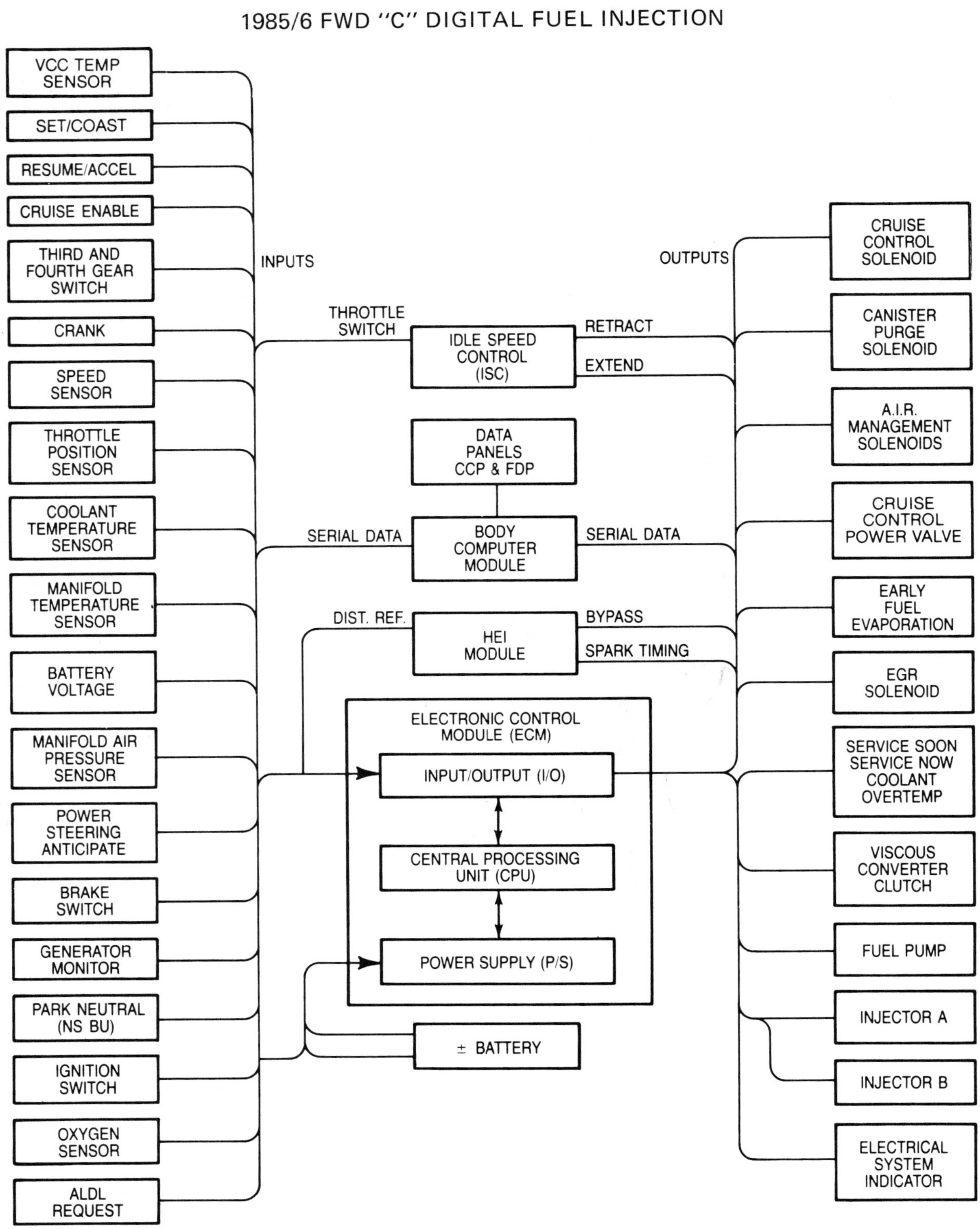

FIGURE 6-30 Electronic control module operating conditions sensed and systems controlled *(Courtesy of Cadillac Motor Car Division).*

UNIT I
General Motors Corporation Systems

7 Multiport Fuel Injection, 1984–1985

OVERVIEW

The earliest use of electronic port fuel injection by General Motors was in 1975 on Cadillac Motor Car Division vehicles, as was discussed in Chapter 1. That system, then called electronic fuel injection, was used until 1980. Port fuel injection was not mass produced by GM again until 1984. The system originally was known as multiport fuel injection (MFI), and has since become known as port fuel injection (PFI). There are two other variations of the PFI system known as sequential fuel injection (SFI) and tuned port injection (TPI), which will be investigated in subsequent chapters.

The MFI or PFI system to be discussed in this chapter was used on 1984 and 1985 vehicles. The PFI system was used on some 3.8-liter Buick V6 and 1.8-liter Pontiac L4 engines in 1984. Use expanded in 1985 to the 2.8-liter engines used by Chevrolet, Pontiac, Buick, and Oldsmobile and to the 3.0-liter engine used by Buick, Oldsmobile, and Pontiac.

Port fuel injection improves distribution of the air/fuel mixture to the cylinders because each cylinder has its own fuel injector. The intake manifold carries only air. Fuel condensation or puddling in the intake manifold is minimized. Fuel economy is improved, emissions are reduced, and overall performance is increased. Because fuel is injected at the intake valve port, cold starting is made easier, and acceleration enrichment and deceleration enleanment can be controlled with more precision (Fig. 7-1). The need for preheating intake air is reduced, while ram tuned intake manifolds can be used.

The port fuel injection system can be broken down into several subsystems: fuel delivery, air induction, sensors, electronic control module, electronic spark timing, electronic spark control, and turbocharger control. The last two subsystems are included in this chapter only because the 1.8-liter L4 engine is turbocharged.

FUEL DELIVERY

The fuel delivery system consists of the fuel tank and cap, an in-tank electric fuel pump, an in-line fuel filter, a fuel pressure regulator, the fuel supply and return lines, a fuel pressure accumulator, the fuel injectors, and the ECM CALPAK or calibration pack (Fig. 7-2).

Fuel Tank

The fuel tank contains a sump area or well where the fuel pump is installed. This sump area contains fuel and helps prevent the pump from running dry during severe maneuvering conditions. The fuel tank cap is a standard pressure-vacuum relief type emission control cap.

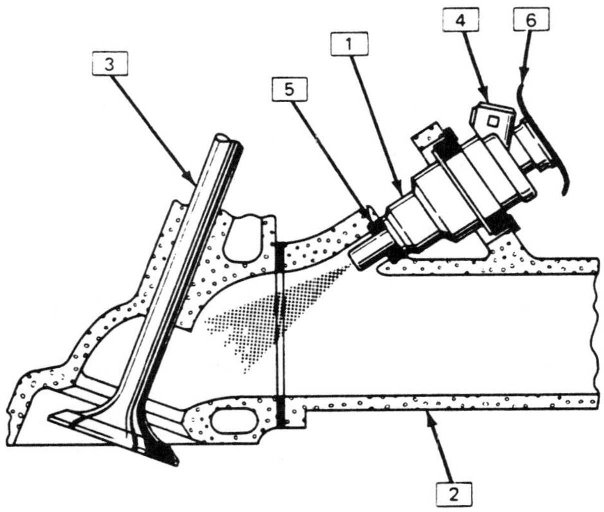

1 FUEL INJECTOR
2 INTAKE MANIFOLD
3 INTAKE VALVE
4 ELECTRICAL TERMINAL
5 "O" RING
6 FUEL RAIL

FIGURE 7-1 Fuel injector *(Courtesy of General Motors Product Service Training).*

In-tank Electric Fuel Pump

The in-tank electric fuel pump is integral with the fuel gauge sending unit (Fig. 7-3). The pump is serviceable separately. The intake or pickup tube of the pump has a 50-micron Saran filter attached. This filters out particles as small as 0.0020 inch and also blocks passage of most water unless the filter becomes completely submerged in water. The electric fuel pump is a high-pressure, positive-displacement roller vane pump. An internal pressure relief valve limits its pressure to 60 to 90 psi. The fuel pump used on the turbocharged engine has a higher flow rate and should not be interchanged.

Fuel Filter

The in-line fuel filter contains a 10 to 20-micron filtering element (Fig. 7-4). This element removes contaminants as small as 0.0004 to 0.0008 inch. The filter is replaced as an assembly. The 1.8-liter L4 fuel filter is located on the engine left side, while the 3.8-liter V6 fuel filter is mounted on the rear frame rail. This filter uses O-ring seals where the fuel lines fasten into the filter.

Fuel Pressure Regulator

The fuel pressure regulator is a vacuum-operated, diaphragm-controlled relief valve (Fig. 7-5). A pressure chamber and vacuum chamber are separated by a dia-

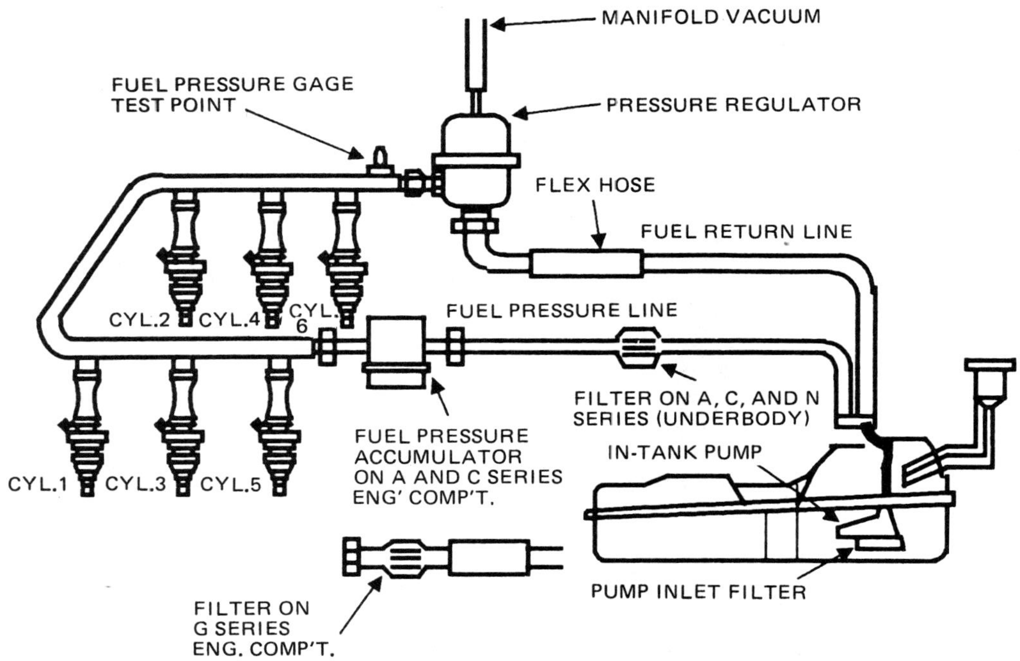

FIGURE 7-2 Fuel system layout *(Courtesy of General Motors Product Service Training).*

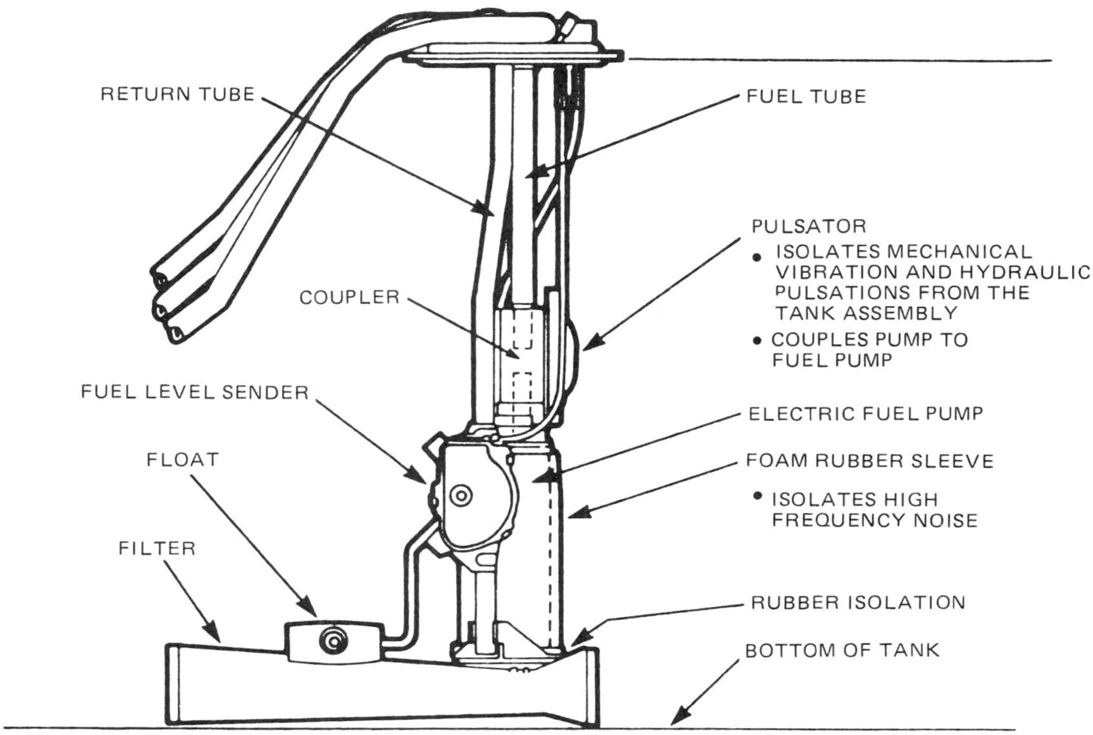

FIGURE 7-3 In-tank roller vane fuel pump *(Courtesy of General Motors Product Service Training).*

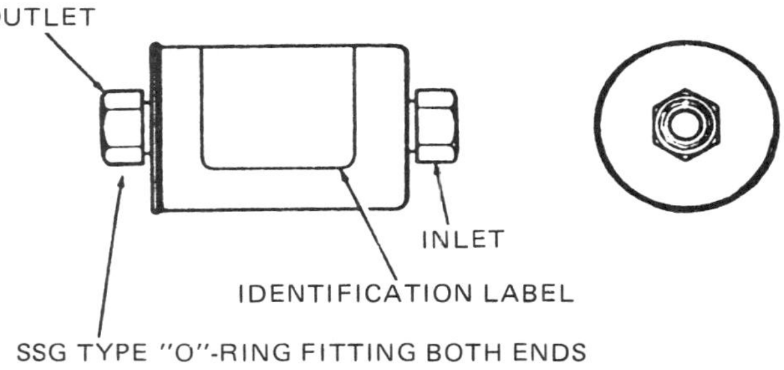

FIGURE 7-4 Typical in-line fuel filter *(Courtesy of General Motors Product Service Training).*

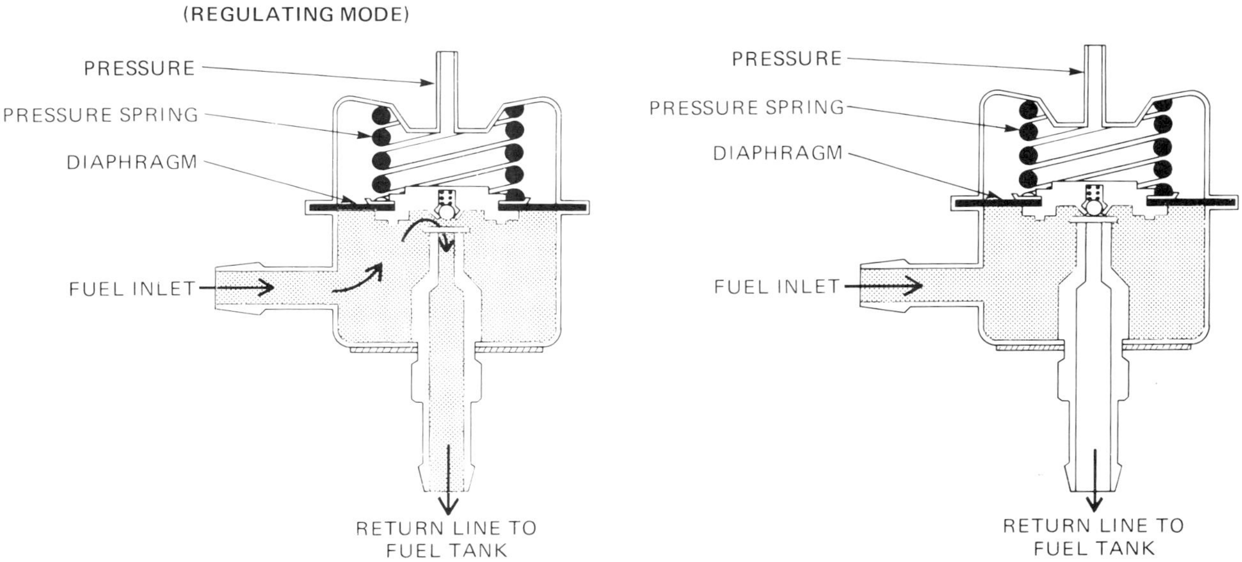

FIGURE 7-5 Fuel pressure regulator *(Courtesy of General Motors Product Service Training).*

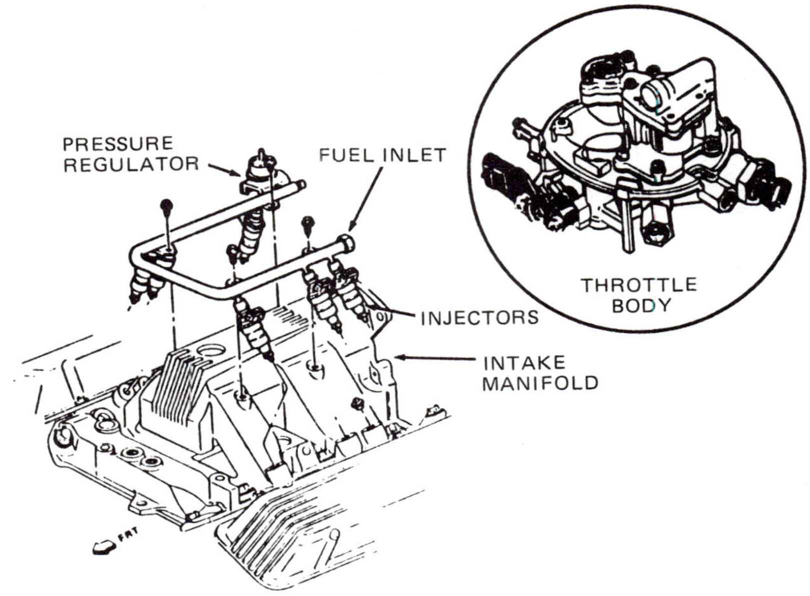

FIGURE 7-6 Fuel rail and throttle body comparison *(Courtesy of General Motors Product Service Training).*

phragm. In addition, the vacuum chamber contains a spring for calibration purposes. Fuel pressure enters the pressure chamber and acts against the spring tension. When fuel pressure rises above spring tension, the diaphragm is moved toward the vacuum chamber. When this happens, a ball valve is unseated, which allows excess pressure to flow into the fuel return line. The pressure regulator is set to maintain 24 to 42 psi. Intake manifold vacuum is connected to the vacuum chamber. This aids the fuel pressure in unseating the ball valve when intake manifold vacuum is high. At idle and low speeds, fuel demand is less and intake manifold vacuum is high. As a result, the pressure regulator maintains fuel pressure near the lower end of the pressure range. When fuel demand increases, as it does during acceleration or at high speeds, intake manifold vacuum reduces and fuel pressure increases. Fuel pressure rises as it tries, by itself, to overcome the regulator spring tension. The fuel pressure regulator is mounted on the fuel rail on the engine (Fig. 7-6).

Fuel Supply and Return Lines

The fuel supply and return lines are either special rubber hose or welded steel tubing. These lines are subjected to high pressures and should be replaced only with high-quality materials. All fuel feed and return line connections are screw types and utilize O-ring seals for system integrity. The ends of the steel lines require a flare that is not the same as for the carbureted flared fuel line.

Fuel Pressure Accumulator

A fuel pressure accumulator may be used on some models as a pressure dampener and sound absorber (Fig. 7-2). On engines using six injectors where all six injectors are pulsed at once, the fuel rail pressure fluctuates with each pulse of the injectors. Because fluids under pressure act as a solid, noises can be produced as a result of the pressure drop and instantaneous buildup. The fuel pressure accumulator is located in the fuel supply line.

Fuel Rail

The fuel rail is mounted on the intake manifold and supplies fuel to the injectors (Fig. 7-7). The fuel pressure regulator is connected to the outlet end of the rail. A fuel pressure test point is provided on the rail. It is similar in appearance to an air-conditioning Schrader valve. The rubber seals on the valve core are special in that they are compatible with gasoline. Fuel pressure in the rail is generally 24 to 50 psi and is maintained by the pressure regulator.

Fuel Injectors

Individual fuel injectors are fastened between the fuel rail and each intake port of the intake manifold. The injectors are mounted so that their nozzles will spray fuel into the intake ports of the cylinder head (Fig. 7-1). Two O-ring seals are used to seal the injectors into the fuel rail and into the intake manifold (Fig. 7-8). If the top seal leaks, gasoline leaks out. If the bottom seal leaks, a vacuum leak is created. Each fuel injector is flow tested and is marked either by part number or by color code as to its flow rate. Injectors should not be interchanged unless the flow rate is equal to what is specified. Port injectors are solenoid-operated pintle valves. When the solenoid coil is energized,

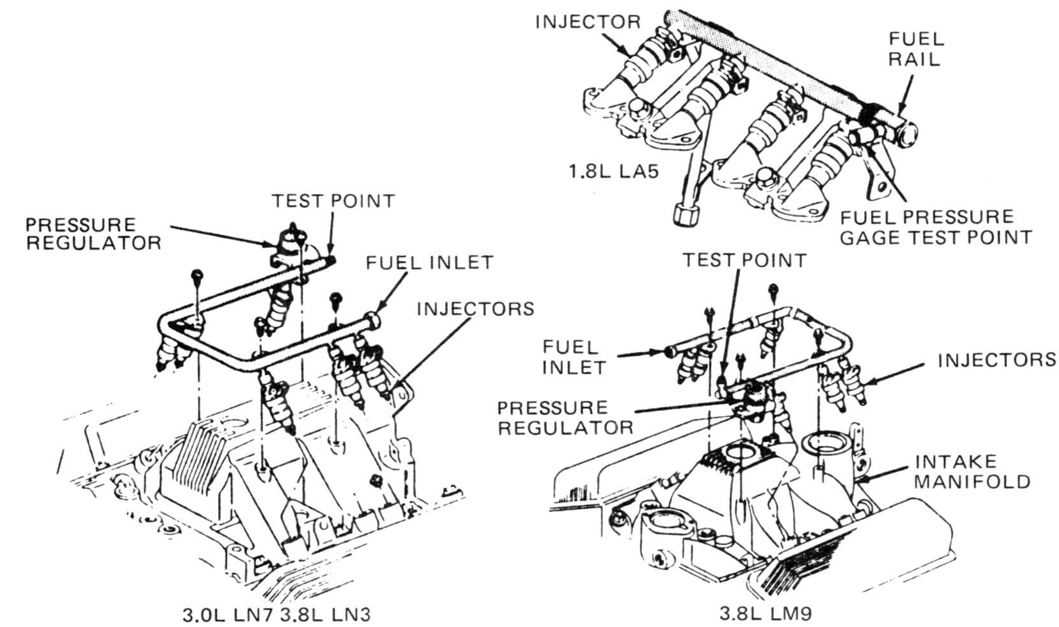

FIGURE 7-7 Fuel rails: 1.8-, 3.0-, and 3.8-liter engines *(Courtesy of General Motors Product Service Training).*

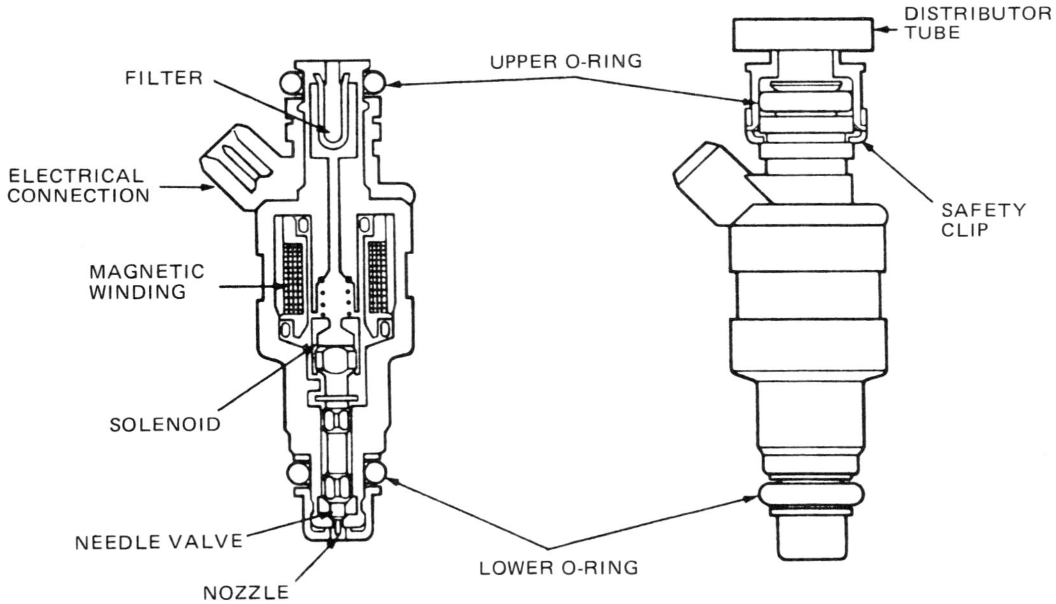

FIGURE 7-8 Typical fuel injector *(Courtesy of General Motors Product Service Training).*

the armature or plunger is attracted by the magnetism produced by the coil. When the plunger moves into the coil, it lifts the pintle or nozzle valve off its seat. This allows fuel to spray through the injector orifice at about 35 psi. The amount of fuel injected is determined by the duration the injector is energized. When the solenoid coil is de-energized, a return spring forces the plunger and pintle valve back to a normal or closed position. Each injector has a two-wire connector. One wire provides a 12-volt source from a 10-ampere fuse in the fuse panel, while the other wire goes to the electronic control module (ECM). The ECM completes and energizes the injector circuit by providing a ground in its injector driver circuit. The length of time the ground is provided is called *pulse width.* This determines the amount of fuel injected. All injectors are energized simultaneously.

Calibration Package

A calibration pack or CALPAK is used in the ECM to provide operating information when ECM voltage falls below 9 volts, cranking voltage falls below 9 volts, the ECM programmable read only memory (PROM) is missing, or the ECM fails to issue operating pulses. If any of these conditions exist, the ECM will operate in a fuel backup circuit mode. The CALPAK provides information necessary to run the engine. In this mode, the engine runs erratically because only the throttle position sensor, coolant temperature sensor, and engine rpm are used to control the fuel pump relay and fuel injector pulse width. The CALPAK is located in the ECM next to the PROM. It is accessible by removing a cover and is serviceable similar to the PROM. If the CALPAK is missing or is not functioning correctly, a code 52 is set in ECM memory.

In-tank Electric Fuel Pump Control

The in-tank electric fuel pump is controlled by a fuel pump relay in the engine compartment (Fig. 7-9). This relay can be energized to turn the fuel pump on by the ECM. The relay is normally open. When the ignition switch is turned to ON, the ECM energizes the relay (and pump) for 2 seconds. If an rpm signal is not received by the ECM, the relay and pump will be shut off. If cranking rpm is sent to the ECM, it will keep the circuit energized. Circuit voltage is provided by the IGN-3 fuse in the fuse panel.

The oil pressure switch also provides voltage to the fuel pump as long as engine oil pressure is above 4 psi. If the fuel pump relay fails, the fuel pump will continue to run via the oil pressure switch circuitry. When the relay is malfunctioning, extended cranking time will be required to build up 4 psi of engine oil pressure in order to complete the fuel pump circuit through the oil pressure switch.

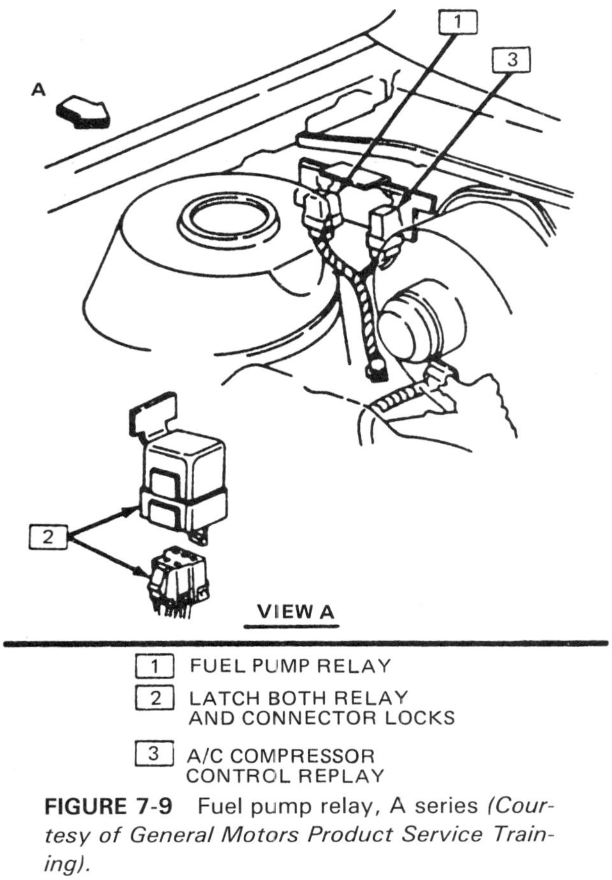

FIGURE 7-9 Fuel pump relay, A series *(Courtesy of General Motors Product Service Training).*

AIR INDUCTION SYSTEM

The port fuel injection air induction system consists of a mass airflow sensor (MAS or MAF), a throttle body, an air cleaner assembly, air ducting, the intake manifold, the idle air control (IAC), and on some models a turbocharger (Fig. 7-10).

Air Cleaner

The air cleaner assembly is not mounted on the engine. It is mounted in the engine compartment and contains a serviceable element.

Mass Airflow Sensor

Air leaving the air cleaner assembly enters the mass airflow sensor on normally aspirated (nonturbocharged) engines (Fig. 7-11). The MAF sensor measures the mass of the air entering the engine. Air mass is affected by temperature, humidity, and altitude or barometric pressure. The MAF sensor measures the temperature of the incoming air with a metal-foil sensing element. This element is heated by current from the MAF sensor module, which is incorporated with the MAF sensor assembly. The module maintains the temperature of the metal-foil sensing element at about 167°F above the temperature of the incoming air. Warmer air requires more current from the module, whereas cooler air requires less current from the module in order to maintain the metal-foil sensing element temperature. Humid air also requires more current, as would air at sea level. Dry air and air at higher altitudes requires less current. The warmer, humid, or sea-level air has more mass than the cooler, drier, higher-altitude air.

The cooling effect of the air passing over the metal-foil sensing element is interpreted by the ECM as air mass. This is more accurate than just measuring air volume because humidity, temperature, and altitude are considered by the MAF sensor. An airflow meter does not do this. When more energy is required to maintain the temperature of the MAF sensor element, the mass of the incoming air

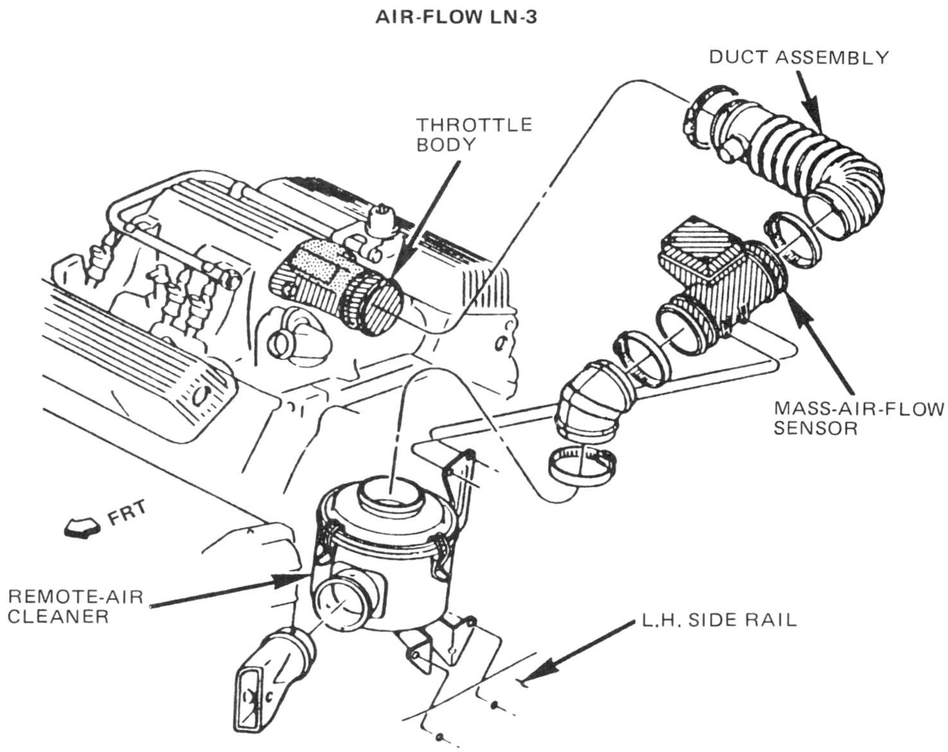

FIGURE 7-10 Air induction *(Courtesy of General Motors Product Service Training).*

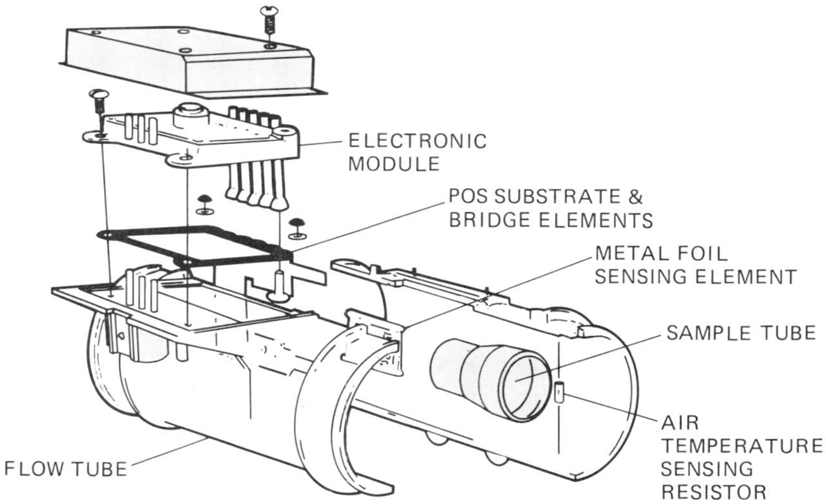

FIGURE 7-11 Mass airflow *(Courtesy of General Motors Product Service Training).*

has increased. This signal is sent to the ECM in the form of a digital signal. The voltage of this signal remains constant, while its frequency changes. An increase in air mass causes an increase in the signal frequency. Problems with the MAF sensor sets a code 33 or 34 in the ECM memory. The ECM considers the MAF sensor signal along with signals from the manifold absolute pressure sensor, engine rpm sensor, coolant temperature sensor, and the EGR vacuum diagnostic switch in order to calculate a fuel injector pulse width. Some models do not use a MAF sensor. An air temperature sensor, to be discussed in the next section of this chapter, is used instead of the MAF sensor.

Air Ducting

The air duct connecting the MAF sensor to the throttle body and engine is made of soft rubber with accordianlike folds. This isolates the MAF sensor from engine vibrations. The PCV clean air connection is integral with the duct. The duct is fastened with clamps at the sensor and at the throttle body. A leak at these points or within the duct (due to cracks) produces false readings at the MAF sensor as air is able to bypass the sensor.

Throttle Body

A single-piece, cast-aluminum housing, throttle plate, and shaft make up the throttle body (Figs. 7-12 to 7-15). Engine coolant is passed through a passage in the throttle body to help prevent icing at the throttle plate. A throttle position sensor (TPS), idle air control motor (IAC), and the throttle body vacuum manifold assembly are attached to the housing. The throttle blade is moved by the shaft and controls the position of the TPS, the volume of air entering the engine, and the amount of vacuum available in the throttle body vacuum manifold. There are about six throttle body models being used; therefore, interchangeability is limited.

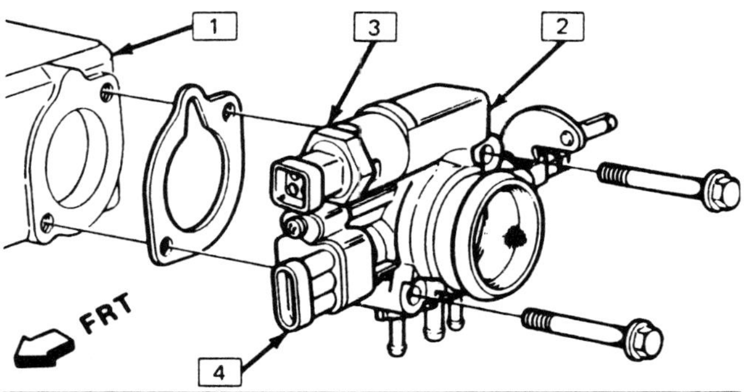

FIGURE 7-12 Throttle body 1.8 liter engine *(Courtesy of General Motors Product Service Training).*

Intake Manifold

The intake manifold is cast aluminum and has tuned runners (Fig. 7-16). This means the passageways have been matched or tuned so that each provides the same resistance to airflow. Because the length, size, and shape of the runners are the same, a denser air mass can be supplied to each intake port. This improves fuel economy and cylinder-to-cylinder distribution.

Idle Air Control Motor

The idle air control (IAC) is used by the ECM to control idle speed (Figs. 7-17 and 7-18). The IAC is actually a reversible electric motor to which the ECM sends pulses. Each pulse causes the IAC armature to step or move. It is sometimes referred to as a stepper motor. When the ECM issues pulses to the IAC, which causes its armature to retract, a valve on the end of the armature unseats and allows air to bypass the closed throttle blade. In this way, the ECM can regulate idle speed in accordance with the rpm signal received from the electronic spark timing module.

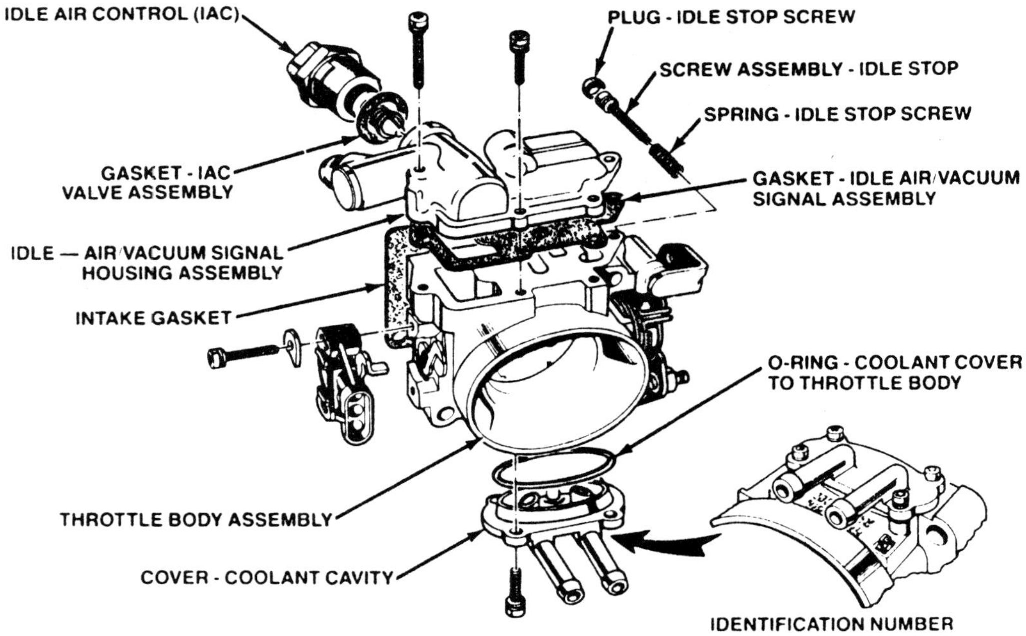

FIGURE 7-13 Engine throttle body 2.8 liter engine *(Courtesy of General Motors Product Service Training).*

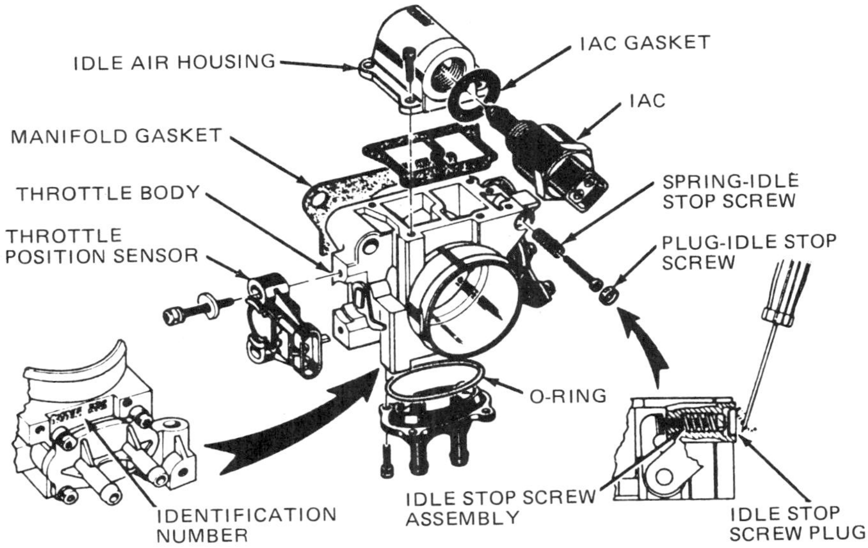

FIGURE 7-14 Throttle body, 3.0-liter engine *(Courtesy of General Motors Product Service Training).*

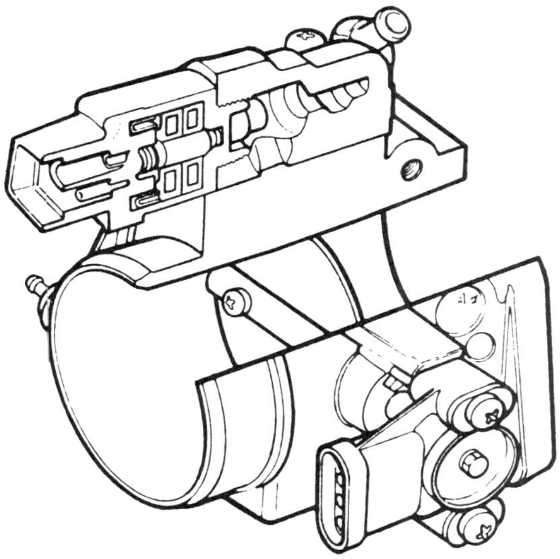

FIGURE 7-15 Throttle body, 3.8-liter engine *(Courtesy of General Motors Product Service Training).*

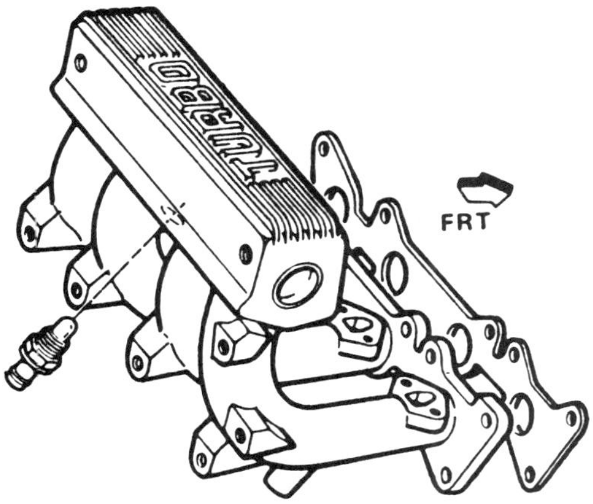

FIGURE 7-16 The 1.8-liter intake and manifold air temperature sensor *(Courtesy of General Motors Product Service Training).*

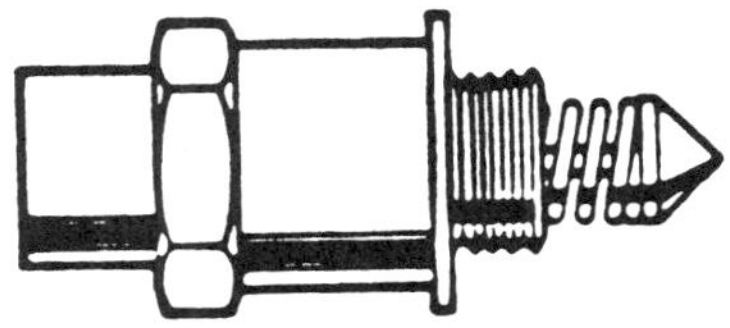

FIGURE 7-17 Idle air control valve *(Courtesy of General Motors Product Service Training).*

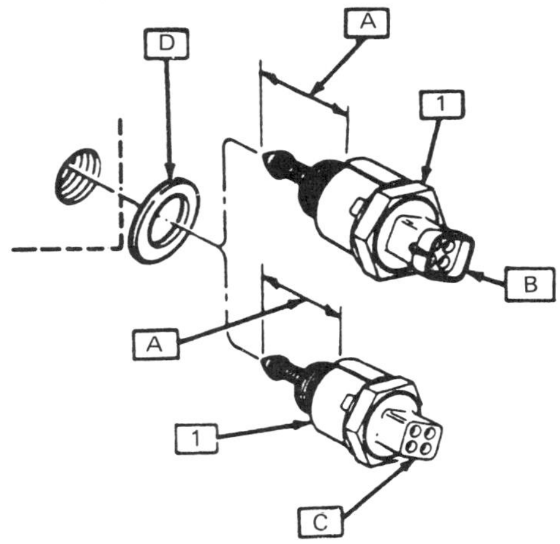

1 IDLE AIR CONTROL VALVE
- A LESS THAN 28mm (1-1/8 IN.)
- B TYPE I (WITH COLLAR)
- C TYPE II (WITHOUT COLLAR)
- D GASKET (PART OF IAC VALVE SERVICE KIT)

FIGURE 7-18 Idle air control valve installation *(Courtesy of General Motors Product Service Training).*

To reduce idle speed, the ECM sends pulses to the IAC that cause its armature to extend and move the valve toward its seat. This reduces the amount of air allowed to bypass the throttle plate, which reduces idle speed. When the ignition is turned off, the ECM resets the IAC by issuing enough pulses to seat the valve. This position is used by the ECM as a reference point. The next time the engine is started, the ECM counts the number of pulses needed to attain the correct idle speed. In this way, the ECM "learns" the motor position for a specified idle speed. Battery voltage, mass airflow, park/neutral switch, air-conditioning clutch signal, coolant temperature, throttle position, engine speed, and vehicle speed information are used by the ECM to calculate IAC motor position for accurate idle speed control.

SENSORS

Throttle Position Sensor

A throttle position sensor (TPS) is mounted on the throttle body and fastens to the end of the throttle shaft (Figs. 7-12 to 7-15). As the throttle blades are opened, the TPS varies in resistance. If a 5-volt reference signal is sent to the TPS by the ECM, the voltage returning to the ECM will be determined by the resistance of the TPS or by throttle blade position. In this way, the ECM is made aware of driver demands for acceleration, deceleration, constant speed, or idle conditions. The TPS has three wire connections, one for the 5-volt reference, one for a ground, and the third is where the throttle position is measured in terms of voltage. When the throttle is closed, the TPS resistance is high, resulting in a low return voltage to the ECM. At wide-open throttle, resistance is at a minimum, so the return voltage to the ECM is high. The TPS can be tested and adjusted if necessary.

Engine Coolant Temperature Sensor

The engine coolant temperature sensor (CTS) is a thermistor whose resistance decreases as the coolant temperature increases (Fig. 7-19). A 5-volt reference signal is sent to the CTS by the ECM. Because the CTS resistance changes with coolant temperature changes, the voltage returning to the ECM will also change. The ECM "views" the return voltage as an indication of engine coolant temperature. Low coolant temperature produces high sensor resistance and a low return voltage to the ECM. At −40°F, the sensor resistance is about 100,000 ohms, while at 266°F, the resistance drops to about 70 ohms. The CTS operation can be checked by using a 10-megohm impedance digital voltmeter.

Manifold Absolute Pressure Sensor

Intake manifold pressure is sensed by the manifold absolute pressure (MAP) sensor (Fig. 7-20). A 5-volt reference signal is provided to the MAP sensor by the ECM. As intake manifold pressure changes, the effective resistance of the sensor changes. As a result, the return voltage to the ECM

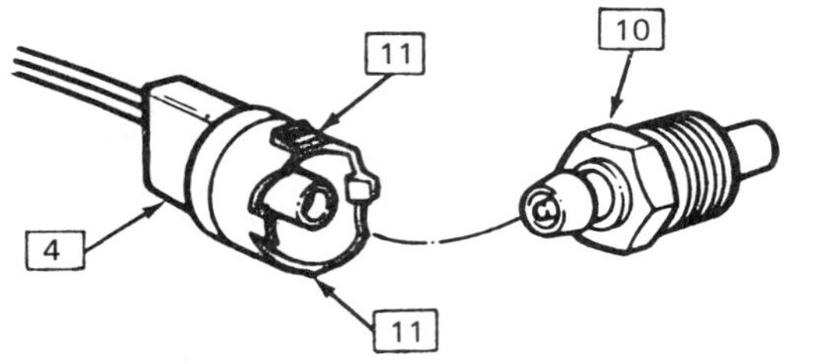

[4] ECM HARNESS—CONNECTOR TO COOLANT TEMPERATURE SENSOR

[10] COOLANT TEMPERATURE SENSOR

[11] DEPRESS TABS TO RELEASE

FIGURE 7-19 Engine coolant temperature sensor *(Courtesy of General Motors Product Service Training).*

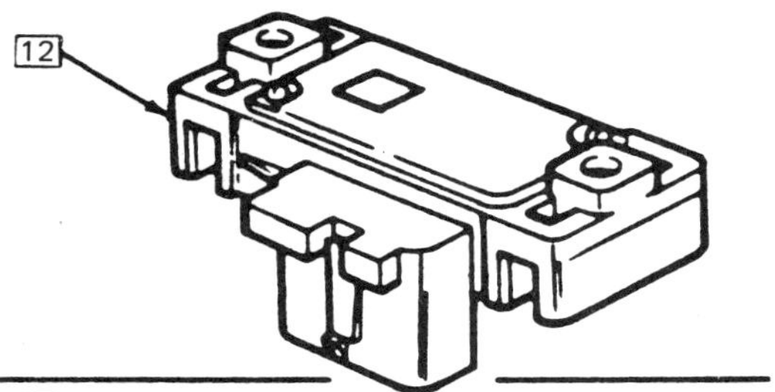

[12] MAP SENSOR

FIGURE 7-20 Manifold absolute pressure sensor *(Courtesy of General Motors Product Service Training).*

also changes. This varying MAP sensor voltage is used by the ECM as an indication of engine speed and load. Low intake manifold pressure (or high vacuum) produces low MAP sensor voltage. High intake manifold pressure (or low vacuum) produces high MAP sensor voltage. The MAP sensor is also used to measure barometric pressure, which allows the ECM to calibrate the fuel system for operation at different altitudes. When the ignition key is initially turned on, the pressure measured by the MAP sensor is the same as barometric pressure. The ECM retains this measurement for engine fuel system operation until the ignition switch is turned off. The MAP sensor can be tested with a 10-megohm impedance digital voltmeter.

Manifold Air Temperature Sensor

Intake manifold air temperature is measured by the manifold air temperature (MAT) sensor (Fig. 7-16). This sensor is a thermistor whose resistance decreases as air temperature increases. Low intake manifold air temperature produces high sensor resistance. At −40°F, the sensor resistance is about 100,000 ohms, while at 266°F the sensor resistance drops to about 70 ohms. A 5-volt reference is sent to the MAT sensor by the ECM. Return voltage to the ECM is determined by the sensor resistance and the manifold air temperature. The return voltage is used by the ECM as an indication of intake manifold air temperature.

Oxygen Sensor

An oxygen sensor (O_2) is used to monitor the oxygen content of the exhaust stream (Fig. 7-21). This sensor is the zirconia type that reacts to the oxygen in the exhaust gases by producing a voltage. When exhaust oxygen content is high, a low voltage is produced, while if exhaust oxygen content is low, a high voltage is produced. Voltage produced by the oxygen sensor varies between 0.1 and 0.9 volts. The ECM uses the oxygen sensor signal to adjust the pulse width for the fuel injectors for emission control purposes.

Park/Neutral Switch

The park/neutral switch provides a signal to the ECM to indicate driver gear range selection. This signal is used to improve idle quality, control ignition timing, control the torque converter clutch circuit, and regulate the idle air control motor.

Crank Signal

A cranking signal is sent to the ECM for control of the fuel system. This signal is provided by the starter solenoid.

Air-conditioning Signal

The air-conditioning system sends a signal to the ECM when its compressor clutch is energized. The ECM uses this signal to regulate the idle air control motor to compensate for added engine load. Idle quality improves as a result.

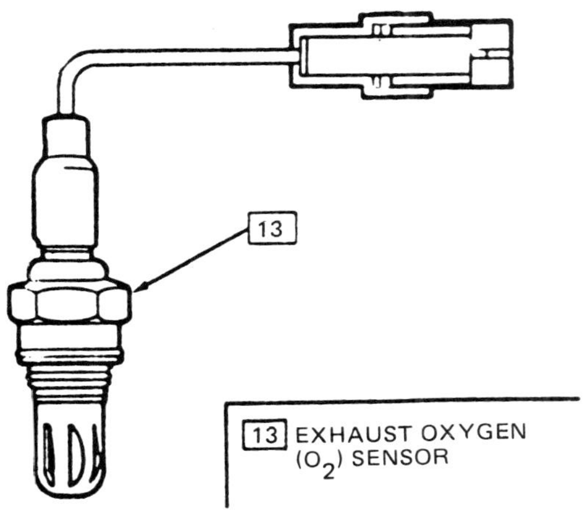

FIGURE 7-21 Exhaust oxygen sensor *(Courtesy of General Motors Product Service Training).*

Engine Speed Signal

An engine rpm signal is provided to the ECM by the electronic spark timing module. This enables the ECM to regulate idle speed and ignition timing.

Vehicle Speed Sensor

A vehicle speed sensor informs the ECM of vehicle speed for correct control of the torque converter clutch system.

ELECTRONIC CONTROL MODULE

Sensor information is sent to the electronic control module where it is used to determine output commands. The ECM processes the sensor input data in accordance with its program and then determines how to change the control of the fuel delivery system or any other device it controls. The oxygen sensor informs the ECM about the results of these changes so that the ECM can make additional compensations, if necessary.

The ECM has the ability to "learn." Each time the power is removed from the ECM or the ignition switch is off for at least 10 seconds, the ECM begins the learning process again. There may be a noticeable change in engine performance until the ECM "learns" over again.

Input Data

The ECM receives data from the air-conditioning system, manifold air temperature sensor, vehicle speed sensor, engine coolant temperature sensor, cranking signal, exhaust gas oxygen sensor, park/neutral switch, throttle position sensor, manifold absolute pressure sensor, engine rpm signal, transmission gear position switch, and system voltage (Fig. 7-22). All but the last two have been discussed earlier in this chapter. The transmission gear position switch is used in some models to provide a signal to the ECM when the transmission is in third or fourth gear. This signal is used to control the torque converter clutch. System voltage is constantly monitored by the ECM. Low voltage affects the entire computer system calibration. As voltage decreases, the ECM attempts to compensate by increasing injector pulse width, raising idle speed, and increasing ignition dwell period.

Output Commands

The output commands issued by the ECM are sent to control the idle air control motor, fuel injectors, electric fuel pump, air-conditioning compressor clutch, engine cooling

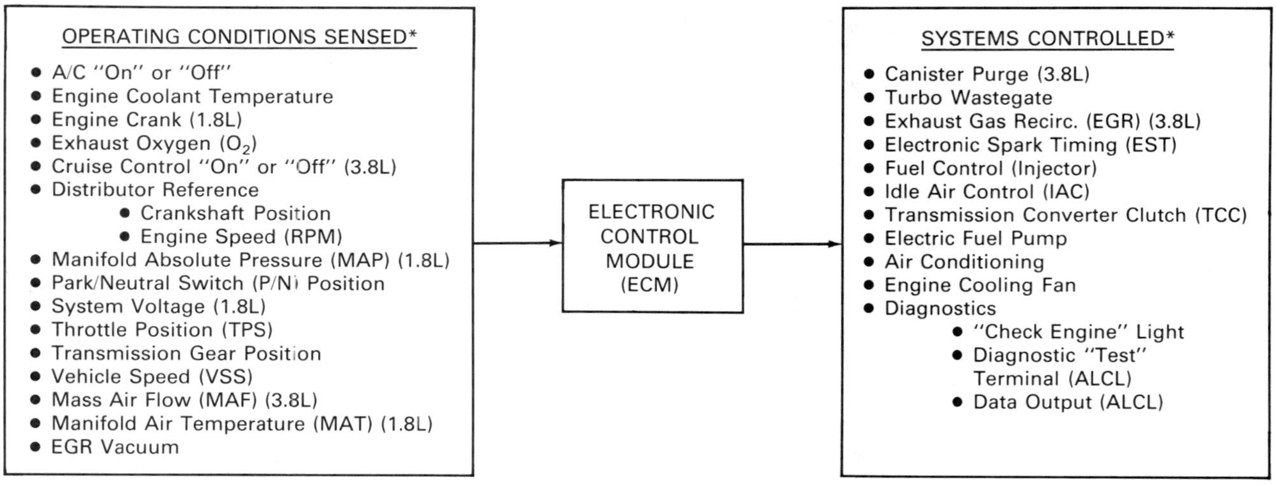

*Not all systems used on all engines.

FIGURE 7-22 Electronic control module operating conditions sensed and systems controlled *(Courtesy of Buick Motor Division).*

fan, turbocharger wastegate, electronic spark timing, torque converter clutch, and the self-diagnostic function (Fig. 7-22). Other model PFI vehicles may have additional data inputs and additional output commands.

ELECTRONIC SPARK TIMING

The high-energy ignition (HEI) system module has circuitry that allows the ECM to assume control of the ignition spark advance. This module has seven terminals instead of the four terminals originally used on HEI modules. Vacuum and centrifugal advance mechanisms are eliminated. To control ignition advance, the ECM must know engine speed, engine load, engine coolant temperature, and crankshaft position. Engine speed and crankshaft position data are provided by the HEI/EST module. Engine load data can be provided by the manifold absolute pressure sensor or the mass airflow sensor. Engine coolant temperature data are provided by the coolant temperature.

A four-wire connector is located in the engine compartment wiring harness between the HEI/EST module and the ECM. The connector wires are labeled A, B, C, and D. Terminal B on some models is the HEI reference or the engine speed and crankshaft position information circuit. Terminal D on some models is a ground wire ensuring a common ground between the module and the ECM. Terminal A on some models, EST, is the circuit by which the ECM controls the ignition timing. Terminal C on some models, bypass, is the circuit the ECM uses to take control of the ignition timing after it receives a 400-rpm signal on circuit B. The ECM issues a 5-volt signal through circuit C, which enables the ECM to control timing. Before the 400-rpm threshold is reached, the ignition system operates as a standard HEI system except that it operates at base timing only.

The ECM provides spark advance in accordance with its program and data from the sensors discussed earlier. In addition, it may provide ignition advance when the engine is cold or when engine load is minimal (based on rpm and low airflow). The ECM may retard or lessen spark advance when the engine is hot or under heavy load (based on high rpm and high airflow).

ELECTRONIC SPARK CONTROL

Some models employ an electronic spark control system (ESC) that is used to inform the ECM control of ignition timing under certain operating conditions. It is used to control detonation, sometimes referred to as spark knock or pinging. The ESC module can cause the ECM to retard ignition timing as much as 20 degrees. This allows the ECM to provide maximum spark advance for best performance and economy.

A piezoelectric device, the knock sensor, is used to detect vibration in the engine block due to detonation. When detonation occurs, the resulting vibration causes the knock sensor to produce a voltage. This voltage signal is sent to the ESC module, which in turn signals the ECM. The ECM retards the ignition timing until the detonation stops. At that point, the ECM begins to advance timing unless detonation begins again. The ECM does not respond to the ESC module signal unless engine rpm is above 1100 on some engines, ECM voltage is above 9 volts, and a code 43 is not stored in the ECM memory. Code 43 is an indication of a problem in the circuitry between the ESC module, knock sensor, and the ECM.

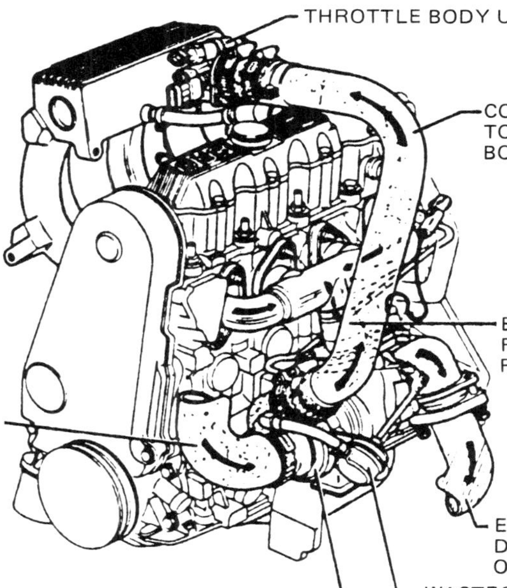

FIGURE 7-23 The 1.8-liter turbo airflow *(Courtesy of General Motors Product Service Training).*

TURBOCHARGER CONTROL

Some PFI engines use a turbocharger to increase engine power by forcing more air into the engine than the engine could take in by itself (Fig. 7-23). When the normal intake of air is exceeded by the turbocharger, it is known as *boost pressure.* Boost pressure must be known as it will affect fuel injector pulse width and ignition timing. Uncontrolled boost results in engine damage. The amount of boost provided by the turbocharger is regulated by a wastegate. The wastegate allows some exhaust to bypass the turbocharger, thereby limiting boost pressure to the engine. The ECM controls the wastegate by means of a wastegate control valve. The ECM controls the wastegate by pulsing the solenoid control valve off and on. The speed or frequency of the pulsing signal determines how much boost is provided by the turbocharger. Should something fail and cause a potential overboost condition, the MAP sensor indicates this to the ECM, and the ECM reduces fuel injector pulse width to prevent damage to the engine. When the ECM pulses the wastegate solenoid control valve on, the turbocharger begins to provide boost. Pulsing the control valve off allows exhaust to bypass the turbocharger and limit boost.

SYSTEM OPERATION

The ECM and PFI system have seven modes of operation: starting, clear flood, run, acceleration, deceleration, battery voltage correction, and fuel cutoff.

Starting Mode

The starting mode is entered when the ignition switch is first turned on. The fuel pump relay and fuel pump are energized by the ECM for about 2 seconds. The fuel rail pressure builds to about 28 to 50 psi. Sensor information is checked by the ECM to determine correct fuel injector pulse width for starting.

Clear Flood Mode

If the engine becomes flooded, the clear flood mode can be entered by depressing the accelerator pedal at least 80% open. The throttle position sensor indicates this to the ECM, while a cranking signal is provided at the same time. The ECM then pulses the injectors to provide a 20 to 1 air/fuel ratio. The ECM remains in the clear flood mode as long as the TPS is at least 80% open and cranking speed is below 600 rpm. If either of these conditions is not met, the ECM enters the starting mode.

Run Mode

The run mode operates in one of two conditions, open loop or closed loop. The run mode remains in open loop until the oxygen sensor output voltage varies between less than 350 millivolts and more than 550 millivolts, the coolant temperature sensor is above about 97°F, and until at least 10 seconds has lapsed since starting the engine. When these conditions are satisfied, the system operates in closed loop. This is when the ECM uses the oxygen sensor data to determine its output control functions. In open loop, the oxygen sensor is ignored by the ECM, and the air/fuel ratio is determined by the coolant and MAP sensors.

Acceleration Mode

The acceleration mode is entered anytime the TPS indicates an opening throttle and the mass airflow sensor indicates an increasing intake of air. The ECM widens the injector pulse width (or increases its frequency) according to the rate at which the throttle opens or the rate of increase in air intake.

Deceleration Mode

The deceleration mode is entered when the ECM receives a closing throttle signal from the TPS and a decreasing airflow signal from the MAF. The ECM shortens the injector pulse width to lean the air/fuel mixture. This is called deceleration enleanment and serves to reduce emissions. The ECM can momentarily stop fuel delivery when the rate of deceleration is very fast.

Battery Voltage Correction Mode

If battery voltage is low, the battery voltage correction mode will be entered. The ECM compensates for the low-voltage condition by increasing the injector pulse width, increasing the ignition dwell period, and raising the idle speed. This is done because low voltage causes the entire system to be out of calibration with the ECM program.

Fuel Cutoff Mode

The fuel cutoff mode is entered whenever the ignition switch is turned off or the ECM stops receiving an engine rpm signal from the HEI/EST module. This not only prevents dieseling or run-on but also prevents flooding or feeding a fire in accidents.

SYSTEM SERVICE

Throttle Position Sensor Service

The throttle position sensor is mounted on the end of the throttle shaft on the side of the throttle body. To check or adjust it, three jumper wires and a 10-megohm impedance digital voltmeter are necessary. Once the adjustment is made a thread locking compound should be added to the TPS mounting screws. On some models, the TPS is not adjustable as routine service.

Idle Speed

Engine idle speed adjustment on some models is accomplished by turning the ignition switch on and grounding the test terminal at the ALCL connector. Wait about 30 seconds and then disconnect the idle air control motor lead. Disconnect the test terminal ground and start the engine. The idle speed should be within specified limits when at operating temperature. If not, turn the adjusting screw until idle speed is acceptable, shut the engine off, and restore all connections.

Injector Balance Test

An injector balance test can be performed to locate a malfunctioning fuel injector (Fig. 7-24). A special tester is used to pulse the questionable injector for 500 milliseconds. The fuel rail pressure will drop proportionally to the amount of fuel injected. A fuel injection pressure gauge must be connected to the fuel rail in order to observe the pressure drop. Before testing the next injector, the fuel rail pressure must be restored to the original beginning pressure. Each injector should provide the same pressure drop when pulsed. Between each injector test, the ignition switch must be off for at least 10 seconds. This allows the ECM to "power down" or complete its power down cycle. If this is not allowed, the ECM may not energize the fuel pump relay and fuel pump when the ignition is turned on. Injectors that cause a pressure drop of more or less than 1.5 psi or 10 kpa as compared to the other injectors should be replaced. Repeated testing of the injectors may cause engine flooding and should be avoided.

Ignition Timing

Ignition timing should be set according to the instructions on the emission control label. The electronic spark timing system must operate in base timing before adjustments are made. This can be accomplished by disconnecting the four-wire distributor connector, a set-timing connector, or by entering the diagnostic mode, depending on the model of car being serviced.

Wastegate Test

A wastegate functional check can be made on turbocharged engines. The procedure is provided in the service manual (Fig. 7-25). A hand-operated pressure/vacuum pump is needed to perform this test. This test checks the operation of the wastegate solenoid control valve, the wastegate actuator, and the electrical control circuits.

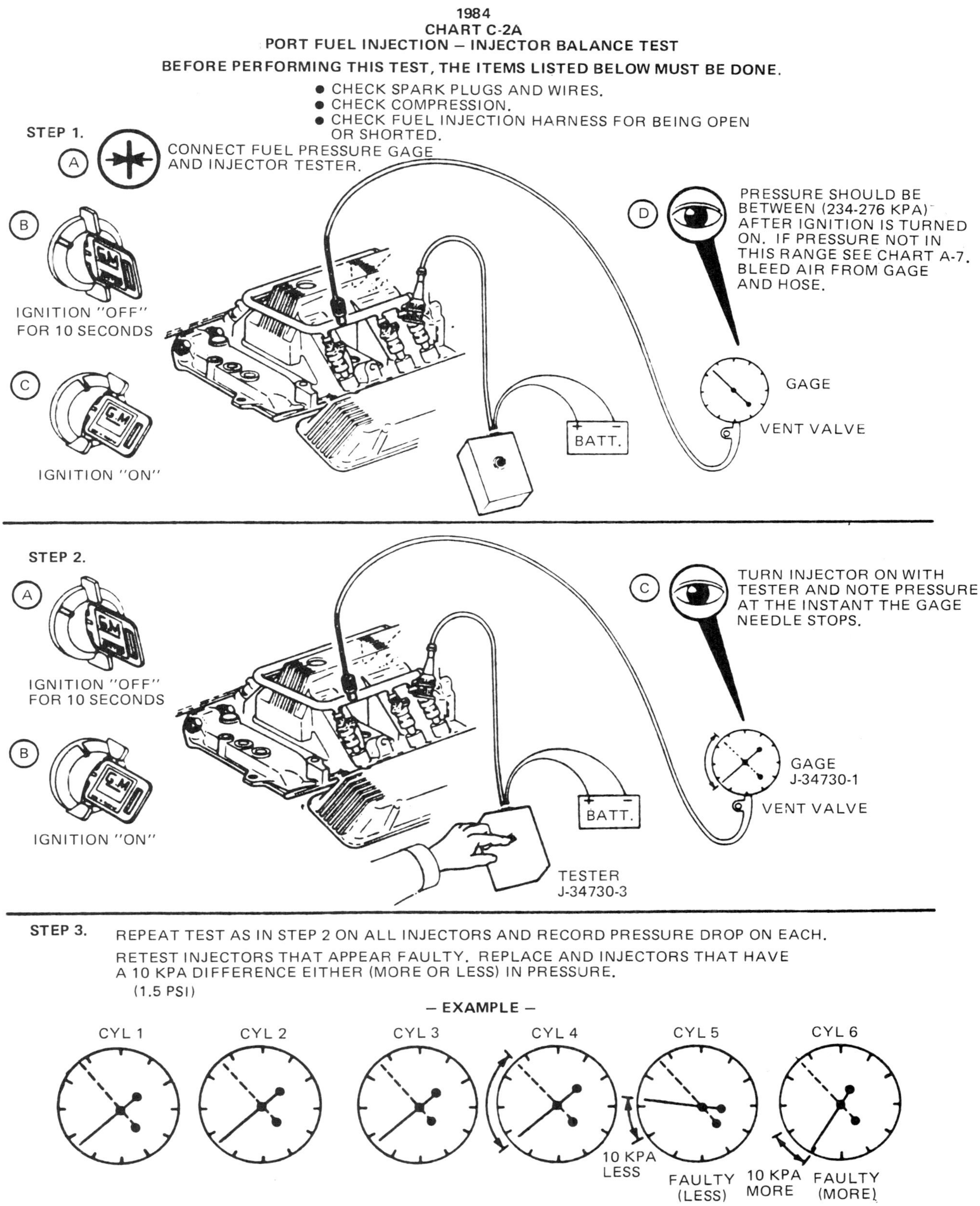

FIGURE 7-24 Injector balance test, 3.8 liter *(Courtesy of Buick Motor Division).*

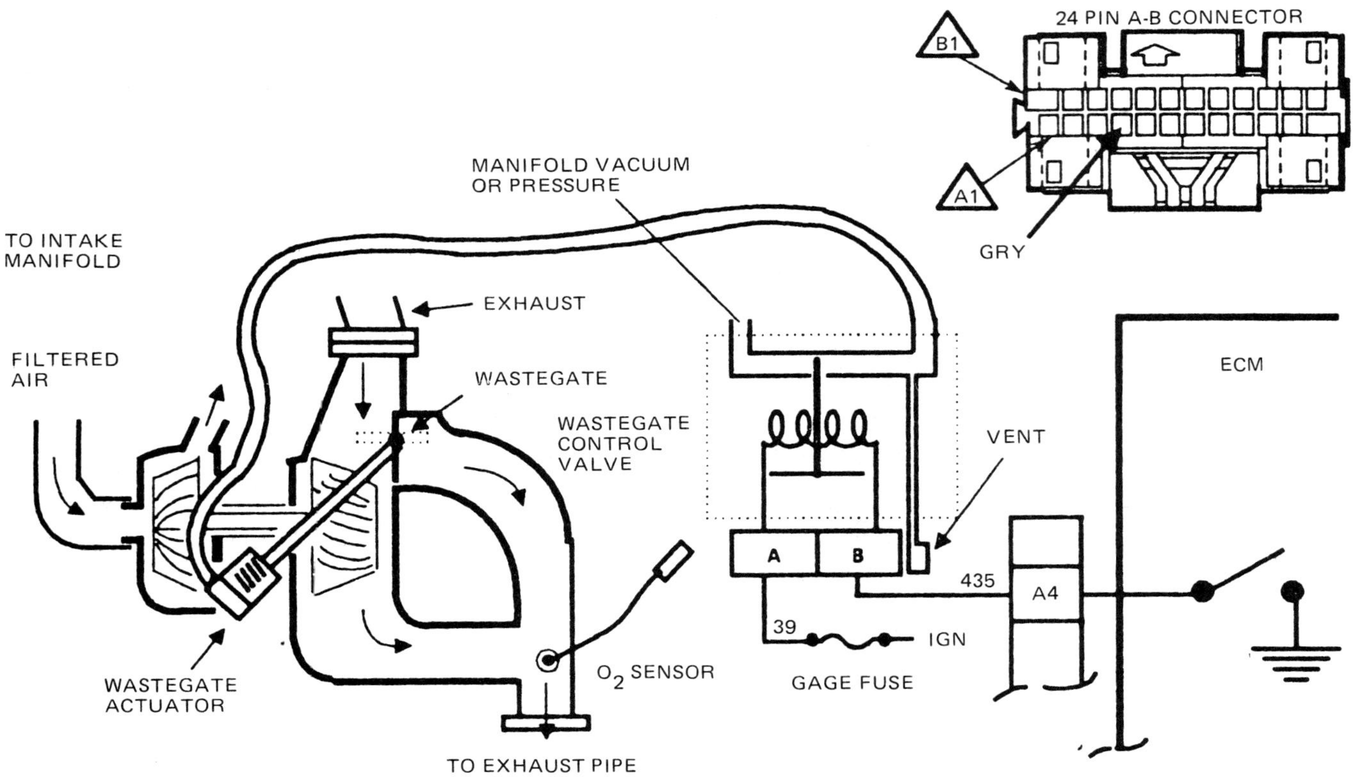

FIGURE 7-25 Wiring diagram, code 31 (1.8 liter) *(Courtesy of Buick Motor Division).*

Electronic Spark Timing Test

The electronic spark timing system can be checked at the same time ignition timing is set. Check ignition timing at 2000 rpm and then check it again with the ALCL connector test terminal grounded (Fig. 7-26). There should be two different readings on the magnetic probe timing meter or timing light. If there is no change, the system is not operating correctly.

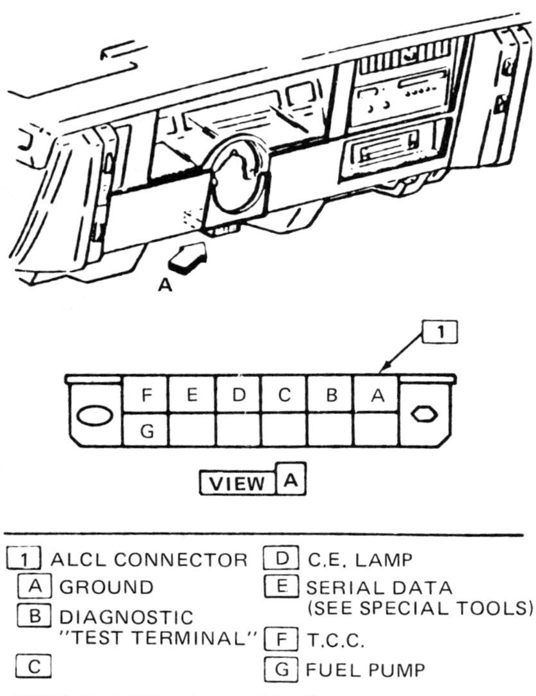

FIGURE 7-26 Assembly line communication link connector *(Courtesy of Buick Motor Division).*

Electronic Spark Control Test

The electronic spark control system can be operationally checked by tapping on the engine block with a ball pein hammer near the knock sensor. The engine should be at operating temperature and running above 1100 rpm when this test is performed. A magnetic probe timing meter or timing light can be used to see if the system causes the ignition spark to retard and then advance when the tapping is stopped.

Fuel Pressure Relief

Whenever the fuel system is going to be opened or a component disconnected, it is necessary to first relieve system fuel pressure. This is safely accomplished by removing the fuel pump fuse and then attempting to start the engine. The engine may start and then run out of fuel. Engage the starter an additional 3 seconds to relieve any residual pressure. With the ignition off, replace the fuse and service the component.

Fuel Pressure Check

Fuel system pressure can be checked by connecting a fuel injection pressure gauge to the Schrader connector on the fuel rail. Care should be taken when connecting or disconnecting the gauge as some fuel may leak out.

The in-line fuel filter should be replaced at the recommended intervals. When O-ring seals are used in the fittings, new seals should be used whenever the fittings are disconnected.

Special tools required for service are shown in Fig. 7-29 and Fig. 7-30.

SYSTEM DIAGNOSIS

Code Display

The PFI system, along with the ECM, has a self-diagnostic capability. An instrument panel warning light will illuminate to advise the driver of a problem in the system. This light provides either a CHECK ENGINE or SERVICE ENGINE SOON display.

While the engine is running, if the warning light comes on, a problem exists in the system and a code has been stored in the ECM memory. If the light goes out with continued driving, the problem has disappeared but a code has still been stored in the ECM memory. An existing problem is referred to as a current or hard code. A problem that appears and then disappears is referred to as a history or intermittent code. Hard codes can be diagnosed by using the decision trees in the manufacturer service manuals. Intermittent codes are generally caused by poor wire connections, loose harness connectors, or poorly tuned engines.

If the ignition switch is turned on, engine not running, the SERVICE ENGINE SOON light should be on steady. This ensures that battery and ignition voltage is available to the ECM. When the test terminal in the ALCL connector is grounded, the warning light should flash once, pause, and then flash twice. This will be repeated three times. This is a code 12, which means the self-diagnostic mode is functioning correctly. Any additional codes are displayed in a similar manner, each being repeated three times. When a code 12 appears again, the end of the code display has been reached. Refer to the service manual decision trees for each code displayed except for code 12. Codes that can be set are shown in Fig. 7-27 and Fig. 7-28.

If no codes were set in ECM memory, the driveability symptom charts in the service manual should be used. The field service mode can be used to aid in diagnosis of this type of problem.

Field Service Mode

To enter the field service mode, start the engine and ground the test terminal in the ALCL connector. The ECM uses the SERVICE ENGINE SOON light to indicate oxygen sensor voltage. While the system is in open loop, the light flashes 2.5 times per second. While in the closed loop, the light flashes once per second. The system flashes open loop after starting for from 30 seconds to 2 minutes depending on the temperature of the oxygen sensor. If the system does not go into the closed-loop mode, raise engine rpm for a few minutes, and if the system remains in open loop, refer to the code 13 decision tree in the service manual. If the light remains out, it is an indication that the exhaust is lean. The code 44 decision tree should be used. If the light remains on, the exhaust is rich and the code 45 decision tree should be used.

When the field service mode is used during a road test, the light may remain on during acceleration. This indicates acceleration enrichment and is a normal result. The light may remain off during deceleration. This indicates deceleration enleanment and is a normal result. At idle, the

DIAGNOSTIC CODES

CODE 13 - OXYGEN SENSOR CIRCUIT	CODE 33 - MAP SENSOR (HIGH)
CODE 14 - COOLANT SENSOR)LOW)	CODE 34 - MAP SENSOR (LOW)
CODE 15 - COOLANT SENSOR (HIGH)	CODE 42 - EST
CODE 21 - TPS (HIGH)	CODE 43 - ESC
CODE 22 - TPS (LOW)	CODE 44 - LEAN EXHAUST INDICATION
CODE 23 - MAT SENSOR (LOW)	CODE 45 - RICH EXHAUST INDICATION
CODE 24 - VEHICLE SPEED SENSOR	CODE 51 - PROM
CODE 25 - MAT SENSOR (HIGH)	CODE 52 - CALPAC
CODE 31 - WASTEGATE SOLENOID	CODE 55 - ECM

FIGURE 7-27 Diagnostic codes, 1.8 liter *(Courtesy of General Motors Product Service Training).*

1984 MFI/SFI DIAGNOSTIC CODES

12	NO REFERENCE PULSES TO ECM
13	OXYGEN SENSOR CIRCUIT FAILED
14	COOLANT READING TOO HIGH
15	COOLANT READING TOO LOW
21	THROTTLE POSITION SENSOR READING TOO HIGH
22	THROTTLE POSITION SENSOR READING TOO LOW
24	VEHICLE SPEED SENSOR FAILED
31	**WASTEGATE ELECTRICAL SIGNAL OPEN OR GROUNDED
32	*EGR ELECTRICAL OR VACUUM CIRCUIT MALFUNCTION
33	*MASS FLOW SENSOR READING TOO HIGH
34	*MASS FLOW SENSOR READING TOO LOW OR NO SIGNAL FROM SENSOR
41	**CAM SENSOR FAILED
42	*ERROR IN DISTRIBUTOR OR C^3 SYSTEM
43	ELECTRONIC SPARK CONTROL FAILURE
44	OXYGEN SENSOR LEAN TOO LONG
51	CALIBRATION PROM ERROR
52	*CALPAK MISSING
55	INTERNAL ECM ERROR (A/D CONVERTER)

*NEW CODES
**S.F.I. CODES ONLY

FIGURE 7-28 Multiport fuel injection codes *(Courtesy of General Motors Product Service Training).*

light may remain on for a longer period, indicating a rich exhaust. This is normal for idle speeds below 1200 rpm.

The PFI system is a further improvement over the throttle body systems in terms of accuracy, economy, and performance. There are, however, variations of PFI that may prove to be even better.

ABBREVIATIONS

A/F Air/fuel (A/F ratio)
AIR Air injection reaction
ALCL Assembly line communication link
BARO Barometric pressure
BATT Battery positive terminal
CCP Controlled canister purge
C.E. Check engine
CO Carbon monoxide
Conv. Converter (catalytic)
Diagnostic "test" term. Lead of ALCL connector that is grounded to get a trouble code
DVM (10 meg.) Digital voltmeter with 10-million ohms resistance
EAC Electric air control
EAS Electric air switching
ECM Electronic control module (controller)
EECS Evaporative emissions control system
EFE Early fuel evaporation
EFI Electronic fuel injection
EGR Exhaust gas recirculation
ESC Electronic spark control
EST Electronic spark timing
FED Federal
HC Hydrocarbons
HEI High-energy ignition
IAC Idle air control
IGN Ignition
I.P. Instrument panel
MAF Mass air flow
MAP Manifold absolute pressure
MAT Manifold air temperature
MFI Multiport fuel injection
mph Miles per hour
N.C. Normally closed
N.O. Normally open
NO_x Oxides of nitrogen
O_2 Oxygen (sensor)
PCV Positive crankcase ventilation
PFI Port fuel injection
P/N Park/neutral
Port Exhaust port
PROM Programmable read-only memory (calibrator)
rpm Revolutions per minute
SFI Sequential fuel injection (used on 3.8L turbo; each injector is individually fired)
TACH Tachometer
TBI Throttle body injection (unit)
TCC Transmission converter clutch
THERMAC Thermostatic air cleaner

SPECIAL TOOLS

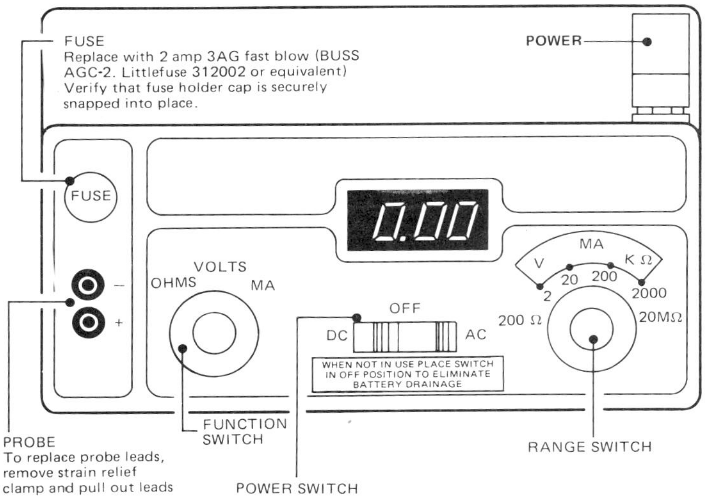

HIGH IMPEDANCE MULTIMETER
J29125-A

VOLTMETER – Voltage Position Measures amount of voltage. Connected parallel to exiting circuit. A digital high impedance voltmeter is used because some circuits require accurate low voltage readings, and some circuits have a very high resistance in the ECM. This meter also accurately measures extremely low current flow. Refer to meter for more information.

- Both function and range switch must be set properly, and the DC or AC position selected. DC is used for most measurements.

OHMMETER – Resistance Position Measures resistance of circuit directly in ohms. Refer to meter for more information.

- [1.] display in all ranges indicates open circuit.
- Zero display in all ranges indicates a short circuit.
- Intermittent connection in circuit may be indicated by digital reading that will not stabilize on circuit.
- Range Switch.

J23738/BT7517

VACUUM PUMP (20 IN. HG. MINIMUM)
Use gage to monitor manifold engine vacuum. Check vacuum sensors, solenoids and valves with hand pump.

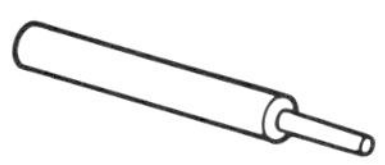

J28742/BT8234-A

WEATHER PACK TERMINAL REMOVER
Used to remove terminals from Weather Pack connectors. Refer to wiring harness service for removal procedure.

J33095/BT8234-A

ECM CONNECTOR TERMINAL REMOVER
Use to extract a terminal from connectors at the ECM.

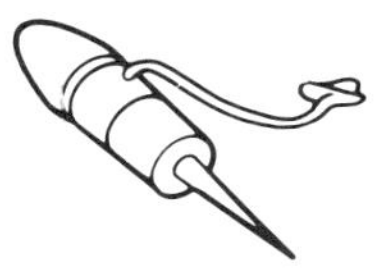

UNPOWERED TEST LIGHT
Used to check wiring for complete circuit. Connect lead wire to good ground. Probe with test prod to connector or component terminal. Bulb will light if voltage is present.

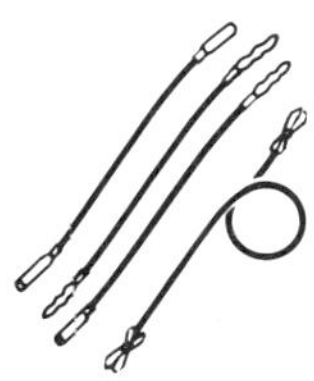

JUMPER WIRES (#16, 18 OR 20 GAGE WIRE)

- Clip jumper wire used to complete a circuit by bypassing an open.
- Set of jumper wires used to insert between Weather Pack connectors to permit access to the connector terminals for circuit checking. Six wires approximately 6" long. Use terminals 12014836 and 12014837. One set - female terminals both ends, one set - male at both ends and four sets - male terminals at one end and female terminals at the opposite end.

TACHOMETER
Use either a crankshaft harmonic balance pickup type or electronic coil trigger signal pickup type.

FIGURE 7-29 Special tools *(Courtesy of General Motors Product Service Training).*

SPECIAL TOOLS

 J33031/BT8130	IDLE AIR CONTROL WRENCH Used to remove or install IAC valve on throttle body.
 J34730-2	INJECTOR TEST LIGHT Used to test injector signal from ECM.
 J34730-1	FUEL PRESSURE GAGE Used to check and monitor fuel line pressure of fuel system.
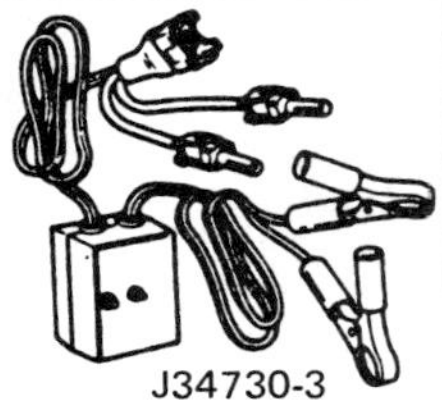 J34730-3	INJECTOR TESTER Used to perform injector balance test chart C-2A.

FIGURE 7-30 Special tools *(Courtesy of General Motors Product Service Training).*

TPS Throttle position sensor
TVS Thermal vacuum switch
V Volt
Vac. Vacuum
VIN Vehicle identification number
VSS Vehicle speed sensor
WOT Wide-open throttle

UNIT I
General Motors Corporation Systems

8 Buick Sequential Fuel Injection 1984–1985

OVERVIEW

Sequential fuel injection is very similar to the multiport fuel injection systems. Distribution of the air/fuel mixture to each cylinder is improved; the flow of fuel from the injector into the cylinder is much better; and the tendency of the fuel to condense before entering the cylinder is minimized. These factors result in reduced emissions and better fuel economy. Fuel delivery can be precisely matched to engine requirements under a wider range of load and speed while fuel economy is maintained. The sequential fuel injection system (SFI) closely monitors engine operation to attain these goals.

The SFI system is currently used on a turbocharged 3.8-liter V-6 engine (Fig. 8-1). Because a turbocharged engine requires even more precise control of the fuel system for its wider operating range from idle to boost conditions and full power, the SFI system was chosen as the best to meet these needs.

The major difference between the multiport fuel injection system and the sequential fuel injection system is that the SFI fuel injectors are pulsed one at a time in the same order in which the spark plugs fire.

SUBSYSTEMS

The sequential fuel injection system has several subsystems, including fuel delivery, air induction, electronic control module, computer-controlled coil ignition, turbocharger control, and electronic spark control.

FUEL DELIVERY

A fuel tank, in-tank electric fuel pump, in-line fuel filter, fuel pressure regulator, fuel lines, and the fuel rail and fuel injectors are part of the SFI fuel delivery subsystem (Fig. 8-2).

Fuel Tank

A special sump area is incorporated into the fuel tank to help prevent the in-tank electric fuel pump from running dry during severe vehicle maneuvering conditions. The tank is equipped with a standard vacuum-pressure relief emission control cap.

In-tank Electric Fuel Pump

The in-tank electric fuel pump supplies fuel for the SFI system (Fig. 8-3). It is a roller vane positive displacement pump that is mounted on the fuel tank sending unit. It is, however, serviceable as a separate part. A Saran sock filter

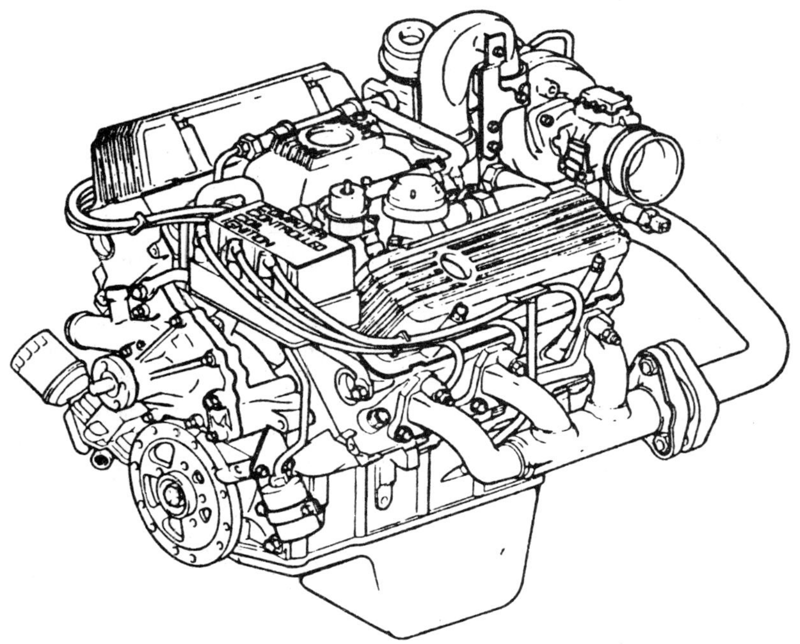

FIGURE 8-1 Sequential fuel injection engine *(Courtesy of General Motors Product Service Training).*

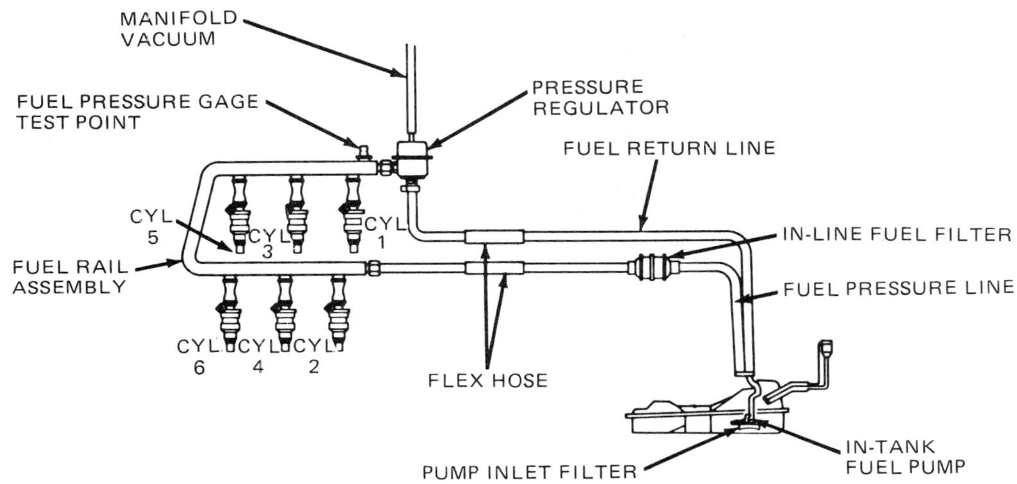

FIGURE 8-2 Fuel System *(Courtesy of General Motors Product Service Training).*

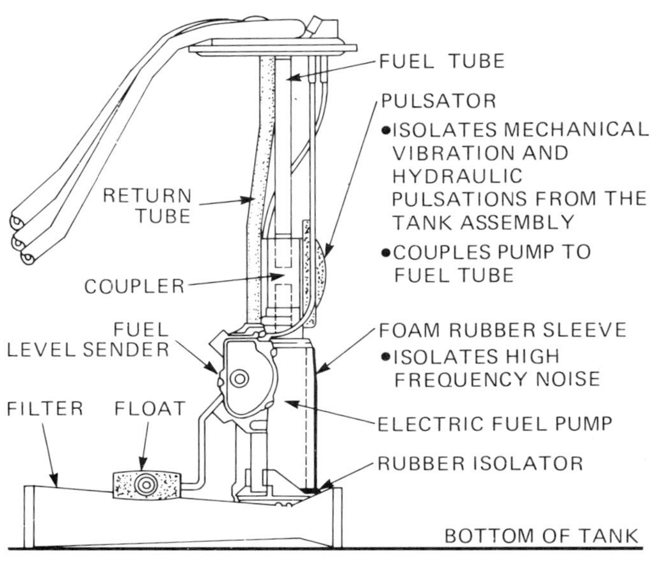

FIGURE 8-3 Pump unit *(Courtesy of General Motors Product Service Training).*

is attached to the end of the fuel pickup tube. This filter prevents passage of particles as small as 50 microns or 0.002 inch. The pump operates at about 3500 rpm and is capable of providing 130 psi, but is limited by the fuel pressure regulator.

Electric Fuel Pump Control

An electrical circuit consisting of a fuse, relay, ECM circuits, and oil pressure switch controls the electric fuel pump. The pump is energized for about 2 seconds when the ignition switch is turned on. The ignition switch signals the electronic control module to energize the fuel pump relay (located on the right front inner fender). The relay then completes a circuit between the ignition 3, 10-ampere fuse and the fuel pump. The electronic control module keeps the relay energized for about 2 seconds unless an rpm signal is received from the computer-controlled coil ignition module. As long as rpm pulses are received, the electronic control module keeps the relay energized.

If the engine rpm signal stops for any reason, the fuel pump is shut off. The oil pressure sending unit has an additional switch to serve as a backup circuit should the fuel pump relay fail. An inoperative relay would prevent the fuel pump from running. With the engine cranking, when about 4-psi oil pressure builds, the switch in the oil pressure sending unit closes and completes a circuit from the ignition 3 10-ampere fuse to the fuel pump. At this point, the engine should start and run; however, prolonged cranking may be necessary to obtain the 4 psi oil pressure needed to turn on the fuel pump.

Fuel Filter

An in-line fuel filter is mounted on the plate that covers the mounting for a previously used mechanical fuel pump on the engine (Fig. 8-4). This is located at the left front corner of the engine. The filtering element prevents passage of particles as small as 10 to 20 microns or 0.0004 to 0.0008 inch. The fuel supply line is a threaded flex hose where it connects to the fuel filter. Rubber O-ring seals are used at all threaded fuel line connections to prevent leakage. Fuel lines are made of steel except where special high-pressure rubber hose is used. This is also true for the fuel return line. Standard fuel line hose should not be used as the pressures developed in the supply line are normally about 28 to 50 psi.

Fuel Pressure Regulator

The fuel pressure regulator maintains pressure in the fuel rail (Fig. 8-5). The fuel rail is mounted on top of the engine and provides fuel under pressure to the fuel injectors. The pressure regulator contains a diaphragm-operated relief valve and a calibrated spring. Fuel pump pressure is applied to one side of the diaphragm while intake manifold vacuum is applied to the other side. The calibrated spring is on the vacuum side of the diaphragm. When there is no vacuum, the fuel pressure must rise and overcome the tension of the spring before excess fuel pressure is relieved. The spring is calibrated to relieve pressure at about 50 psi. When manifold vacuum is maximum, it relieves fuel pressure sooner with the spring resisting it. The vacuum overcomes the spring and fuel pressure is limited to about 28 psi. Varying vacuum with changing engine operating conditions causes the fuel pressure to range between 28 and 50 psi.

Fuel Injectors

The fuel injectors are mounted between the fuel rail and the intake manifold (Fig. 8-6). A rubber O-ring seals the connection between the fuel rail and the fuel injector and another O-ring seals the fuel injector to the intake mani-

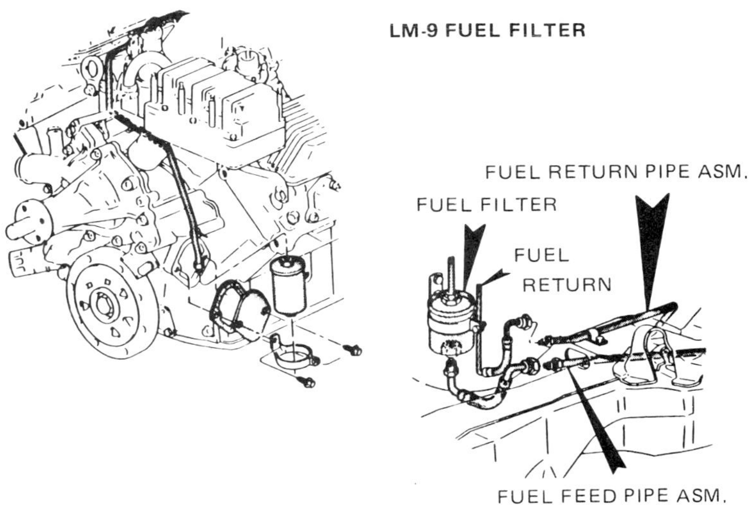

FIGURE 8-4 Fuel filter *(Courtesy of General Motors Product Service Training).*

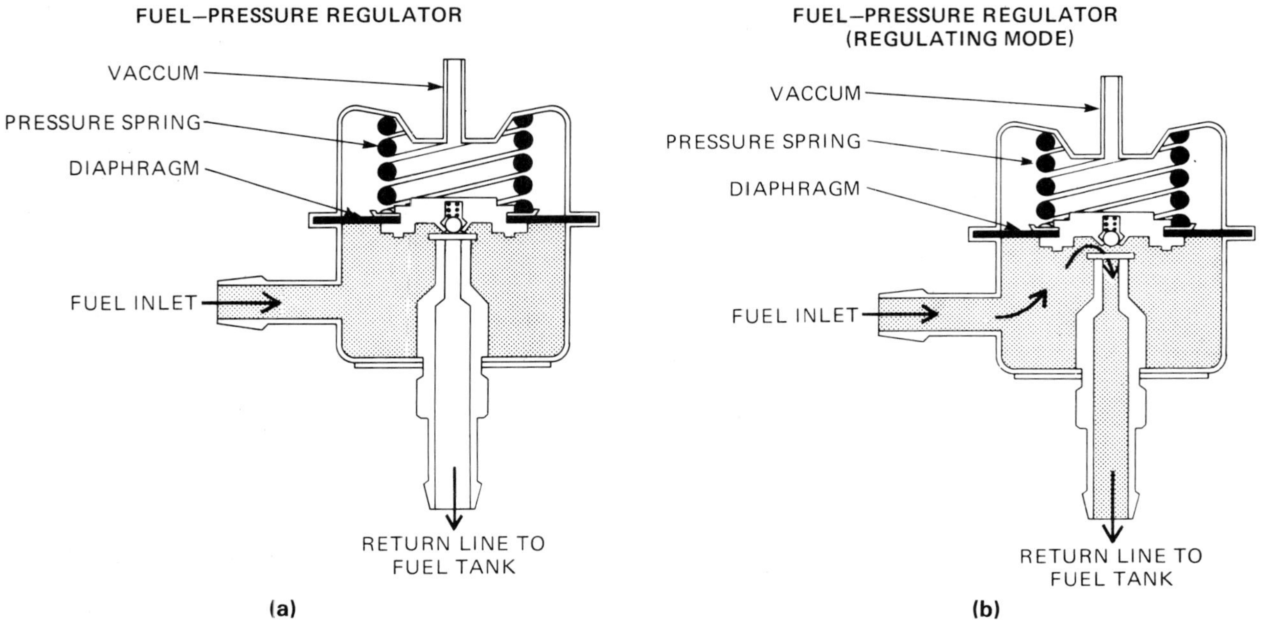

FIGURE 8-5 (a) Pressure regulator (b) Regulating mode *(Courtesy of General Motors Product Service Training).*

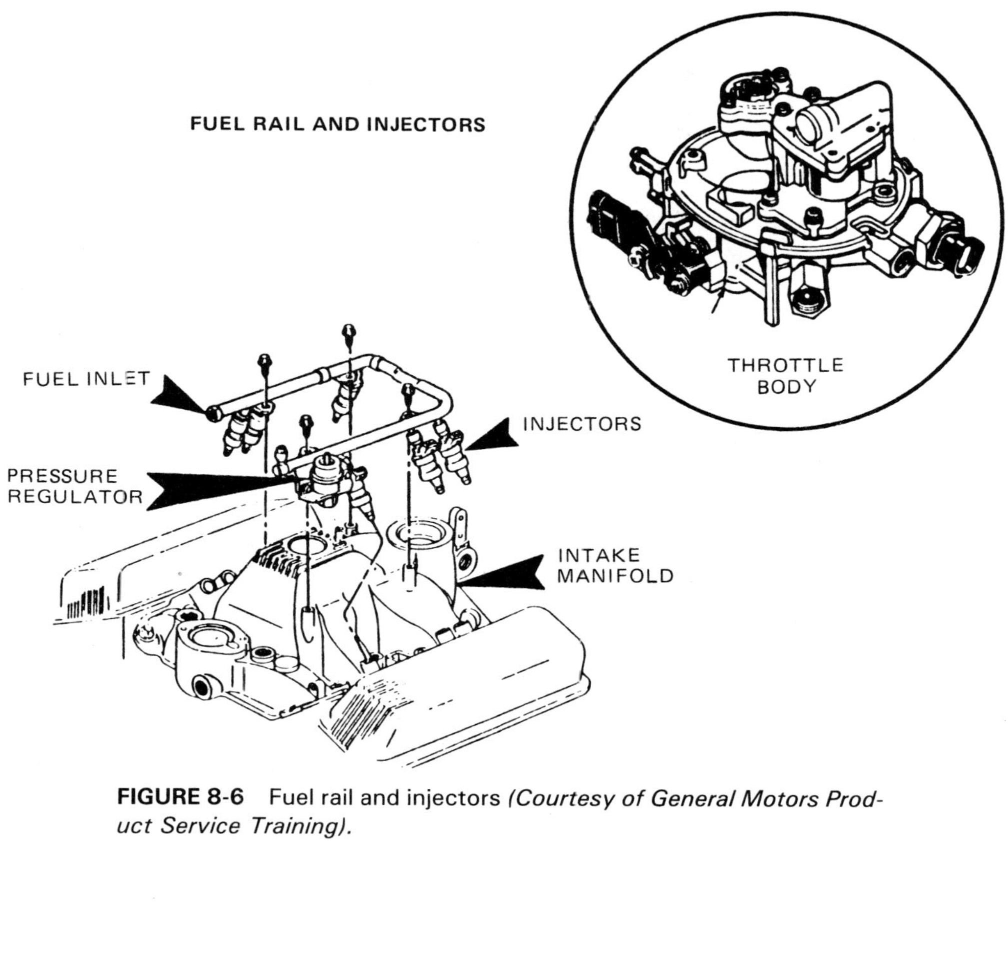

FIGURE 8-6 Fuel rail and injectors *(Courtesy of General Motors Product Service Training).*

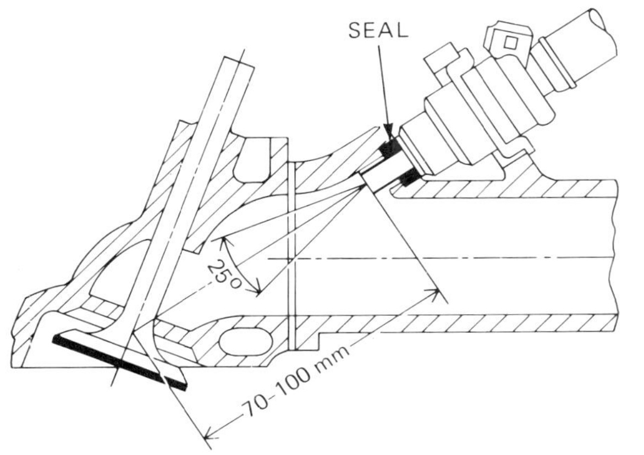

FIGURE 8-7 Injector mounting *(Courtesy of General Motors Product Service Training).*

fold. The O-ring at the top of the fuel injector prevents fuel leaks and the O-ring at the bottom of the fuel injector prevents vacuum leaks (Fig. 8-7). The injectors are marked with a GM/Bosch part number and also have a color-coded top to indicate the injector fuel flow rate. The SFI system uses a black color code. Injectors with different color codes should not be interchanged as their flow rate would not match the program calibration in the electronic control module.

Each fuel injector is basically a solenoid-operated pintle valve (Fig. 8-8). When current is passed through the solenoid winding, the resulting magnetic field pulls the solenoid armature and pintle valve. This action moves the pintle valve off its seat and fuel sprays into the intake valve port. When the current flow is interrupted, a return spring closes or seats the pintle valve and fuel spray stops. Each injector has a two-wire connector. One wire provides voltage from a 10-ampere ECM fuse. The other wire connects to the electronic control module, which completes the injector circuit by providing a ground. The making and breaking of this ground pulses the injector. The amount of time or duration that the ground is provided is known as pulse width. The pulse frequency and pulse width determine how much fuel is sprayed into the intake valve port. The fuel injectors in the SFI system are pulsed in the same order in which the spark plugs fire (1-6-5-4-3-2). The connectors for the injectors must not be interchanged or fuel will be injected at the wrong time. The brown wire at each injector is the source voltage wire, and the other wire (each is a different color) enables the ECM to provide a ground. The injector ground wire color code is as follows:

Cylinder	*Wire Color*
1	Black
2	Dark Green
3	Pink
4	Red
5	White
6	Light green

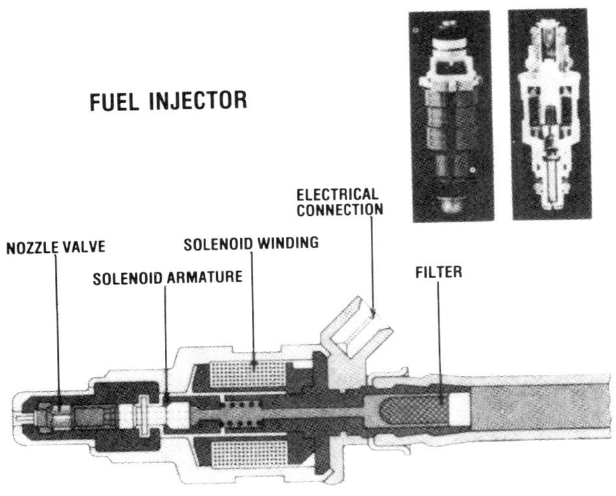

FIGURE 8-8 Injector cutaway *(Courtesy of General Motors Product Service Training).*

Fuel injectors can be tested for leaks and correct flow rate. The solenoid windings can also be checked for shorts or grounds. These tests will be described later in this chapter.

AIR INDUCTION

Air Cleaner

An air cleaner assembly, a mass airflow sensor, a throttle body, the intake manifold, and a turbocharger make up the air induction system (Fig. 8-9).

The air cleaner assembly is mounted alongside the engine. This remote mounting is necessary to provide room for mounting the turbocharger on top of the engine. Air ducting connects the air cleaner to the mass airflow sensor (MAF). The ducting is soft rubber, which reduces transfer of vibrations to the MAF sensor.

Mass Airflow Sensor

The mass airflow sensor provides data to the ECM about the mass of the air entering the engine (Fig. 8-10). Air temperature, barometric pressure, and humidity affect air mass. The MAF senses these variables and uses them to make calculations of the air mass, which are then used by the ECM to control the air/fuel mixture. The MAF has a screen to break up the airflow through it, an air temperature sensing resistor, a heated-foil sensing element, and an electronic module (Fig. 8-11). The module attempts to keep the metal-foil sensing element heated at a temperature of about 160°F above the temperature of the incoming air. The module's efforts to do this are interpreted as the air

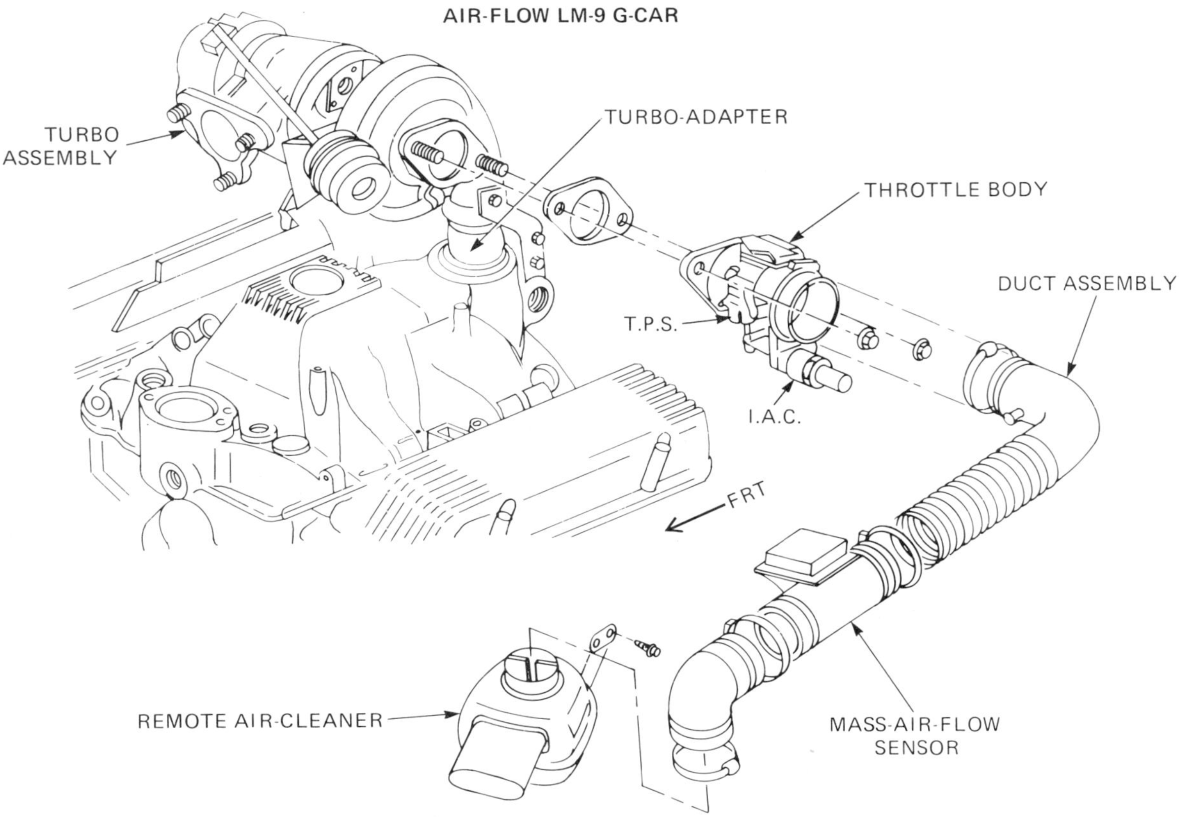

FIGURE 8-9 Air induction *(Courtesy of General Motors Product Service Training).*

FIGURE 8-10 Mass airflow sensor *(Courtesy of General Motors Product Service Training).*

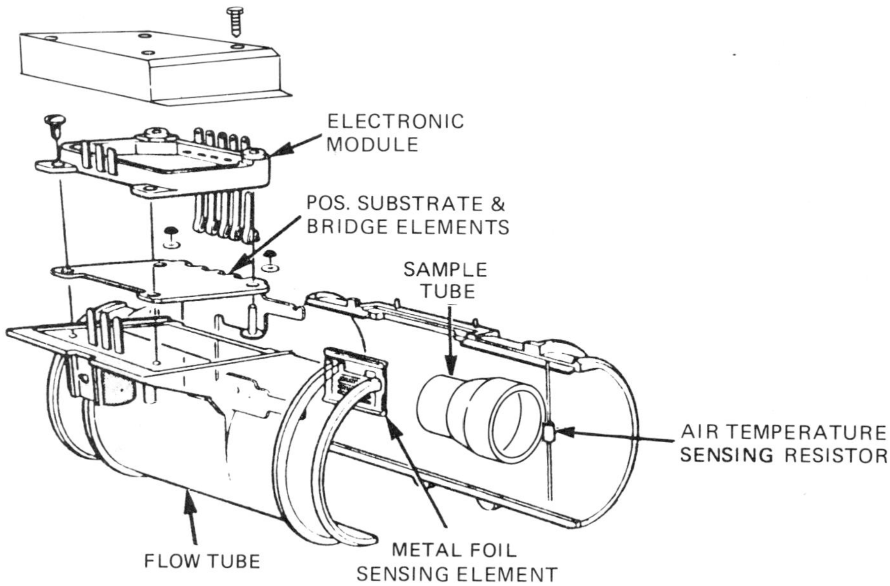

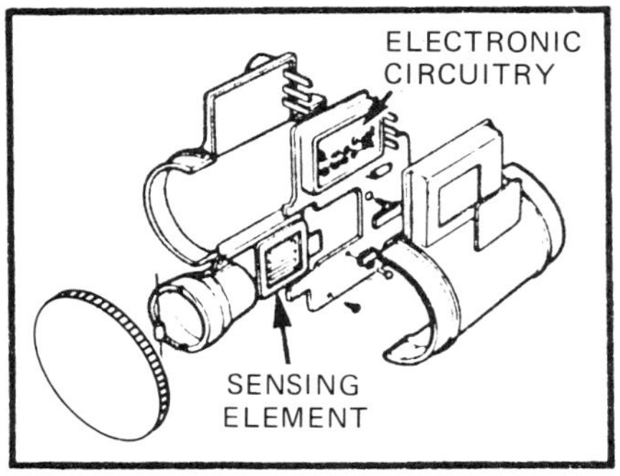

FIGURE 8-11 Exploded view of mass airflow sensor *(Courtesy of General Motors Product Service Training).*

mass. An air mass signal is provided to the ECM, which uses these data along with coolant temperature, engine rpm, engine load, and oxygen sensor data, to calculate the air/fuel mixture.

Testing the MAF sensor requires the use of the diagnostic charts in the service manual. The MAF sensor voltage signal to the ECM remains constant; however, the signal frequency increases as the air mass increases. The MAF replaces the manifold air temperature sensor, the manifold absolute pressure sensor, and the barometric pressure sensor used in other systems.

Throttle Body

A cast-aluminum housing incorporates the throttle shaft and plate, throttle position sensor, idle air control motor, and vacuum manifold assembly into the SFI system throttle body (Fig. 8-12). The throttle shaft and plate regulate the volume of air allowed to enter the engine. Attached to the end of the throttle shaft and mounted on the side of the throttle body housing is the throttle position sensor. The ECM uses this sensor to monitor throttle position as determined by the vehicle driver. The vacuum manifold assembly provides hose connections for various vacuum-actuated devices or accessories.

The idle air control (IAC) motor is used by the ECM to regulate idle speed. It is an electric motor that varies the size of an air passage that bypasses the throttle plate. The IAC is a reversible motor that has a valve attached to its armature. The ECM controls the IAC by means of counts or pulses of voltage. The motor armature and valve moves a predetermined amount for each pulse. When the IAC motor retracts the valve from the air bypass passage, more air is allowed to bypass the throttle plate. This causes the idle speed to increase. To decrease the idle speed, the ECM

FIGURE 8-12 Throttle body cutaway *(Courtesy of General Motors Product Service Training).*

must pulse the IAC motor to move the valve into the passage and restrict bypass airflow.

Each time the ignition switch is turned off, the ECM seats the IAC motor valve. The ECM uses this position as a reference to be able to "know" how far to open the passage for restarting purposes. System voltage, air-conditioning system operation, mass airflow, vehicle speed, engine speed, throttle position, and coolant temperature influence the ECM control of idle speed.

Intake Manifold

A special "tuned" aluminum intake manifold is used with the SFI system to increase air mass density at the intake valve ports. The length and size and shape of the air passages in the manifold have been matched or tuned to achieve this goal. This manifold carries only air because the injectors spray fuel into the intake valve ports.

The SFI system is used in conjunction with a turbocharger mounted on the intake manifold by means of an adapter (Fig. 8-13). This adapter is sealed to the intake manifold and to the turbocharger with O-ring seals to prevent air leakage. The throttle body is mounted on the end of the turbocharger. Turbocharger operation and control will be discussed later in this chapter.

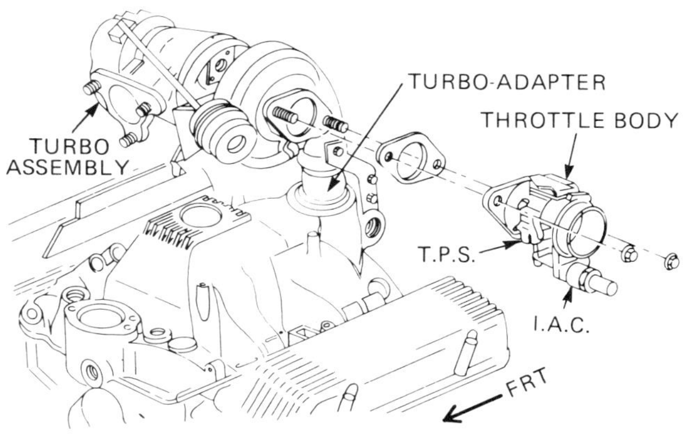

FIGURE 8-13 Sequential fuel injection turbo *(Courtesy of General Motors Product Service Training).*

SENSORS

Engine Coolant Temperature Sensor

The SFI system utilizes a mass airflow sensor, throttle position sensor, engine coolant temperature sensor, vehicle speed sensor, engine speed sensor, and an exhaust gas oxygen sensor. The mass airflow sensor has been covered earlier in this chapter. The engine coolant temperature sensor (CTS) is a thermistor whose resistance decreases as tem-

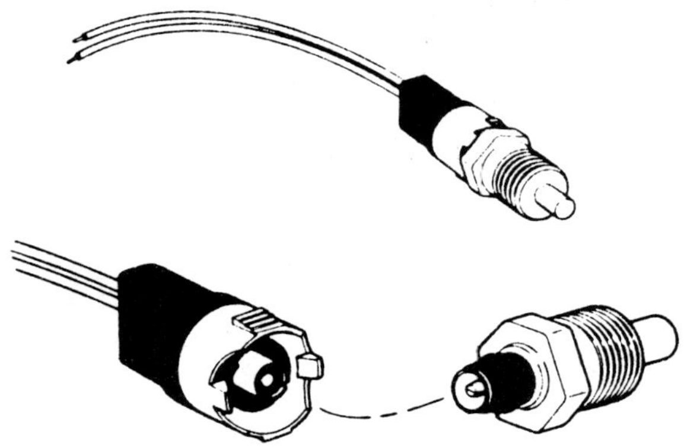

FIGURE 8-14 Coolant sensor (typical) *(Courtesy of General Motors Product Service Training).*

perature increases (Fig. 8-14). The range of resistance spans from about 100,000 ohms at −40°F to about 70 ohms at 266°F. The ECM provides a 5-volt source to the CTS and monitors the return voltage on another wire. The sensor resistance determines the return voltage. The ECM interprets the return voltage as engine coolant temperature.

Vehicle Speed Sensor

The vehicle speed sensor can be one of two designs. If the vehicle has a standard instrument cluster, the vehicle speed sensor will be mounted on the back of the speedometer head (Fig. 8-15). Reflective blades, driven by the speedometer cable, reflect light from a light-emitting diode (LED) to a photo-transistor. The photo-transistor produces an on-off voltage signal when the reflected light strikes it. This signal is sent to a buffer amplifier where it is conditioned for use by the ECM. The speed or frequency of the signals is interpreted to be vehicle speed. Vehicles equipped with digital display instrument clusters utilize a permanent-magnet generator in the transmission/transaxle (Fig. 8-16). The output shaft drives the generator, which produces a signal to the buffer amplifier. The buffer amplifier sends a conditioned signal to the ECM as an indication of vehicle speed. There is no speedometer cable or head in this sensor. The signal produced by the buffer amplifier also operates the digital speedometer display.

Engine Speed Signal

An engine speed signal is provided by the computer-controlled coil ignition system through its crankshaft sensor. This sensor and system will be discussed in detail later in this chapter.

Oxygen Sensor

Exhaust gas oxygen content is measured by an oxygen sensor (O_2) (Fig. 8-17). This sensor is mounted in the exhaust stream where it can react to the oxygen in the exhaust gases. The oxygen sensor is a zirconium oxide type that consists of a thimble-shaped layer of zirconia oxide coated on both sides with thin layers of platinum. The inside of the thimble is exposed to fresh air and the outside is exposed to the exhaust gases. As oxygen comes in contact with the zirconia oxide, a voltage is produced. The difference between the inner layer and outer layer voltages represents the difference between the oxygen content of the fresh air versus the exhaust gases. This voltage signal is sent to the ECM to enable it to make corresponding adjustments in the air/fuel ratio. The oxygen sensor output voltage may go as high as 900 millivolts for a rich exhaust and as low as 100 millivolts for a lean exhaust. This allows the ECM to compensate with the fuel injector pulse width.

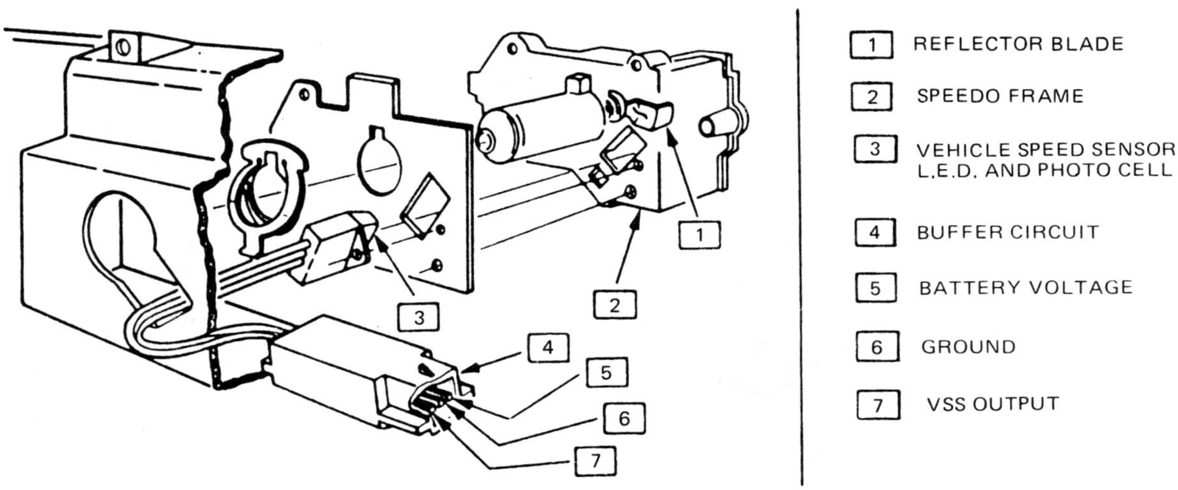

FIGURE 8-15 Vehicle speed sensor installation *(Courtesy of General Motors Product Service Training).*

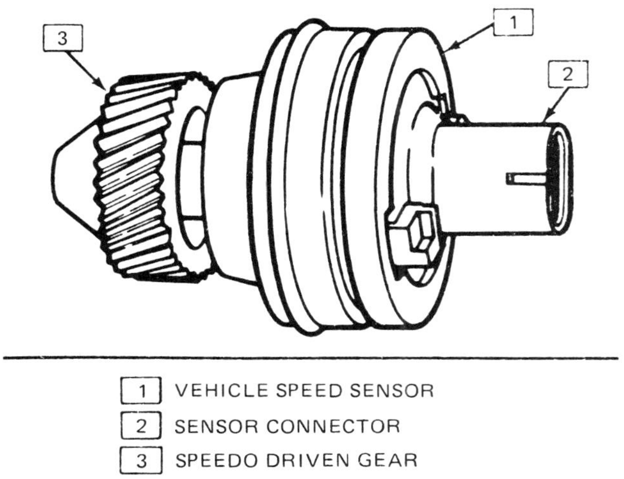

FIGURE 8-16 Vehicle speed sensor *(Courtesy of Oldsmobile Division).*

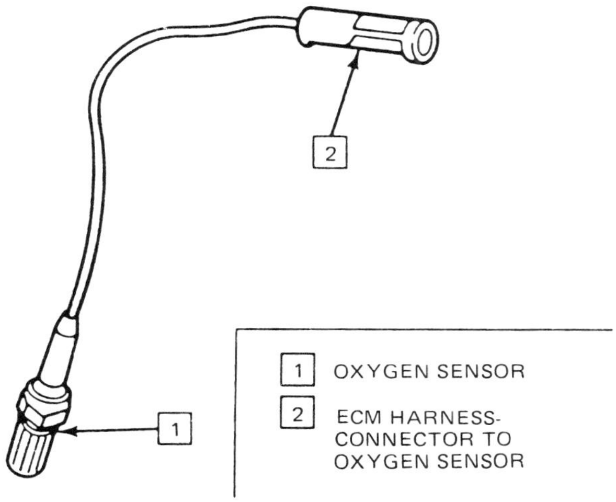

FIGURE 8-17 Exhaust oxygen (O_2) sensor *(Courtesy of Oldsmobile Division).*

Throttle Position Sensor

The throttle position sensor (TPS) is a variable-resistor potentiometer type that provides a voltage signal to the ECM as an indication of throttle position (Fig. 8-12). The ECM provides a 5-volt source to the TPS, which is grounded. A third wire is connected to a movable arm that moves with the throttle shaft. This arm contacts the TPS winding and senses varying voltage as it sweeps across the winding. The varying voltage signal is sent to the ECM. At closed throttle the voltage signal is about 400 millivolts, and at wide-open throttle the signal is about 5 volts. This signal is used by the ECM to help determine the air/fuel ratio. TPS adjustment and testing will be covered later in this chapter.

ELECTRONIC CONTROL MODULE

An electronic control module (ECM) is used to process many incoming data signals, make a decision, and then issue appropriate commands to controlled devices (Fig. 8-18). The ECM has a header-type connector to electrically connect its two circuit boards to the wiring harness. Its program and circuitry not only control the fuel injectors, but also can control the air-conditioning system, turbocharger wastegate, ignition system, emission controls, and torque converter clutch (Fig. 8-19).

The ECM has the ability to "learn," but the learning process must start over again everytime the battery is disconnected or the ignition switch is turned off for more than 10 seconds. While the learning process is taking place, a noticeable difference in performance may be experienced. To "teach" the computer, the engine must be at operating temperature and the vehicle should be driven with moderate acceleration, at part throttle, and at idle conditions until performance returns.

FIGURE 8-18 Electronic control module (SFI) *(Courtesy of General Motors Product Service Training).*

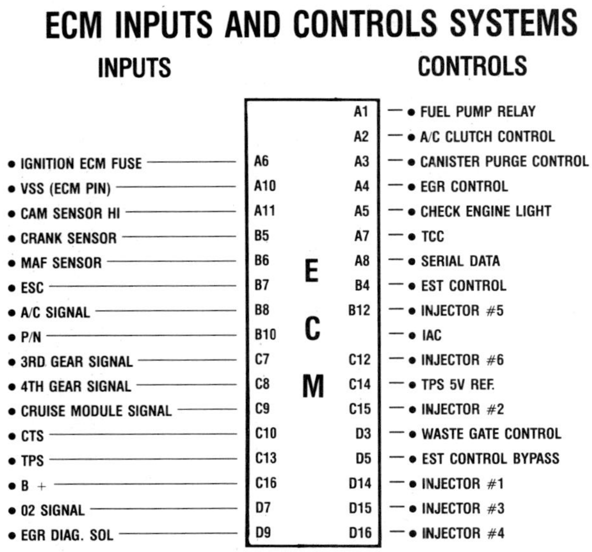

FIGURE 8-19 Electronic control module inputs and outputs *(Courtesy of General Motors Product Service Training).*

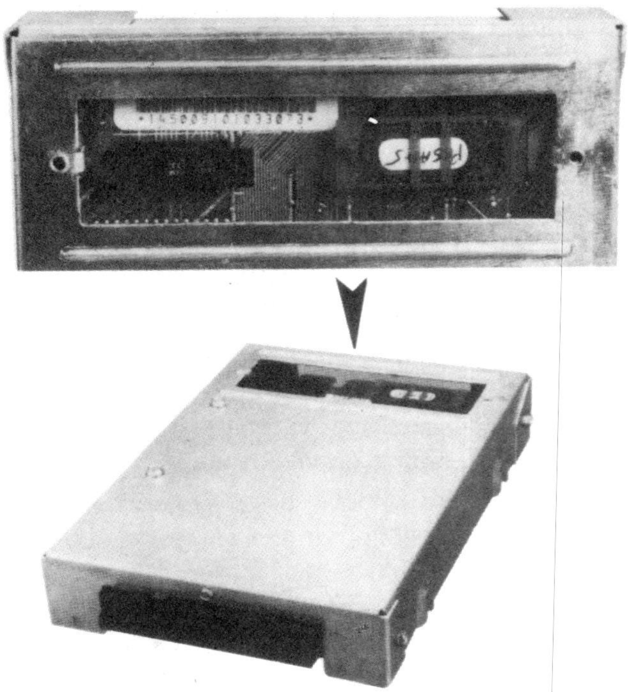

FIGURE 8-20 Electronic control module calibration package *(Courtesy of General Motors Product Service Training).*

MODES OF OPERATION

Calibration Package

There are eight modes of operation designed into the ECM. A calibration pack (CALPAK) is a resistor network resembling the serviceable PROM (Fig. 8-20). The CALPAK provides data for cold start cranking fuel and for a fuel backup circuit should a malfunction in the normal fuel control circuitry occur. The ECM operates in the fuel backup circuit mode if the ECM voltage drops below 9 volts; if cranking voltage drops below 9 volts, if the programmable read only memory (PROM) integrated circuit is malfunctioning or missing; or if the ECM circuitry fails to issue computer operating pulses. The fuel backup circuit is powered by the ignition switch and senses engine coolant temperature, throttle position, and engine rpm. It controls the fuel pump relay and the fuel injectors. The engine runs erratically while the computer is in this mode.

Starting Mode

The starting mode is entered when the ignition switch is turned on. The ECM energizes the fuel pump relay for 2 seconds, and the fuel pump builds fuel pressure to about 28 to 50 psi. The ECM, meanwhile, is receiving data from the coolant temperature sensor, throttle position sensor, and mass airflow sensor. It will determine an air/fuel ratio ranging from 1.5 to 1 at −30°F to 14.7 to 1 at 200°F.

Clear Flood Mode

The clear flood mode is used to start the engine if it should become flooded. If the accelerator pedal is depressed about three-quarters of the way to the floor, the TPS will signal the ECM, and if a cranking signal is received at the same time, the clear flood mode will be entered. The ECM pulses the fuel injectors to provide a 20 to 1 air/fuel ratio. This continues until the TPS is less than about 60% open or engine rpm is above 500. When the throttle is released, the ECM returns to the starting mode.

Run Mode

The run mode can operate under one of two conditions known as open loop and closed loop. Open-loop operation begins when the engine is first started. The ECM ignores the oxygen sensor signal and relies on the remaining engine sensor signals for fuel ratio delivery. When the oxygen sensor varying voltage ranges from about 350 to 550 millivolts, the coolant temperature sensor is above 95°F, and a programmed amount of time has elapsed since the engine was started, the ECM can enter the closed-loop run mode. If any of these conditions are not met, even after the engine has been running, the system will return to the open-loop mode of operation. In closed loop, the ECM uses the oxygen sensor signal to determine output commands.

Acceleration Mode

During the acceleration mode, the ECM receives signals from the TPS and MAF indicating an increasing fast rate of change. It responds by increasing the fuel injector pulse width and, therefore, the amount of fuel injected.

Deceleration Mode

The deceleration mode is entered when the ECM receives signals from the TPS and MAF that indicate a decreasing rate of change. It responds by decreasing the fuel injector pulse width. If the decreasing rate of change is rapid enough, the ECM can momentarily shut off the fuel injectors.

Battery Voltage Correction Mode

The ECM has the capability of compensating for low battery voltage. This is known as the battery voltage correction mode. The ECM responds to low battery voltage by increasing the fuel injector pulse width, the idle speed, and the ignition dwell period.

Fuel Cutoff Mode

The eighth mode of operation is the fuel cutoff mode. The injectors are shut off whenever the ignition switch is turned off. No fuel is injected and dieseling is prevented. If the ECM does not receive engine rpm signals, it will shut the injectors off. This prevents flooding and also prevents a hazardous situation if the vehicle is involved in an accident.

Input Data

In addition to the modes of operation, the ECM monitors signals from the vehicle speed sensor, camshaft sensor, crankshaft sensor, mass airflow sensor, exhaust gas recirculation diagnostic switch, oxygen sensor, throttle position sensor, coolant temperature sensor, electronic spark control module, air-conditioning system switch, park/neutral switch, cruise control module, and third- and fourth-gear switches (Fig. 8-19). In accordance with its program, the ECM uses this data input to issue output commands to the fuel pump relay, the six fuel injectors, the electronic spark timing module, the turbocharger wastegate control solenoid, the idle air control motor, the torque converter clutch, the CHECK ENGINE light, the exhaust gas recirculation solenoid, the canister purge control solenoid, and the air-conditioning compressor clutch.

Self-diagnostic Mode

A self-diagnostic capability is built into the ECM circuitry. A total of 19 trouble codes can be set and stored in the ECM memory to aid the service technician in diagnosing problems (Fig. 8-21). The self-diagnostic mode will be described in the diagnosis section of this chapter.

COMPUTER-CONTROLLED COIL IGNITION

The SFI, turbocharged, 3.8-liter, V-6 engine does not have an ignition distributor. The computer-controlled coil ignition system (C^3I) replaces it. This system includes a new coil pack, a camshaft sensor, a crankshaft sensor, and a C^3I module (Fig. 8-22). The coil pack consists of three ignition coils. When each coil discharges, it fires two spark plugs at the same time. The spark plugs are in cylinders that are opposite each other in the firing order. Cylinders

1984 MFI/SFI DIAGNOSTIC CODES

12 NO REFERENCE PULSES TO ECM
13 OXYGEN SENSOR CIRCUIT FAILED
14 COOLANT READING TOO HIGH
15 COOLANT READING TOO LOW
21 THROTTLE POSITION SENSOR READING TOO HIGH
22 THROTTLE POSITION SENSOR READING TOO LOW
24 VEHICLE SPEED SENSOR FAILED
31**WASTEGATE ELECTRICAL SIGNAL OPEN OR GROUNDED
32 *EGR ELECTRICAL OR VACUUM CIRCUIT MALFUNCTION
33**MASS FLOW SENSOR READING TOO HIGH
34**MASS FLOW SENSOR READING TOO LOW OR NO SIGNAL FROM SENSOR
41**CAM SENSOR FAILED
42 *ERROR IN DISTRIBUTOR OR C^3 SYSTEM
43 ELECTRONIC SPARK CONTROL FAILURE
44 OXYGEN SENSOR LEAN TOO LONG
45 OXYGEN SENSOR RICH TOO LONG
51 CALIBRATION PROM ERROR
52 *CALPAK MISSING
55 INTERNAL ECM ERROR (A/D CONVERTER)

*NEW CODES
**S.F.I. CODES ONLY

FIGURE 8-21 Sequential fuel injection codes *(Courtesy of General Motors Product Service Training).*

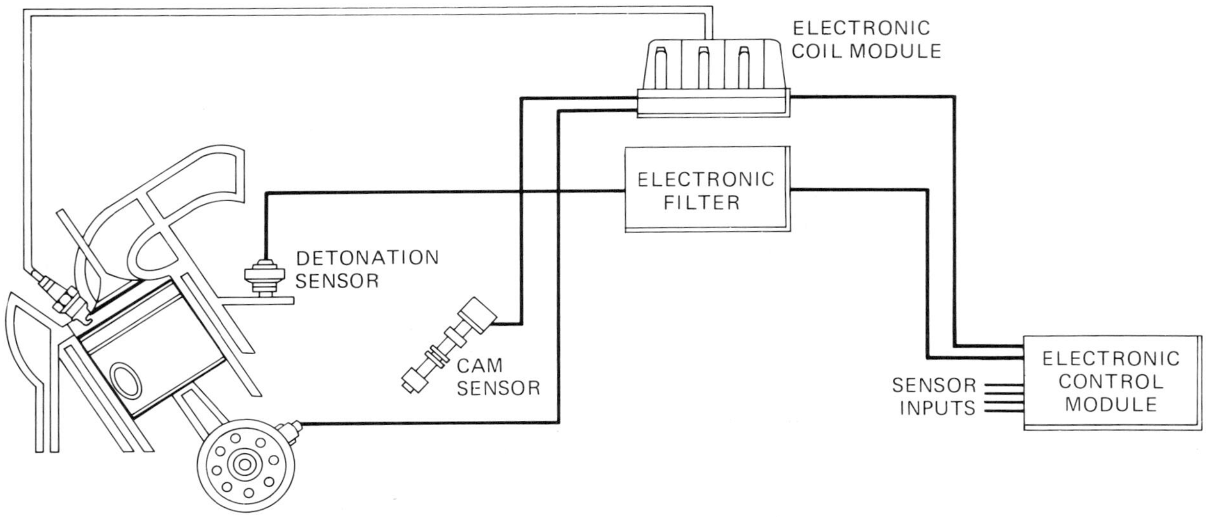

FIGURE 8-22 Computer-controlled coil ignition overview *(Courtesy of General Motors Product Service Training).*

1 and 4 are fired by one coil, cylinders 2 and 5 are fired by a second coil, and cylinders 3 and 6 are fired by the third coil. For example, when cylinder 1 is in its compression stroke and ready to be fired, cylinder 4 will be on its exhaust stroke. The spark at 4 spark plug will be wasted. The spark plug that fires on the exhaust stroke requires only about 2 to 3 kilovolts to fire; therefore, the majority of the coil output voltage is available to fire the spark plug on the compression stroke.

The 10-ampere ECM fuse supplies ignition voltage to the C[3]I module. The module is connected to the ECM by a wiring harness. The ignition coils are powered by a separate 7.5-ampere fuse in the fuse panel. A 0.7-ohm resistor wire is located in this power feed wire to limit coil current when surrounding ambient temperatures are high. The module also provides a terminal for tachometer operation.

The C[3]I module supplies voltage to both the crankshaft and camshaft sensors (Fig. 8-23). Both sensors are grounded through the module. A protection circuit inside the module shuts the current off to the coil driver circuits if the engine is not started within 1 to 2 seconds after the ignition switch is turned on. This prevents circuit overheating and deterioration. The system will be reactivated when it receives a cranking signal.

Crankshaft and Camshaft Sensors

The crankshaft and camshaft sensors are Hall-effect switches (Fig. 8-24). A plate with three vanes is attached to the crankshaft harmonic balancer. As the vanes pass through a slot in the crankshaft sensor, the Hall-effect switch is triggered off and on. This digital signal represents engine rpm and crankshaft position and is sent to the C[3]I module, which sends it to the ECM. At about 400 rpm, the ECM switches control of the ignition timing from the ignition module to its own electronic spark timing circuitry. Base ignition timing is set by the crankshaft sensor and is not adjustable. The engine will not run without the crankshaft sensor.

The camshaft sensor is also a Hall-effect switch that fits into the traditional distributor mounting hole (Fig. 8-25). A gear on the camshaft drives it. The camshaft sensor signal determines the firing sequence. If this signal is

CAM/CRANK SENSOR

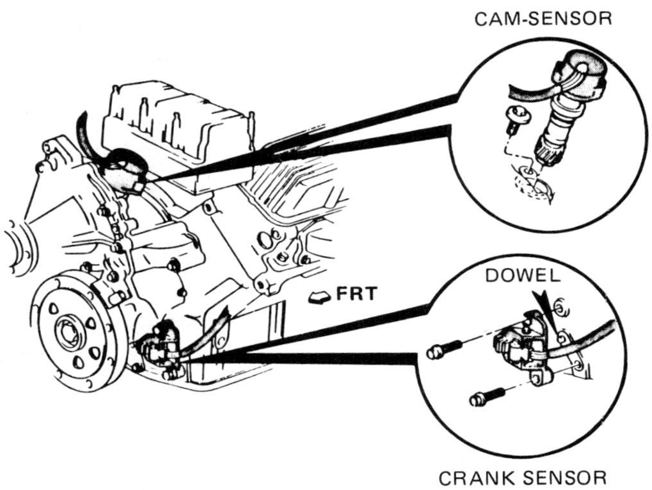

FIGURE 8-23 Cam and crank sensors *(Courtesy of General Motors Product Service Training).*

CRANKSHAFT SENSOR

FIGURE 8-24 Crank sensor *(Courtesy of General Motors Product Service Training).*

FIGURE 8-25 Cam sensor *(Courtesy of General Motors Product Service Training).*

not received by the C³I module, all injectors will be pulsed simultaneously. The camshaft sensor is adjustable.

TURBOCHARGER CONTROL

A turbocharger consists of an engine exhaust-driven turbine mounted on a common shaft with a compressor (Fig. 8-9). The engine exhaust drives the turbine and shaft. When the shaft speed is great enough, the pressure produced by the compressor begins to exceed that required by the engine. The compressor begins to pressurize the intake manifold. This is called *boost pressure.* During boost conditions, more air enters the cylinder on its intake stroke because it enters under pressure. The turbocharger can, however, produce too much pressure. This may result in engine damage. A control, called a *wastegate,* is used to limit turbocharger boost by allowing exhaust gases to bypass the turbine. This limits turbine, shaft, and compressor speed and also boost pressure.

Wastegate Control

The SFI system utilizes a wastegate solenoid to limit turbocharger boost (Fig. 8-26). The solenoid is controlled by the ECM. The wastegate actuator, a vacuum diaphragm, is operated by intake manifold pressure. Pressure is applied to the actuator diaphragm, which moves a valve in the exhaust stream to direct exhaust gas flow through the turbine. The pressure applied to the diaphragm can be vented to atmosphere through the wastegate solenoid. When this happens, the actuator moves the valve to a position that

FIGURE 8-26 Wastegate overboost *(Courtesy of General Motors Product Service Training).*

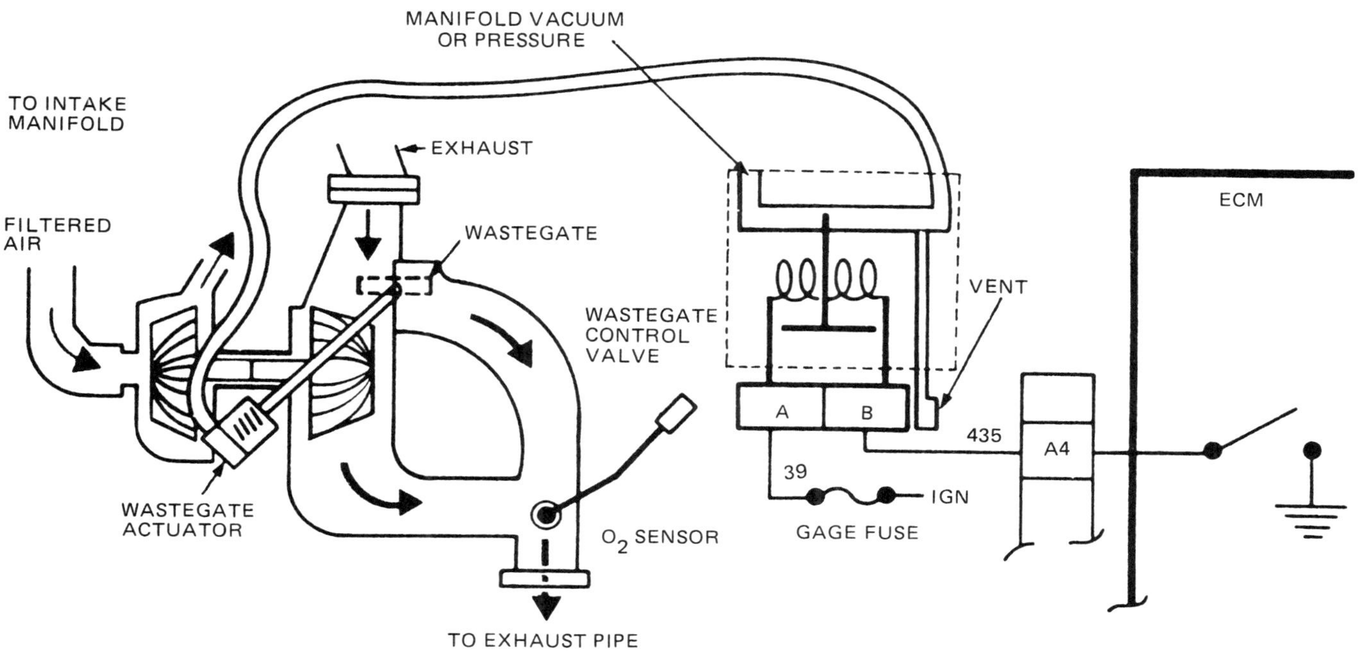

allows the exhaust gases to bypass the turbine. When the ECM energizes the wastegate solenoid, boost is limited.

ELECTRONIC SPARK CONTROL

An electronic module and a knock sensor are the only components of the electronic spark control (ESC) system. During boost conditions, the engine's tendency to knock is increased. This is because of the varying octane levels of fuels available today in combination with the higher compression under boost conditions. The ESC system controls damaging knock by retarding the ignition timing a maximum of about 20 degrees.

Knock Sensor

The knock sensor is a piezoelectric device that senses vibrations and produces a voltage signal. Sometimes referred to as a detonation sensor, the knock sensor is located in the engine block at the rear of the intake manifold (Fig. 8-27). When detonation occurs, vibrations are produced and the sensor provides a voltage signal to the ESC module. The ESC module turns off an 8- to 10-volt signal to the ECM, which then begins to retard the timing until the detonation stops. The ECM then advances the timing again unless detonation is sensed. The ECM retards the timing if the engine rpm is over 1100, supply voltage to the ECM is more than 9 volts, and the ESC system is operational.

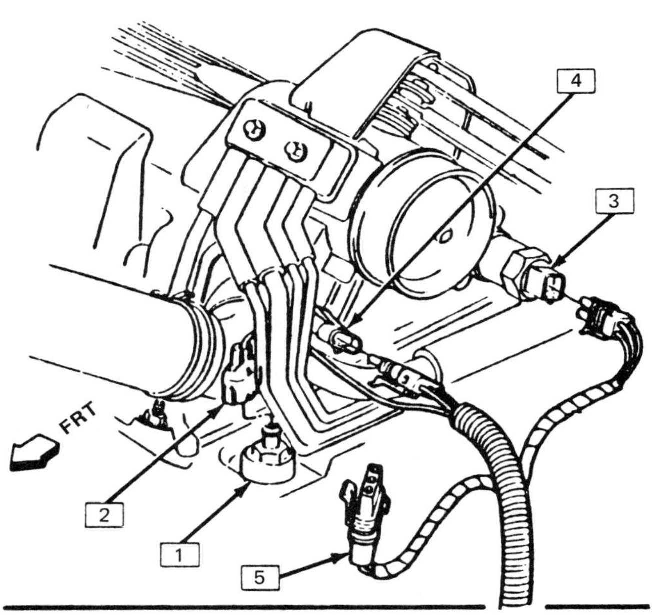

1 KNOCK SENSOR
2 KNOCK SENSOR CONNECTOR
3 IAC MOTOR
4 T.P.S.
5 MASS AIR FLOW SENSOR CONNECTOR

FIGURE 8-27 Knock sensor *(Courtesy of Oldsmobile Division).*

SYSTEM SERVICE

Injector Balance Test

It is important to have each fuel injector provide the same amount of fuel. If a driveability symptom appears that indicates unequal injection, a fuel injector balance test can be performed. The procedure involved is similar to a cylinder power balance test done with an engine analyzer. A special injector tester, a fuel injection pressure gauge, and the necessary adapters are required to perform this test (Fig. 8-28).

After connecting a fuel injection pressure gauge to the Schrader fitting on the fuel rail, turn the ignition switch on and off two or three times to cycle the fuel pump and build up fuel pressure. Disconnect all fuel injectors from the wiring harness. The fuel injector tester is a push-button timer that energizes one injector for about 500 milliseconds. When an injector is energized, a pressure drop will be noticed on the pressure gauge. This drop should be recorded. The same procedure should be used for the remaining injectors. Fuel pressure should be restored before each injector is tested. The ECM needs to power down each time also or the fuel pump may not cycle. It takes about 10 seconds for the ECM to power down. When all injectors have been tested, the pressure drop readings should be compared. Any injector that has a 1.5-psi larger drop or a 1.5-psi smaller drop should be replaced. A difference of 1.5 psi means that the injector is injecting too much fuel (larger drop), or too little fuel (smaller drop). During the test, the injectors should not be pulsed more than three times each as engine flooding could result.

Idle Speed Adjustment

Idle speed adjustment should not be necessary unless the throttle body is being replaced or has been tampered with by an inexperienced person. The engine should be at operating temperature and the wheels should be blocked as the adjustment is done with the transmission in drive. When this adjustment is completed, the throttle position sensor adjustment should be checked. Detailed steps for adjusting idle speed are in the manufacturer shop manual.

Throttle Position Sensor Adjustment

The TPS should be adjusted whenever idle speed is adjusted or whenever it is replaced or removed for any reason. The TPS adjustment requires special jumper leads and

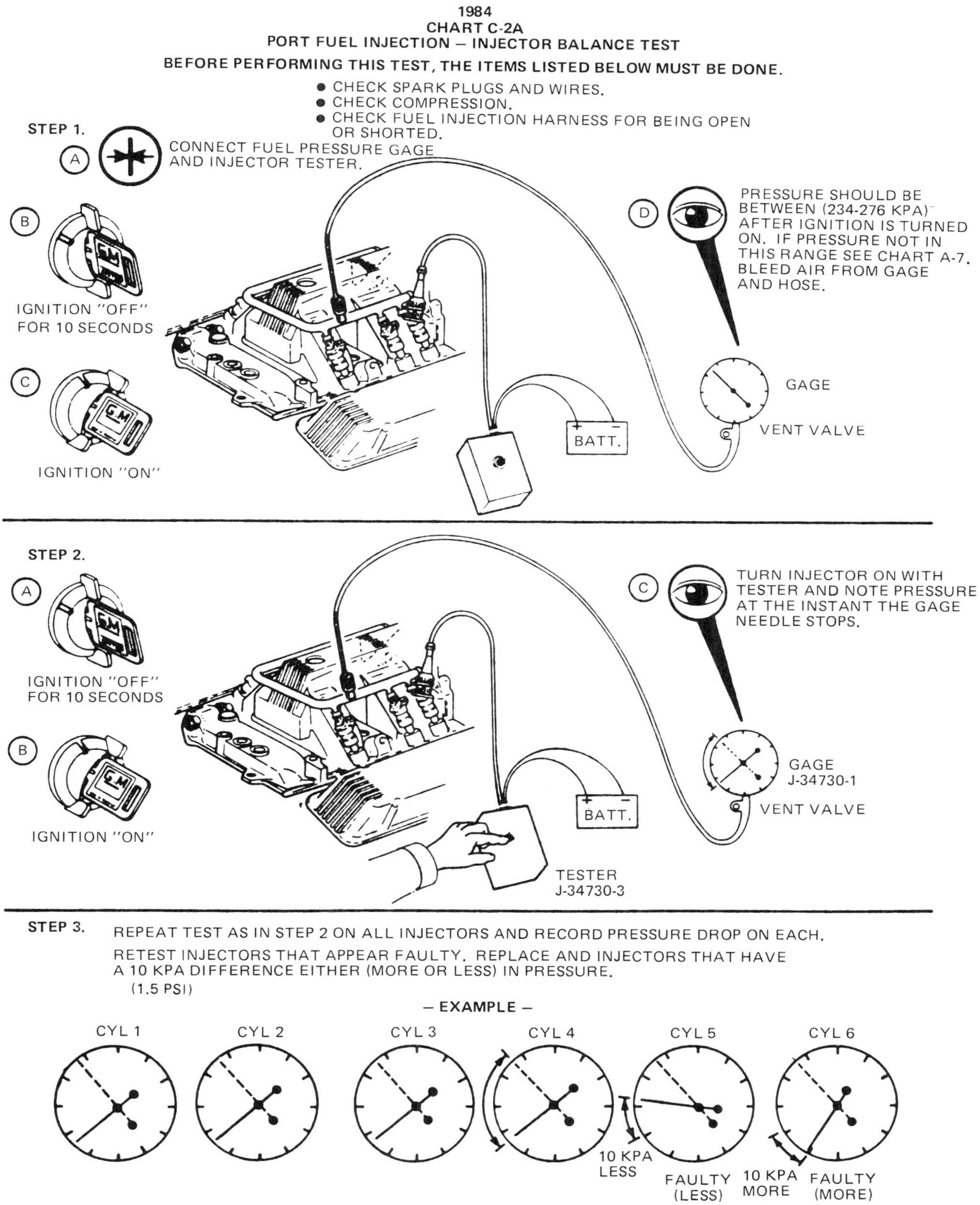

FIGURE 8-28 Injector test *(Courtesy of General Motors Product Service Training).*

a 10-megohm impedance digital multimeter. Adjust the sensor as detailed in the service manual. The calibration reading on the digital multimeter is very important as all systems do not have the same calibrations.

Camshaft Sensor Adjustment

The camshaft sensor must be reset anytime it has been removed from the engine. A WeatherPack terminal removal tool and a digital voltmeter are necessary to adjust the sensor. It is adjusted by turning it in a manner similar to how the distributor used to be turned to adjust ignition timing. Follow the procedure given in the shop manual for correct adjustment.

Crankshaft Sensor Gap Adjustment

The crankshaft sensor is fastened into a holding bracket. The holding bracket clamps onto the sensor by means of a screw that, when tightened, squeezes the bracket onto the sensor. When loosened, the sensor can slide within the bracket and can be moved closer to or farther from the harmonic balancer on the front of the crankshaft. Refer to the shop manual for adjustment specifications. This adjustment does not affect ignition timing as timing is controlled by the ECM. It does, however, prevent damage to the sensor and ensures an accurate crankshaft position signal.

Fuel Filter Service

The in-line fuel filter is mounted on the engine and uses an O-ring seal in the screw-type line fitting. New seals should be used anytime the fittings are disconnected. Fuel line hoses should be inspected anytime the filter is serviced.

SYSTEM DIAGNOSIS

The SFI system ECM has the capability of monitoring its own circuits and, should a malfunction occur, the ECM will illuminate a telltale lamp on the instrument panel. At the same time, a code number is stored or *set* in the ECM memory. This code can be used by the service technician to help diagnose the problem. Nineteen codes can be stored by the ECM.

Codes

There are generally two kinds of problems that can occur in the circuitry. A temporary or intermittent problem is one that happens and then corrects itself or "goes away." A loose connector is an example of an intermittent problem. In this case, the ECM illuminates the SERVICE ENGINE SOON telltale light on the instrument panel when the problem occurs. The light goes out when the problem disappears. A code, however, will remain in the ECM memory so that the technician can determine which circuit had a temporary problem. A problem that occurs and remains causes the ECM to illuminate the SERVICE ENGINE SOON telltale, but this time it remains lit. This is known as a permanent or hard problem. A shorted sensor is an example of a hard problem. A code is stored in the ECM memory so that the technician can retrieve it to help in the diagnostic procedure.

Self-diagnostic Mode

The ECM enters its self-diagnostic mode when a jumper is placed between the correct two terminals in the assembly line communications link (ALCL) connector under the instrument panel (Fig. 8-29). The ECM causes the SERVICE ENGINE SOON telltale to flash once, pause, and then flash twice when the ignition key is turned on. This will be repeated three times and is a code 12. Code 12 indicates that the self-diagnostic mode is functional and capable of displaying any additional codes. Additional codes follow, each being repeated three times, in ascending order. If there are no additional codes, or when the last code has been shown, the process will begin again with

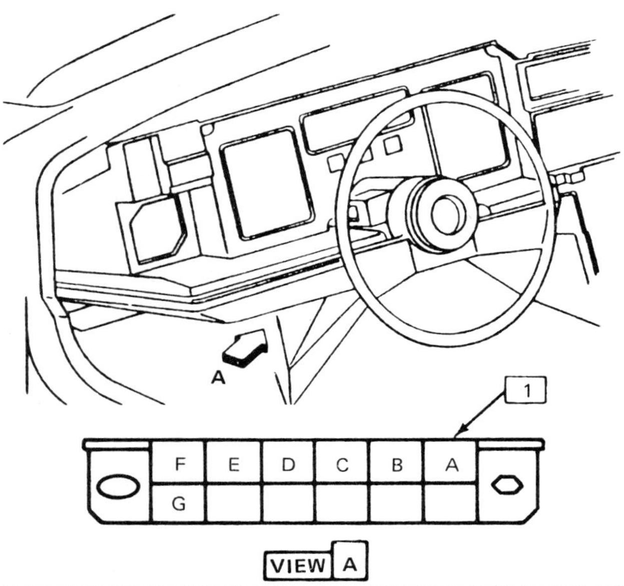

FIGURE 8-29 Assembly line communication link connector *(Courtesy of Oldsmobile Division).*

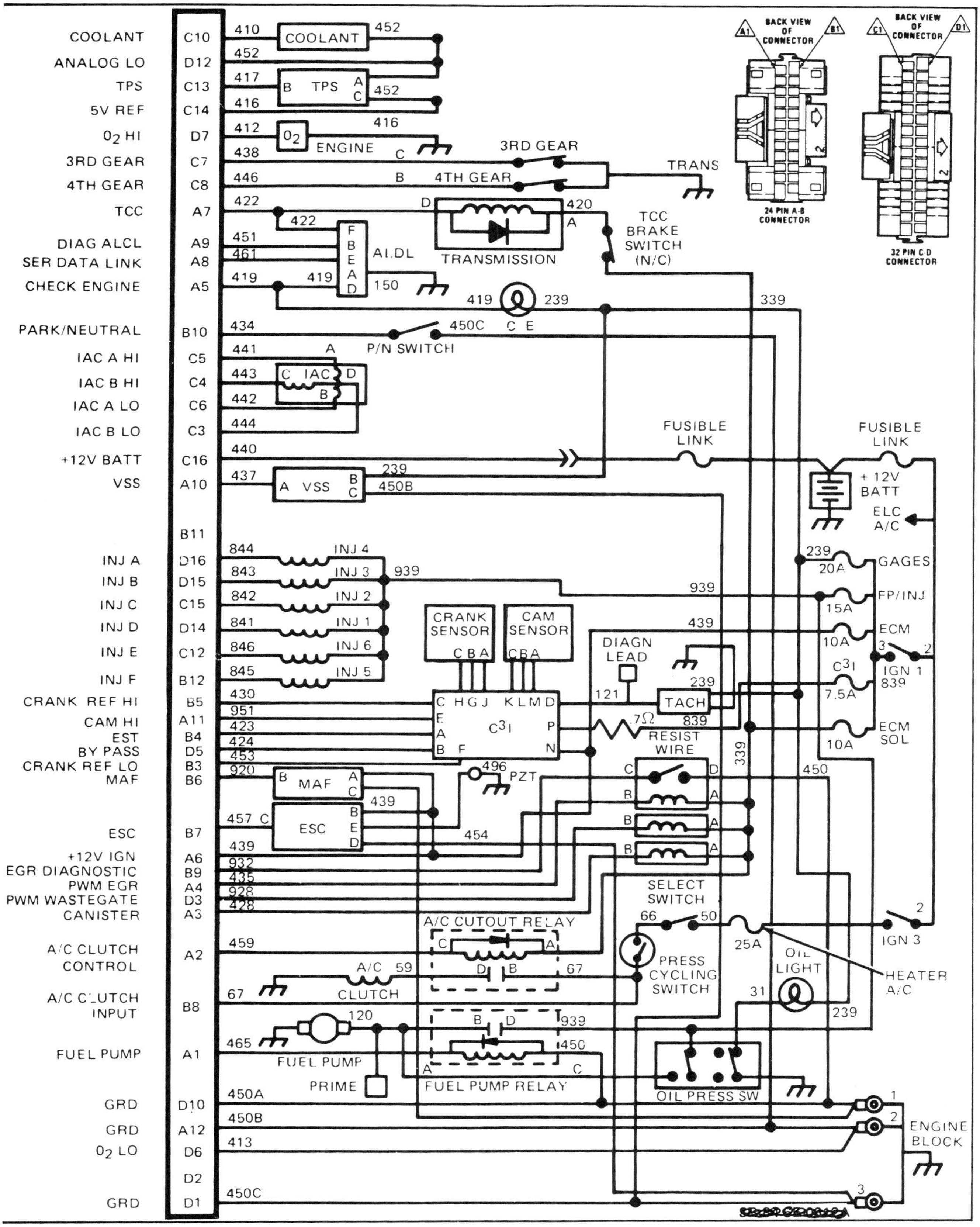

FIGURE 8-30 Electronic control module terminals *(Courtesy of General Motors Product Service Training).*

code 12. This cycle continues until the ignition switch is turned off or the jumper lead is removed.

The service manual contains decision trees or diagnostic charts for each code. Also included are charts to aid in diagnosing problems when no telltale light is illuminated and no codes are displayed. A diagram of the ECM terminals is shown in Fig. 8-30.

Field Service Mode

When a driveability problem exists but no codes are stored in memory, the ECM field service mode can be entered. With the engine running and a jumper lead connected between the correct terminals on the ALCL connector, the ECM responds to the oxygen sensor voltage. The SERVICE ENGINE SOON telltale then indicates open-loop operation, closed-loop operation, lean exhaust, or rich exhaust. Open-loop operation is indicated when the telltale flashes about 2.5 times per second. Closed loop is indicated when the telltale flashes about 1 time per second. A lean exhaust is indicated when the telltale stays off. A rich exhaust is indicated when the telltale stays on.

When road testing the vehicle while in the field service mode, the telltale may remain on during acceleration. This indicates a rich exhaust and is a normal condition. During deceleration, the telltale may remain off indicating a lean exhaust. This is also a normal condition. At idle, if the engine speed is below 1200 rpm, the light may remain on for a long period of time. This is normal due to low speed.

While the sequential fuel injection system is different from most other systems, it is still not the answer for fuel delivery on all engines.

ABBREVIATIONS

A/F Air/fuel (A/F ratio)

AIR Air injection reaction

ALCL Assembly line communication link

BARO Barometric pressure

BATT Battery positive terminal

CCP Controlled canister purge

C.E. Check engine

CO Carbon monoxide

Conv. Converter (catalytic)

Diagnostic "test" term. Lead of ALCL connector that is grounded to get a trouble code

DVM (10 meg.) Digital voltmeter with 10-million ohms resistance

EAC Electric air control

EAS Electric air switching

ECM Electronic control module (controller)

EECS Evaporative emissions control system

EFE Early-fuel evaporation

EFI Electronic fuel injection

EGR Exhaust gas recirculation

ESC Electronic spark control

EST Electronic spark timing

FED Federal

HC Hydrocarbons

HEI High-energy ignition

IAC Idle air control

IGN Ignition

I.P. Instrument panel

MAF Mass airflow

MAP Manifold absolute pressure

MAT Manifold air temperature

MFI Multiport fuel injection

mph Miles per hour

N.C. Normally closed

N.O. Normally open

NO_x Oxides of nitrogen

O_2 Oxygen (sensor)

PCV Positive crankcase ventilation

PFI Port fuel injection

P/N Park/Neutral

Port Exhaust port

PROM Programmable read-only-memory (calibrator)

rpm Revolutions per minute

SFI Sequential fuel injection (used on 3.8-liter turbo; each injector is individually fired)

TACH Tachometer

TBI Throttle body injection (unit)

TCC Transmission converter clutch

THERMAC Thermostatic air cleaner

TPS Throttle position sensor

TVS Thermal vacuum switch

V Volt

Vac. Vacuum

VIN Vehicle identification number

VSS Vehicle speed sensor

WOT Wide-open throttle

UNIT I
General Motors Corporation Systems

9
Chevrolet and Pontiac Tuned Port Fuel Injection, 1985

OVERVIEW

The tuned port fuel injection system (TPI) replaced the crossfire fuel injection system described in an earlier chapter. The name "tuned port fuel injection" describes the tubular runners connecting the cylindrical air plenum to the intake ports of the cylinder heads (Fig. 9-1). Each runner is shaped, by volume and length, the same. This matching is called *tuning*. The runners are also very smooth on the inside and are curved very gradually to provide a minimum of resistance to airflow. Fuel is injected into the intake ports where each runner connects to the cylinder heads (Fig. 9-2). Each intake port has its own fuel injector. Air mass rather than volume is used to calculate desired air/fuel ratios for engine operation. The tuned port fuel injection system consists of subsystems that include fuel delivery, air induction, ignition management, and emission control. These systems are interrelated, but they will be discussed separately for better understanding.

FUEL DELIVERY

The fuel delivery system includes the fuel tank and cap, in-tank electric fuel pump, in-line fuel filter, fuel filter, fuel supply line, fuel rail, fuel pressure regulator, fuel return line, fuel injectors, and a cold start fuel injector.

Fuel Tank

The fuel tank is a standard stamped steel tank except that it has a sump area where the in-tank electric fuel pump is located. The sump area is surrounded by baffles to prevent fuel from running away from the pump during cornering or severe maneuvering conditions. The fuel tank cap is a standard vacuum-pressure relief emission control cap.

In-tank Electric Fuel Pump

The in-tank electric fuel pump supplies fuel to the fuel rail and injectors and then to the pressure regulator (Fig. 9-3). The pump is a positive-displacement roller vane type. It is mounted on the fuel gauge sending unit but is serviceable as a separate part. A pulsator mounted on the pump/sending unit assembly reduces mechanical and hydraulic vibrations. This reduces noise that would otherwise echo within the fuel tank. The pump runs at about 3500 rpm and is capable of providing a maximum of 60 to 90 psi. The pump pressure far exceeds the maximum demand of the injectors. Excess pressure is returned to the fuel tank by the pressure regulator. The constant cycling of fuel keeps it cool and reduces the tendency to form vapor bubbles that cause vapor lock. The inlet of the fuel pump is filtered by a woven plastic filter capable of removing contaminants as small as

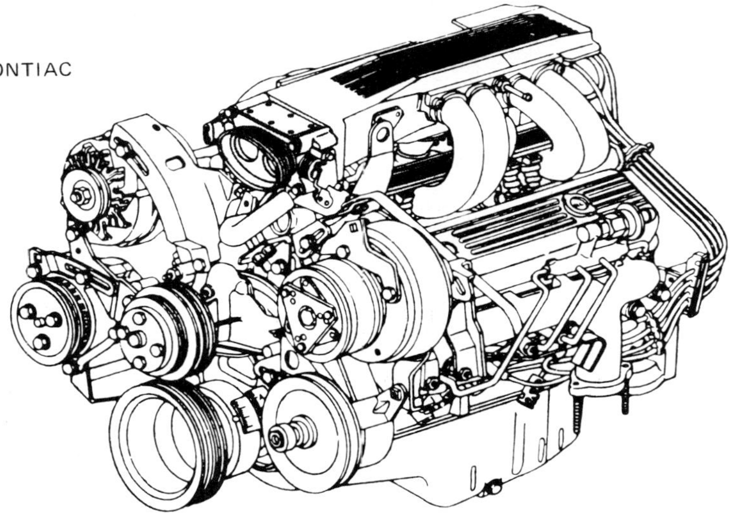

FIGURE 9-1 The 5.0 and 5.7-liter 90-degree V-8 tuned port injection engine *(Courtesy of General Motors Product Service Training).*

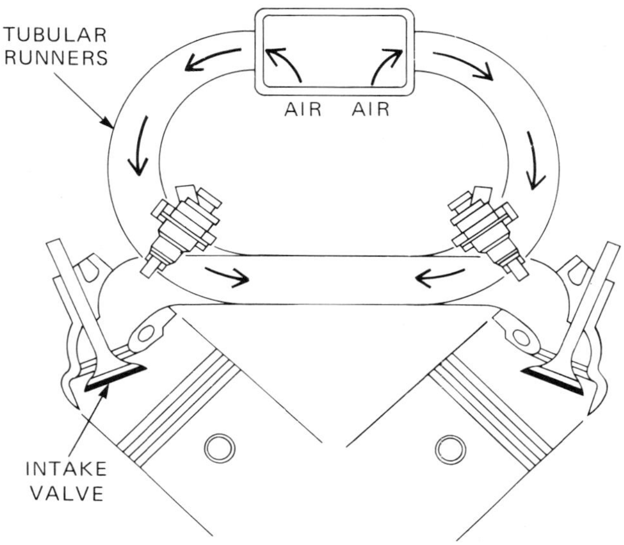

FIGURE 9-2 The 5.0- and 5.7-liter tuned port injection system *(Courtesy of Chevrolet Motor Division).*

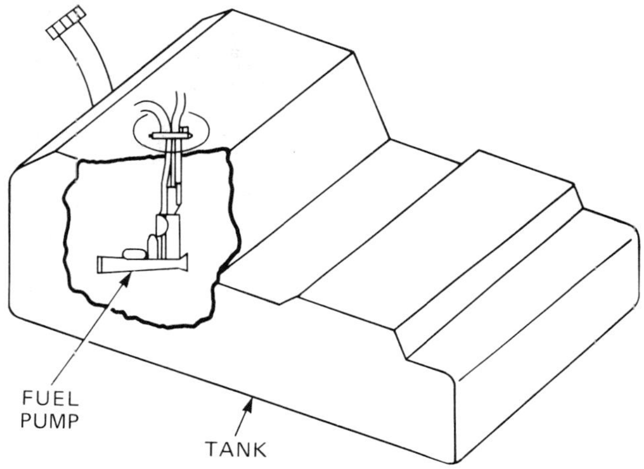

FIGURE 9-3 Port fuel injection fuel pump *(Courtesy of Chevrolet Motor Division).*

50 microns or 0.002 inch. The filter is serviceable as a separate part.

Fuel Filter

An in-line fuel filter is located on the frame rail under the floor pan (Fig. 9-4). It has screw connections where the fuel lines fasten to it. These fittings are sealed to the fuel lines with O-ring seals made of a fluoroelastomer rubber (Fig. 9-5). Other rubber materials should not be used as leaks and potential fire hazard may result. The filter is a 10- to 20-micron element and is serviced by replacement only.

Fuel Supply Line

The fuel supply line connects the in-tank electric fuel pump to the in-line fuel filter and the filter to the fuel rail. This line may be made of special fuel line rubber hose, steel with inverted flared ends, or flexible rubber hose covered with a woven braided stainless steel mesh. The special synthetic rubber hose is a fluoroelastomer that not only withstands the higher pressure of fuel injection systems, but also is more resistant to deterioration caused by contaminants and additives in the fuel. A rubber hose connects the fuel pump to the steel line leading to the fuel filter. A steel

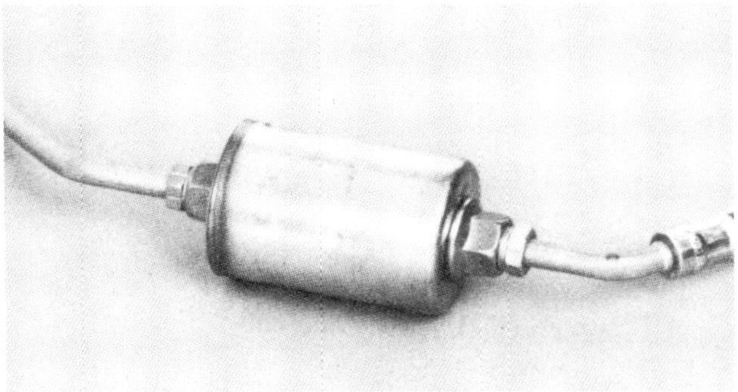

FIGURE 9-4 In-line fuel filter *(Courtesy of Chevrolet Motor Division).*

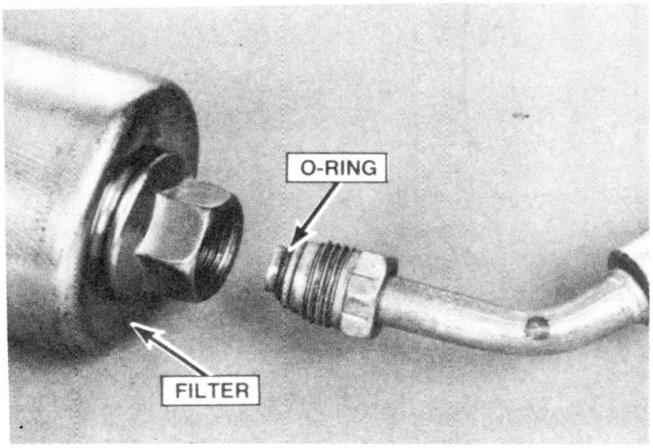

FIGURE 9-5 In-line fuel filter *(Courtesy of Chevrolet Motor Division).*

line connects the fuel filter to a flexible rubber hose. This flexible hose is in the engine compartment. The other end of this hose connects to the fuel rail.

Fuel Rail

The fuel rail is fitted with the fuel injectors and supplies them with pressurized fuel from the pump (Figs 9-6 and 9-7). The injectors are fastened to the rail with clips. To remove an injector, when looking from the bottom of the injector toward its top, the clip should be rotated clockwise (Fig. 9-8). The injector is sealed to the rail with an O-ring

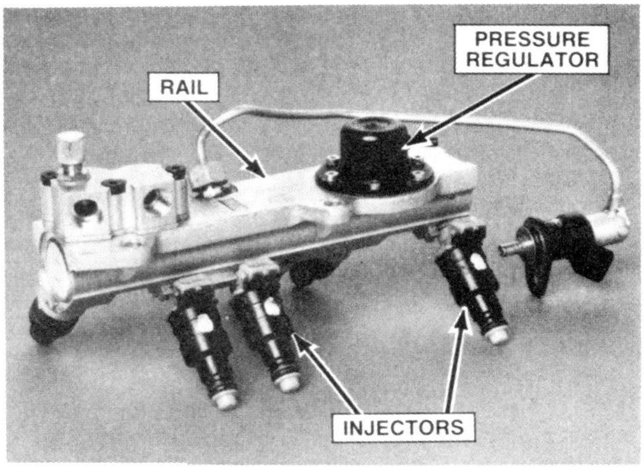

FIGURE 9-6 Fuel rail and injectors, 2.8-liter V-6 *(Courtesy of Chevrolet Motor Division).*

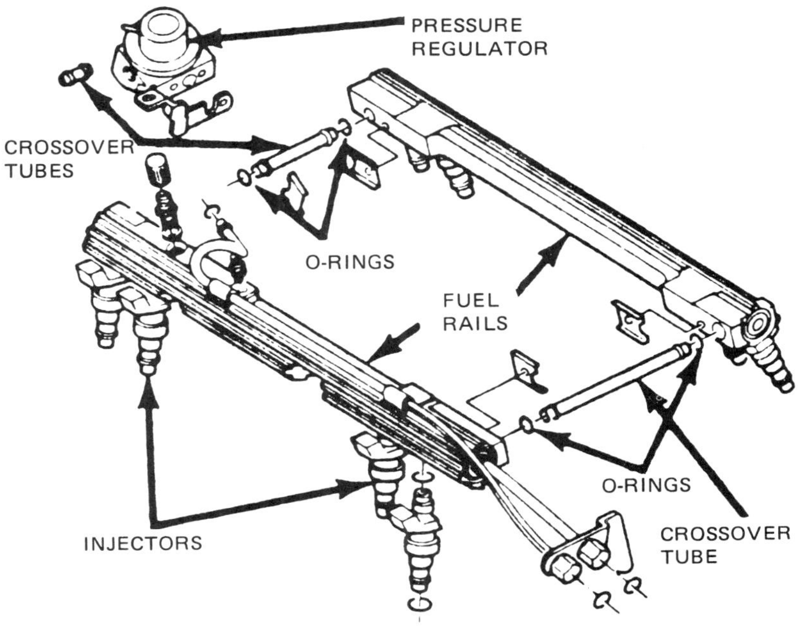

FIGURE 9-7 Fuel rail assembly, 5.0- and 5.7-liter V-8s *(Courtesy of Chevrolet Motor Division).*

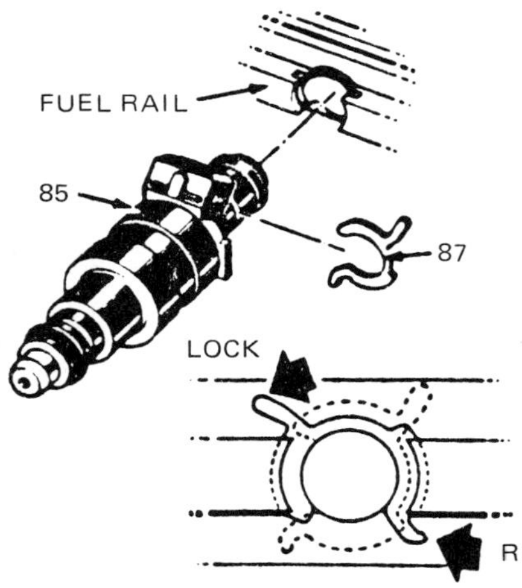

85 INJECTOR - PORT
87 CLIP-INJECTOR RETAINER

FIGURE 9-8 Port injector with injector retainer clip *(Courtesy of Chevrolet Motor Division).*

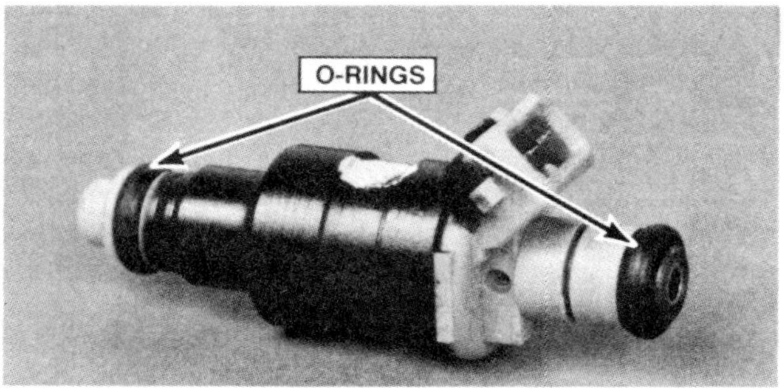

FIGURE 9-9 Injector O-rings *(Courtesy of Chevrolet Motor Division).*

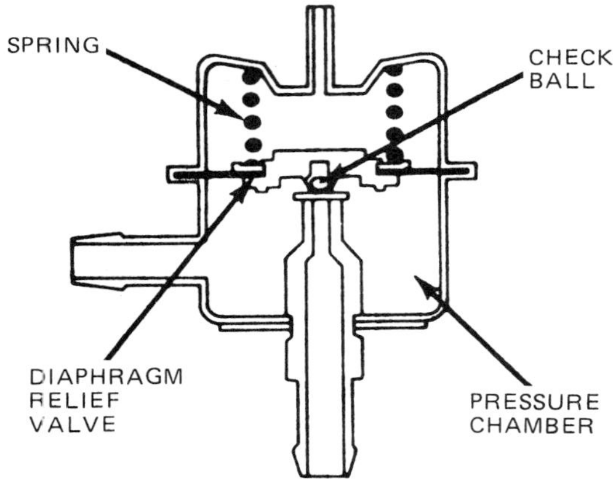

FIGURE 9-10 Fuel pressure regulator operation *(Courtesy of Chevrolet Motor Division).*

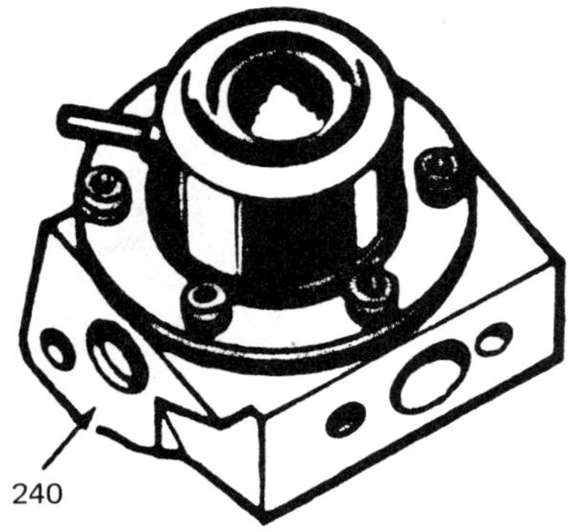

240 PRESSURE REGULATOR AND BASE ASSEMBLY

FIGURE 9-11 Pressure regulator and base assembly *(Courtesy of Chevrolet Motor Division).*

seal. This prevents fuel leaks (Fig. 9-9). The fuel rail is cast aluminum and has an identification number that must be known for correct replacement. The front end of the rail is where the fuel supply and return lines are connected. The rear end of the rail has a Schrader valve fitting for checking fuel pressure and also a fuel pressure regulator.

Fuel Pressure Regulator

The function of the fuel pressure regulator is to maintain about a 35-psi drop across the fuel injectors (Figs. 9-10 and 9-11). The regulator contains a diaphragm that moves a valve in response to spring tension and intake manifold pressure changes. Fuel pressure acts on the diaphragm on the opposite side. For fuel pressure to move the diaphragm, it must overcome the calibrated spring tension. When intake manifold pressure increases, it is harder for the fuel to unseat the regulator valve. When manifold pressure decreases, it is easier for the fuel pressure to unseat the regulator valve. When the regulator valve is seated, fuel pressure rises in the rail. If the valve unseats, fuel rail pressure is lower. The calibrated spring maintains a specified fuel pressure and the intake manifold pressure varies fuel rail pressure in response to engine load. The fuel pressure regulator is serviced as a separate part.

When the pressure regulator valve unseats, fuel in the rail is allowed to escape into the fuel return line. This excess fuel is returned to the fuel tank, where it empties into the sump area. This assures a good supply of fuel to the pump during low-fuel-level conditions. The fuel return line materials are the same as those for the supply line.

Fuel Injectors

Eight individual fuel injectors are mounted between the fuel rail and the cylinder intake ports (Figs. 9-6 and 9-7). The injectors are sealed to the intake manifold by an O-ring seal. Both the upper seal in the rail, discussed earlier, and the lower seal in the intake manifold should be replaced whenever the injector is removed. The lower seal prevents air or vacuum leaks. Both seals also insulate the injector from heat and vibration. Each injector has an identification number that must be known for correct replacement. In prior years, the plastic tops of the injectors were color coded to indicate their flow rates. This is no longer true, so the identification number is an important factor during replacement.

The injector is a solenoid-operated pintle valve that when energized, causes the pintle valve to rise off its seat (Fig. 9-12). This allows fuel to be sprayed into the intake port (Fig. 9-13). A spring above the pintle valve returns it to its seat when the solenoid is de-energized. Each injector has a wiring harness connector with two terminals. Some models use a retention clip that must be released before removing the connector. One terminal supplies voltage from a fuse in the fuse panel, while the other terminal connects the injector to the electronic control module, which provides the ground for circuit operation.

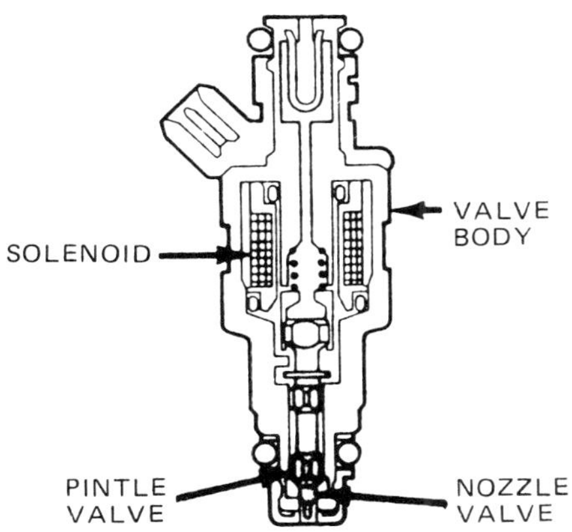

FIGURE 9-12 Typical fuel injector *(Courtesy of Chevrolet Motor Division).*

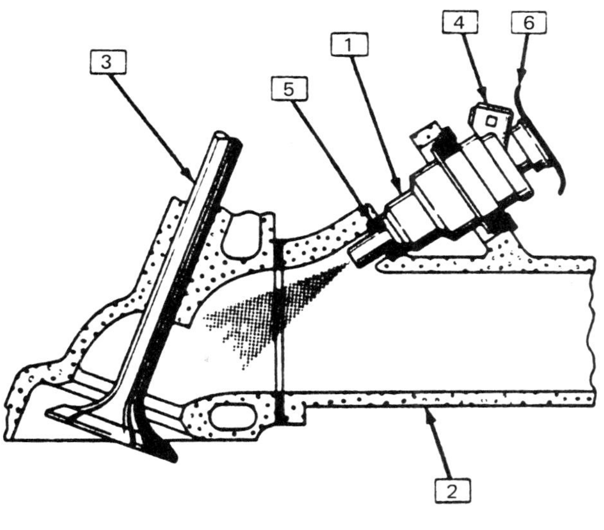

1 FUEL INJECTOR
2 INTAKE MANIFOLD
3 INTAKE VALVE
4 ELECTRICAL TERMINAL
5 "O" RING
6 FUEL RAIL

FIGURE 9-13 Fuel injector *(Courtesy of Chevrolet Motor Division).*

Cold Start Fuel Injector

A cold start fuel injector is located at the left side of the intake manifold (Fig. 9-14). A fuel line from the fuel rail is connected to the cold start fuel injector by means of a banjo-type connection (Fig. 9-15). A 12-volt supply is provided by the ignition switch when it is in the START position. This voltage is supplied to a thermo-time switch screwed into the engine cooling system. When the temperature is below about 95°F, the 12-volt source is applied to the cold start fuel injector, which then sprays additional fuel into the intake manifold for easier cold starting. The circuit is deactivated when the ignition switch is returned to the ON position or whenever engine coolant temperature is above about 95°F. The thermo-time switch contains a heating coil and a bimetal switch. The bimetal switch is opened by the heating coil, which takes about 8 seconds at 68°F engine coolant temperature. As engine coolant temperature goes up, the cold start fuel injector ON times goes down. When installing the cold start fuel injector onto the fuel line, it should be turned clockwise until seated and then backed off one full turn. The lock tang should be bent over to hold it in position.

AIR INDUCTION

The air induction system consists of an air cleaner, air ducting, a mass airflow sensor, throttle body, intake manifold, and idle air control motor (Fig. 9-16).

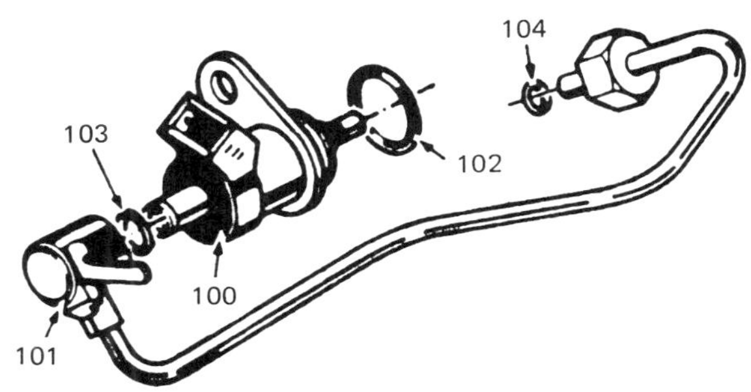

100 VALVE - COLD START
101 TUBE AND BODY ASSEMBLY
102 O-RING SEAL - VALVE
103 O-RING SEAL - BODY
104 O-RING SEAL - TUBE

FIGURE 9-14 Cold start valve *(Courtesy of Chevrolet Motor Division).*

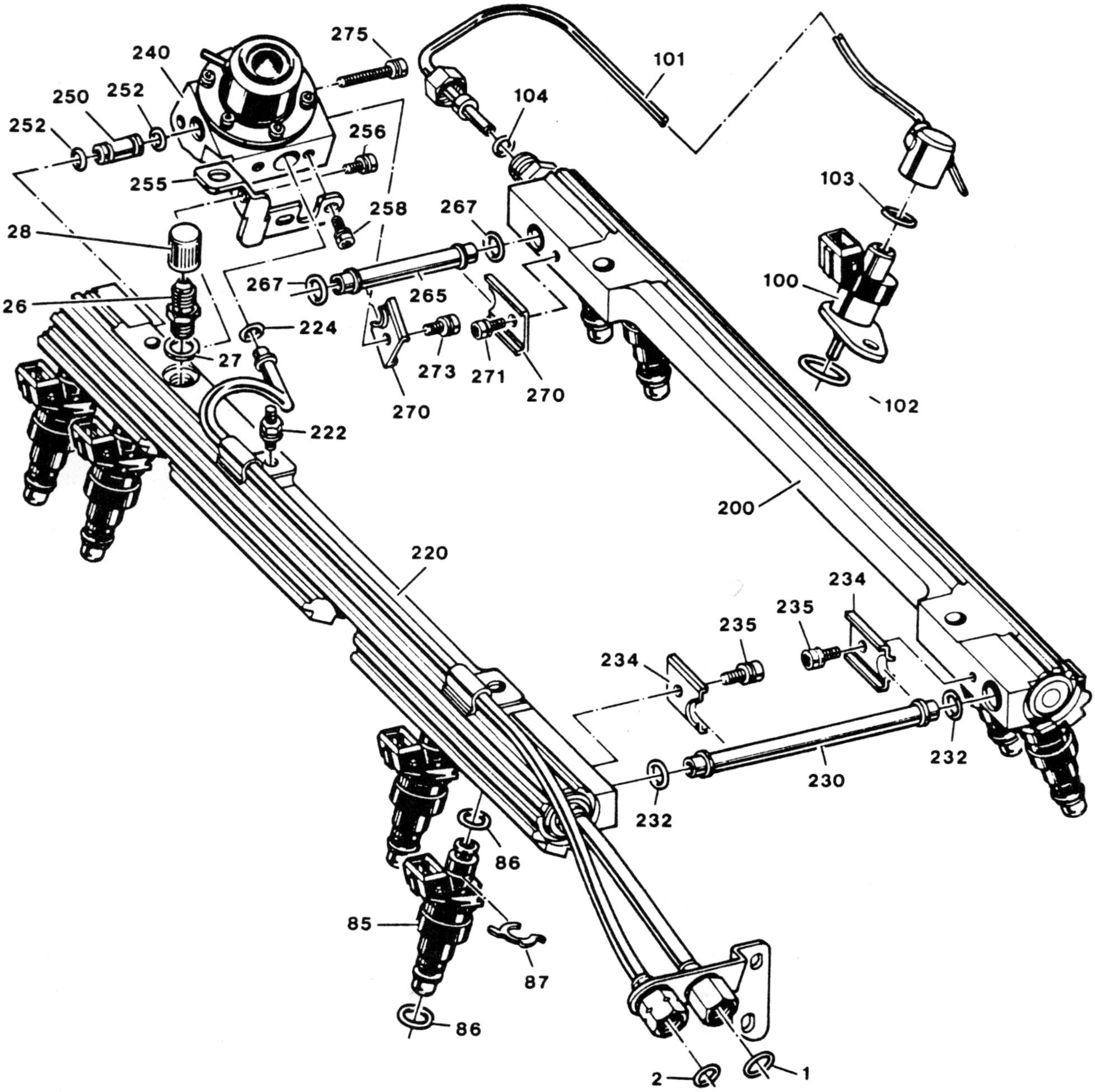

FIGURE 9-15 Fuel rail assembly, 5.0 liter *(Courtesy of Chevrolet Motor Division).*

Air Cleaner

The air cleaner assembly is not mounted on top of the engine. It is remotely mounted under the hood and consists of a replaceable paper element.

Air Ducting

Flexible ducting connects the air cleaner to the mass airflow sensor. Another flexible duct connects this sensor to the throttle body. Air leaks at either of these ducts will affect overall system operation. Duct-clamp tightness and cracks in the ducting should be checked periodically.

Mass Airflow Sensor

The mass airflow sensor is a thermal measurement device that uses a heated-metal-foil sensing element to measure air flow (Fig. 9-17). This sensor is different from those used on other port fuel injection systems. It is capable of

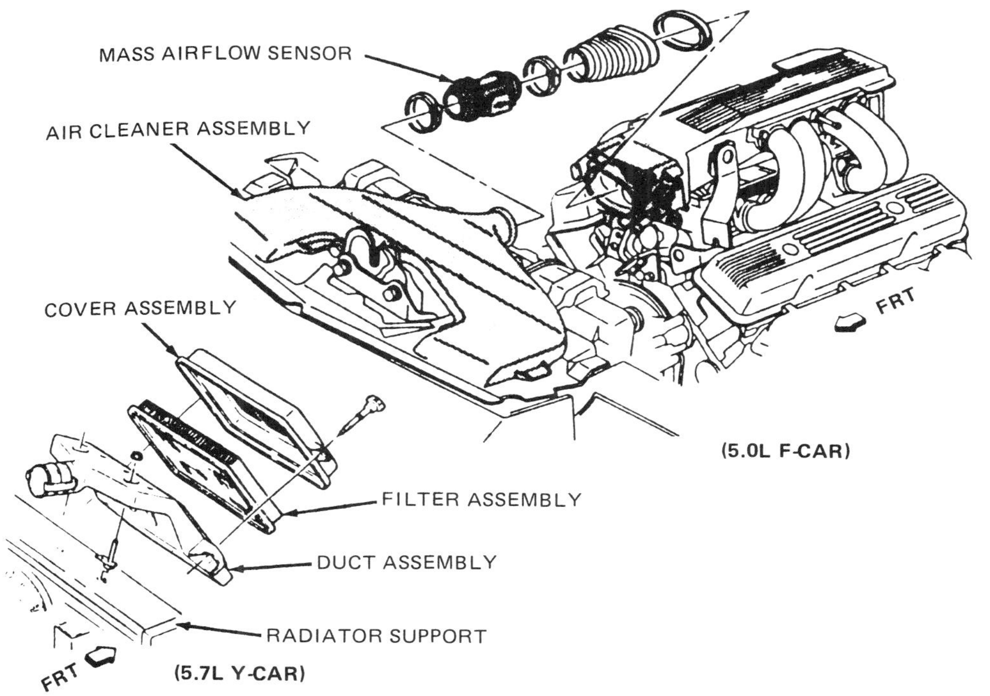

FIGURE 9-16 Air filtering system for 5.0- and 5.7-liter engines *(Courtesy of General Motors Product Service Training).*

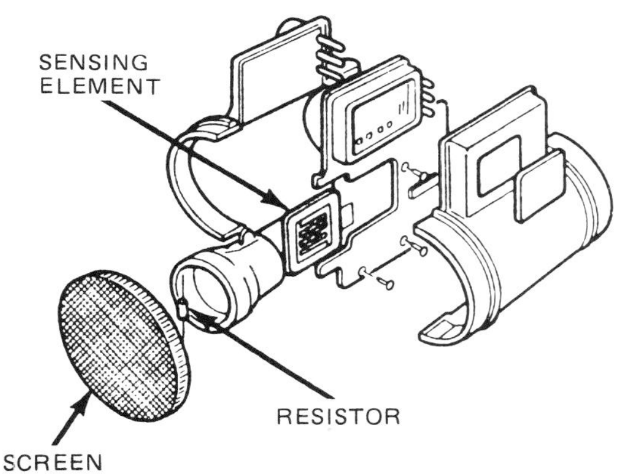

FIGURE 9-17 Mass airflow sensor for 2.8-liter V-6 engine (Courtesy of Chevrolet Motor Division).

measuring airflow. The metal-foil sensing element is heated to a temperature of about 170°F above the temperature of the incoming air. When the incoming air cools the element, the element resistance drops and causes a corresponding increase in current flow. When the current flow increases, the supply voltage decreases. This signal is seen by the ECM as a change in voltage drop and is used as an indication of airflow. An increase in airflow causes an increase in the voltage drop seen by the ECM. The electronic control module uses the signal to calculate fuel injector pulse width or ON time and ignition timing. Engine coolant temperature, engine speed, manifold air temperature, and the mass airflow sensor signals enable the computer to make the necessary calculations.

Throttle Body

The throttle body is mounted on the air plenum at the front of the engine (Fig. 9-18). The throttle position sensor and the idle air control motor are part of the throttle body assembly. A throttle shaft controls the movement of the throttle plates. The throttle position sensor is a variable resistor that is connected to the end of the throttle shaft. The resistance of the sensor signal circuit is high when the throttle blades are closed. This causes a low-voltage signal to be sent to the ECM. Signal voltage varies from 0.5 volt at closed throttle to about 4.5 volts at wide-open throttle. The throttle body also contains a passage through which engine coolant flows and can warm the throttle blade area to prevent icing. The idle air control motor is operated by the ECM to regulate the amount of air allowed to bypass the throttle plates. This bypass air determines idle speed. The IAC motor is a reversible motor that reacts to pulses of power from the ECM (Fig. 9-19). The ECM can pulse the motor to cause it to move in steps in either direction. The IAC is sometimes referred to as a stepper motor.

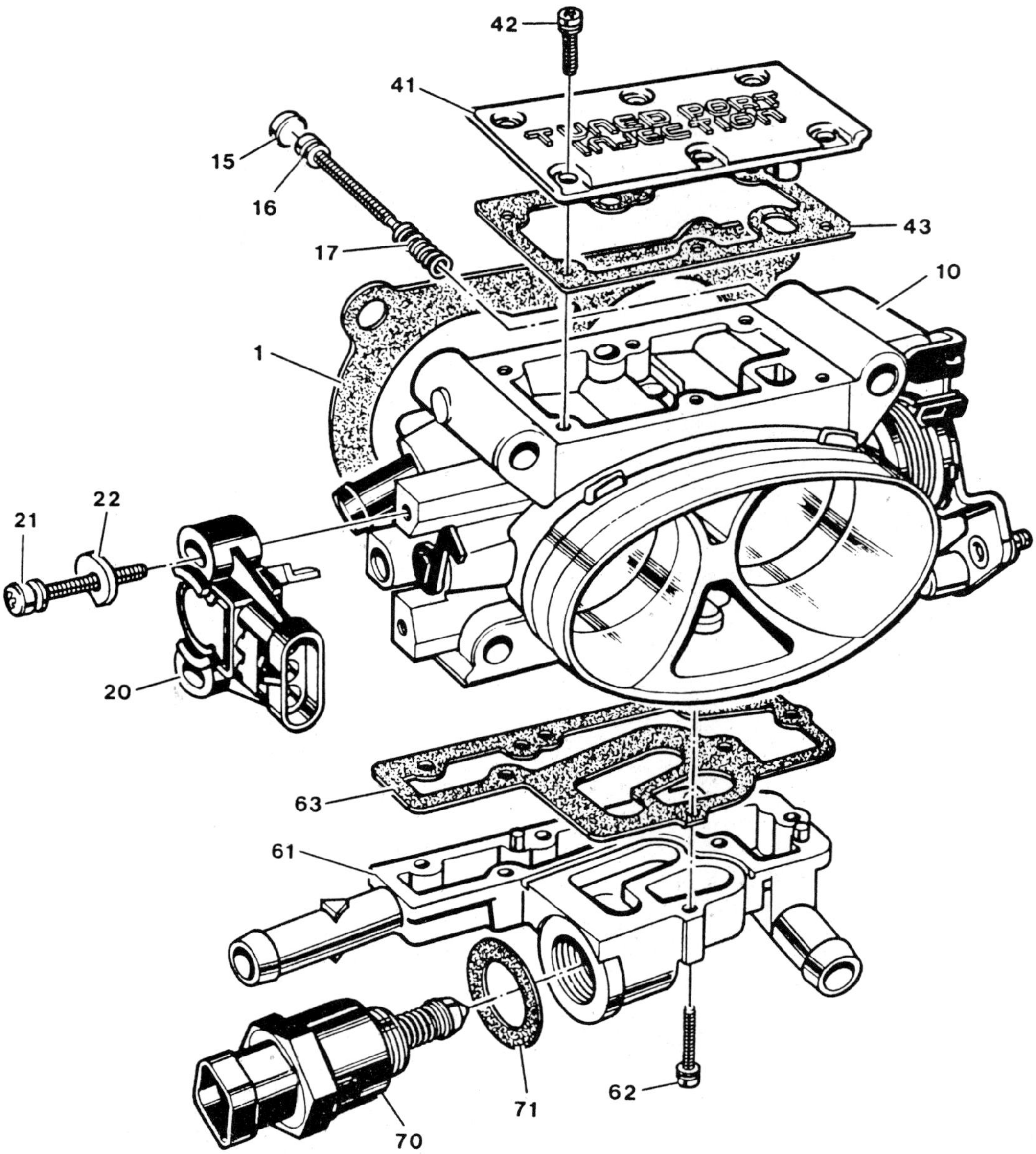

FIGURE 9-18 Throttle body, 5.0 liter *(Courtesy of Chevrolet Motor Division).*

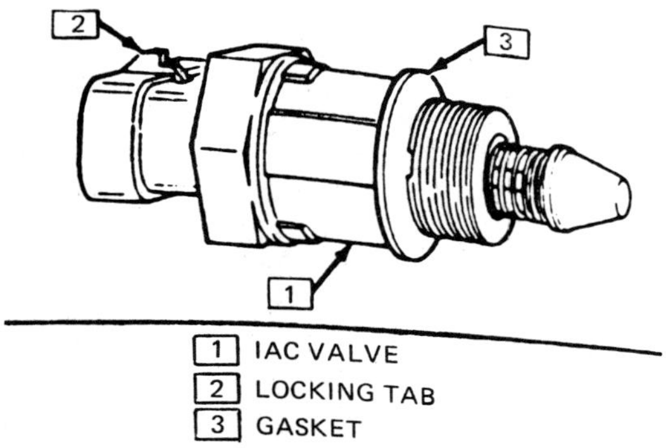

FIGURE 9-19 Idle air control valve *(Courtesy of Chevrolet Motor Division).*

IGNITION MANAGEMENT

Electronic Spark Timing

Two circuits, the electronic spark timing circuit and the electronic spark control circuit, along with the ECM, control ignition timing. The electronic spark timing circuit (EST) consists of a seven-terminal high-energy ignition module and the ECM. The distributor pickup coil provides pulses to the ignition module, which in turn relays these signals to the ECM as an engine speed reference and crankshaft position reference. During engine cranking, these same signals are used by the ignition module to operate the ignition primary circuit. This is known as the HEI cranking mode of operation. Once engine speed reaches about 400 rpm, the ECM activates a circuit in the ignition module that allows the ECM to assume control of the ignition primary circuit. At this point the ignition system is in the EST mode of operation. Information from the mass airflow sensor, engine coolant temperature sensor, and engine speed are used by the ECM to control ignition timing. Ignition timing is advanced by the ECM when coolant temperature is low or when engine load and airflow are low. As these factors increase, less spark advance is provided.

Electronic Spark Control

The electronic spark control circuit consists of a knock or detonation sensor (Fig. 9-20), an ESC module, and the related ECM and EST circuits. Detonation in an engine can occur as a result of several variables but is damaging regardless of the cause. The knock sensor responds to detonation because of the resulting vibrations in the engine block. The vibrations cause the sensor, which is a piezoelectric device, to produce a voltage signal to the ESC module. The ESC module then sends a reduced voltage signal, which is normally 8 to 10 volts, to the ECM. This causes the ECM to retard the timing until the detonation stops. At this point, the knock sensor signal stops and the ESC module signal voltage rises to its normal 8 to 10 volts. The ECM will gradually advance the timing.

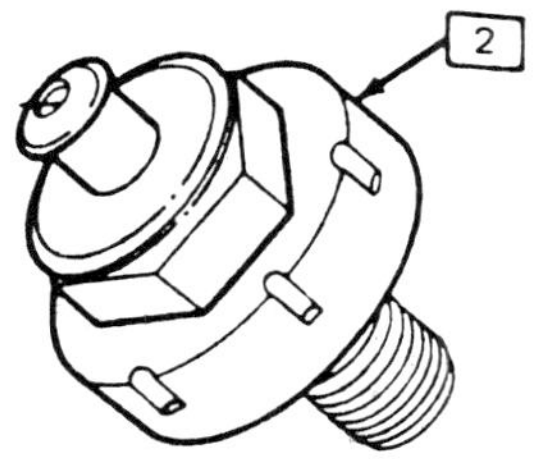

2 ESC KNOCK SENSOR

FIGURE 9-20 Electronic spark control system *(Courtesy of Chevrolet Motor Division).*

EMISSION CONTROL

Air Management

In addition to the ECM responsibilities already discussed, three emission control systems are operated. An air injection reaction system is regulated by the ECM through control of the air management valve. This valve directs fresh air either to the exhaust manifold, the catalytic converter, or the air divert silencer.

Exhaust Gas Recirculation

The exhaust gas recirculation valve is also controlled through use of an electrically operated solenoid-vacuum valve. The ECM activates the EGR solenoid to block the valve from opening. The valve operates as a normal EGR valve when the ECM deactivates the solenoid.

Evaporative Emission Control

The evaporative emission control system functions much the same as it has for past models. It collects, stores, and recycles fuel vapors, preventing them from escaping into the air. The ECM controls a purge valve that allows engine vacuum to clear the system of fuel vapors. When the purge valve solenoid is energized, no purging takes place. When the ECM de-energizes the solenoid, purging begins.

In addition, the ECM controls the automatic transmission torque converter clutch, the manual transmission overdrive unit, and the engine coolant fan.

MAF Burn-off Feature

Another feature of the TPI system is related to the mass airflow sensor. Contaminants in the air may cause a film or residue buildup on the surface of the heated-metal-wire sensing element. This may result in inaccurate readings from the sensor. To avoid potential problems, a *burn-off* cycle has been designed into the mass airflow circuitry. When the engine is in the closed-loop mode of operation and the ignition switch is turned off, a burn-off module is triggered that causes 12 volts to be applied to the metal-wire sensing element for about 1 second. This causes the element to get red hot, about 1000°F, and any film or residue is burned off.

SENSORS AND SWITCHES

For the electronic control module to make high-speed calculations and issue commands to the devices it controls, it must rely on information from a variety of sensors and switches. Each sensor and switch provides a range of data to the ECM known as *parameters.* Parameters sensed by the ECM include manifold air temperature, mass airflow, air-conditioning system operation, engine coolant temperature, engine detonation, vehicle speed, transmission gear position, exhaust gas oxygen content, crankshaft position, engine speed, throttle position, system voltage, and park/neutral switch position.

Manifold Air Temperature Sensor

The manifold air temperature sensor (MAT) is a thermistor whose resistance varies inversely with the temperature.

When temperature is about 265°F, the sensor resistance is about 70 ohms, whereas when the temperature is about −40°F, its resistance is about 100,000 ohms. The MAT sensor is supplied with a 5-volt source from the ECM. The resistance of the sensor determines how much of this voltage is returned to the ECM. This return voltage is interpreted by the ECM to be manifold air temperature. When a malfunction occurs in the sensor or its circuit, a code is set in the ECM memory. The MAT sensor signal is used to delay operation of the EGR valve.

Mass Airflow Sensor

The mass airflow sensor has been discussed earlier in this chapter. It, too, provides a varying voltage signal that ranges from 0.5 volt at idle to about 4.7 volts at wide-open throttle. The ECM uses the MAF signal to calculate air mass. A malfunction in the sensor or its circuit sets a code in the ECM memory.

Air-conditioning Signal

To control idle speed and prevent undesirable changes in it, the ECM must know when the air-conditioning system is cycling on. Because the air-conditioning compressor places additional load on the engine, the ECM needs to cause the idle air control motor to back out and allow more air to bypass the throttle plates. This would normally raise idle speed, but, because of the compressor, the rise is offset and idle speed is maintained. A slight delay in the activation of the compressor clutch gives the ECM time to anticipate and compensate for the engine load. The delay time is about 200 milliseconds.

Engine Coolant Temperature Sensor

The engine coolant temperature sensor is a thermistor whose resistance varies inversely with the temperature. Low coolant temperature produces a high resistance. At about −40°F, the sensor resistance is about 100,000 ohms, whereas at about 266°F the sensor resistance is about 70 ohms. The ECM supplies a 5-volt source for the sensor and monitors what returns. The sensor resistance determines the return voltage signal. The ECM interprets this signal into a value for engine coolant temperature. The coolant temperature sensor and manifold air temperature sensor are not the same and should not be interchanged. A malfunction in the sensor or its circuit sets a code in the ECM memory.

Detonation Sensor

Engine detonation is monitored through the detonation or knock sensor, as described earlier. Because this sensor is related to the electronic spark control system, a malfunction in the sensor or the ESC module and related circuitry sets a code in the ECM memory.

The vehicle speed sensor on some models is located in the transmission where the former speedometer drive was mounted. It is a permanent-magnet generator that produces a varying ac voltage. This voltage increases with vehicle speed. The ac voltage signal is sent to a buffer amplifier where the signal is conditioned for use by the instrument cluster display and by the ECM. The ECM uses this vehicle speed signal to control the automatic transmission torque converter clutch. A malfunction in the vehicle speed sensor, buffer amplifier, or related circuitry sets a code in the ECM memory. Some models use a vehicle speed sensor that is located behind the speedometer head. Operation is similar to those found in earlier chapters.

Fourth-gear Switch

Transmission gear position is monitored by the ECM through a fourth-gear switch in the automatic transmission. The manual transmission provides gear position information through its first-gear switch. This information is used by the ECM to control the automatic transmission torque converter clutch and the manual transmission overdrive unit.

Oxygen Sensor

The ECM is provided data about exhaust gas oxygen content by the oxygen (O_2) sensor. This sensor is mounted in the exhaust stream and produces a voltage signal related to the comparison of the oxygen content of fresh air with that of the engine exhaust. This signal enables the ECM to make corrections in the air/fuel mixture. A high oxygen content produces a low voltage, about 100 millivolts. A high oxygen content means the air/fuel mixture is too lean. A low oxygen content or rich air/fuel mixture produces a higher voltage of about 900 millivolts. A malfunction in the oxygen sensor or its circuit for a specified period of time sets a code in the ECM memory.

Engine Speed Signal

Crankshaft position and engine speed information are provided by the ignition module. The ECM uses these data to control ignition timing, as discussed earlier.

Throttle Position Sensor

Throttle position data are provided to the ECM for computing fuel delivery demanded by the driver. It is a potentiometer-type variable resistor that provides a low-voltage signal of about 0.5 volt at closed throttle and a higher voltage signal of about 5 volts at wide-open throttle. This sensor not only can malfunction but can also be out of adjustment. A trouble code is set in the ECM memory when the sensor malfunctions or a problem occurs in its circuit.

System Voltage Monitor

System voltage is monitored by the ECM. When system voltage drops, the ECM operation is affected. The ECM program enables it to compensate for low-voltage conditions. For example, low voltage causes injector pulse width to narrow, resulting in a leaner mixture. The ECM program allows for a wider pulse width under these conditions in an effort to offset the end result.

Park/Neutral Switch

The park/neutral switch position is monitored by the ECM for control of ignition timing, torque converter clutch, exhaust gas recirculation valve, and idle speed.

ELECTRONIC CONTROL MODULE

The fuel delivery system is controlled by the electronic control module (ECM). The ECM contains all the circuitry necessary for controlling not only the TPI system, but also emission controls and other fuel economy devices. The ECM also contains a programmable read-only memory (PROM) that contains information for the ECM to enable it to control various devices (such as fuel injectors) with a high degree of accuracy.

Calibration Package

Another memory, CALPAK, is a calibrated network of resistors. The CALPAK enables the ECM to operate the fuel system when the normal fuel system circuitry malfunctions. The CALPAK is part of the fuel backup circuit (FBC) in the ECM. The CALPAK calibrates the FBC and provides fuel for cold start and limp-home operation. The ECM operates in the FBC mode if the ECM voltage is less than 9 volts, the cranking voltage is less than 9 volts, the PROM is malfunctioning or missing, or the ECM fails to issue computer operating pulses. When in the FBC mode, the engine runs erratically. The FBC relies on the throttle position sensor (TPS), the engine coolant temperature sensor (CTS), the engine speed signal, and the power from the ignition switch for its operation. With this information, the FBC controls the fuel pump relay and fuel injectors. Both the PROM and CALPAK are serviceable components.

To function accurately, the ECM must measure and be aware of engine operating conditions. Engine data sensors are used to provide up-to-the-minute information about these conditions. Each sensor has a range of operation or parameter. The ECM monitors all these parameters. If a sensor goes out of range due to some fault, the ECM indicates this to the driver by illuminating a telltale light on the instrument panel. At the same time, the ECM stores or sets a fault code, representing the failed sensor or sensor circuit, in its memory. As an aid in diagnosing the problem, the service technician can read the stored fault code at a later time by causing the ECM to enter its self-diagnostic mode of operation.

SYSTEM OPERATION

There are seven modes of operation that the ECM can enter in response to engine operating conditions. The starting mode, clear flood mode, run mode, acceleration mode, deceleration mode, battery voltage correction mode, and fuel cutoff mode are used to control the fuel delivery system.

Starting Mode

When the ignition switch is turned on, the ECM energizes the electric fuel pump relay and fuel pump for about 2 seconds. This pressurizes and purges the fuel system. At the same time, engine coolant temperature is checked and throttle position is received from the TPS. The ECM can determine from this information how long to pulse the fuel injectors. The cold start fuel injector, which is not controlled by the ECM, provides additional fuel for starting when temperatures are very low. This is significant because the eight fuel injectors cannot be pulsed long enough to provide sufficient fuel under very cold conditions. This is known as the starting mode.

Clear Flood Mode

The clear flood mode is activated when the accelerator pedal is pushed completely to the floor, causing a wide-open throttle signal from the TPS, and cranking speed is below about 400 rpm. During this time, the ECM provides a very short pulse to the fuel injectors. If throttle position returns to about 80% open, the system deactivates and the starting mode is reentered.

Run Mode

The run mode can operate under one of two conditions: open or closed loop. When engine speed is above about 400 rpm after starting, the run mode operates in the open-loop condition. While in open loop, the ECM does not use the oxygen sensor signal in its calculations for determining the air/fuel ratio. Data signals from the coolant and mass airflow sensors are used instead. When the oxygen sensor reaches its operating temperature, its voltage signal varies in response to the exhaust gas oxygen content. At the same time, the coolant sensor has been indicating a rising temperature and a timer circuit in the ECM has been keeping track of how long the engine has been running since it started. When specific values for the oxygen sensor, coolant sensor, and timer circuit have been reached, the system enters closed-loop operation. These specifications or values are stored in ECM PROM discussed earlier. The ECM calculations for the air/fuel mixture are now based on the oxygen sensor signals, along with signals from the rest of the engine sensors.

Acceleration Mode

When the throttle position sensor and MAF sensor indicate an opening throttle, the acceleration mode is activated. The ECM pulses the fuel injectors longer based on the rate of throttle opening and the mass airflow sensor signal.

Deceleration Mode

When the throttle position sensor and MAF sensor indicate a closing throttle, the deceleration mode is entered. The ECM pulses the injectors for a shorter time based on the rate of throttle closing and the mass airflow sensor signal. The ECM can actually shut the injectors off momentarily if the deceleration rate is very fast.

Battery Voltage Correction Mode

When system voltage drops below a specified level, the battery voltage correction mode is activated. While in this mode, the ECM can compensate for weak ignition spark by increasing fuel delivery, idle speed, and ignition dwell period.

Fuel Cutoff Mode

The fuel cutoff mode is entered when the engine speed signal is interrupted. When the ignition switch is turned off, the HEI reference signal is stopped and the ECM will not pulse the injectors. This prevents ''after run'' or ''dieseling.'' If the engine stalls, an rpm signal is not available. Once again, the ECM will not pulse the injectors. This prevents flooding. If the engine stalled due to an accident, a potential fire hazard is avoided.

SYSTEM SERVICE

Fuel Pressure Relief

When servicing the fuel delivery system, it is necessary to relieve fuel pressure before opening the system. This can be done by using the special fuel injection pressure gauge with bleed-off valve. Attach the gauge to the Schrader connector on the fuel rail and bleed pressure into an approved gasoline can carefully.

Minimum Idle Speed

Minimum idle speed is adjustable through the idle stop screw. To set minimum idle speed, follow the detailed instructions in the service manual. The throttle position sensor adjustment should be checked after performing the minimum idle speed adjustment.

Throttle Position Sensor Adjustment

Adjustment of the throttle position sensor requires the use of special jumper leads and a high-impedance digital multimeter (Figs. 9-21 and 9-22). Specific instructions for adjusting the TPS are given in the service manual and should be followed carefully.

Idle Air Control Motor Service

Whenever the idle air control motor is replaced, its adjustment should be checked before installation (Figs. 9-23 and 9-24). Failure to do so may result in damage to the pintle valve. Follow service manual instructions carefully.

Ignition Timing Adjustment

To set initial ignition timing, install a timing light, or preferably a magnetic timing meter, and disconnect the timing connector in the wiring harness under the hood. Set timing according to specification. This procedure causes the ECM to set a code in its memory because the electronic spark timing system was disabled when the timing connector was opened. For this reason, when ignition timing has been set, it is also necessary to clear the ECM memory. This will be discussed in the next section of this chapter.

FIGURE 9-21 Adjusting the throttle position sensor *(Courtesy of Chevrolet Motor Division).*

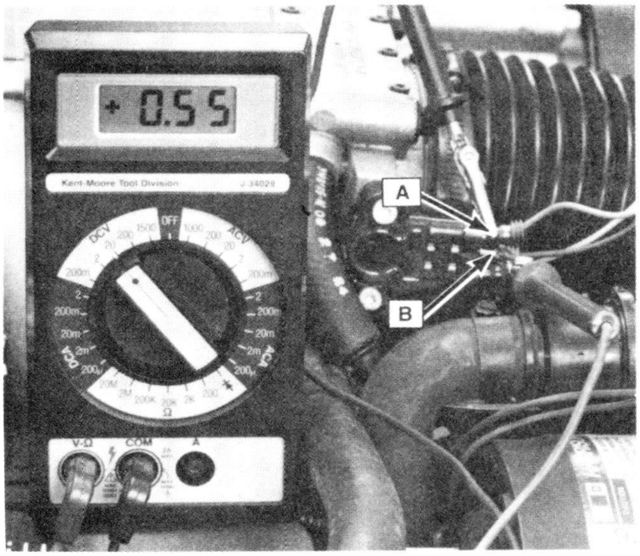

FIGURE 9-22 Connecting a voltmeter to the throttle position sensor *(Courtesy of Chevrolet Motor Division).*

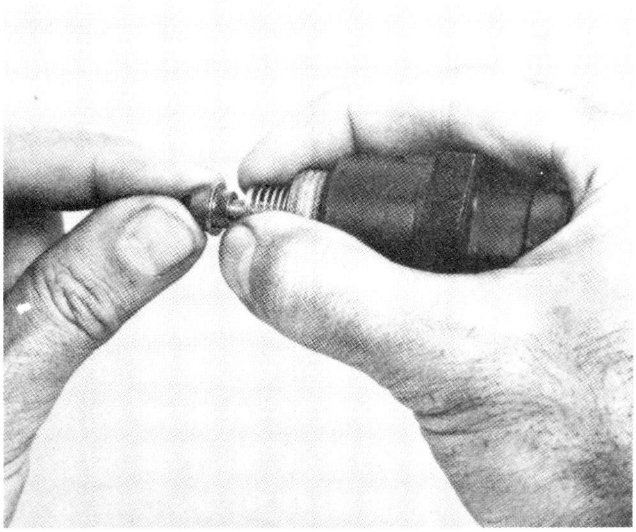

FIGURE 9-23 Adjusting a collarless idle air control motor *(Courtesy of Chevrolet Motor Division).*

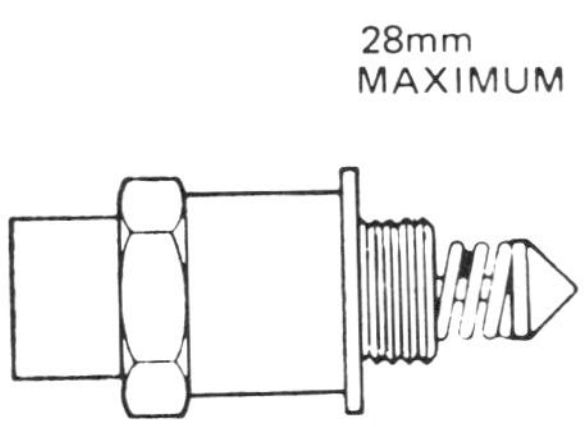

FIGURE 9-24 Idle air control motor specifications *(Courtesy of Chevrolet Motor Division).*

Fuel Injector Balance Test

By using a special fuel injection tester, a fuel injector balance test can be performed (Fig. 9-25). This test helps the service technician to determine if an injector is sticking, closed, restricted, or leaking. Attach a special fuel injection pressure gauge to the pressure Schrader fitting on the fuel rail (Fig. 9-26). Install the injector tester harness to one fuel injector and the battery. Pressurize the fuel system by turning the ignition switch on, then off for 10 seconds, and then on again. Record the fuel pressure, activate the tester, and record the pressure drop. The tester energizes the fuel injector for 500 milliseconds, resulting in a drop in fuel pressure. This procedure should be repeated for each fuel injector. When finished, a comparison of the fuel pressure drops can be made. If an injector drops less pressure, it is not allowing as much fuel to be injected. An internal restriction is present. No drop in pressure indicates the injector is inoperative. Too great a pressure drop indicates a leak internally. The fuel system should be pressurized between each test. A pressure drop difference of more than about 1.5 psi (10 kpa), indicates that the injector should be replaced.

FIGURE 9-25 Testing injector operation *(Courtesy of Chevrolet Motor Division).*

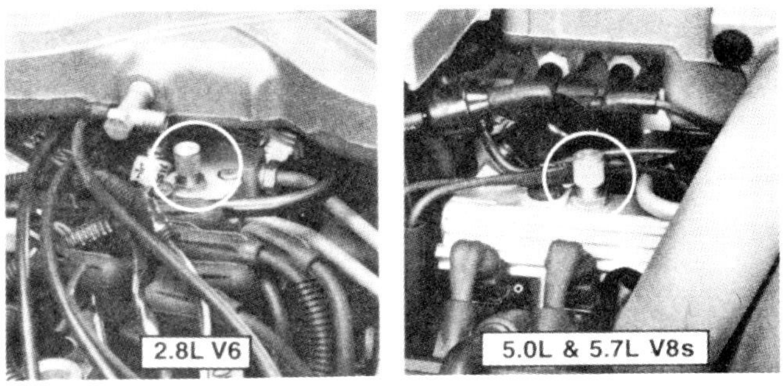

FIGURE 9-26 Fuel pressure taps *(Courtesy of Chevrolet Motor Division).*

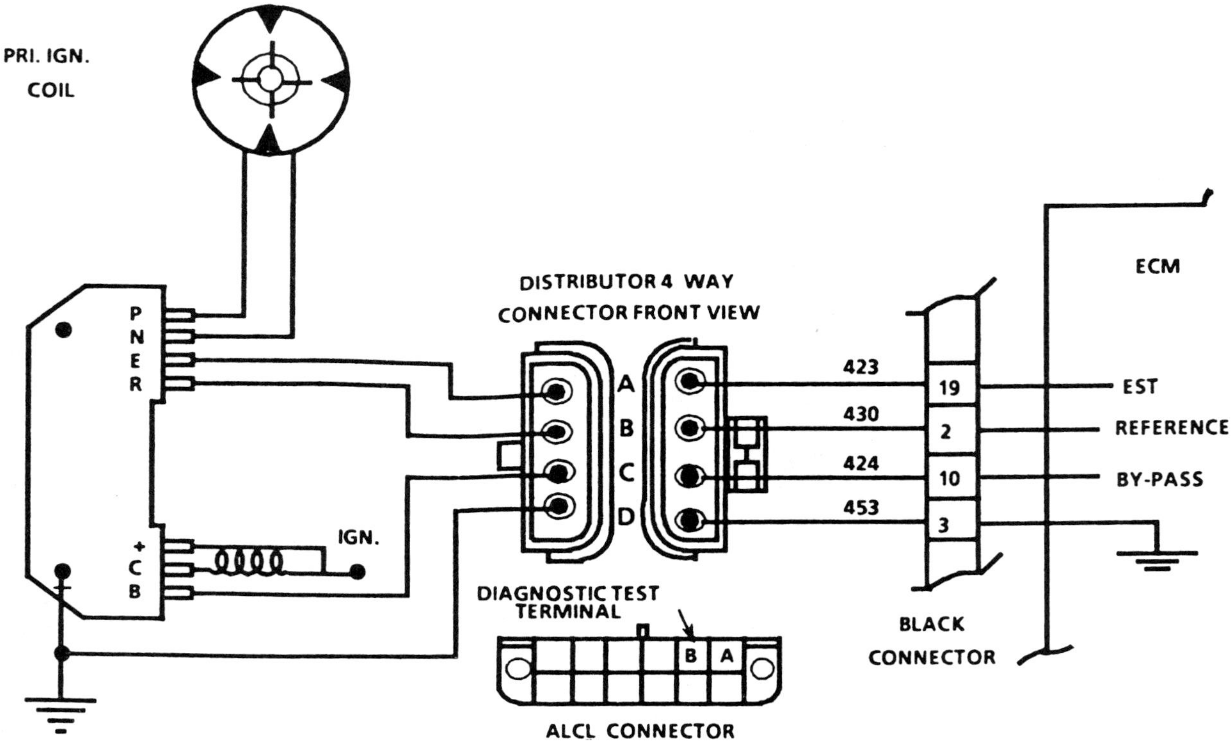

FIGURE 9-27 Assembly Line Communication Link connector and distributor 4-way connector *(Courtesy of Chevrolet Motor Division).*

Electronic Spark Timing Test

The electronic spark timing system can be checked for operation by checking ignition timing at 2000 rpm and then grounding the test terminal in the assembly line communications link (ALCL) connector under the instrument panel (Fig. 9-27). The timing should change. This indicates correct operation of the system.

Electronic Spark Control Test

The electronic spark control system can be checked for operation by checking ignition timing at 2000 rpm and then tapping on the engine block near the detonation sensor with a ball peen hammer. The timing should change. When the tapping stops, the timing should change again. This indicates correct operation of the system. Special tools required for service are shown in Fig. 9-28, Fig. 9-29, and Fig. 9-30.

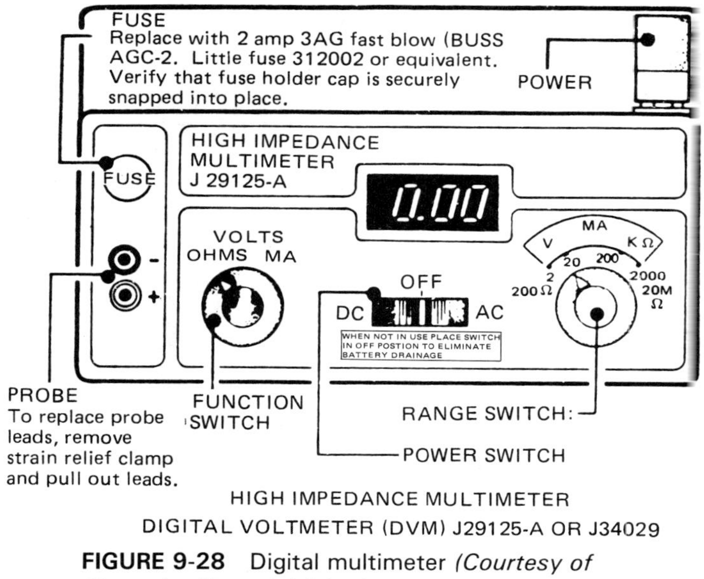

FIGURE 9-28 Digital multimeter *(Courtesy of Chevrolet Motor Division).*

SYSTEM DIAGNOSIS

Self-diagnostic Mode

As discussed earlier, the ECM has a memory in which codes are stored when malfunctions occur in specific circuits. These codes are an aid to the service technician in diagnosing problems with system operation. To gain access to these codes, the technician must ground the test terminal in the assembly line communications link (ALCL) connector in the passenger compartment. Grounding this terminal causes the computer to enter either the diagnostic mode or the field service mode.

 **J23738/BT7517**	**VACUUM PUMP (20 IN. HG. MINIMUM)** Use gauge to monitor manifold engine vacuum. Check vacuum sensors, solenoids and valves with hand pump.
 **J28742/BT8234-A**	**WEATHER-PACK TERMINAL REMOVER** Used to remove terminals from Weather-Pack connectors. Refer to wiring harness service for removal procedure.
 **J33095/BT8234-A**	**ECM CONNECTOR TERMINAL REMOVER** Used to extract a terminal from connectors at the ECM.
	**UNPOWERED TEST LIGHT** Used to check wiring for complete circuit. Connect lead wire to good ground. Probe with test prod to connector or component terminal. Bulb will light if voltage is present.
	**JUMPER WIRES (#16, 18 or 20 GAUGE WIRE)** • Clip jumper wire used to complete a circuit by by-passing an open. • Set of jumper wires used to insert between Weather-Pack connectors to permit access to the connector terminals for circuit checking. Six wires approximately 6 inches long. Use terminals 12014836 and 12014837. One set – female terminals both ends; one set – male at both ends; and four sets – male terminals at one end and female terminals at the opposite end.
	**TACHOMETER** Use either a crankshaft harmonic balance pickup type or electronic coil trigger signal pickup type.
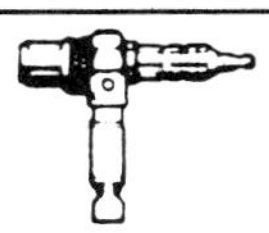 **J26792/BT7220-1**	**HEI SPARK TESTER** Used to check HEI spark voltage. Also called ST125.
 **J34730-A**	**PORT FUEL INJECTION DIAGNOSTIC KIT** Used to diagnose and service port fuel injection systems. The kit includes: • Fuel Pressure Gauge – to check fuel pump pressure and compare injectors for equal fuel distribution. • Test Light – to check electrical impulses to an injector. • Injector Tester – to evaluate an injector.

FIGURE 9-29 Special tools *(Courtesy of Chevrolet Motor Division).*

The diagnostic mode is entered by turning the ignition switch on and then grounding the ALCL connector test terminal. The ECM then causes the CHECK ENGINE light to flash in a sequence. One flash, followed by a pause, and then two flashes is a code 12. This code indicates that there are no engine speed reference pulses being received by the ECM. Since the engine is not running, this is to be expected. Therefore, a code 12 means the self-diagnostic mode is functional. The code 12 flashes are repeated three times. If any other codes are set in the ECM memory, each is displayed as a series of flashes of the instrument panel telltale light. Each code is repeated three times before the

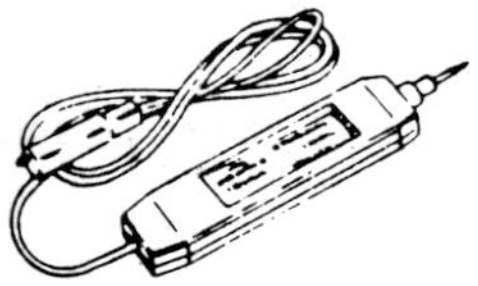 J34636	CIRCUIT TESTER Used to check all relays and solenoids before connecting them to a new ECM. Measures the circuit resistance and indicates pass or fail via green or red LED. Amber LED indicates current polarity. Can also be used as a nonpowered continuity checker.
 J28687-A/BT8220	OIL PRESSURE SWITCH WRENCH Used to remove or install oil pressure gauge switch on engine.
 J33095/BT8234-A	ECM CONNECTOR TERMINAL REMOVER Used to remove terminal from Micro-Pack connectors. Refer to wiring harness service for removal procedure.
 J29533-A/BT8127	OXYGEN SENSOR WRENCH Used to remove or install the oxygen sensor.
 J33031/BT8130	IDLE AIR CONTROL WRENCH Used to remove or install IAC valve on throttle body.
 J34730-2	INJECTOR TEST LIGHT Used to test injector signal from ECM.
 J34730-1	FUEL PRESSURE GAUGE Used to check and monitor fuel line pressure of fuel system.
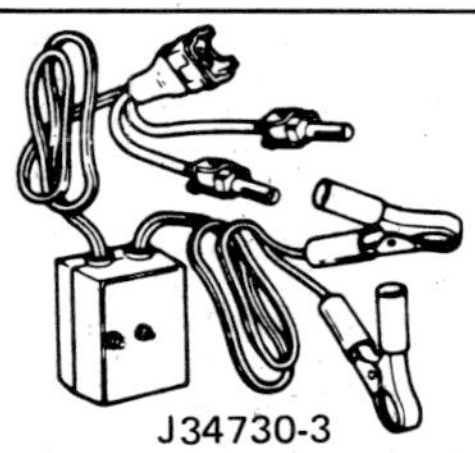 J34730-3	INJECTOR TESTER Used to perform injector balance test chart C-2A.

FIGURE 9-30 Special tools *(Courtesy of Chevrolet Motor Division).*

next code is displayed. When all stored codes have been displayed, code 12 appears again and the system begins to display the code sequence again. The system continues to do this until either the ALCL connector ground is removed or the ignition switch is turned off. While in the self-diagnostic mode, the ECM also energizes all controlled relays (except the fuel pump relay) and solenoids.

When codes are set in the ECM memory, accessed through the self-diagnostic mode, the service technician is made aware of a circuit problem identified by the code number. The service manual contains charts to define symptoms and code numbers. It also contains individual charts for each code fault and specific instructions for testing affected circuits, sensors, switches, relays, or solenoids.

Field Service Mode

The field service mode can be activated by grounding the ALCL connector test terminal while the engine is running. The ECM causes the CHECK ENGINE light to flash at one of two speeds. A flash every 2.5 seconds means the computer system is in open-loop operation. A flash every second means the system is in closed loop. If the air/fuel mixture is too lean, the light will remain out most of the time. If the mixture is too rich, the light will remain on most of the time. There are charts in the service manual to accompany any of these conditions.

Clearing Codes

When circuit repairs have been made, it becomes necessary to erase or clear the ECM memory. This can be accomplished by disconnecting the ECM harness from the positive battery cable pigtail near the battery for at least 10 seconds. Operate the vehicle to see if the code resets. If it does, the malfunction still exists.

The tuned port fuel injection system is yet another improved version of the multiport fuel injection system used on earlier models.

ABBREVIATIONS

A/F Air/fuel (A/F ratio)
AIR Air injection reaction
ALCL Assembly line communication link
ATS Air temperature sensor
BARO Barometric pressure
BATT Battery positive terminal
CCP Controlled canister purge
C.E. Check engine
CO Carbon monoxide
Conv. Converter (catalytic)
Diagnostic "Test" Term. Lead of ALCL connector that is grounded to get a trouble code.
DVM (10 meg.) Digital voltmeter with 10-million ohms resistance
EAC Electric air control
EAS Electric air switching
ECM Electronic control module (controller)
EECS Evaporative emissions control system
EFE Early fuel evaporation
EFI Electronic fuel injection
EGR Exhaust gas recirculation
ESC Electronic spark control
EST Electronic spark timing
FED Federal
HC Hydrocarbons
HEI High-energy ignition
IAC Idle air control
IGN Ignition
I.P. Instrument panel
MAF Mass air flow
MAP Manifold absolute pressure
MAT Manifold air temperature
MFI multiport fuel injection
mph Miles per hour
NO_x Oxides of nitrogen
O_2 Oxygen (sensor)
PCV Positive crankcase ventilation
PFI Port fuel injection
P/N Park/Neutral
PROM Programmable read-only memory (calibrator)
rpm Revolutions per minute
TACH Tachometer
TCC Transmission converter clutch
THERMAC Thermostatic air cleaner
TPI Tuned port injection
TPS Throttle position sensor
TVS Thermal vacuum switch
V Volt
VAC Vacuum
VIN Vehicle identification number
VSS Vehicle speed sensor
WOT Wide-open throttle

UNIT II
Ford Motor
Company Systems

10
Central Fuel Injection/Electronic Engine Control III (1980–1982) and IV (1983–1985)

OVERVIEW

The Ford central fuel injection system used in conjunction with the electronic engine control system III was developed to provide lower emissions and increased fuel economy while performance remained good. The electronic engine control III (EEC-III) system is the result of an evolutionary process that began in 1978 with the EEC-I system. The EEC-I system was an open-loop, electronically controlled carburetor-type fuel system. From this, the EEC-II system evolved. The EEC-II system was a closed-loop, electronically controlled carburetor system that was much more sophisticated and sensitive to engine operating conditions. The EEC-III system is very similar to its predecessor except that it is also used in combination with a central fuel injection system. The EEC-IV system represents further refinements, use of integrated circuitry, more sensitive responsiveness to engine operating conditions, and faster reaction to changes in engine conditions than the three earlier systems. It is also used in combination with central fuel injection (CFI).

This chapter will discuss the EEC-III and EEC-IV systems as they relate to central fuel injection. Both of these systems utilize an electronic control assembly (ECA) or computer to monitor engine operating conditions through sensor circuits (Fig. 10-1). It controls devices or actuators in order to adjust the air/fuel ratio, ignition timing, exhaust gas recirculation flow rate, and thermactor airflow. The central fuel injection system provides fuel and air induction systems that are also controlled by the electronic control assembly.

The fuel delivery, air induction, engine sensors, and electronic engine control systems will be individually discussed in detail.

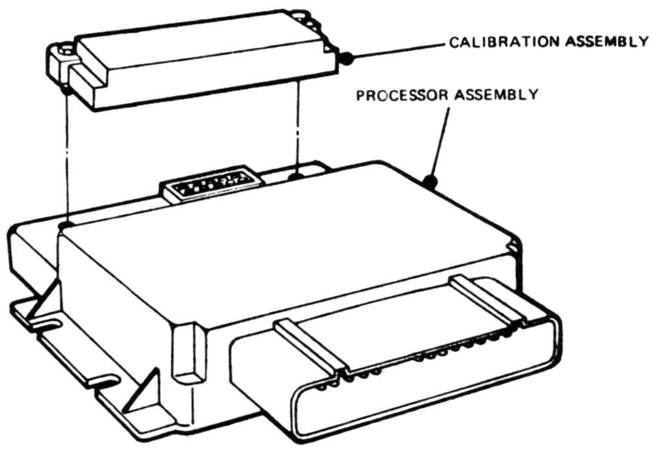

FIGURE 10-1 Electronic control assembly *(Courtesy of Ford Motor Company).*

FUEL DELIVERY

The fuel delivery system includes the fuel tank, an in-tank electric fuel pump, fuel supply line, in-line fuel filter, fuel charging assembly, and fuel return line.

Fuel Tank

The central fuel injection system has a new-design fuel tank (Fig. 10-2). It is made of stamped steel and has a sump area that acts as a fuel reservoir for the in-tank electric fuel pump. The tank requires a standard vacuum-pressure relief emission control cap.

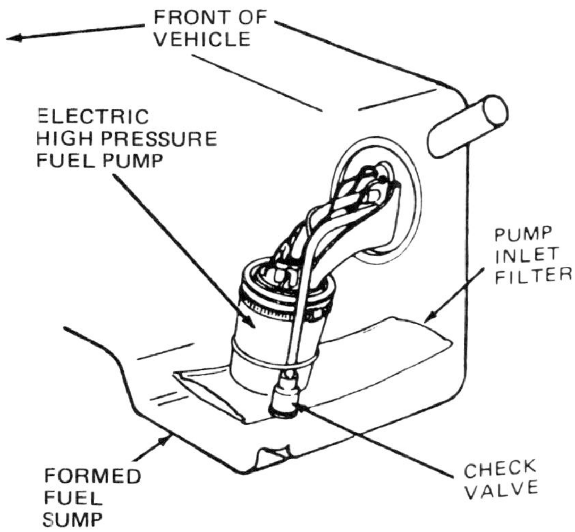

FIGURE 10-2 Fuel tank and pump assembly *(Courtesy of Ford Motor Company).*

In-tank Electric Fuel Pump

The in-tank electric fuel pump is mounted in the sump area of the fuel tank with the fuel inlet at the bottom of the tank (Fig. 10-3). This mounting ensures an adequate supply of fuel even during severe vehicle maneuvering conditions. The in-tank pump provides high regulated pressure of about 39 psi. It is capable, however, of producing about 80 to 120 psi. The pressure is limited by an internal relief valve. A woven plastic fuel filter is mounted on the fuel pump inlet. This nylon element prevents dirt and contaminants from entering the pump. It will, however, allow small amounts of water to pass.

The fuel pump has two controls. One control is operated by a vacuum switch, which was discontinued in early 1981. The other control is operated by the electronic control assembly. When the ignition switch is off, the vacuum-controlled relay is closed and the ECA relay is open. When

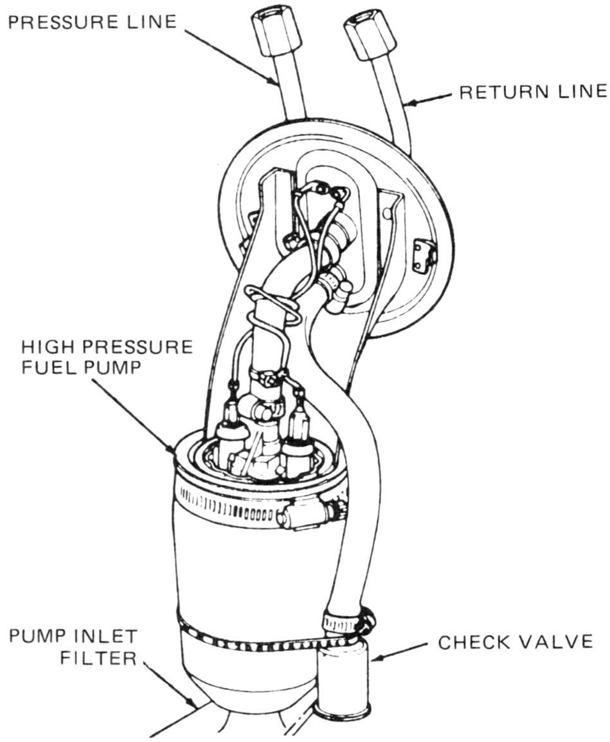

FIGURE 10-3 Electric fuel pump (vehicles equipped with EFI) *(Courtesy of Ford Motor Company).*

the ignition switch is turned on, both relays are closed and the fuel pump runs to purge and pressurize the fuel delivery system. If a cranking signal is not received by the ECA within 1 second, the ECA relay is de-energized to stop the fuel pump. When the engine cranks and starts, the vacuum switch that controls the vacuum relay closes and causes the vacuum relay to open. This causes power for the fuel pump to be fed through a resistance wire, which reduces fuel pump operating voltage. When engine load causes a large enough decrease in vacuum, the switch opens, and the relay de-energizes and provides full battery voltage to the fuel pump. The ECA can also control the fuel pump by opening its relay when the engine stops or the engine speed is less than 120 rpm.

In addition, an inertia switch is used in the fuel pump circuit (Fig. 10-4). It is located in the luggage compartment on the left hinge support of most models. The function of the switch is to open the fuel pump circuit and stop the pump in the event of a collision. The fuel pump shuts off even if the engine continues to run. Eventually, the engine will also stop. To restart the engine, inspect the fuel system for damage and leaks. The inertia switch must be manually reset by depressing both buttons on the early-style switch or a single button on the later-style switch.

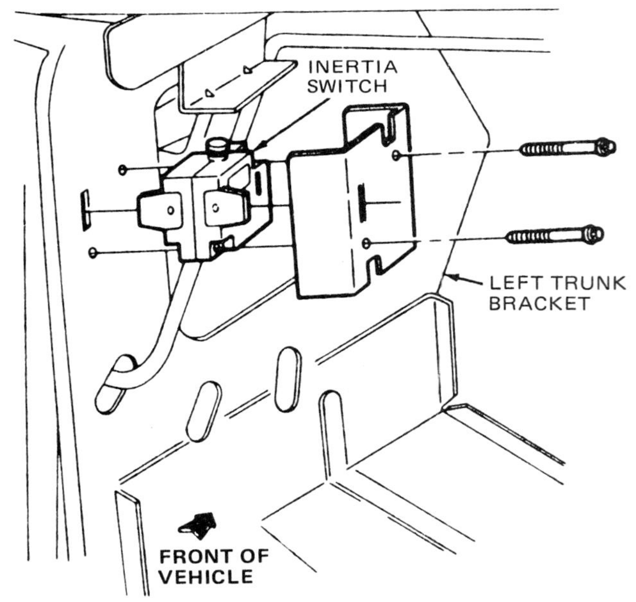

(a)

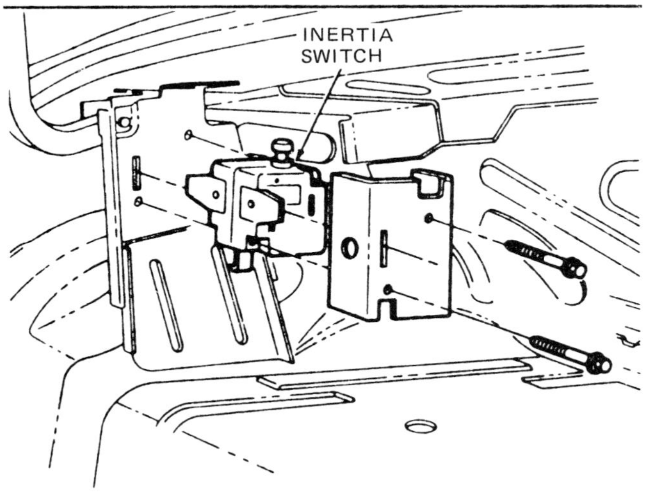

(b)

FIGURE 10-4 Inertia switch location. (a) Ford and Mercury. (b) Lincoln Continental and Continental Mark VI *(Courtesy of Ford Motor Company).*

Fuel Supply Line

Fuel supply lines are made of either steel tubing or rubber hose. Because of higher fuel pressures, connections are threaded fasteners. All hoses are serviced as assemblies. Steel tubing can be replaced by cutting and forming sections from stock. Standard fuel line hose and clamps should not be used as a replacement. Fuel supply lines may remain pressurized for a long time after the engine has been stopped. Procedures for relieving pressure during system servicing are provided in the service manual and will be included later in this chapter.

Fuel Filter

An in-line fuel filter is located under the vehicle floor pan near the right rear wheel well (Fig. 10-5). It is mounted on the frame rail and is a replaceable cartridge. Its function is to remove particles and contaminants that may otherwise cause damage to the fuel injectors. The filter and housing is protected by a metal stone shield (Fig. 10-6). A second-

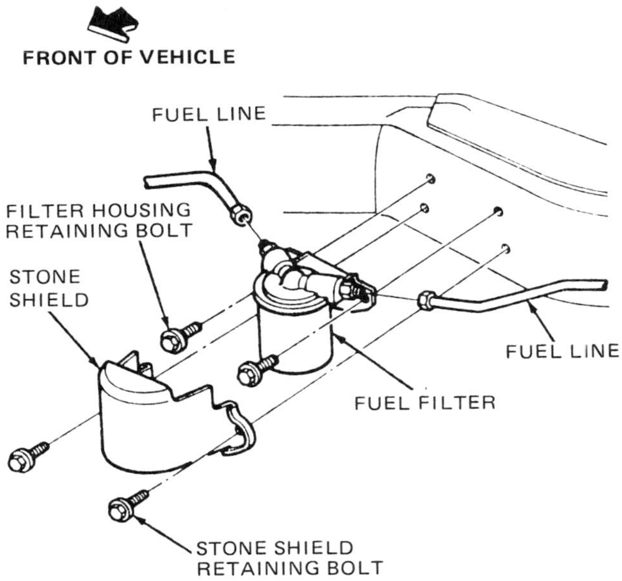

FIGURE 10-5 Primary fuel filter removal and installation *(Courtesy of Ford Motor Company).*

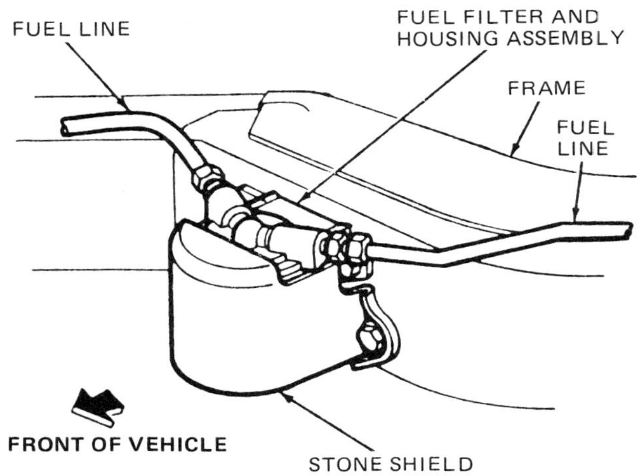

FIGURE 10-6 Primary fuel filter stone shield *(Courtesy of Ford Motor Company).*

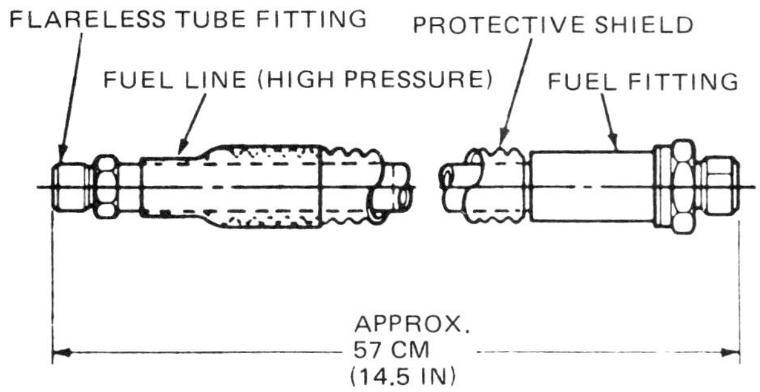

FIGURE 10-7 Secondary fuel filter and line assembly *(Courtesy of Ford Motor Company).*

ary in-line filter is a small, serviceable filter located in the lower engine compartment (Fig. 10-7). This filter has about 15 inches of flexible line permanently attached and is serviced as an assembly. The purpose of this filter is to remove particles resulting from servicing the rest of the system. Particles can also be loosened during the purging process whenever the system is pressurized. This filter should never need replacement.

Fuel Charging Assembly

The fuel charging assembly is a throttle body that includes two vertically mounted fuel injectors, fuel inlet and return fittings, fuel pressure regulator, a fuel pressure relief valve and cap, throttle positioner assembly, throttle position sensor, thermostat housing assembly, and vacuum hose connections (Fig. 10-8).

Fuel Injectors

Two fuel injectors are vertically mounted above the throttle plates and are connected in parallel with the fuel pressure regulator. When fuel pressure supplied by the fuel pump exceeds that required for engine operation, the fuel pressure regulator valve unseats and allows the excess pressure to enter the fuel return line. The fuel injectors are electric solenoids operated by the electronic control assembly (Fig. 10-9). When the ECA energizes the injector, the solenoid action raises a pintle valve off its seat and allows fuel to spray into the incoming air charge. Since the fuel pressure is constant and the injector orifice is fixed, the amount of fuel sprayed is determined by the duration the injector is

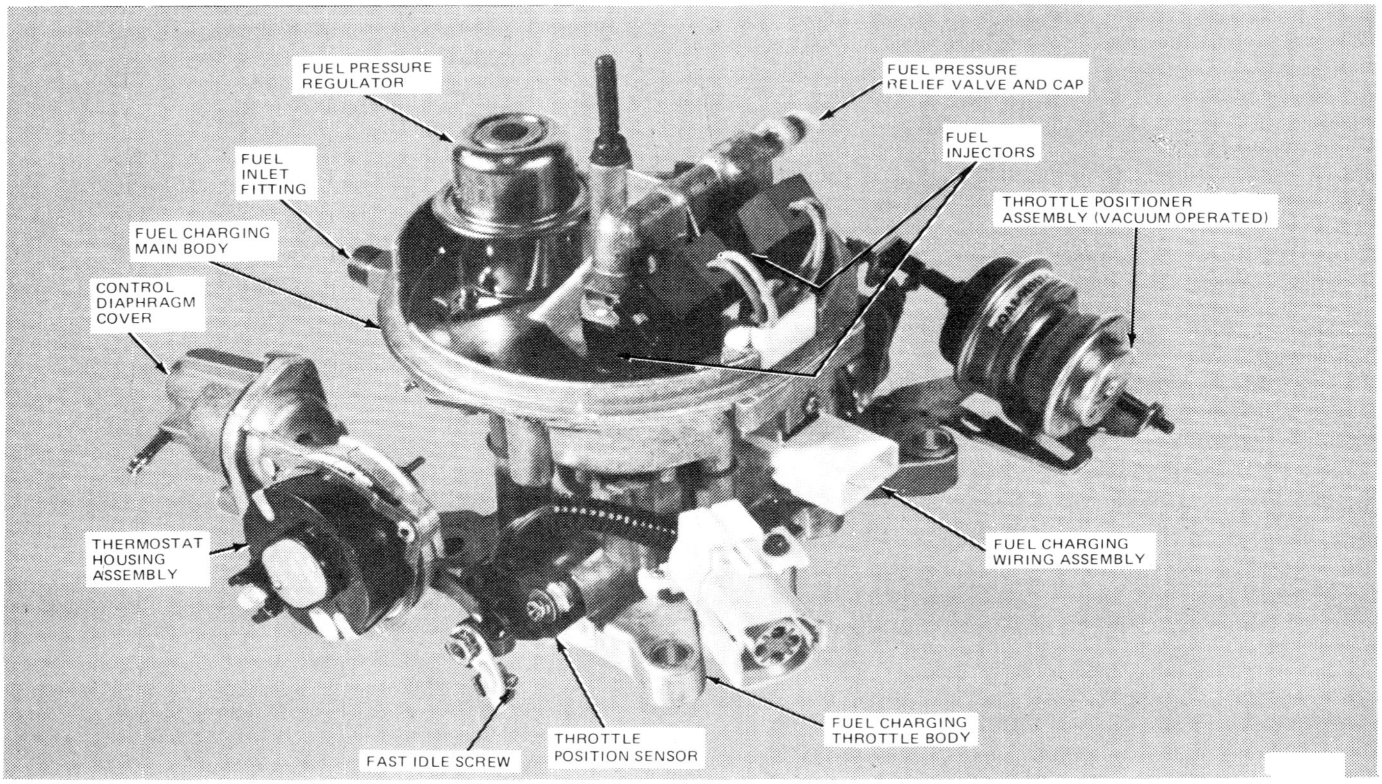

FIGURE 10-8 Electronic fuel injection: fuel charging assembly, three-quarter front view *(Courtesy of Ford Motor Company).*

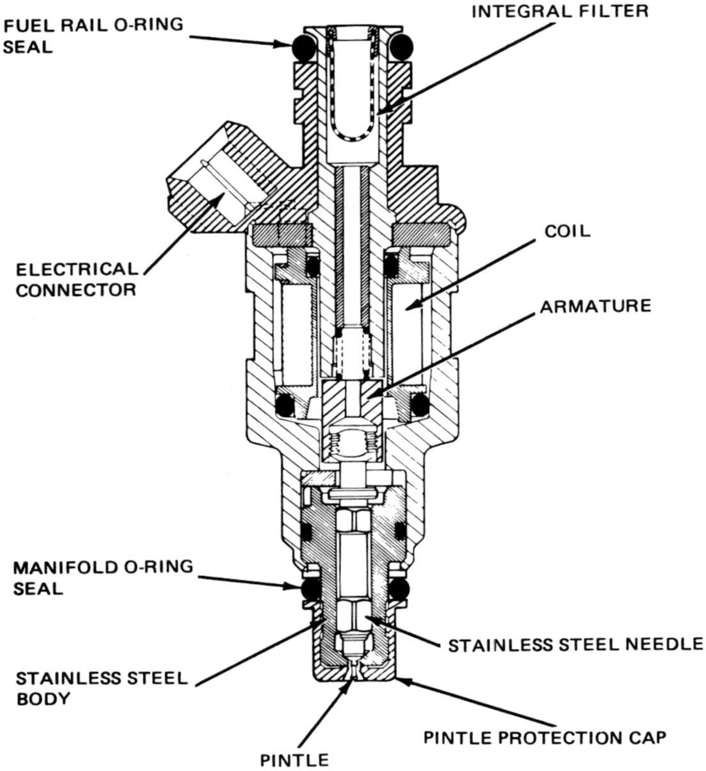

FIGURE 10-9 Fuel injector *(Courtesy of Ford Motor Company).*

energized. This is known as *pulse width.* A pulse width of 10 milliseconds provides more fuel than a pulse width of only 6 milliseconds.

The fuel inlet and return fittings are threaded connections from the fuel pump and filter to the fuel return line.

Fuel Pressure Regulator

A fuel pressure regulator is mounted on the fuel charging assembly near the rear of the air horn surface (Fig. 10-10). A function of the regulator is to smooth out the pressure drops that occur in the fuel supply line whenever the injectors are pulsed. Another function of the regulator is to maintain fuel system pressure whenever the engine or fuel pump is shut off. This prevents formation of bubbles in the fuel line due to engine heat and allows for quicker hot starts and stable idle conditions after a hot start. The pressure regulator maintains about 39 psi pressure to the fuel injectors.

Fuel Pressure Tap

The fuel pressure relief valve and cap are sometimes referred to as a fuel pressure diagnostic valve (Fig. 10-8). This valve is located at the top of the fuel charging assembly and is essentially a Schrader-type valve. It has a threaded connection for a fuel pressure gauge and for re-

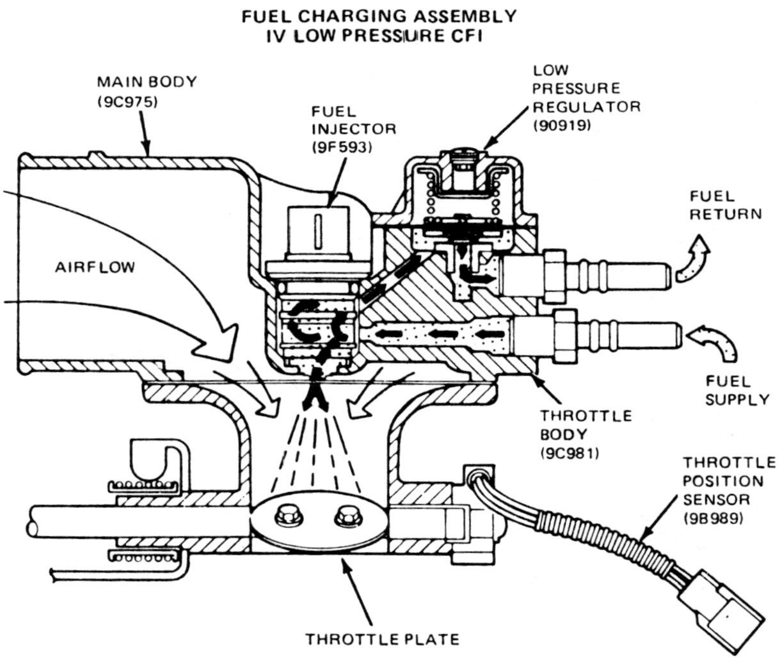

- LOW PRESSURE
- SINGLE LOW PRESSURE INJECTOR
- INTEGRAL PRESSURE REGULATOR
- TWO PIECE ALUMINUM CASTING WITH NO CROSS-OVER CHANNELS
- IDLE SPEED CONTROL
- THROTTLE POSITION SENSOR

FIGURE 10-10 Low-pressure fuel charging assembly, EEC-IV, 2.3-liter HSC *(Courtesy of Ford Motor Company).*

lieving system pressure. When necessary, air can be bled from the system at this point also.

Throttle Position Sensor

One of the engine sensors is mounted on the fuel charging assembly at the end of the throttle shaft (Fig. 10-11). The throttle position sensor sends a signal to the ECA to indicate idle, acceleration, steady speed, and deceleration conditions. This signal is used by the ECA to determine the air/fuel ratio, ignition timing, and exhaust gas flow rate. The throttle position sensor will be discussed further in the sensor section of this chapter.

FIGURE 10-11 Throttle position sensor removal *(Courtesy of Ford Motor Company).*

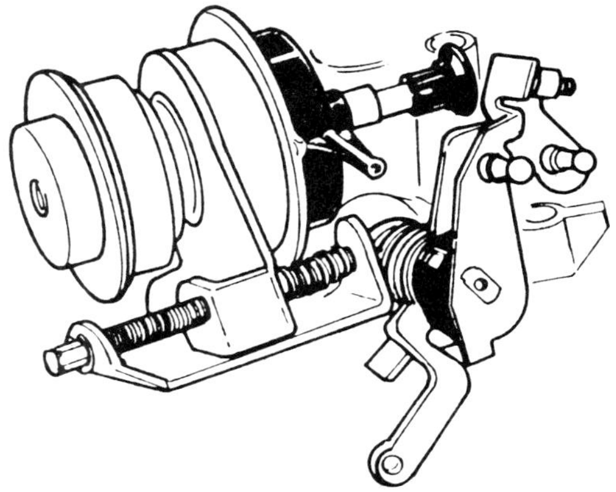

FIGURE 10-12 Throttle kicker *(Courtesy of Ford Motor Company).*

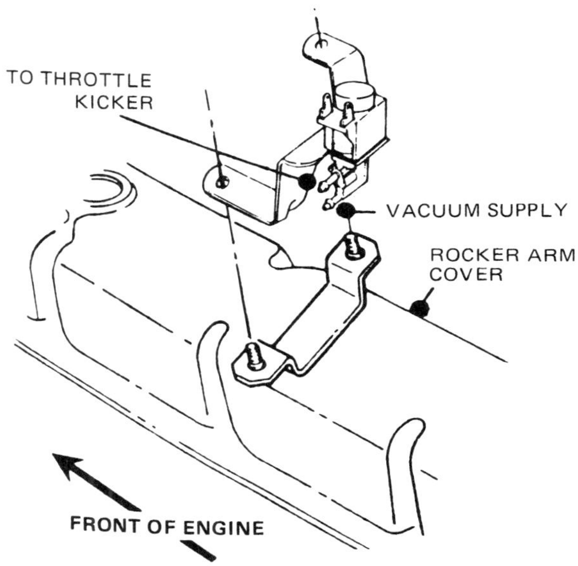

FIGURE 10-13 Throttle kicker solenoid *(Courtesy of Ford Motor Company).*

Throttle Position/Throttle Kicker Assembly

A throttle position assembly or throttle kicker is mounted on the throttle linkage side of the fuel charging assembly (Fig. 10-12). The throttle kicker is operated by vacuum from the throttle kicker solenoid (Fig. 10-13). Intake manifold vacuum is connected to the throttle kicker solenoid. The ECA controls the throttle kicker solenoid. When the ECA energizes the solenoid, vacuum applies a diaphragm in the throttle kicker. The diaphragm moves and causes an attached shaft to extend and open the throttle plates. This causes an increase in idle speed. When a higher idle speed is no longer needed, the ECA de-energizes the solenoid, which blocks vacuum to the throttle kicker. A return spring causes the diaphragm to move back and retract its attached shaft. Higher idle speeds are needed when the air-conditioning system is on, during engine warm-up, and during overheating conditions. The ECA receives an engine speed signal from the crankshaft sensor, engine coolant temperature sensor, and throttle position sensor in order to control the throttle kicker solenoid and throttle kicker.

Cold Engine Speed Control

A thermostat housing assembly, also known as a cold engine speed control, is mounted on the same side of the fuel charging assembly as the throttle position sensor (Fig. 10-14). This thermostatic spring operates a fast idle cam to provide higher engine speed during cold idle operation. The thermostatic spring is heated by an electric resistive element that uses a 7.3-volt source from the alternator stator. The fast idle cam has multiple positions that automat-

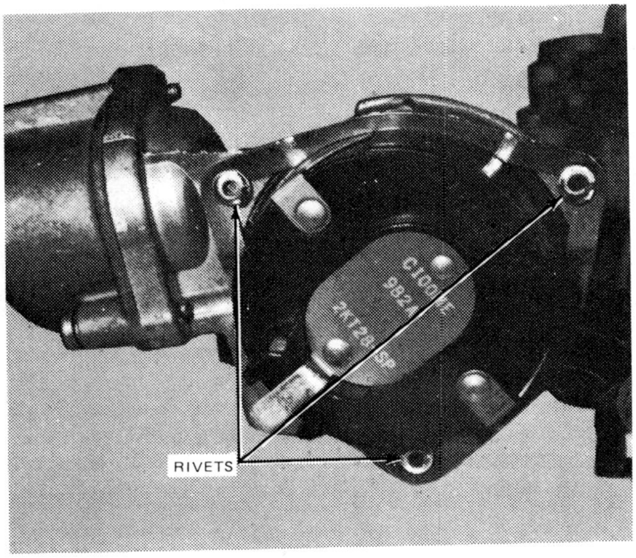

FIGURE 10-14 Choke cap rivets *(Courtesy of Ford Motor Company).*

ically allow for a decreasing idle speed as the engine warms (Fig. 10-15). The automatic kickdown feature is accomplished by the ECA control of a pulldown control on the thermostat housing. The ECA can mechanically move the fast idle cam to a lower step through operation of the pulldown control as the engine reaches specified temperature levels. The fuel charging assembly also has provisions for various vacuum hose connections.

Fuel Return Line

A steel fuel line connects the fuel charging assembly to the fuel tank for the return of excess system fuel pressure. The return line is connected to the fuel charging assembly and electric fuel pump by threaded fasteners. It empties into the sump area of the fuel tank and has a check valve at the end. The check valve prevents siphoning of the fuel into the return line should the vehicle become involved in a rollover accident.

AIR INDUCTION

The air induction system consists of the air cleaner assembly, throttle body, and intake manifold.

Air Cleaner Assembly

The air cleaner assembly contains a replaceable pleated paper element and a replaceable positive crankcase ventilation system filter (Fig. 10-16). A controlled combustion system or preheated intake air system is used in the duct attached to the housing.

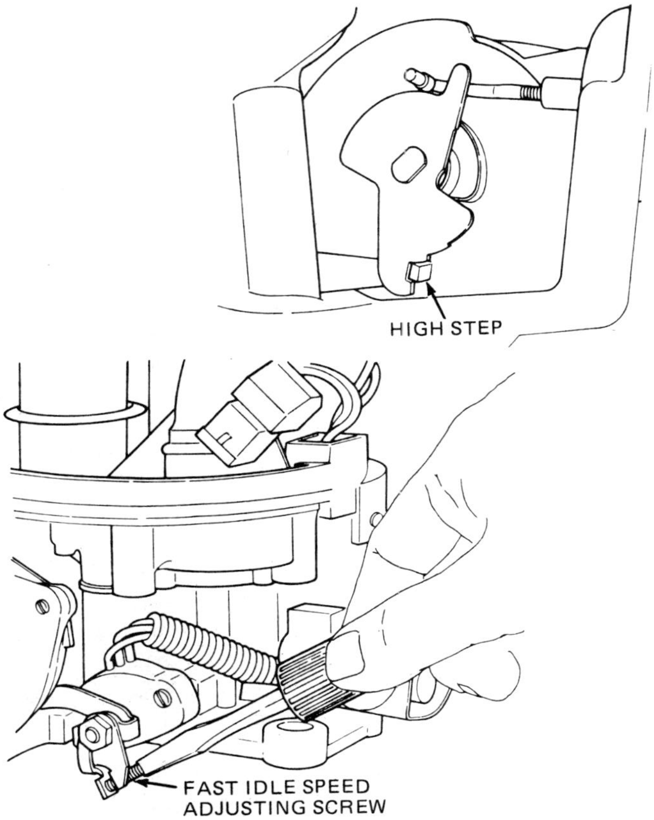

FIGURE 10-15 The 5.0-liter engine with electronic fuel injection and fast idle rpm adjustment *(Courtesy of Ford Motor Company).*

Throttle Body

The throttle body is part of the fuel charging assembly (Fig. 10-17). It regulates air volume entering the engine according to vehicle driver demand on the accelerator pedal. The throttle body is die-cast aluminum and is mounted in a conventional way on the intake manifold.

Intake Manifold

The intake manifold is of standard design. It has runners that carry the air/fuel mixture to the intake valve ports of the cylinder heads.

ENGINE SENSORS

Input Data

To control various devices or actuators, the electronic control assembly relies on input data from engine sensors. These sensors provide data over a range of operation sometimes referred to as *parameters.* Engine parameters sensed

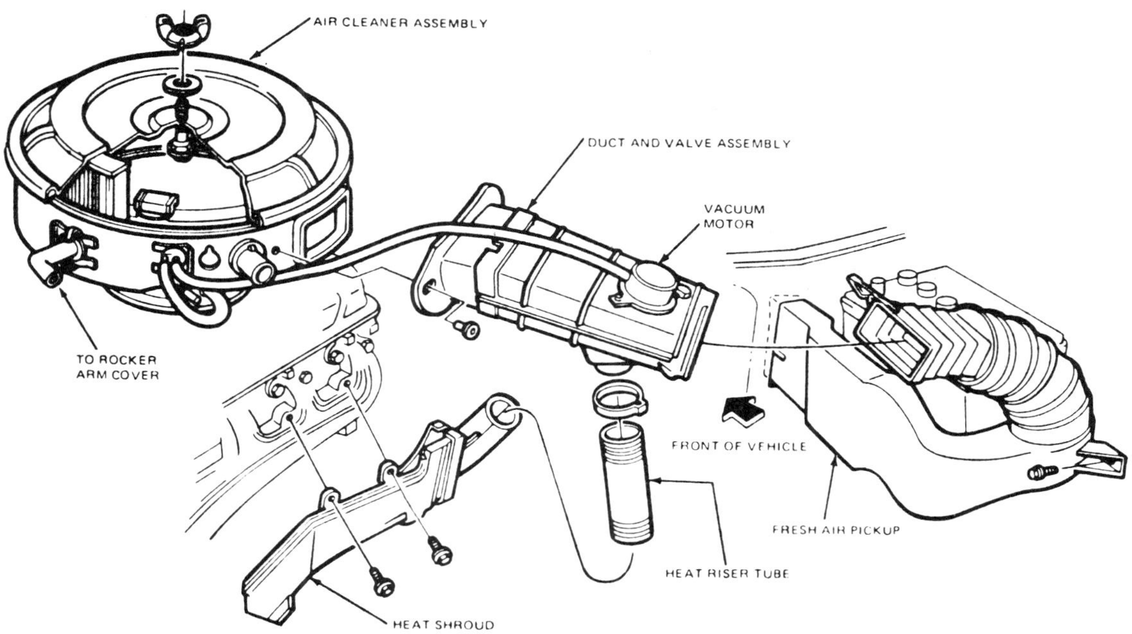

FIGURE 10-16 Typical air cleaner and duct system, V-8 *(Courtesy of Ford Motor Company).*

include crankshaft position, barometric and manifold absolute pressure, engine coolant temperature, exhaust gas oxygen content, throttle position, exhaust gas recirculation valve position, air charge temperature, and a distributor pickup sensor that is used only on engines without a crankshaft position sensor.

Crankshaft Position Sensor

Crankshaft position and engine speed data are produced by the crankshaft position sensor (CPS). This information is used by the ECA to control the EGR valve, canister purge valve, fuel injectors, and ignition timing. The CPS, which is used on the 5.0-liter V-8 only, is mounted at the front of the engine block and is in direct alignment with a crankshaft pulse ring that is attached to the crankshaft harmonic balancer (Fig. 10-18). On eight-cylinder engines, the pulse ring has four teeth. When the pulse ring spins with the crankshaft, an electrical signal is produced in the CPS as each tooth passes the sensor. The signal is produced because the tip of the sensor contains a permanent magnet (Fig. 10-19). When the pulse ring tooth passes near the sensor, the magnetic field increases in intensity and induces a voltage in the sensor coil winding as the tooth goes past (Fig. 10-20). This activity is similar to the reluctor and pickup coil action in ignition distributors. The CPS signal is sent to the ECA where it is used to compute ignition timing, engine speed, when to trigger the fuel injectors, and when to allow EGR valve and canister purge valve operation. The pulse produced in the CPS coil is an ac voltage and is generated at 10° BTDC due to the position of the pulse ring on the harmonic balancer.

B/Map Sensor

The barometric and manifold absolute pressure sensors (B/MAP) are combined into a single assembly (Fig. 10-21). The barometric pressure sensor measures atmospheric pressure and changes it into an electrical signal. The ECA receives this signal and uses it to compute and control ignition timing, fuel injector pulse width, and EGR valve and throttle kicker solenoid operation at specific altitudes. The manifold absolute pressure sensor measures intake manifold pressure and converts it into an electrical signal. The ECA receives this signal and uses it to calculate and control ignition timing, fuel injector pulse width, and EGR valve and throttle kicker solenoid operation according to engine load. The B/MAP sensor is located in the engine compartment on the right front fender well. It is a device that changes in resistance value with changes in pressure. The ECA supplies an 8- to 10-volt source for the sensor assembly, which in accordance with its resistance will return a specified voltage representing altitude and manifold pressure via two signal lines.

Engine Coolant Temperature Sensor

A thermistor is used to measure engine coolant temperature (Fig. 10-22). As the sensor temperature increases, its resistance decreases. At operating temperature, the sensor

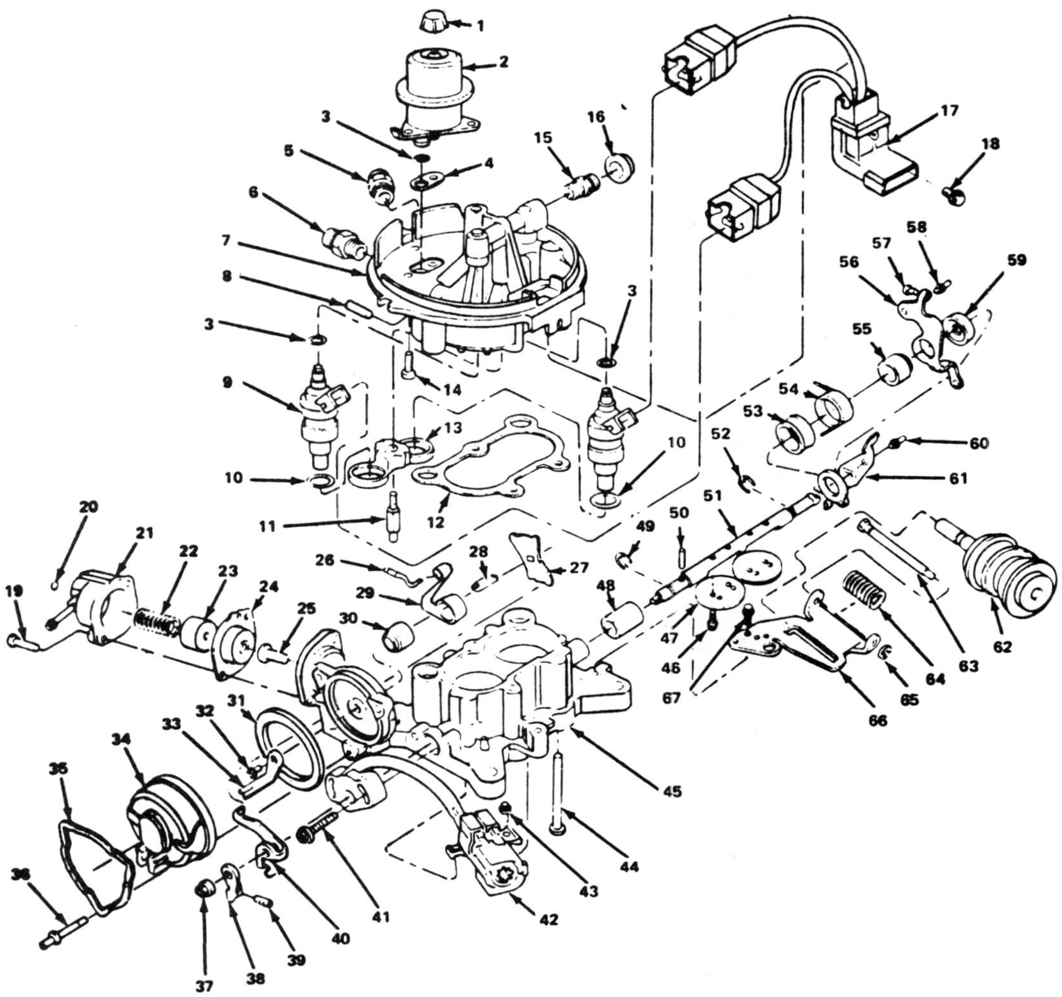

ITEM **PART NAME**

1. PLUG-FUEL PRESSURE REGULATOR ADJUSTING SCREW
2. REGULATOR ASSEMBLY—FUEL PRESSURE
3. SEAL—5/16 × 070 "O" RING
4. GASKET—FUEL PRESSURE REGULATOR
5. CONNECTOR—1/4 PIPE TO 1/2 20
6. CONNECTOR—1/8 PIPE TO 9/16 16
7. BODY—FUEL CHARGING MAIN
8. PLUG—1/16 × 27 HEADLESS HEX
9. INJECTOR ASSEMBLY—FUEL
10. SEAL—5/8 × 103 "O" RING
11. SCREW—FUEL INJECTOR RETAINING
12. GASKET—FUEL CHARGING BODY
13. RETAINER—FUEL INJEJCTOR
14. SCREW M5 0 × 20 0 PAN HEAD
15. VALVE ASSEMBLY—DIAGNOSTIC VALVE
16. CAP—FUEL PRESSURE RELIEF VALVE
17. WIRING ASSEMBLY—FUEL CHARGING
18. SCREW—M3.5 × 1.27 × 12.7 PAN HEAD
19. SCREW & WASHER—M4 × 70 × 20 00
20. BALL—LEAD SHOT 26 24 DIA
21. COVER ASSEMBLY—CONTROL DIAPHRAGM
22. SPRING—CONTROL MODULATOR
23. RETAINER—PULLDOWN DIAPHRAGM
24. DIAPHRAGM—PULLDOWN CONTROL
25. ADJUSTER—PULLDOWN CONTROL
26. ROD—FAST IDLE CONTROL
27. CAM—FAST IDLE
28. SHAFT—CHOKE HOUSING
29. POSITIONER—FAST IDLE CONTROL ROD
30. BUSHING—CHOKE HOUSING
31. GASKET—THERMOSTAT HOUSING
32. SCREW & WASHER—M3.5 × 06 × 6 PAN HEAD
33. LEVER—CHOKE THERMOSTAT
34. HOUSING ASSEMBLY—THERMOSTAT

ITEM **PART NAME**

35. RETAINER—HOUSING ASSEMBLY
36. POP RIVER—1/8 × 517
37. NUT & WASHER ASSEMBLY—7 6H HEX
38. LEVER—FAST IDLE CAM ADJUSTER
39. SCREW—NO 10 32 × 50 SET SLOTTED HEAD
40. LEVER—FAST IDLE
41. SCREW & WASHER—M4 0 7 × 22 0 PAN HEAD
42. THROTTLE POSITION SENSOR
43. SCREW—M4 × 7 × 14 0 HEX WASHER TAP
44. SCREW—M5 × 0 7 × 55 0
45. BODY—FUEL CHARGING—THROTTLE
46. SCREW—M3 × 05 × 74 HEX WASHER HEAD
47. PLATE—THROTTLE
48. REARING—THROTTLE CONTROL LINKAGE
49. "E" RING—7/32 RETAINING
50. PIN—SPRING COILED
51. SHAFT—THROTTLE
52. "C" RING—THROTTLE SHAFT BUSHING
53. BEARING—THROTTLE CONTROL LINKAGE
54. SPRING—THROTTLE RETURN
55. BUSHING—ACCELERATOR PUMP OVER TRAVEL SPRING
56. LEVER—TRANSMISSION LINKAGE
57. SCREW—M4 × 07 × 76
58. PIN—TRANSMISSION LINKAGE LEVER
59. SPACER—THROTTLE SHAFT
60. BALL—THROTTLE LEVER
61. LEVER—THROTTLE
62. POSITIONER ASSEMBLY—THROTTLE
63. SCREW—1/4 28 × 2 53 HEX HEAD ADJUSTING
64. SPRING—THROTTLE POSITIONER RETAINING
65. "E" RING—RETAINING
66. BRACKET—THROTTLE POSITIONER
67. SCREW—M5 × 8 × 14 0 HEX WASHER TAP

FIGURE 10-17 Electronic fuel injection fuel charging assembly (exploded view) *(Courtesy of Ford Motor Company).*

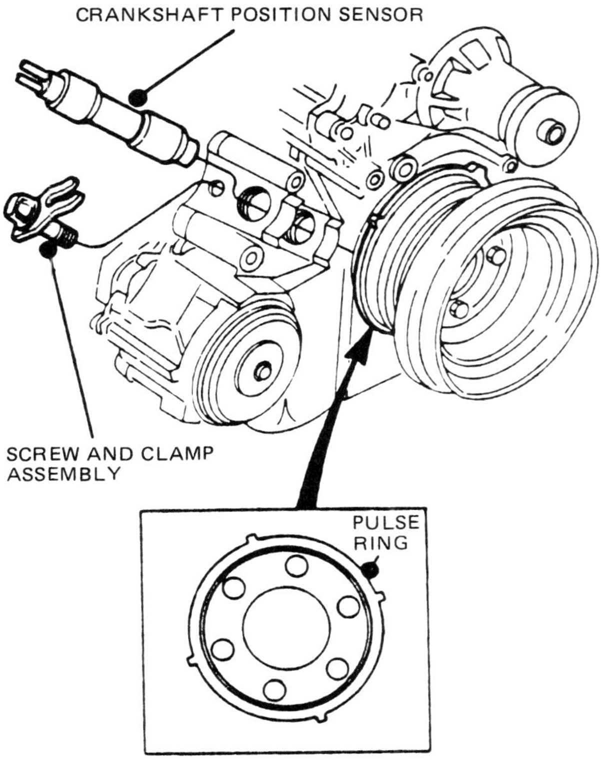

FIGURE 10-18 Crankshaft position sensor installation *(Courtesy of Ford Motor Company).*

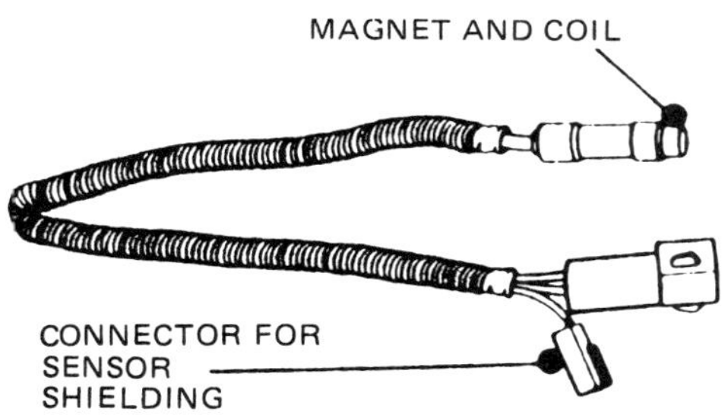

FIGURE 10-19 Crankshaft position sensor *(Courtesy of Ford Motor Company).*

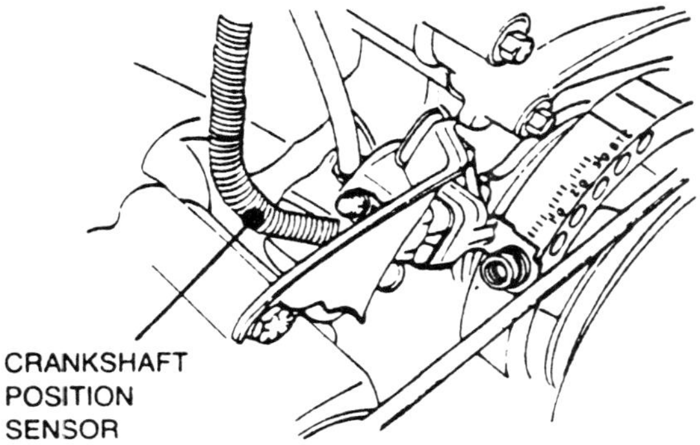

FIGURE 10-20 Crankshaft position sensor mounting *(Courtesy of Ford Motor Company).*

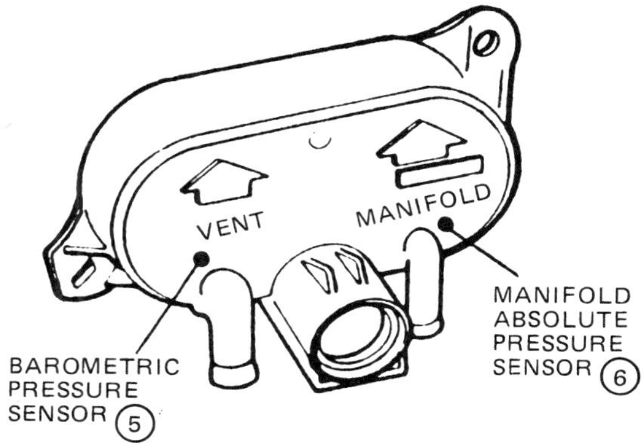

FIGURE 10-21 Barometric and manifold absolute pressure sensor *(Courtesy of Ford Motor Company).*

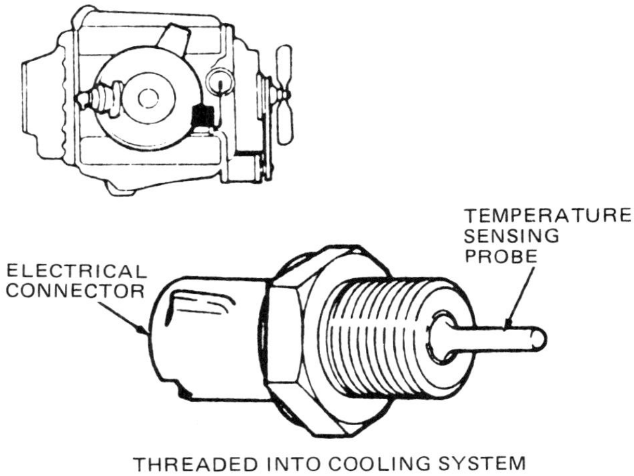

FIGURE 10-22 Engine coolant temperature sensor *(Courtesy of Ford Motor Company).*

should provide a 0.75- to 2.6-volt signal to the ECA. The CTS is located at the front of the intake manifold in the heater outlet fitting at the right valve cover. When the engine coolant temperature is low, the voltage signal from the sensor to the ECA is high. When engine coolant temperature is high, the CTS signal is low. The ECA uses the CTS signal to control ignition timing, exhaust gas recirculation valve operation, and the air/fuel ratio or injector pulse width. The sensor also affects operation of the thermactor (air pump) system, canister purge, and throttle kicker.

Exhaust Gas Oxygen Sensor

An exhaust gas oxygen (EGO) sensor measures the oxygen content of the exhaust stream (Fig. 10-23). It generates a voltage that varies with the amount of oxygen in the exhaust. This sensor is mounted in the right exhaust manifold (Fig. 10-24). It contains two electrodes made of zirconia oxide coated with platinum to protect them against the high

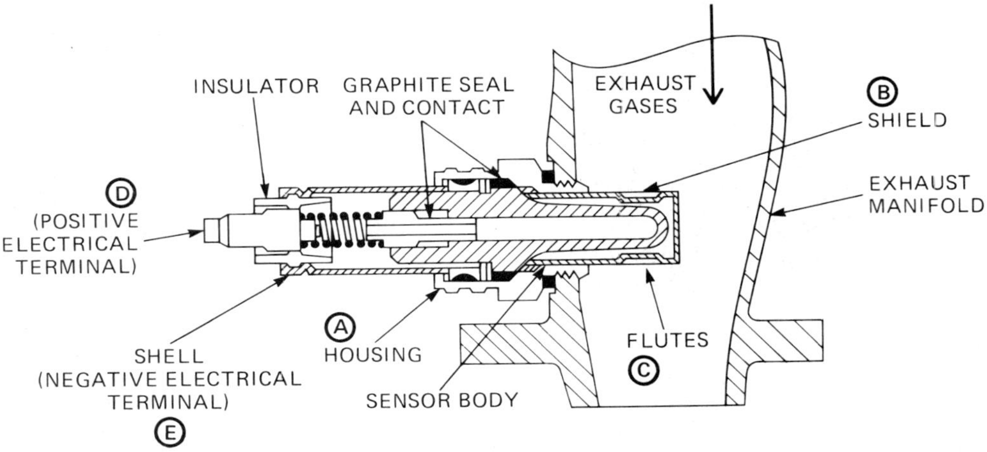

FIGURE 10-23 Exhaust gas oxygen sensor assembly *(Courtesy of Ford Motor Company).*

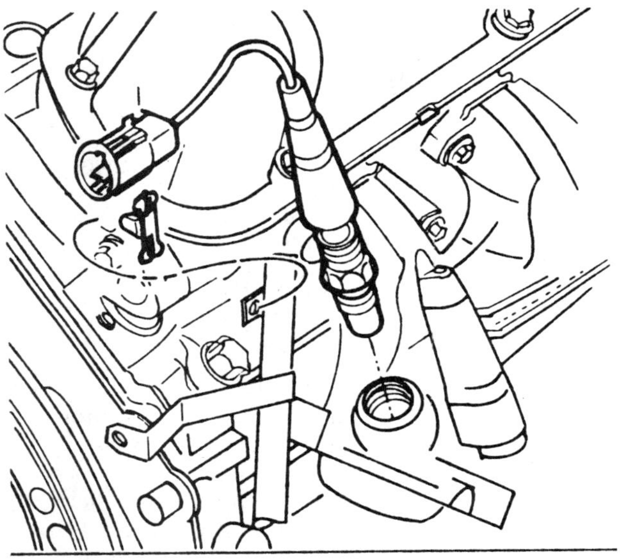

FIGURE 10-24 Exhaust gas oxygen sensor installation *(Courtesy of Ford Motor Company).*

exhaust temperatures. When oxygen in the exhaust stream comes into contact with one electrode, a voltage is produced. The other electrode is exposed to oxygen from outside fresh air and produces a voltage. By comparison, the ECA can determine the oxygen content of the exhaust stream. Low exhaust oxygen content produces a high sensor voltage signal. High exhaust oxygen content produces a low sensor voltage signal. Low exhaust oxygen content also means the engine air/fuel mixture entering the cylinders is too rich. The ECA then shortens the fuel injector pulse width to lean the mixture. High exhaust gas oxygen content means the engine air/fuel ratio entering the cylinders is too lean. The ECA responds by widening and enrichening the fuel injector pulse width.

The EGO sensor enables the ECA to see the results of its efforts to control the air/fuel mixture. The EGO sensor must reach an operating temperature of about 600°F before it can produce a voltage large enough to be recognizable by the ECA. Until this temperature is reached, the EEC system operates in open loop. When the operating temperature is reached, the EEC system operates in closed loop. These two modes of operation will be discussed later in this chapter.

Throttle Position Sensor

A throttle position sensor (TPS) is used to indicate to the ECA the driver's demand on the throttle. It is mounted on the side of the throttle body on the end of the throttle shaft (Fig. 10-11). A varying voltage signal sent to the ECA by the TPS is used for control of ignition timing, EGR flow rate, injector pulse width, and thermactor air. The TPS is a variable resistor whose resistance value changes with the angle of the throttle plates. The ECA supplies an 8- to 10-volt source to the sensor, which transmits throttle position data over a return signal line back to the ECA. The TPS indicates to the ECA one of three modes of operation: closed throttle, part throttle, or wide-open throttle. The closed-throttle signal represents idle or deceleration to the ECA, depending also on engine speed and B/MAP signals. Part-throttle signals indicate normal operation for steady speed or acceleration, depending on the rate of change of the signal. Wide-open-throttle signals indicate the need for maximum acceleration where the ECA may cause the EEC system to enter the open-loop mode of operation in order to provide full fuel enrichment. The TPS is adjustable on early EEC-III systems only. This will be discussed in the system service section of this chapter.

EGR Valve Position Sensor

An exhaust gas recirculation valve position sensor monitors the amount of movement of the EGR valve pintle. The exhaust gas recirculation valve is located in the intake manifold beneath the fuel charging assembly. It controls the oxides of nitrogen (NO_x) emissions by recirculating exhaust gases into the intake manifold air/fuel mixture. The EGR valve position sensor is attached to the end of the EGR pintle. This sensor is supplied with an 8- to 10-volt source from the ECA. As the position of the EGR valve pintle changes, the position sensor resistance changes, and the return signal to the ECA also changes. With these data, the ECA can accurately control the EGR valve solenoid, which controls how much exhaust gases are allowed to pass through the EGR valve into the intake manifold.

Air Charge Temperature Sensor

An air charge temperature sensor is used to provide data to the ECA concerning the incoming air/fuel mixture temperature (Fig. 10-25). It is located in the intake manifold runner for cylinder 7. The ECA provides an 8- to 10-volt source for the sensor, which is a thermistor-type variable resistor. As the air charge temperature rises, the sensor resistance decreases and the signal to the ECA is a lower voltage. When air charge temperature is low, the sensor resistance is high and the signal to the ECA is a higher voltage.

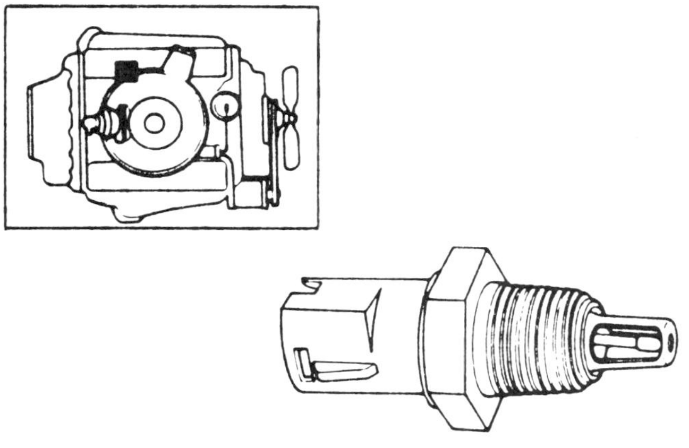

FIGURE 10-25 Air charge temperature sensor *(Courtesy of Ford Motor Company).*

Distributor Pickup Sensor

A distributor pickup sensor is used on the 3.8-liter V-6 EEC-IV engine. It serves the same function as the crankshaft position sensor on the 5.0-liter V-8 engine. Instead of the pickup coil pulse triggering the ignition module directly for control of spark advance, the pickup coil pulse is sent to the ECA, which has circuitry to control spark advance. The conventional centrifugal and vacuum advance mechanisms are no longer necessary. The ECA also uses the pickup coil pulses to determine engine speed.

ELECTRONIC ENGINE CONTROL SYSTEM

The electronic engine control III system consists of an electronic control assembly (ECA), sensor and switch inputs, and controlled devices or actuators on the command or output side of the system (Fig. 10-26).

Electronic Control Assembly

The electronic control assembly is a computer that has a microprocessor and a calibration unit. It receives input data from sensors and switches discussed earlier in this chapter. These data are used to make high-speed calculations in accordance with the computer program. The microprocessor assembly and calibration assembly take part in this decision-making process. As a result, a command or output signal is issued to some device or actuator that is controlled by the computer. The electronic control assembly, then, is the "brain" of the EEC-III system.

Input signals are received by the microprocessor assembly. It converts analog sensor signals into digital signals, temporarily stores them for analysis, chooses an operating strategy, and makes high-speed computations for control of ignition spark, EGR flow, air/fuel ratio, canister purge, throttle kicker, and other output signals. The calibration assembly is attached to the top side of the microprocessor assembly by two screws (Fig. 10-1). It is a memory that provides calibration information so that the microprocessor can operate the EEC system on a particular vehicle. It can also recall from memory data stored at an earlier time. The EEC system is provided with power through its own separate electrical relay (Fig. 10-27). The relay is energized whenever the ignition switch is in the start or run position. The relay connects power to the electronic control assembly. A diode in the relay prevents damage to the ECA if a reversed polarity condition occurs.

The EEC system uses eight sensor inputs and various switch inputs depending on vehicle application. Because these have been discussed earlier, the next system topic

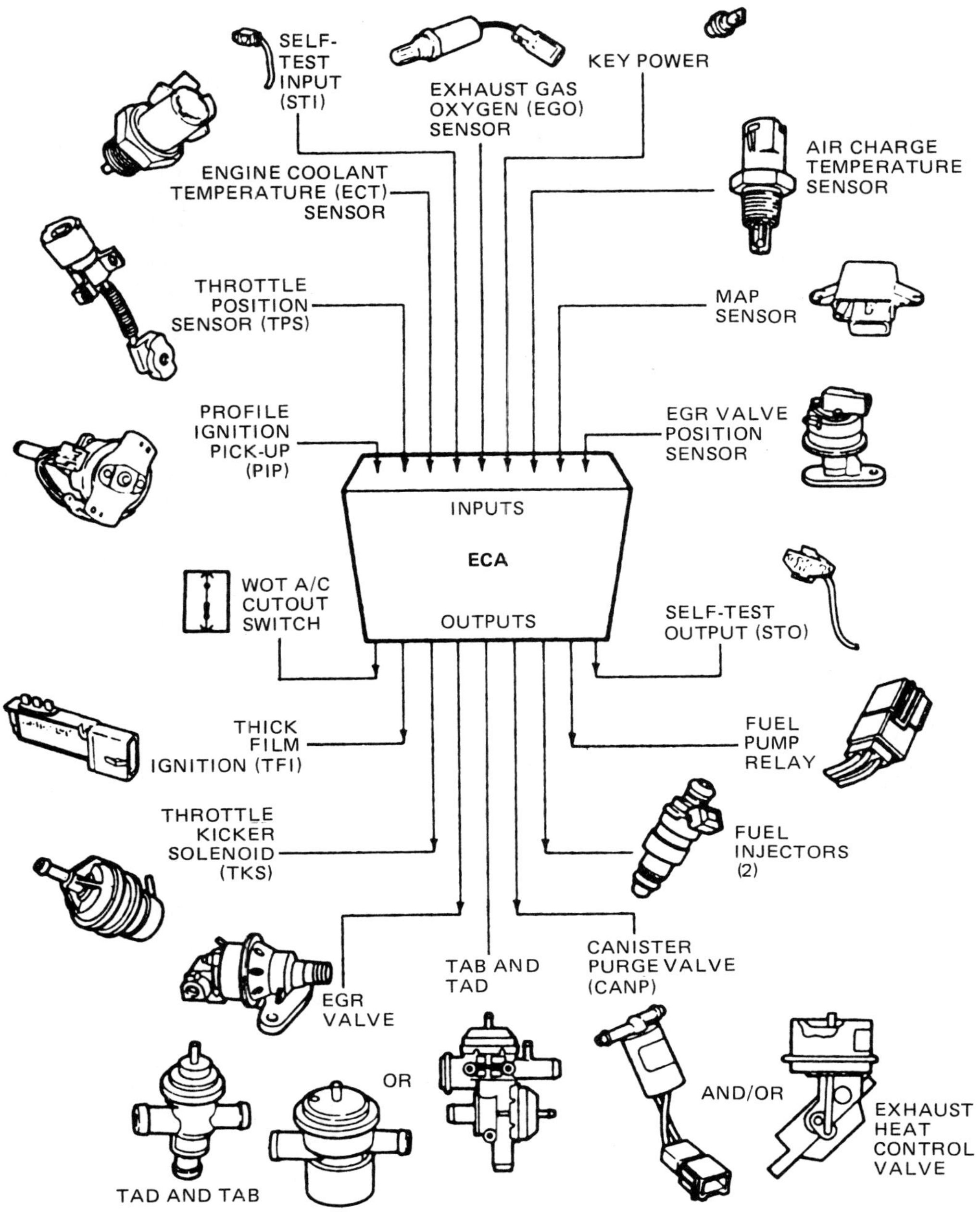

FIGURE 10-26 Electronic control assembly inputs and outputs *(Courtesy of Ford Motor Company).*

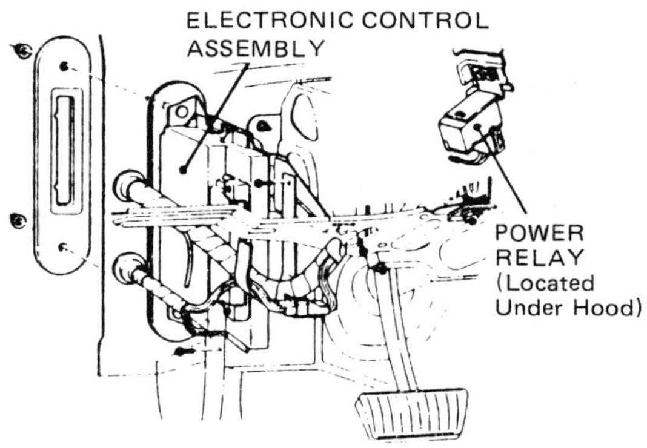

FIGURE 10-27 Electronic control assembly for Ford/Mercury and Lincoln/Mark VI *(Courtesy of Ford Motor Company).*

involves those devices or actuators that are controlled by the ECA.

Output Commands

The ECA output circuitry controls the throttle kicker solenoid (TKS), EGR control solenoid (EGRV and EGRC), thermactor air diverter (TAD) and thermactor air bypass (TAB) solenoids, canister purge solenoid (CANP), fuel injectors, and fuel pump relay (Fig. 10-26).

Throttle Kicker Solenoid

The ECA uses the throttle kicker solenoid (TKS) to control idle speed for various engine operating conditions (Fig. 10-28). The throttle kicker solenoid is activated by the ECA when engine coolant temperature is below a specified value. This improves cold start performance. When engine coolant temperature is above a specified value, the ECA energizes the TKS to provide additional engine cooling. When the air-conditioning system is turned on, the TKS is activated to offset the load of the compressor on the engine. This improves idle quality and prevents stalling. The TKS is energized at specified altitudes to improve idle quality. It is also activated whenever the engine is restarted to prevent stalling upon transmission engagement. In each case, the TKS controls vacuum to the throttle kicker actuator. The actuator is a vacuum-operated diaphragm and shaft that extends when the TKS is activated and allows vacuum to move the diaphragm.

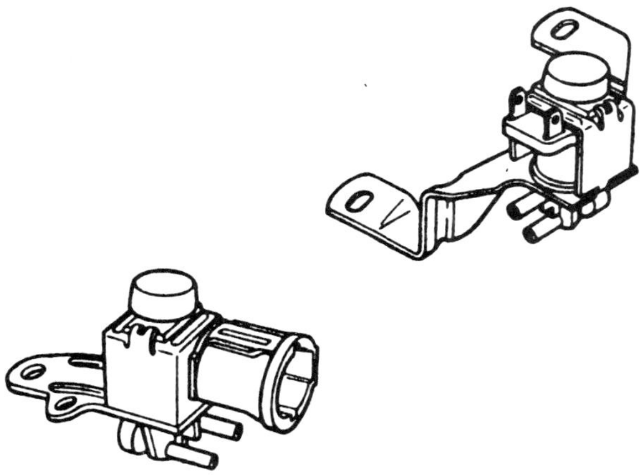

FIGURE 10-28 Throttle kicker solenoid (Courtesy of Ford Motor Company).

EGRC and EGRV Solenoids

Two solenoids are used to regulate vacuum to the EGR valve (Fig. 10-29). The EGR control (EGRC) solenoid is used by the ECA to allow or block vacuum to the EGR valve. This solenoid blocks vacuum to the EGR valve until a specified engine operating temperature is reached and when engine speed is above a specified rpm. When these conditions are satisfied, the ECA can control EGR flow rate through the use of the EGR vent (EGRV) solenoid. By triggering or pulsing the EGRV, the vacuum to the EGR diaphragm can be regulated or modulated. By doing this, a specific EGR valve position can be achieved for a particular set of engine operating conditions. The ECA receives information concerning the EGR valve position from the EGR valve position sensor discussed earlier in the Engine Sensors section of this chapter and can then control the EGRV solenoid to obtain correct exhaust gas recirculation. The EGRC and EGRV solenoids are serviceable separately. Testing information is provided in the service manual.

Thermactor Emission Control

The ECA also controls the thermactor air emission control system. This system consists of an air pump that blows fresh air into either the exhaust manifold, catalytic converter, or the atmosphere. The ECA uses a thermactor air diverter (TAD) solenoid and a thermactor air bypass (TAB) solenoid to direct airflow from the thermactor pump (Fig. 10-30). When the engine is started, the ECA energizes the TAD and TAB solenoids, which then direct fresh air into the exhaust manifold. This helps reduce carbon monoxide and hydrocarbons during engine warmup and also causes the EGO sensor to reach its operating temperature sooner.

When the engine reaches normal operating tempera-

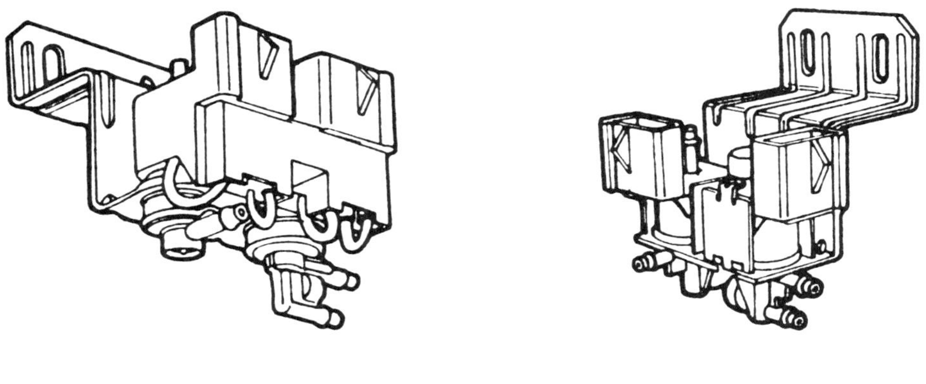

FIGURE 10-29 Normally open and normally closed valve assembly combinations (EGR) *(Courtesy of Ford Motor Company).*

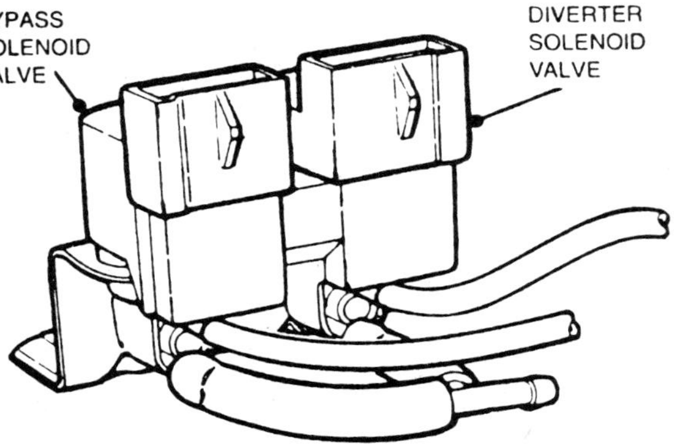

FIGURE 10-30 Thermactor air bypass and diverter solenoid valves *(Courtesy of Ford Motor Company).*

ture, the ECA de-energizes the TAD solenoid and leaves the TAB solenoid energized, which then directs fresh air into the catalytic converter. This enables the catalytic converter to not only reduce carbon monoxide and hydrocarbon levels, but also oxides of nitrogen levels. When the ECA de-energizes the TAB and TAD solenoids, fresh air is diverted to the atmosphere. This is necessary when idling for a long period or if the EGO sensor signal does not indicate rich or lean conditions for a period of time. This helps protect the catalytic converter.

Canister Purge

The canister purge (CANP) solenoid, which was used until 1983, is part of the evaporative emission control system. This solenoid is controlled by the ECA and determines when vapors stored in the charcoal canister are drawn into the intake manifold air/fuel charge. When the ECA energizes the solenoid, vapors are purged from the canister. This occurs only when engine coolant temperature is above a specified cold temperature or below a specified overheat temperature, engine speed is above a specified rpm, and after a timer circuit has expired following engine startup. Vapors will not be purged at closed throttle. Testing of the solenoid and related circuits is detailed in the service manual.

Fuel Control

The ECA controls the fuel injectors in the fuel charging assembly. When the injectors are energized, fuel sprays into the throttle bores. The amount of time the injectors are energized is known as pulse width and determines the air/fuel ratio. Based on input data from engine sensors and switches, the ECA can calculate and issue an appropriate pulse width for accurate air/fuel mixture control.

As discussed earlier, the ECA controls the fuel pump relay and ignition module. The EEC system uses a Duraspark III ignition module and a Duraspark II ignition coil. No advance mechanisms are needed as the ECA determines spark advance and signals the ignition module when to break the coil primary circuit. The distributor only sends spark to the appropriate spark plug.

ELECTRONIC CONTROL ASSEMBLY MODES OF OPERATION

The EEC computer has two general modes of operation: open loop and closed loop.

Open Loop

The open-loop mode of operation is activated whenever the EGO sensor signal is too small to be recognized by the ECA, the engine is cold, a wide-open-throttle signal is received by the ECA, or the ECA enters the limited-operation strategy mode. The LOS mode will be discussed in the next section.

Closed Loop

As long as the EGO sensor signal is present and the engine is at operating temperature, the closed-loop mode of operation is activated by the ECA within about 60 seconds of starting the engine.

During open-loop conditions, the ECA determines air/fuel ratio without using the EGO sensor signal. During closed loop, the EGO sensor provides a feedback signal that enables the ECA to accurately control the air/fuel ratio.

ENGINE OPERATING MODES

Cold Start-up

During the cold startup mode, the engine runs at fast idle. When the ignition switch is turned on, the ECA energizes the fuel pump relay and fuel pump for 1 second unless a cranking signal is received. When a cranking signal is present, the fuel pump runs continuously and the fuel injectors begin to deliver fuel based on engine sensor input data to the ECA. The timing of injection is based on the distributor signal. The electric choke coil provides fast idle through rotation of the fast idle cam when the driver depresses the accelerator pedal prior to cranking the engine. The amount of fuel delivered is determined by the ECA, which relies on the engine coolant temperature sensor and B/MAP sensor signals for its calculations. The ECA does not energize the EGR and CANP solenoids until operating temperature and closed-loop operation are achieved. Spark advance is controlled by the ECA, which bases its calculations on engine speed, load, coolant temperature, B/MAP, and EGR flow rate. The system is in open-loop operation.

As the engine coolant temperature rises to about 100°F, the ECA reduces the fuel enrichment, the electric choke coil heats to allow the fast idle cam to drop to a lower step when the accelerator pedal is depressed, the EGR and CANP solenoids are energized, and the EGO sensor signal has increased enough to be used by the ECA for closed-loop operation.

Acceleration

Acceleration at wide-open throttle when the engine is at operating temperature causes the ECA to respond by reentering the open-loop mode. The wide-open-throttle operating condition is sensed by the throttle position sensor. The ECA widens the fuel injector pulse width for fuel enrichment. Canister purge and EGR flow are stopped and additional spark advance is provided.

Hot Curb Idle

During hot curb idle closed-throttle conditions, the ECA controls the fuel injectors in the closed-loop mode. If the EGO sensor signal stops for longer than 15 seconds, the ECA enters the open-loop mode. After at least two EGO sensor signals are received, the ECA reenters the closed-loop mode. The canister purge and EGR flow should be stopped.

The ECA operates in the closed-loop mode when the engine is at operating temperature and the vehicle is at a steady highway speed. This mode is known as *hot cruise.* The EGO sensor is providing feedback data and the canister purge and EGR flow are activated. Spark advance is controlled by the ECA according to its programmed schedule.

Deceleration

When decelerating, the cannister purge and EGR flow are stopped, and the ECA computes a fuel injector pulse width based on engine speed, engine coolant temperature, B/MAP, and EGO sensor data. If the computed pulse width is less than 2 milliseconds and the engine speed is 1500 rpm or more, the ECA shuts off the fuel injectors. They remain off until the engine speed drops to about 1200 rpm. Spark advance is determined by the ECA program.

The limited-operation strategy (LOS) mode is entered when a malfunction occurs in the ECA or certain sensor data circuits. While in the LOS mode, the output commands of the ECA are interrupted and the engine operates at base ignition timing of 10° BTDC. Sensor data are ignored by the ECA. The fuel injectors provide a fixed air/fuel ratio. The result of operating in the LOS mode is poor fuel economy and poor performance, higher emissions, and a noticeable decrease in engine power. The vehicle can, however, be driven so that repairs can be made.

SYSTEM SERVICE

In-tank Electric Fuel Pump Service

The in-tank electric fuel pump is serviced as an assembly. Control circuitry and fuel lines should be checked before the pump is condemned. The fuel pump is normal if it provides about 35 to 45 psi, pumps about 10 ounces of fuel in 10 seconds, and fuel pressure remains at a minimum of 30 psi immediately after the pump is shut off. Pressure can be checked at the Schrader valve on the fuel charging assembly. A low pressure reading may also be caused by a fuel line leak, pressure regulator malfunction, or leaking fuel injector. Before disconnecting fuel lines, residual pressure must be relieved. This can be done at the pressure Schrader valve on the fuel charging assembly. Care must be taken to prevent possible fire hazards. A special diagnostic pressure tool is available for pressure testing and also pressure relieving.

The in-line fuel filter is located on the vehicle frame rail near the right rear wheel. Fuel pressure must be relieved before the filter cartridge is unscrewed and discarded. It is a spin-on type of filter and is protected by a sheet metal stone shield.

Engine Sensor Service

The engine sensors can be checked with an accurate digital ohmmeter. The service technician should refer to the appropriate shop manual for checking procedures and specifications.

ECA-controlled Solenoids and Relays

Computer-controlled solenoids can also be checked with a digital ohmmeter. The EGRC and EGRV solenoids, the throttle kicker solenoid, and the TAB and TAD solenoids should be checked according to the appropriate service manual.

The fuel pump relay coil should be tested as described in the service manual.

Throttle Position Sensor Service

Whenever the throttle position sensor is removed or replaced, it is necessary to calibrate its position if it is the adjustable type. With the throttle in the hot curb idle position and off the fast idle cam, the TPS signal line should read 1.6 to 2.4 volts on a high-impedance digital voltmeter with the ignition switch on. The TPS can be adjusted to obtain this reading. Appropriate wiring diagrams and service manuals should be referred to for correct identification of wire terminals.

Idle Speed Adjustment

To set idle speed for both fast idle and curb idle, instructions provided in the service manual should be closely followed. When the adjustments are completed, the TPS calibration should be checked and adjusted if necessary.

SYSTEM DIAGNOSIS

The EEC system computer has a self-diagnostic capability. Before checking EEC circuits, all other possibilities involving engine basics should be considered.

The tools and equipment necessary for EEC system diagnosis include a high-impedance digital multimeter, electronic fuel injection adapter harness, fuel injection pressure gauge, tachometer, hand vacuum pump, jumper wires, vacuum gauge, timing light, ignition spark tester, and an EEC-III tester.

Self-test Mode

A self-test can be performed, which when used in combination with the *quick test,* will provide accurate diagnosis. Connect the EEC-III tester (Fig. 10-31). Engine sensors and actuators are tested by the ECA microprocessor during the self-test. Related sensor and actuator circuitry and connectors are also tested. The self-test mode can be entered by first starting the engine. (This is actually part of the quick test.) The engine should be run at about 1800 rpm until the upper radiator hose is hot and pressurized. Set the EEC-III tester to A-8 and apply vacuum to the B/MAP vent

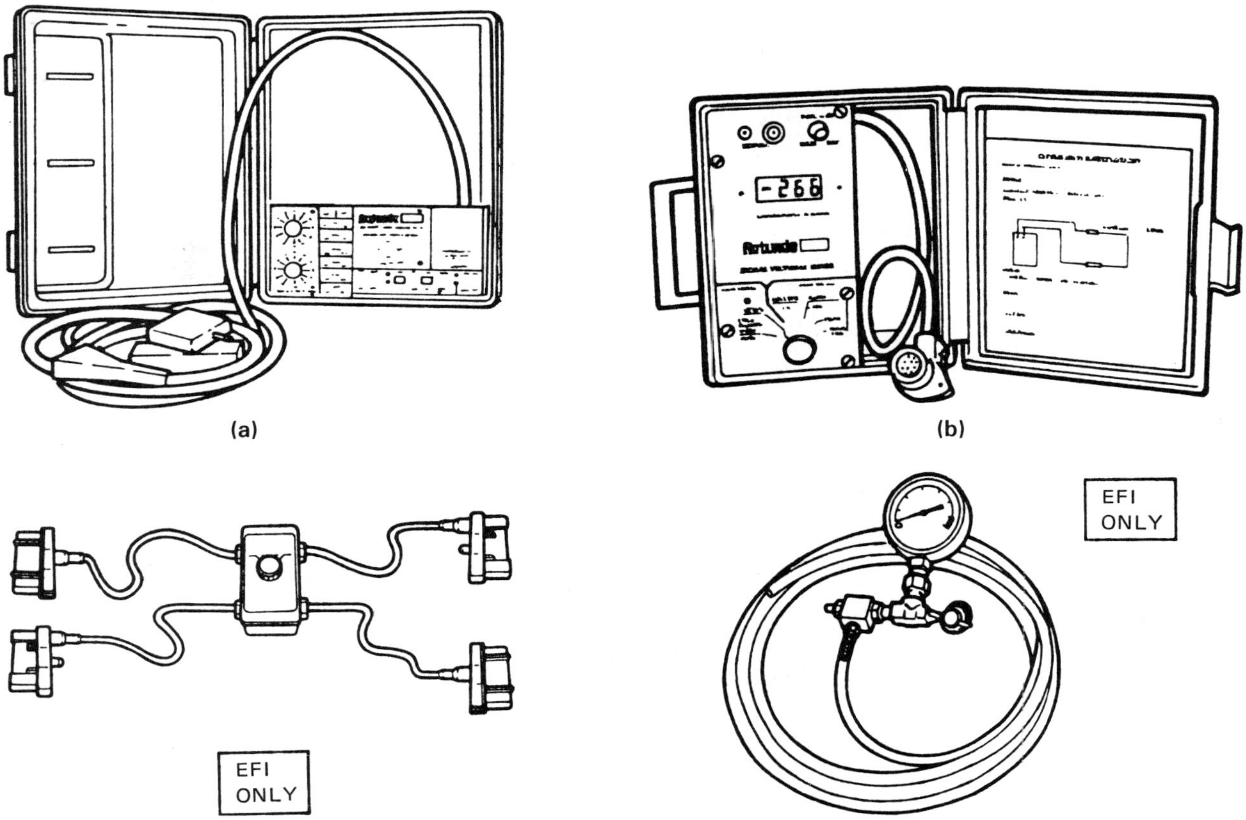

FIGURE 10-31 (a) EEC-III tester. (b) DVOM Rotunda T78L-50-DVOM or T79L-50-DVOM or equivalent. (c) Electronic fuel injection adapter harness T80L-5-EFI or equivalent (EFI only). (d) Electronic fuel injection pressure gauge T80L-9974-A or equivalent (EFI only). (e) Tachometer. (f) Rotunda pump 21-0014 or equivalent. (g) Jumper wire. (h) Vacuum gauge. (i) Timing light. (j) Spark tester. *(Courtesy of Ford Motor Company).*

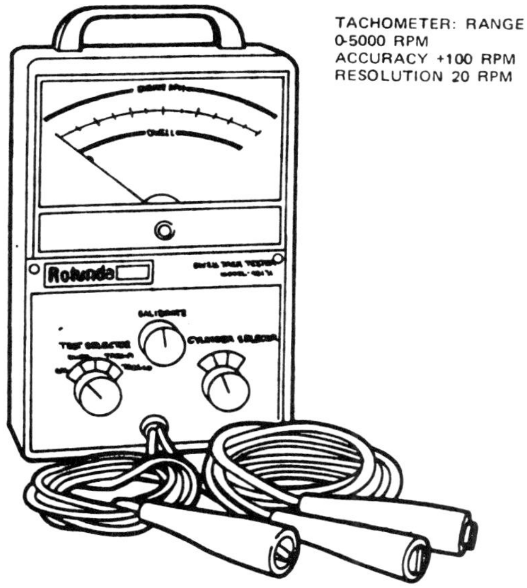

(e)

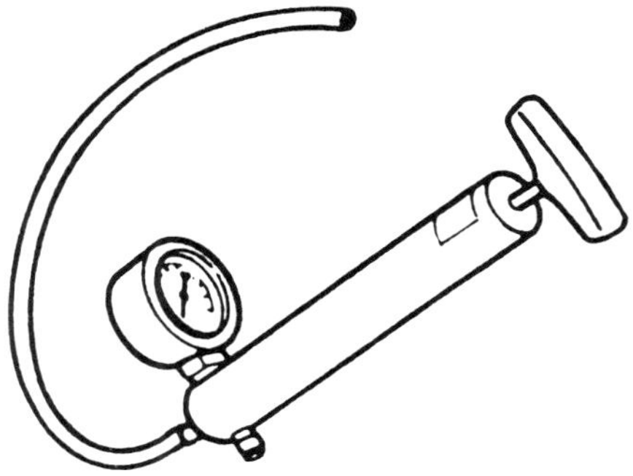

(f)

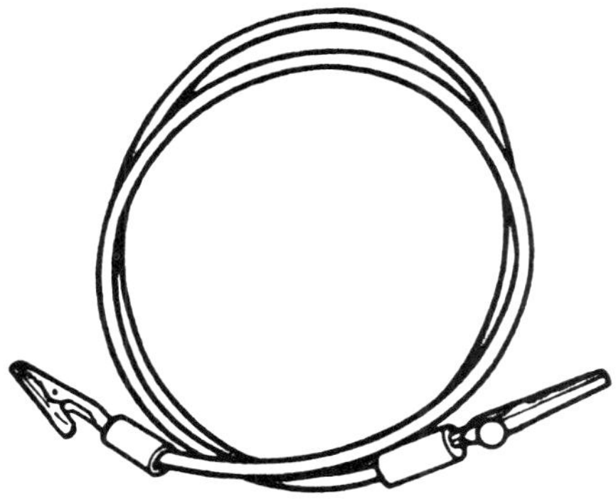

(g)

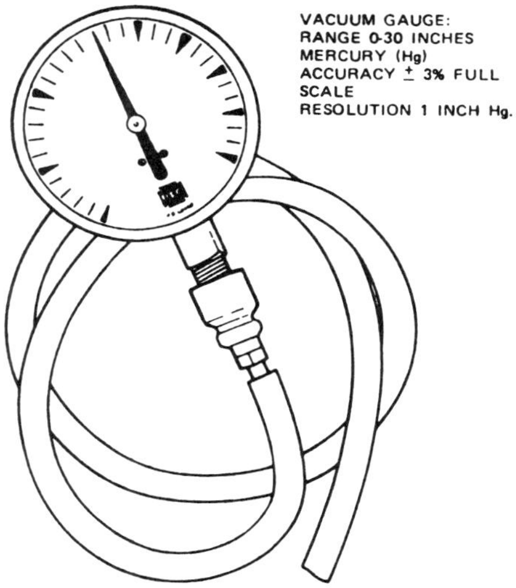

(h)

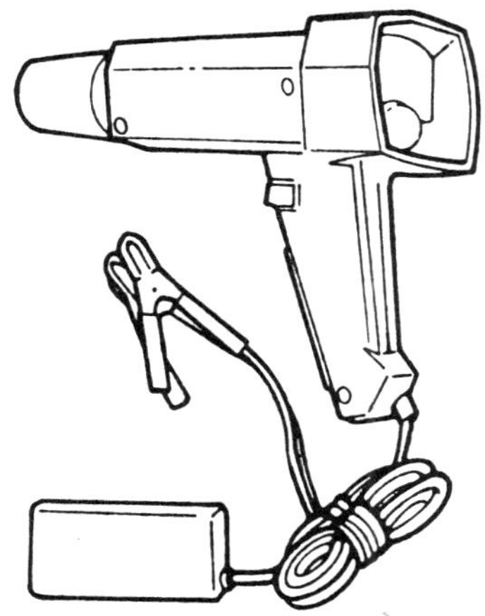

(i)

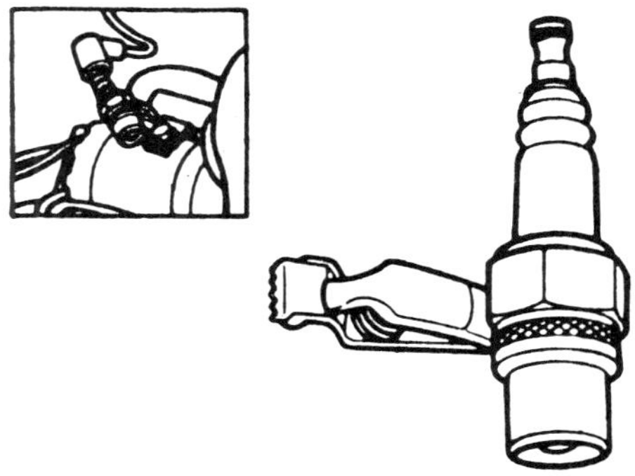

(j)

FIGURE 10-31 (Continued)

port until the voltmeter reading is less than 2.85 volts. Hold the vacuum for 8 seconds and then release it. The self-test will begin by lighting the TKS light, TAD light, TAB light, and CANP light (Fig. 10-32). Initially, the TKS light will come on, go off, pause for 2 seconds, and come on again, and the throttle kicker actuator will extend for about 1.5 minutes. When the TKS light goes out, the tester will display service codes stored in the ECA memory.

Service Codes

Service codes are displayed on the TAD and TAB lights simultaneously. The pulsing of the lights represents digits of the two-digit codes. The lights are off for 5 seconds between codes. A code 21, for example, is represented when the TAD and TAB lights come on for 0.5 second, then off for 0.5 second, then on for 0.5 second, then pause

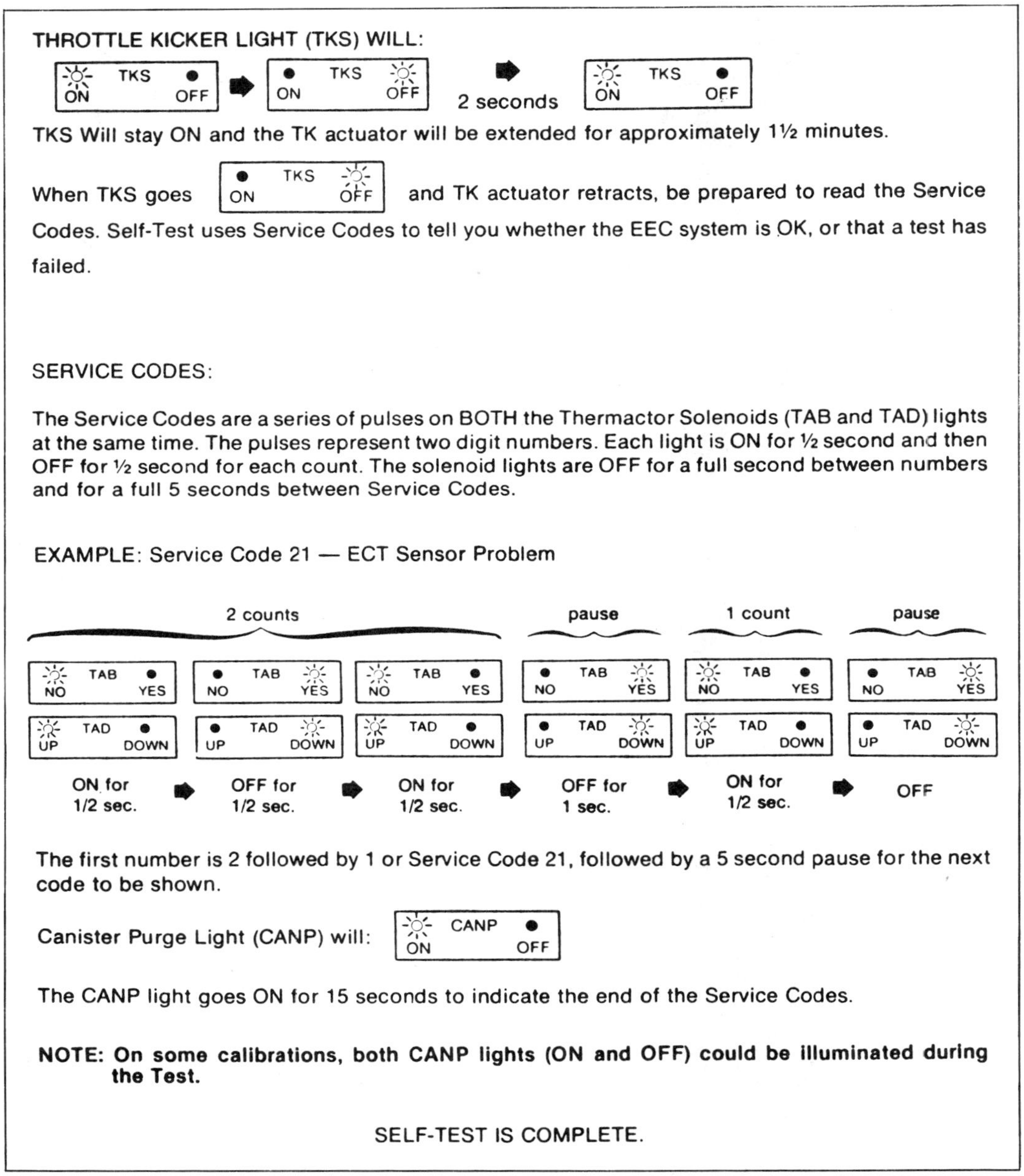

FIGURE 10-32 Tester lights *(Courtesy of Ford Motor Company).*

for 1.0 second, then on for 0.5 second, and then off for 5 seconds. This could be interpreted as a flash, pause, flash, longer pause, flash, 5 second pause. Code 21 is how the flashes are defined. When all codes have been displayed, the CANP light will come on for 15 seconds. There are 13 codes that can be stored in memory.

Quick Test

Once the EEC system has been determined as the cause of the problem, the quick test can be performed. Service manual procedures should be followed closely for the quick test as there are two parts to the test: key on, engine off and key on, engine running. When the quick test reveals a particular malfunction, a *pinpoint test* can be performed. The quick test checks overall operation of the EEC system and the pinpoint test checks a particular EEC circuit. Pinpoint test procedures are also provided in the service manual.

For problems not related to the EEC system, the service manual contains diagnostic routines. These are step-by-step procedures for checking and testing components of the engine not controlled by the ECA.

EEC-IV AND CENTRAL FUEL INJECTION

This chapter has described the CFI system as used in combination with the EEC-III system for 1980 through 1982. In 1983 through 1985, the EEC-IV system was used with CFI. This section will highlight the few differences between the EEC-III and EEC-IV systems.

The ECA reference voltage for engine sensors is 5 volts instead of 8 to 10 volts as was used in the EEC-III system. This ECA has the capability of monitoring and controlling seven functions in less than one revolution of the crankshaft. The calculation capability of the microprocessor has been increased to about 1 million per second. Internal circuitry of the ECA was reduced to two microcircuits instead of the four or five used in earlier computers. The calibration unit is an integral part of the ECA and is not serviceable separately as on the EEC-III system. The EEC-IV wiring harness has edgeboard connectors where it fastens to the ECA on some models. Later systems use socket and pin connectors.

The EEC-IV system still has a self-diagnostic ability and can also store intermittent codes. The same tools and testers are needed for this system except for the EEC-III tester. Instead, either an analog voltmeter or self-test automatic readout tester is required (Fig. 10-33). Either can be connected to the self-test connector of the ECA. Codes are read on the analog voltmeter by counting the swings of the pulsing meter needle. The self-test automatic readout (STAR) tester displays codes digitally. A high impedance digital voltmeter should be used for testing circuits, Fig. 10-34.

System operation, service, and diagnosis are the same except for these differences.

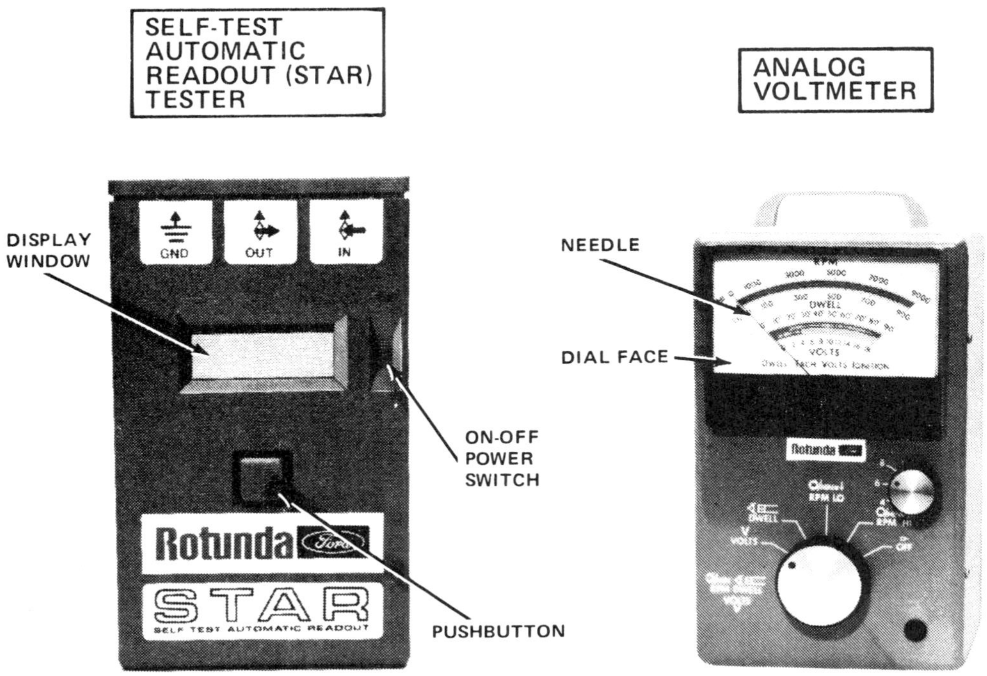

FIGURE 10-33 Self-test automatic readout (STAR) tester and analog voltmeter *(Courtesy of Ford Motor Company).*

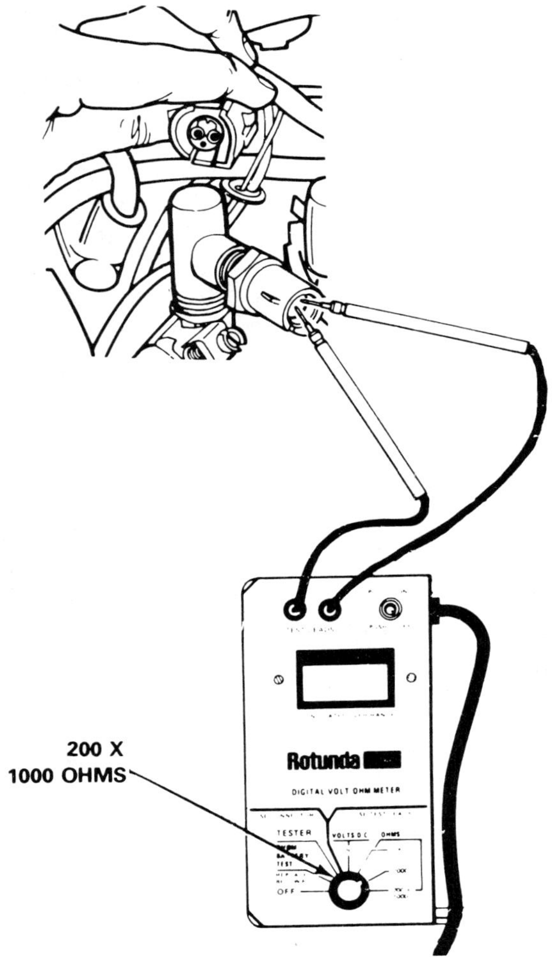

FIGURE 10-34 Digital volt-ohmmeter *(Courtesy of Ford Motor Company).*

ABBREVIATIONS

ACT Air charge temperature
BMAP Barometric and manifold absolute pressure
BP Barometric pressure
CANP Canister purge
CP Crankshaft position
ECA Electronic control assembly
ECT Engine coolant temperature
EEC Electronic engine control
EFI Electronic fuel injection
EGO Exhaust gas oxygen
EGR Exhaust gas recirculation
EGRC EGR control
EGRV EGR vent
EVP EGR valve position
IAT Intake air temperature
IM Ignition module
LOS Limited operational strategy
MAP Manifold absolute pressure
MCU Microprocessor control unit
TAB Thermactor air bypass
TAD Thermactor air diverter
TKS Throttle kicker solenoid
TP Throttle position
VREF Reference voltage
WOT Wide-open throttle

UNIT II
Ford Motor
Company Systems

11 Multipoint Fuel Injection, Electronic Engine Control IV, 1983–1985

OVERVIEW

Electronic engine control IV (EEC-IV) was developed and produced in 1983 and is still in use today. The first phase was electronic engine control I, which was used in 1978. It was a completely solid-state module using a digital microprocessor, custom-designed integrated circuits, and seven sensors. This system, used in conjunction with a three-way catalytic converter in 1979–1980, became known as EEC-II. This second-generation system was a feedback system in which the air/fuel mixture was controlled through the use of an electronic carburetor. When this system was combined with central fuel injection, it became EEC-III and was used from 1980 to 1982.

As these systems developed, they became more responsive to changes in engine operation, less complex, and less expensive to produce. Rather than use these systems in all models, Ford selects certain engines and models where the benefits are most pronounced.

The EEC-IV system was developed for use in 1983 only on selected engines and models. It is still in use today. The EEC-IV system is designed to control engine operation over a wide range of driving conditions. The computer or microprocessor unit is referred to as an electronic control assembly (ECA). It is capable of controlling, electronically, seven engine operating functions and processing up to 1 million engine control commands per second. An engine control command is a signal sent by the ECA to an actuating device such as a relay or solenoid.

The biggest improvement that the EEC-IV system has over its predecessors is that it has only two microcircuits, where earlier versions had four or five. This means less complexity, lower costs, and yet increased monitoring capability and memory storage. The two microcircuits control fuel metering, air induction, ignition timing, and emission control devices. The ECA also has a self-diagnostic ability that can recall circuit malfunctions for the service technician.

The EEC-IV system used in conjunction with the multipoint fuel injection system provides for improved emissions, engine performance, cold starting and drive-away, fuel economy, idle and fast idle control, and "no touch" starting. This last improvement means that the driver does not have to depress the accelerator pedal to start the engine.

The EFI/EEC-IV system can be broken down into the fuel, air, and electronic engine control subsystems.

FUEL DELIVERY

The fuel delivery system consists of the fuel tank and cap, high-pressure electric fuel pump, in-line fuel filter, fuel supply line, fuel charging manifold assembly, and a fuel return line (Figs. 11-1 and 11-2).

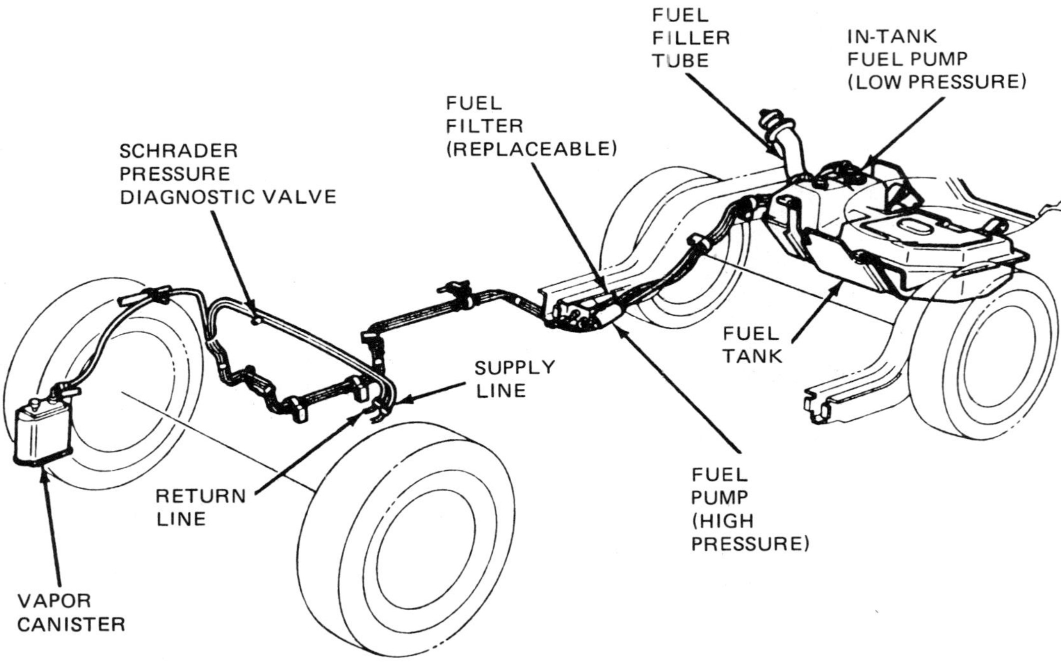

FIGURE 11-1 Fuel delivery system *(Courtesy of Ford Motor Company).*

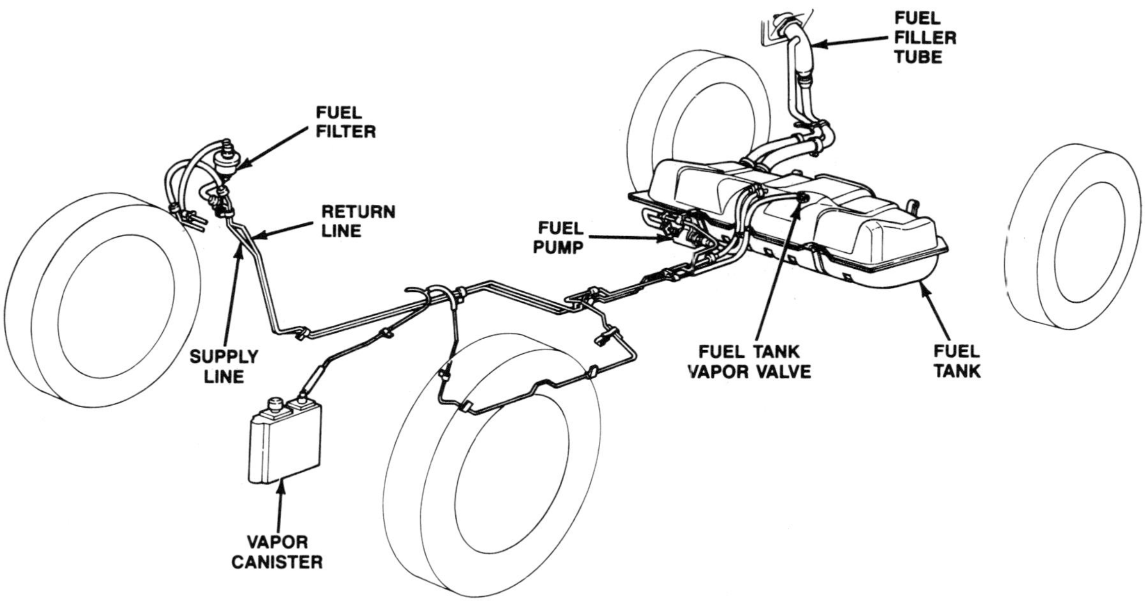

FIGURE 11-2 Fuel pump and fuel delivery lines *(Courtesy of Ford Motor Company).*

Fuel Tank

The fuel tank is stamped steel and is a new design. It has an internal pump cavity where the fuel gauge sending unit and fuel pickup tube are located. The cavity ensures adequate fuel supply during severe maneuvering conditions and low fuel level. The fuel tank cap is a standard pressure-vacuum relief emission control cap. Vacuum is relieved when it reaches −0.5 psi, and pressure is relieved when it reaches 1.8 psi. Correct cap replacement is very important as EFI system operation will be affected.

High-pressure Electric Fuel Pump

The high-pressure electric fuel pump is located outside and in front of the fuel tank (Fig. 11-3). Normal working pressure is about 39 psi, but the pump is protected by an internal relief valve that limits maximum pressure to about 120 psi. A woven plastic (nylon) filter is attached to the end of the pickup tube to protect the pump from contaminants. This filter will allow water to pass through. The fuel pump is controlled by a relay that is activated by the ECA when the ignition switch is in the CRANK or RUN position. A timer circuit in the ECA causes the relay and pump to be de-energized if a cranking signal is not received within 1 second. The activation of the fuel pump before cranking allows for purging and pressurization of the fuel system.

The ECA monitors engine speed for fuel pump control. It opens the ground circuit for the fuel pump relay if there is no engine speed signal or if the signal indicates an engine speed of less than 120 rpm. In addition, an inertia switch is used in the fuel pump circuit to open the circuit in the event of a collision (Fig. 11-4). The fuel pump shuts off even if an engine speed signal is still provided. The engine stops when it runs out of fuel. This prevents possible fire hazard conditions. The inertia switch must be manually reset by pushing a button on the switch. It is located behind the left side trim panel behind the rear seat

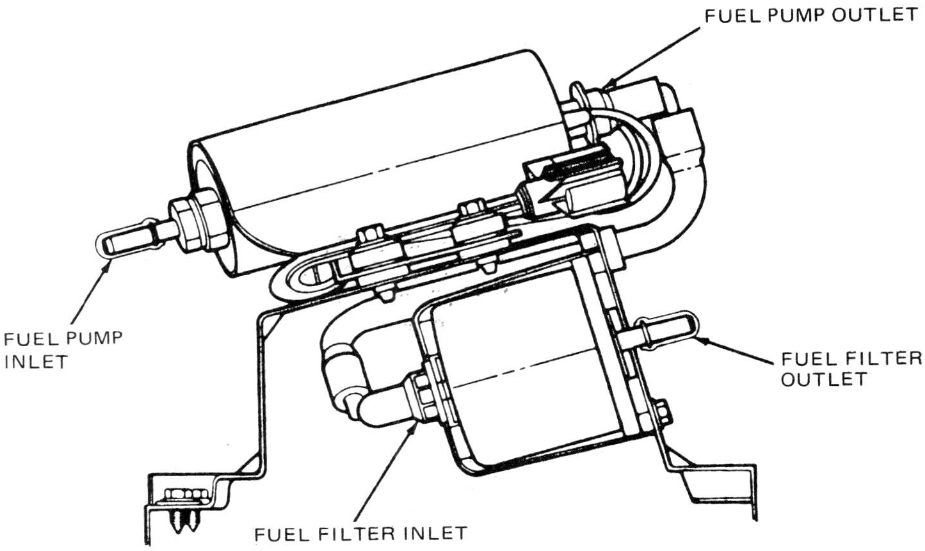

FIGURE 11-3 Rail-mounted high-pressure fuel pump and filter assembly (typical) *(Courtesy of Ford Motor Company).*

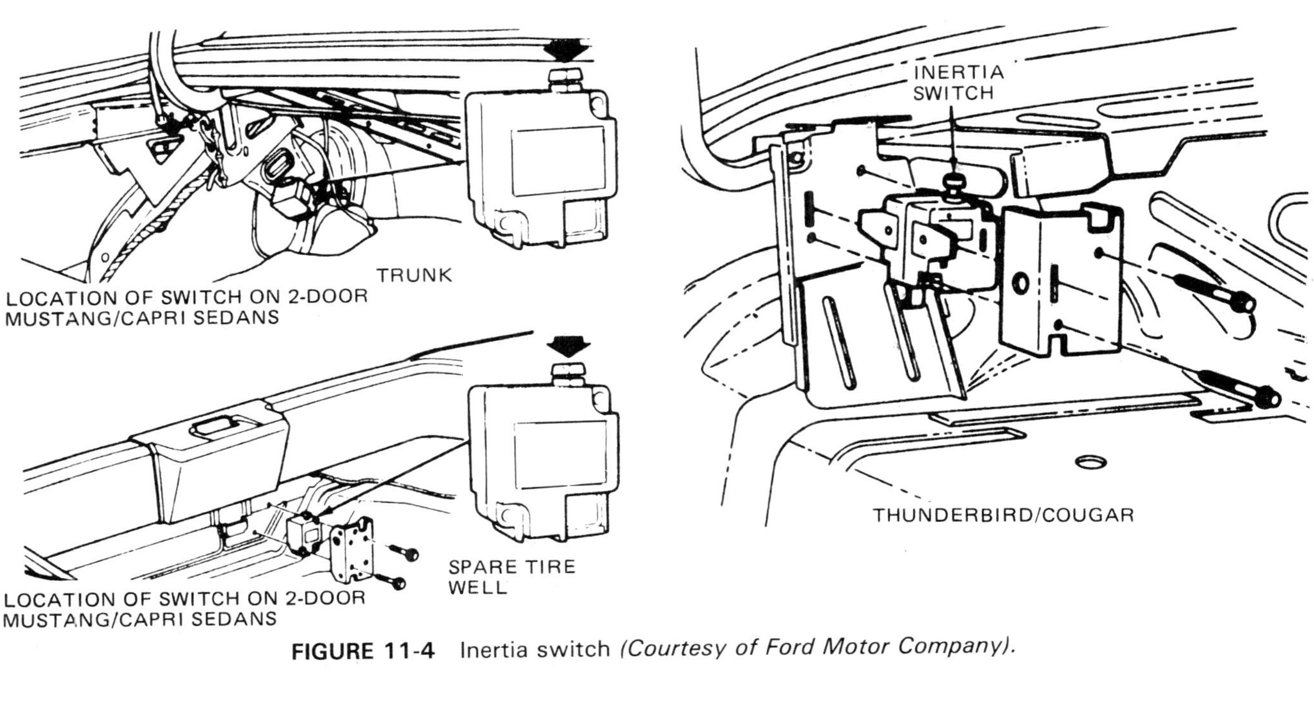

FIGURE 11-4 Inertia switch *(Courtesy of Ford Motor Company).*

on Escort and Lynx. The EXP and LN7 inertia switch is located inside the left rear quarter-panel shock tower access door. The fuel pump relay is located in the engine compartment and, on some models, under the instrument panel.

The 2.3-liter engine multipoint fuel injection system utilizes two electric fuel pumps. One is located in the fuel tank and supplies the second high-pressure pump mounted on the right-side frame rail of the vehicle. The fuel pumps are controlled by the ECA in a similar manner as for the 1.6-liter engine. This relay is located above the ECA in the passenger compartment.

Fuel Filter

A one-piece construction-type in-line fuel filter provides fine filtration to protect the small orifices of the fuel injectors (Fig. 11-5). It is serviced as an assembly and is located in the right rear corner of the engine compartment. Quick disconnect fittings with internal O-ring seals are used at the filter. The filter is secured to the cowling with a spring-loaded clip on 1.6-liter engines. The 2.3-liter engine fuel filter is mounted on a bracket fastened to the underbody ahead of the right rear wheel.

Fuel Supply Line

The fuel supply line is under high pressure during normal operation and remains pressurized for a long time after the engine is shut off. Nylon hose and steel tubing are used with push connect fittings. When servicing the fuel lines or any component of the fuel system, the system pressure must be relieved. This procedure is discussed in the system service section of this chapter. Only steel lines should be used to replace bad sections of the fuel supply or return lines. Threaded fasteners should be used to couple the sections together. Replacement hoses should be serviced as assemblies. Normal fuel line hose and clamps should not be used as a potential fire hazard may be created. The push connect fittings are also serviceable.

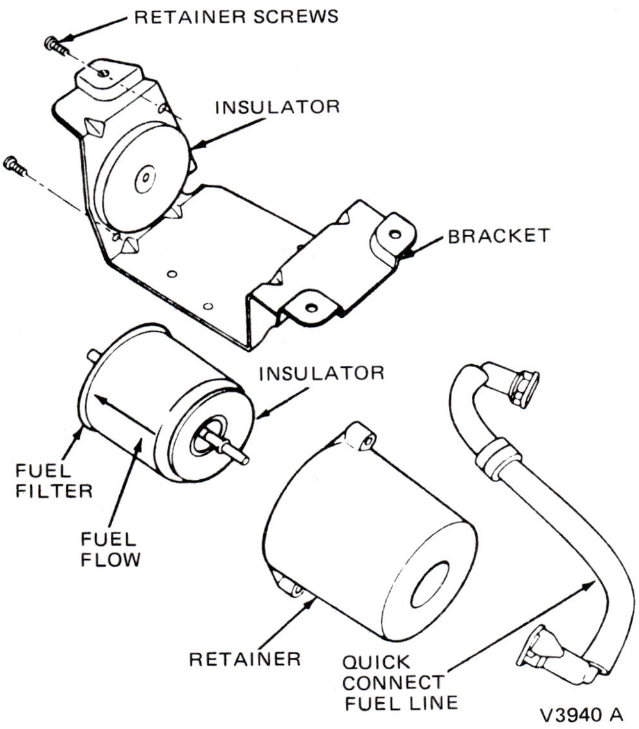

FIGURE 11-5 Primary fuel filter *(Courtesy of Ford Motor Company).*

Fuel Charging Manifold Assembly

The fuel charging manifold assembly consists of four fuel injectors, a pressure regulator, a fuel supply manifold assembly, a throttle body assembly, a throttle air bypass valve, and a two-piece air intake manifold (Fig. 11-6). The fuel injectors are electrically operated solenoid valves that meter and atomize the fuel entering the engine cylinders (Fig. 11-7). They are mounted in the lower intake manifold and are angled so their tips are aimed into the cylinder head intake ports. A signal from the ECA energizes the injector coil, which produces a magnetic field. The field causes a pintle valve to be raised off its seat and fuel sprays into the intake valve port. The amount of time the injector coil is energized determines how much fuel is injected. This is known as fuel injector pulse width. When the ECA stops its signal, a return spring causes the pintle valve to seal against its seat. The shape of the pintle valve tip causes the fuel to swirl and atomize as it is injected. The fuel injectors are sealed to the manifold by O-ring seals. These are located at the bottom of each injector. Poor sealing in this area would result in a vacuum or air leak. The pintle valve tip is protected by a plastic cap. If this cap is damaged, fuel atomization will be affected.

Fuel Pressure Regulator

A fuel pressure regulator is mounted at one end of the fuel supply manifold assembly or fuel rail (Fig. 11-8). The regulator maintains 39- to 40-psi pressure supply to the fuel injectors under normal operating conditions. When idling or under high vacuum conditions, the regulator reduces this pressure to about 30 psi. A diaphragm in the regulator is operated by spring pressure and vacuum on one side and fuel pressure on the other side. When fuel pressure and manifold vacuum overcome the spring pressure, the diaphragm moves and opens the pressure relief valve. This action allows excess fuel pressure to enter the fuel return line. In this way, fuel injector pressure is maintained in accordance with engine demand.

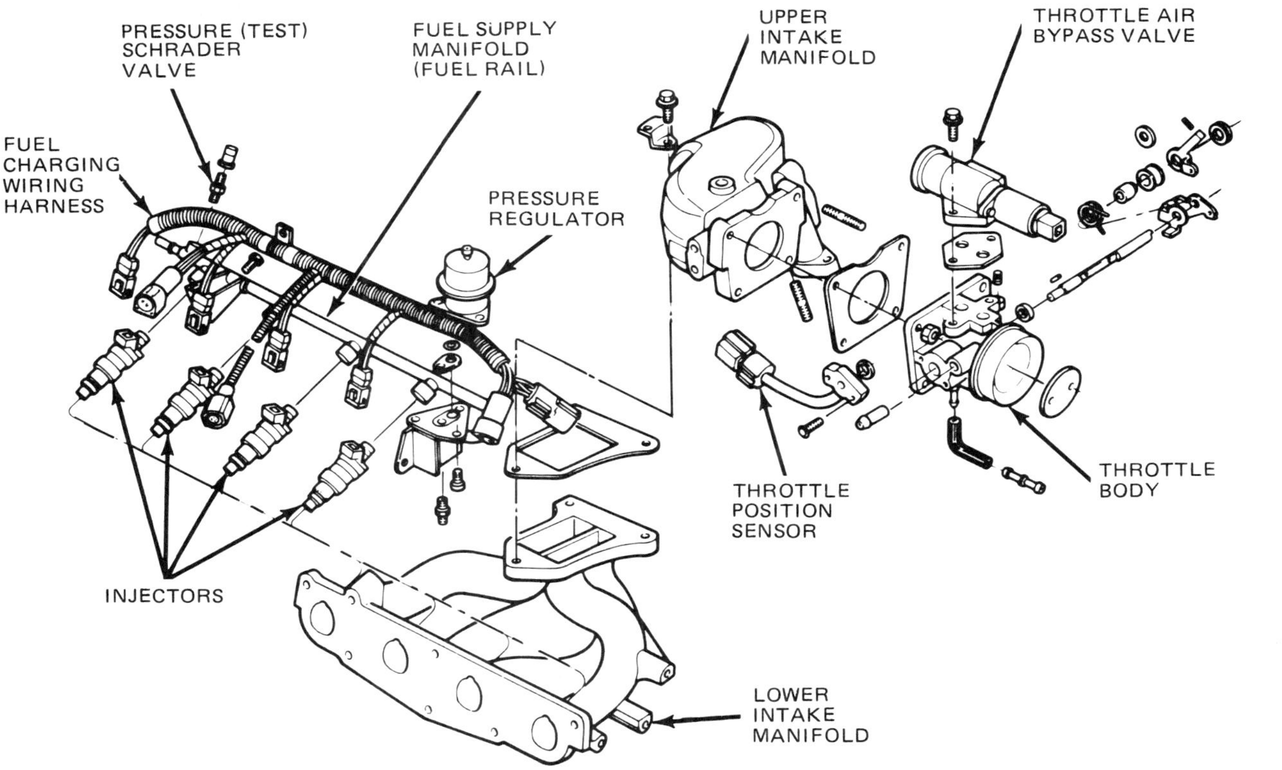

FIGURE 11-6 Fuel charging manifold assembly *(Courtesy of Ford Motor Company).*

FIGURE 11-7 Fuel injector *(Courtesy of Ford Motor Company).*

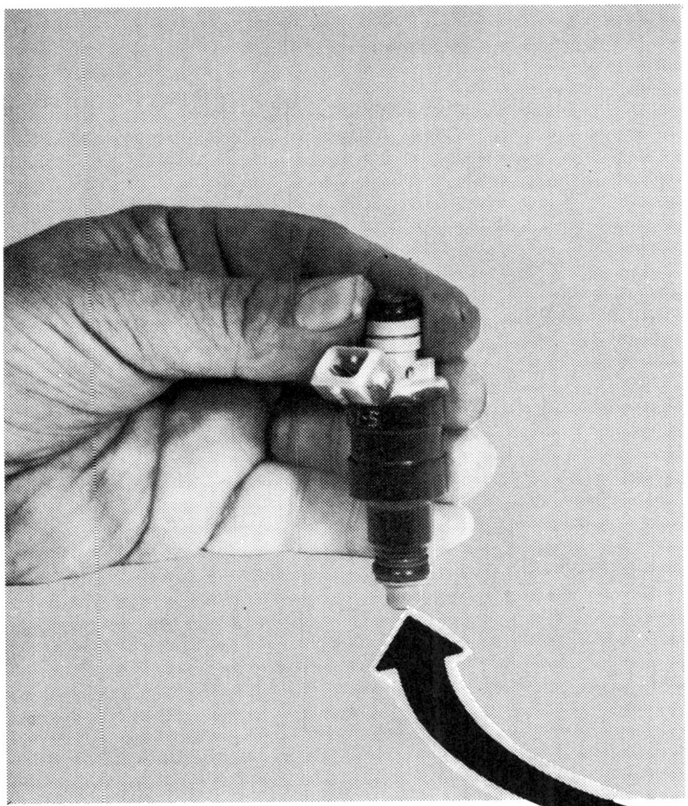

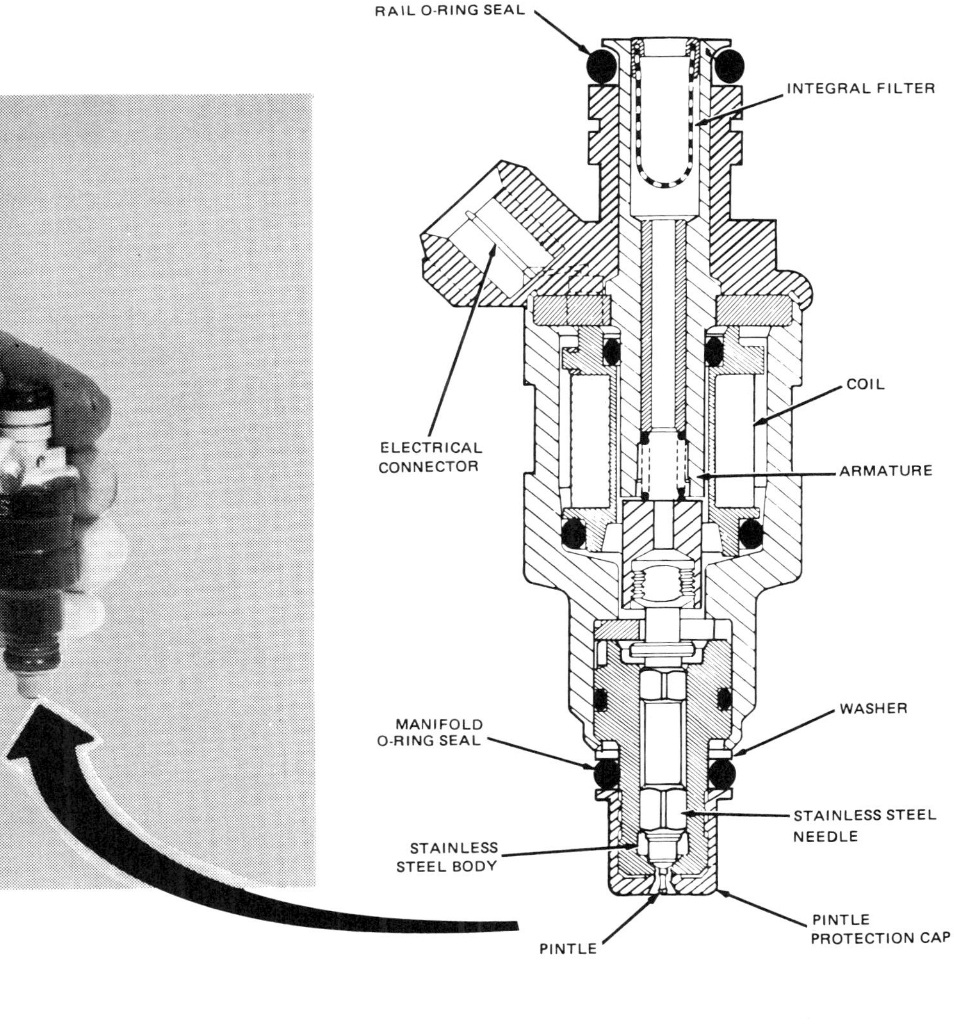

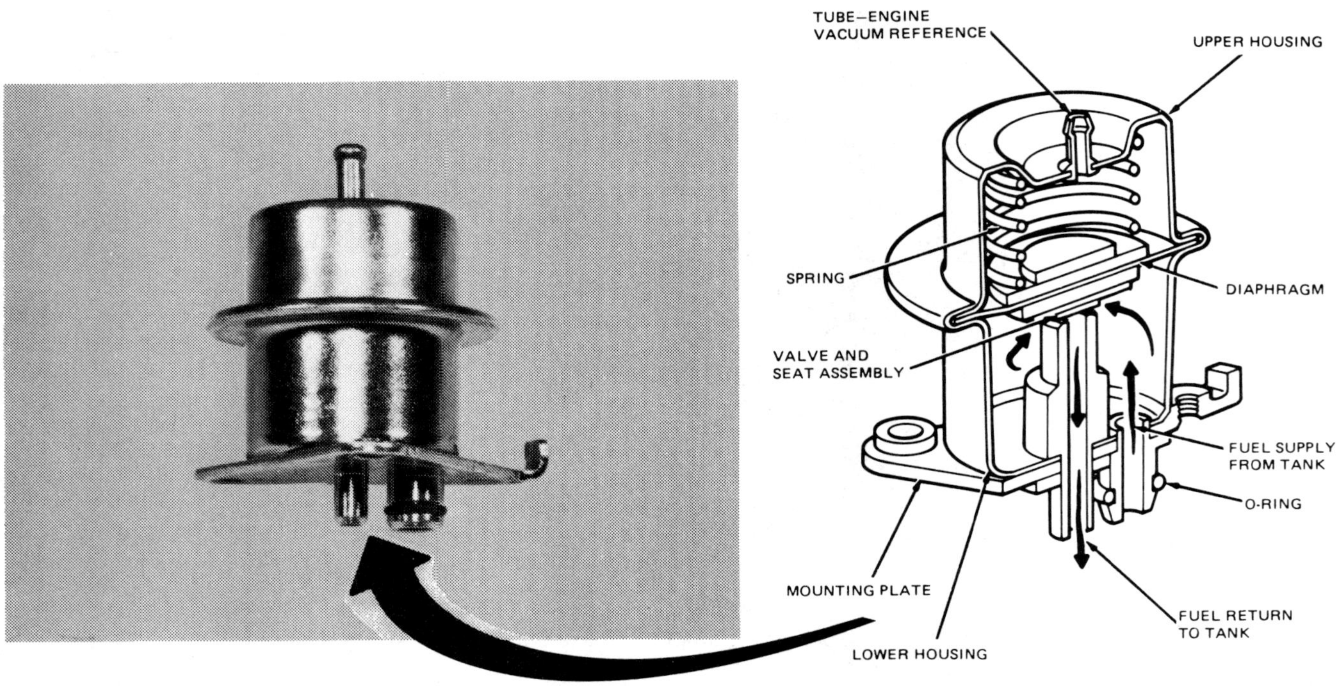

FIGURE 11-8 Fuel pressure regulator *(Courtesy of Ford Motor Company).*

Fuel Supply Manifold

The fuel supply manifold or fuel rail supplies high-pressure fuel to the fuel injectors (Fig. 11-9). It is a preformed steel tube that incorporates mounts for the fuel injectors and pressure regulator and has a fuel pressure Schrader valve for diagnostic gauge connection and pressure bleed-down. Mounting brackets on the rail minimize vibration and also provide tight fuel injector connections.

Throttle Body Assembly

The throttle body assembly controls airflow entering the engine (Fig. 11-10). It is a single-piece aluminum casting

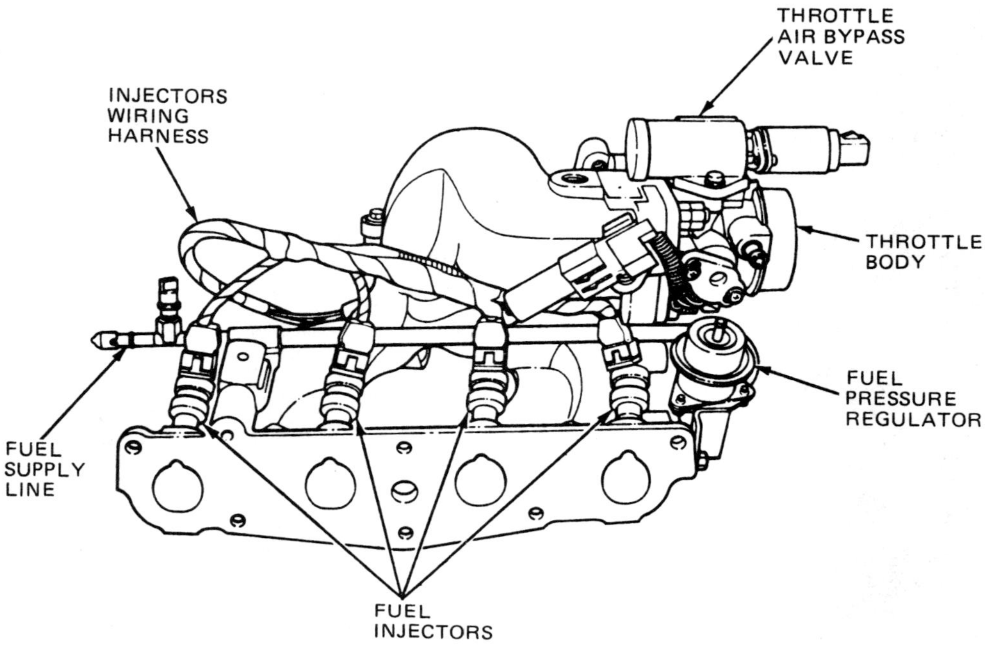

FIGURE 11-9 Fuel charging manifold assembly *(Courtesy of Ford Motor Company).*

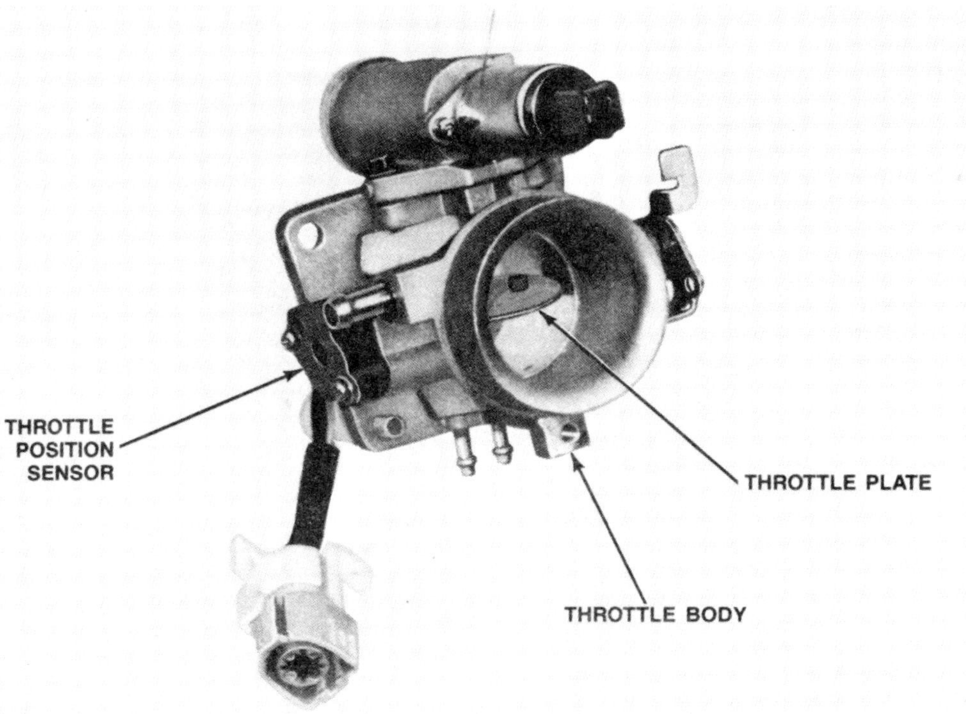

FIGURE 11-10 Throttle body *(Courtesy of Ford Motor Company).*

that provides vacuum hose connections, a mount for the throttle position sensor, and a mount for the throttle air bypass valve. It has a single bore and throttle plate, an adjusting screw for minimum idle, a PCV hose connection, and an air channel that bypasses the throttle plate for idle speed control.

Air Intake Manifold

The air intake manifold, so called because it conducts only air to the cylinders, is a two-piece aluminum casting (Fig. 11-11). Its runner lengths are tuned for best engine torque and power output. The manifold has mounting surfaces for the throttle body, fuel supply manifold, and accelerator pedal brackets. Vacuum taps are provided for engine accessories. The fuel injectors mount into machined bores near where the manifold fastens to the cylinder head. These bores are accurately machined to prevent air or fuel leaks and are angled so the injector tips point toward the intake valve in the cylinder head.

Fuel Return Line

The fuel return line begins at the pressure regulator on the fuel charging manifold assembly and ends in the sump cavity of the fuel tank. Information pertaining to the fuel supply line also applies to the return line.

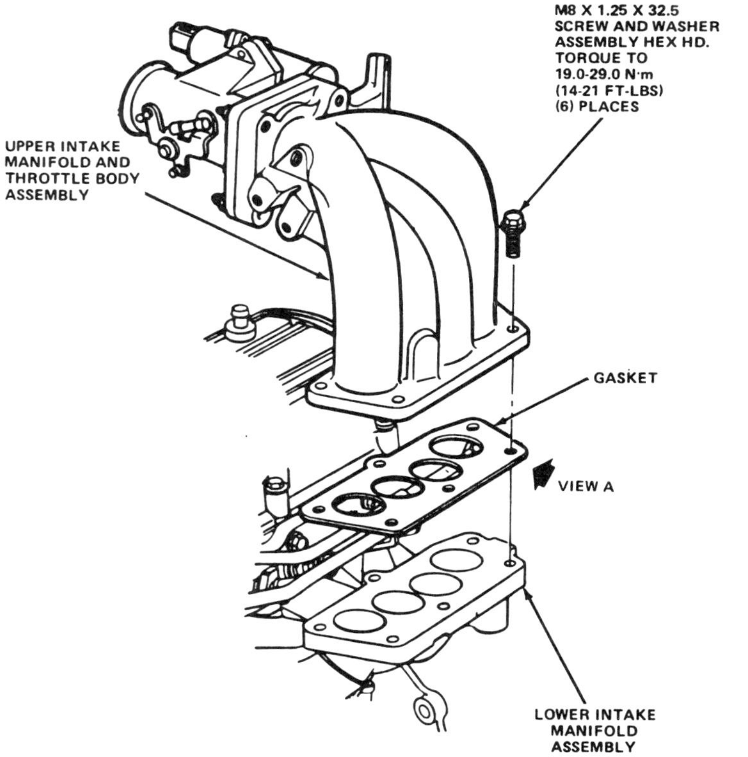

FIGURE 11-11 Upper and lower intake manifold, installation reference and accelerator controls *(Courtesy of Ford Motor Company).*

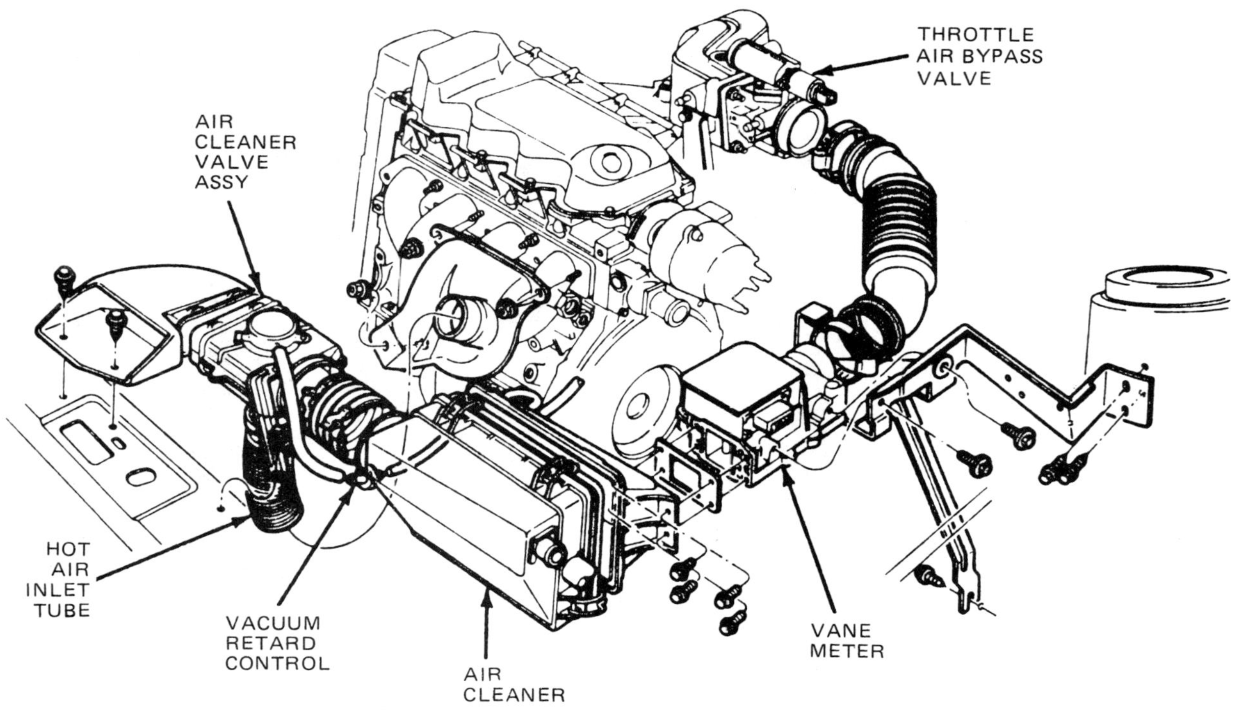

FIGURE 11-12 Air subsystem *(Courtesy of Ford Motor Company).*

AIR INDUCTION

Included as part of the air induction system is an air cleaner, vane meter, throttle body and throttle air bypass valve, and air intake manifold (Fig. 11-12).

Air Cleaner

The air cleaner consists of ducting, air preheater, and air cleaner housing and filter element. The ducting is flexible and conducts fresh, cool air from in front of the radiator support into the air preheater duct valve assembly. The function of the preheater is the same as on other fuel systems. The air cleaner housing and filter element are not mounted in the conventional location on the throttle body. The housing base bolts to the vane meter, and the element and housing cover clip to the base. The filter element is paper and is serviceable as a separate part.

Vane Airflow Meter

The vane meter (Fig. 11-13) is located between the air cleaner and the throttle body. Flexible ducting connects it to the throttle body. Two sensors are located in the vane meter. A vane airflow sensor measures the amount of air entering the engine (Fig. 11-14). A vane air temperature sensor measures the temperature of the incoming air (Fig. 11-15). The airflow entering the vane meter causes a vane in the meter to pivot on a shaft. When airflow increases, the vane pivots farther. As the vane pivots, it causes a potentiometer connected to its shaft to vary a signal to the ECA. The ECA provides a 5-volt source to the vane airflow sensor, whose resistance, determined by airflow and vane position, affects how much voltage is available on the sensor signal line. A high volume of air causes the potentiometer to have low resistance, resulting in a high voltage on the signal line.

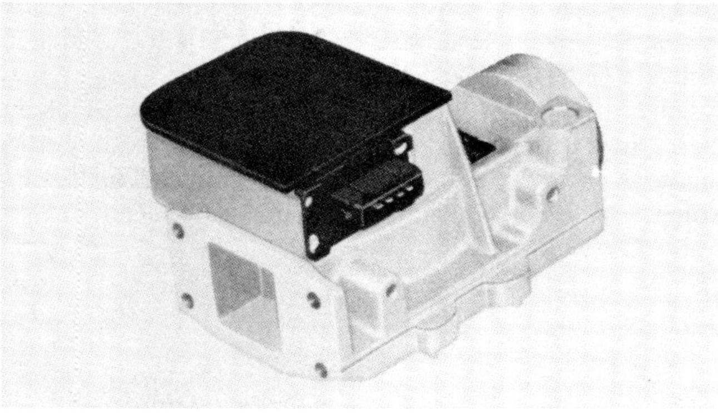

FIGURE 11-13 Vane meter *(Courtesy of Ford Motor Company).*

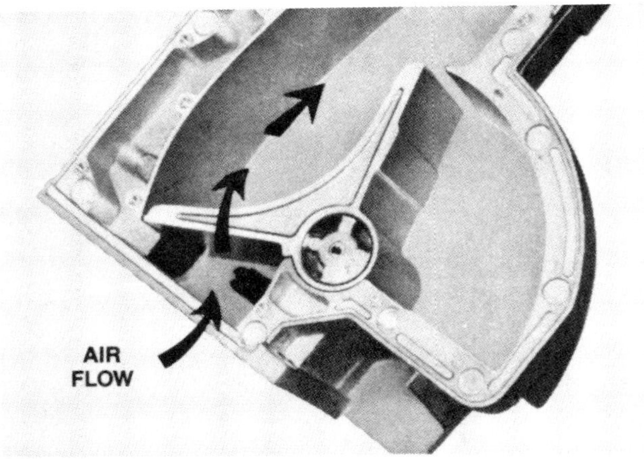

FIGURE 11-14 Vane airflow sensor air vane *(Courtesy of Ford Motor Company).*

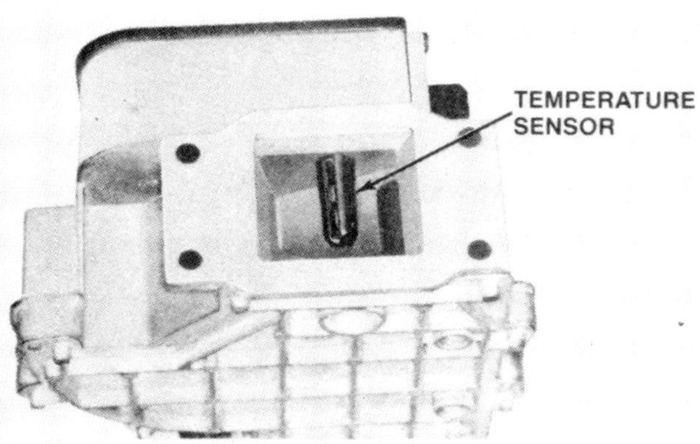

FIGURE 11-15 Vane air temperature sensor *(Courtesy of Ford Motor Company).*

Vane Air Temperature Sensor

The volume of air entering the engine must be measured in terms of air mass in order to achieve accurate air/fuel ratios. Since the mass of the air is dependent on pressure and temperature, a temperature sensor is used to measure air temperature in front of the air vane. In some respects, the vane air temperature sensor serves the same function as the air charge temperature sensor used on other EEC systems; however, in this example, its signal is used by the ECA to compute mass airflow.

Throttle Air Bypass Valve

The throttle body was discussed in the fuel delivery section of this chapter; however, the throttle air bypass valve was only mentioned. This electromechanical valve is controlled by the ECA (Fig. 11-16). It responds to the ECA commands for control of idle speed during cold and warm operation. It consists of a solenoid that positions an air

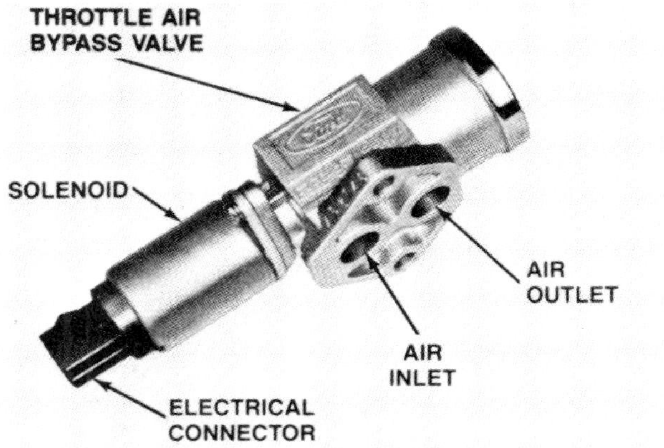

FIGURE 11-16 Throttle air bypass valve solenoid *(Courtesy of Ford Motor Company).*

metering valve. The TAB valve is located on the throttle body in an air bypass tube. When the valve is open, air is allowed to bypass the closed throttle plate through the bypass tube. The amount of current directed to the TAB solenoid by the ECA determines the position of the valve and the amount of air allowed to bypass the throttle plate. Through use of the TAB valve, the ECA can regulate idle speed for cold or warm engine operation and can compensate for increased engine load caused by such things as the air-conditioning compressor or transmission or transaxle. The TAB valve serves the same function as the fast idle cam, antidieseling solenoid, throttle kicker, and other throttle-positioning mechanisms. Fast idle is provided by the ECA and TAB valve when engine coolant is below 160°F. In addition, the use of the TAB valve enables the engine to be started without touching or depressing the accelerator pedal. During engine cranking, the ECA provides a 100% duty cycle signal (100% current) to the TAB valve to allow maximum air bypass for "no touch" starting.

The air intake manifold was discussed earlier in this chapter.

ELECTRONIC ENGINE CONTROL SYSTEM

The electronic engine control IV system includes an electronic control assembly (ECA), sensors and switches, and actuators and relays.

Electronic Control Assembly

The electronic control assembly receives input data from a variety of sensors and switches (Fig. 11-17). The ECA, which is a microcomputer, uses the input data in combination with its program, makes high-speed calculations,

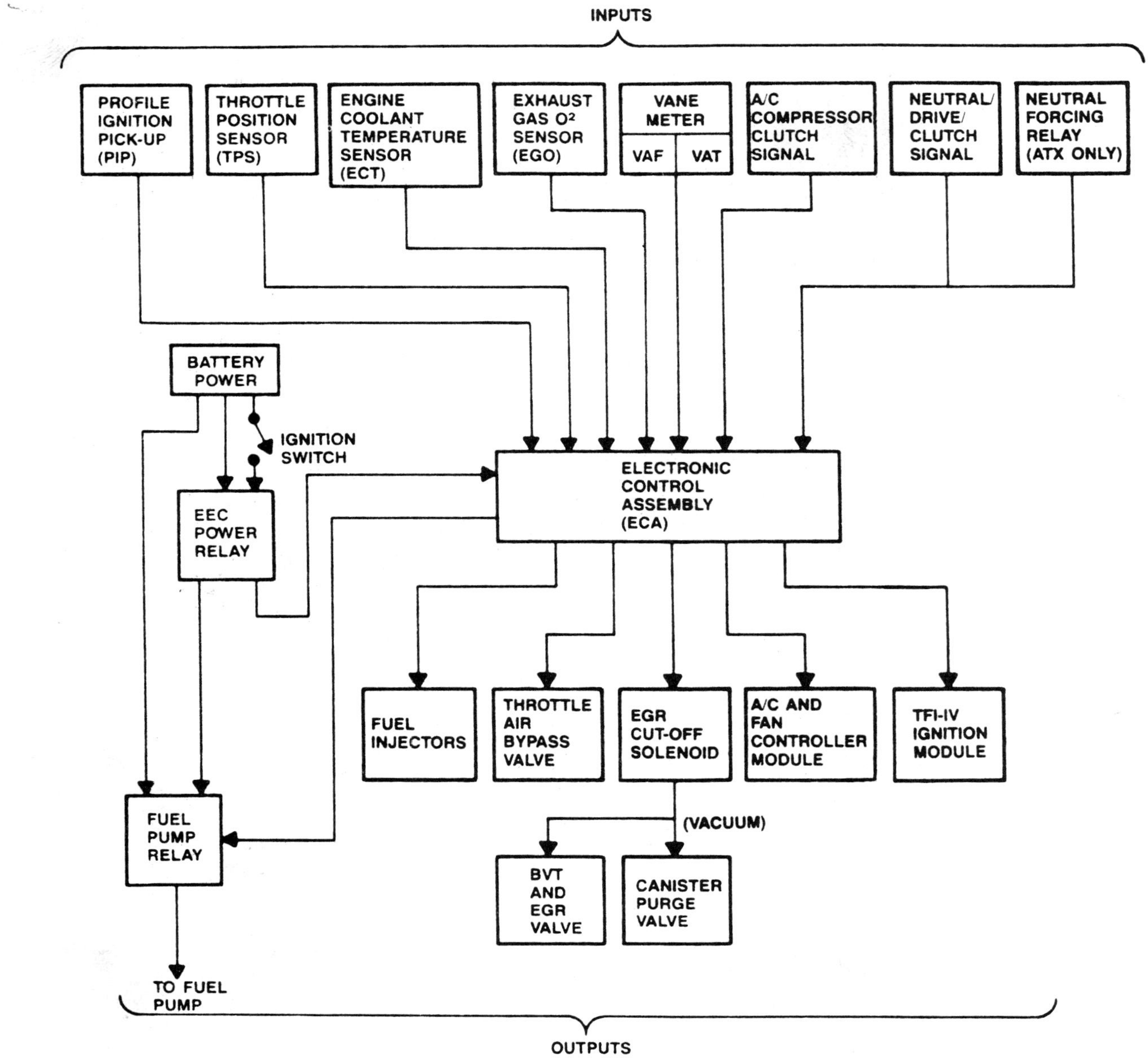

FIGURE 11-17 EFI-EEC IV system inputs and outputs *(Courtesy of Ford Motor Company).*

and issues command signals to its actuators, relays, and solenoids. The calibration module for the ECA is inside the computer. This is a new feature of the EEC-IV system. The connectors for the ECA are pin-and-socket type. The ECA is located in the passenger compartment under the instrument panel.

EEC-IV SYSTEM INPUTS

Input Data

Input data to the ECA are provided by the profile ignition pickup, throttle position sensor, engine coolant temperature sensor, exhaust gas oxygen sensor, vane meter, air-conditioning clutch switch, neutral/drive/clutch switch, and neutral forcing relay.

Profile Ignition Pickup

The profile ignition pickup (PIP) provides crankshaft position data (Fig. 11-18). It replaces the previously used crankshaft position sensor. Located in the distributor, the PIP is a Hall-effect switch that has an armature with four metal vanes with an opening or window between each vane. The vanes rotate past a stator assembly and pass between a permanent magnet and a coil or transducer. When a vane is between the magnet and transducer, voltage signal is produced and sent to the ECA. This signal represents the crankshaft position at 10° BTDC. The ECA can calculate

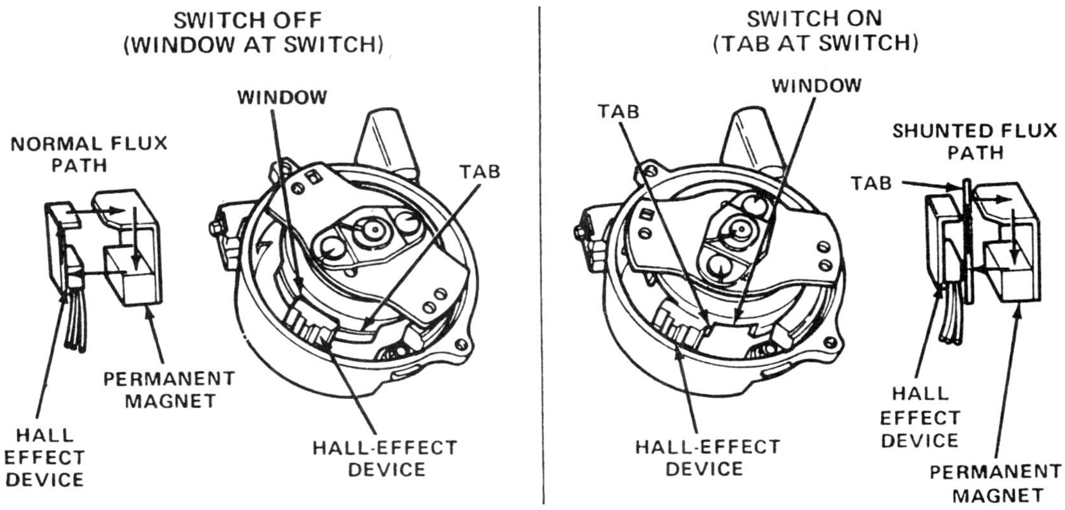

FIGURE 11-18 Ignition pickup operation *(Courtesy of Ford Motor Company).*

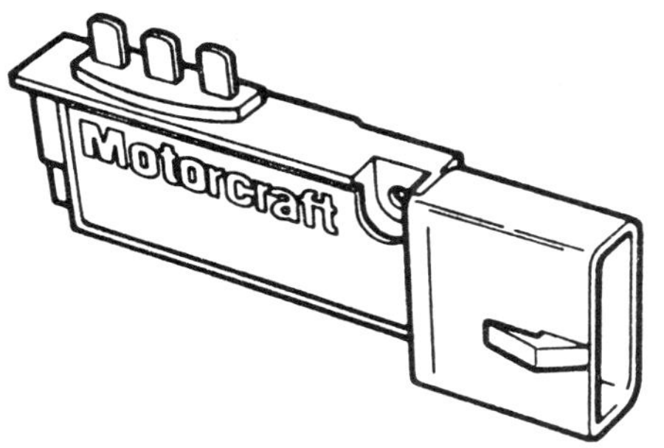

FIGURE 11-19 Thick-film ignition module *(Courtesy of Ford Motor Company).*

ignition timing and will send the appropriate signal to the thick-film ignition (TFI) module (Fig. 11-19). The ignition module controls primary circuit current and produces secondary voltage in the ignition coil.

Throttle Position Sensor

The throttle position sensor (TPS) is located on the throttle body at the end of the throttle shaft (Fig. 11-10). The ECA sends a 5-volt reference to the TPS. As the throttle position changes, the TPS, which is a variable resistor, changes in resistance value. At wide-open throttle, the TPS resistance is lowest, and its signal voltage to the ECA will be about 4.75 volts. At closed throttle, the TPS resistance is highest, and its signal to the ECA will be about 1 volt. The TPS is not adjustable and must be replaced if it is out of calibration. The ECA program compensates for minor production differences. The TPS signal is used by the ECA to select wide-open-throttle, part-throttle, or closed-throttle modes of operation. It also causes the ECA to provide fuel enrichment for wide-open throttle, provide additional spark advance at wide-open throttle, shut off EGR, canister purge and air conditioning for 30 seconds while at wide-open throttle, provide cold start kickdown or fast idle speed reduction, and shut off fuel during deceleration. The EEC-IV TPS affects spark advance only during wide-open-throttle conditions.

Engine Coolant Temperature Sensor

An engine coolant temperature sensor is located in the heater supply tube at the back of the 1.6-liter engine. The coolant temperature sensor is located in the lower intake manifold of the 2.3-liter engine. It is a thermistor whose resistance changes inversely as the coolant temperature changes (Fig. 11-20). At low temperatures, the ECT sensor resistance is high, while at high temperatures, the sensor

FIGURE 11-20 Engine coolant temperature sensor *(Courtesy of Ford Motor Company).*

resistance drops. The ECA provides a 5-volt reference to the ECT sensor, and the return signal is determined by the sensor resistance and coolant temperature. The ECA uses this return signal to control the air/fuel ratio, idle speed, spark advance, and EGR flow and canister purge. During cold engine operation, the ECT sensor data causes the ECA to enrich the air/fuel mixture for optimum cold performance.

Exhaust Gas Oxygen Sensor

By knowing about the exhaust gas oxygen content, the ECA is better able to precisely control and adjust the air/fuel ratio. The exhaust gas oxygen sensor is located in the exhaust manifold (Fig. 11-21). The end of the EGO sensor extends into the exhaust stream. Inside the louvered protective shield are two electrodes made of zirconia oxide. One electrode is exposed to the engine exhaust and the other is subjected to fresh air. When oxygen comes in contact with the zirconia, a voltage is produced. The voltage produced on the electrode exposed to fresh air is relatively constant, whereas the voltage produced on the electrode exposed to exhaust gas oxygen varies with the air/fuel ratio. The difference between the two voltages is monitored by the ECA. A signal greater than 600 millivolts indicates a rich air/fuel ratio, and a signal of less than 400 millivolts is the result of a lean air/fuel ratio. The voltage signal may vary from about 100 to 900 millivolts in actual operation. Because of such small voltage, an ohmmeter should never be used to test this sensor. If a voltmeter is used to check its operation, a 10-megohm high-impedance voltmeter must be used to prevent damage to the sensor.

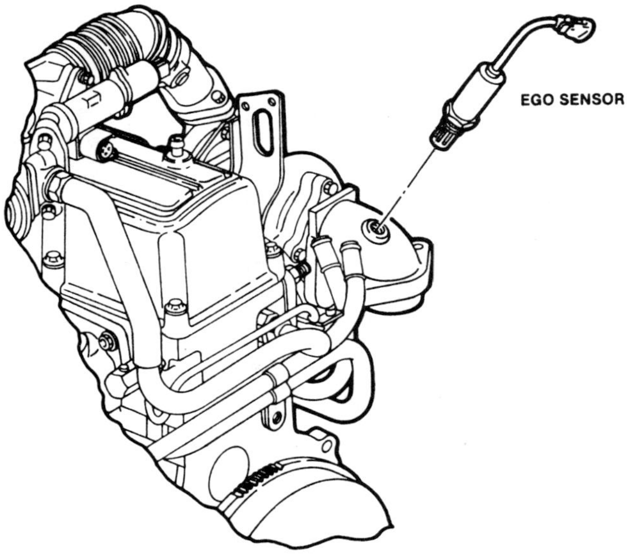

FIGURE 11-21 Location of exhaust gas oxygen sensor *(Courtesy of Ford Motor Company)*.

Vane Airflow Meter

The vane meter, consisting of a vane airflow sensor (VAF) and a vane air temperature sensor (VAT), was discussed in the air induction section of this chapter. It is mounted on a bracket near the left shock tower in the engine compartment of the 1.6-liter engine. The vane meter for the 2.3-liter engine is much larger and is located under the air cleaner on a bracket between the right shock tower and the radiator cradle.

Air-conditioning Clutch Switch

An air-conditioning clutch switch provides a signal to the ECA whenever battery voltage is supplied to the air-conditioning compressor clutch. This enables the ECA to compensate for the additional engine load by pulsing the throttle air bypass valve in order to maintain a stable idle speed. When the compressor clutch cycles off, the ECA again pulses the TAB valve so that engine idle speed remains steady.

Clutch-engaged Switch and Transaxle Neutral Switch

Vehicles equipped with manual transaxles have a clutch-engaged switch and a transaxle neutral switch to provide data to the ECA. These switches are wired in parallel and indicate load or no-load conditions to the ECA. The clutch-engaged switch is located on the clutch pedal lever and closes when the clutch is disengaged. The transaxle neutral switch is mounted on the shift linkage. It is closed only when the transaxle is in neutral. These switch signals enable the ECA to better control idle speed and prevent potential engine stall.

Neutral Start Switch

Vehicles equipped with automatic transaxles have a neutral start switch to indicate to the ECA whether or not the transaxle is in gear. A reference voltage on the switch signal line means the switch is open and the transaxle is either in reverse or drive. Low-voltage signals mean the transaxle is in park or neutral. The ECA uses these signals to control idle speed.

Neutral Forcing Relay

The neutral forcing relay is used in conjunction with automatic transaxles only (Fig. 11-22). While cranking the engine, this relay is energized to cause the neutral start switch signal to be less than 1 volt. This causes the ECA to see a no-load condition for better fast idle control. Without the neutral forcing relay, the neutral start switch signal

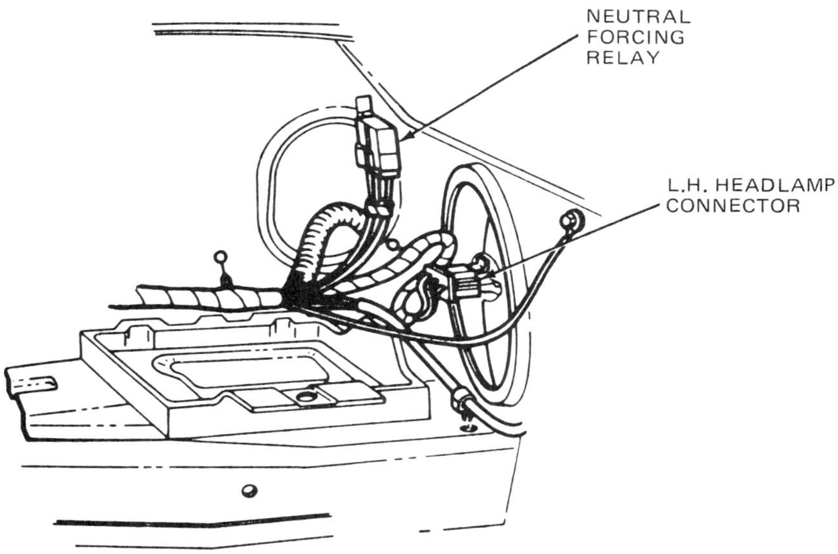

FIGURE 11-22 Neutral forcing relay *(Courtesy of Ford Motor Company).*

would be about 5 volts because of the battery voltage being applied through the ignition switch. The ECA would incorrectly interpret this as a load condition, and fast idle speed would not be controlled correctly.

Detonation Sensor

In addition to these sensor and switch inputs, the 2.3-liter turbocharged engine has a detonation or engine knock sensor. It is located in the lower intake manifold toward the rear of the engine (Fig. 11-23). It is a piezoelectric device that produces a voltage signal when excessive vibrations caused by detonation are present. A signal from the knock sensor causes the ECA to retard the ignition timing until the knock disappears. It then advances the timing.

Barometric Air Pressure Sensor

The 2.3-liter engine system also uses a barometric air pressure sensor (BAP). This sensor signal is used to allow the

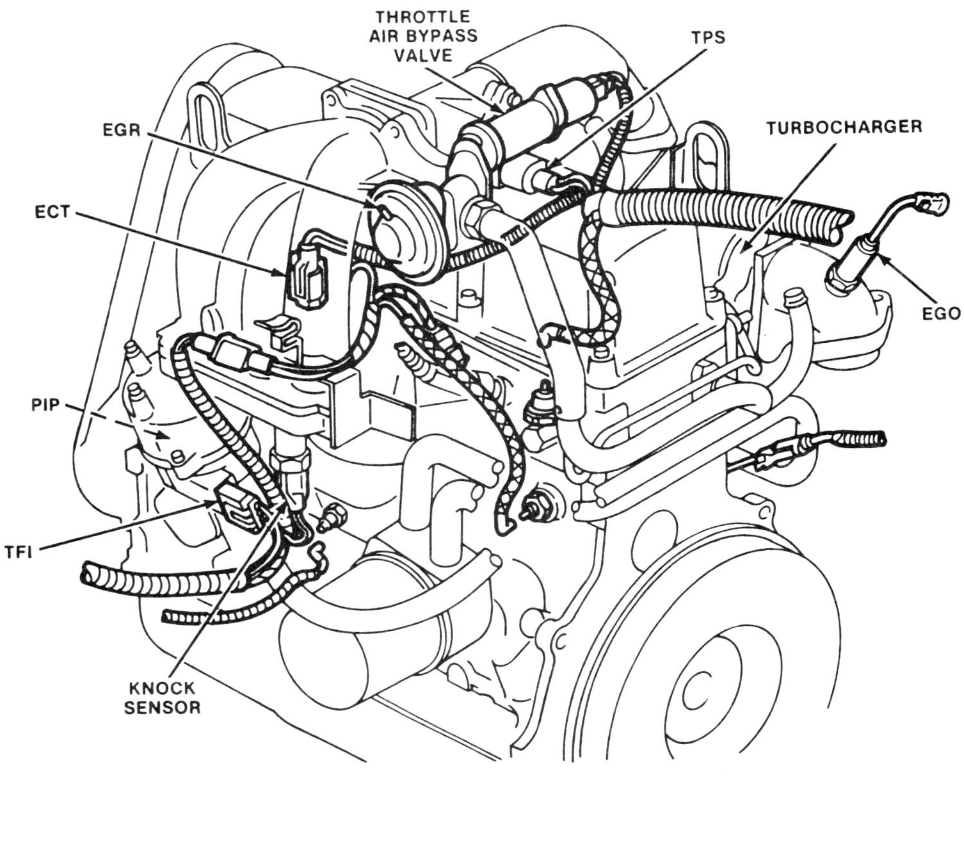

FIGURE 11-23 Turbocharger engine *(Courtesy of Ford Motor Company).*

ECA to compensate for changes in altitude. The ECA also uses the BAP signal to control air/fuel ratio, spark advance, idle speed, and the EGR solenoid. The BAP signal is based on atmospheric pressure. It is located on the right inside fender in the engine compartment.

EEC-IV SYSTEM OUTPUTS

As a result of receiving sensor and switch input data, the ECA makes high-speed computations according to its program and issues command signals or pulses to the devices or actuators and solenoids that it controls. These control devices include the EGR solenoid, canister purge solenoid, ignition timing module, throttle air bypass valve, air-conditioning and cooling-fan module, fuel injectors, and fuel pump relay.

EGR Shut-off Solenoid

The ECA issues either an ON or OFF signal to control the EGR shut-off solenoid (Fig. 11-24). When the signal is off, vacuum to the EGR valve is blocked. The EGR solenoid is commanded off during a cold start, closed-throttle, or wide-open-throttle operation. The solenoid will be on at all other times. The EGR solenoid is mounted on the left side of the cowling in the engine compartment of 1.6-liter engines and is mounted on the right shock tower in the engine compartment of 2.3-liter engines.

Canister Purge Solenoid

The canister purge solenoid is controlled by the ECA for regulation of when stored vapors are drawn from the charcoal canister. The solenoid blocks purge when it is deenergized. When the ECA pulses the solenoid ON, vapors are drawn from the canister. The purge solenoid is located in the engine compartment in the vapor line near the charcoal canister.

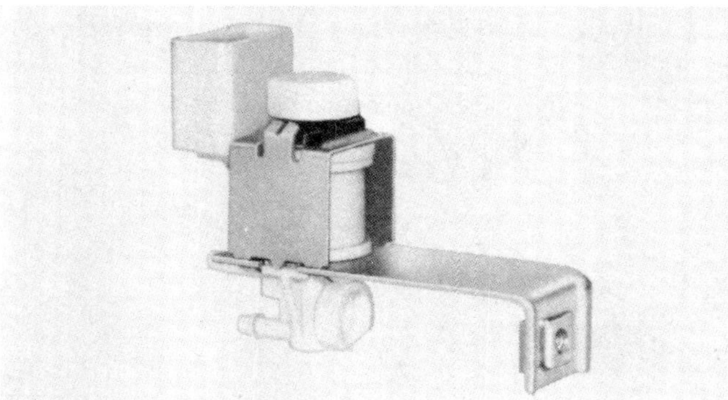

FIGURE 11-24 Exhaust gas recirculation shut-off solenoid *(Courtesy of Ford Motor Company).*

Thick-film Ignition Module

The ignition timing module is of the thick-film type and is similar to earlier versions (Fig. 11-19). The EEC-IV thick-film integrated (TFI) module has six terminals, whereas earlier versions had only three. The additional three terminals provide PIP signals to the ECA, trigger signals for control of timing from the ECA, and a common ground signal between the TFI module and the ECA. The TFI module is molded into a thermoplastic case and is mounted on the base of the distributor.

Throttle Air Bypass Valve

The throttle air bypass valve is used by the ECA to control idle speed, as discussed earlier.

Air-conditioning and Cooling-fan Module

The ECA commands the air-conditioning and cooling-fan module for control of the air-conditioning compressor clutch and engine cooling system electric fan. It is used on the 1.6-liter engine and is located under the right side of the instrument panel. The ECA uses the module to deactivate the air-conditioning compressor clutch during wide-open-throttle operation. The same ECA command also causes the module to de-energize the engine cooling fan if coolant temperature is below 221°F. When coolant temperature is above this specification, an engine coolant temperature switch provides a ground for the fan circuit and overrides the ECA signal even during wide-open-throttle conditions. If wide-open-throttle operation lasts for longer than 30 seconds, the ECA signal to the control module is interrupted by a timer circuit in the ECA. This allows the air-conditioning compressor clutch and coolant fan to operate normally. If, during wide-open-throttle operation, the accelerator pedal is returned to part throttle and then to wide-open throttle, the timer will reset and another 30 seconds will have to pass before normal operation of the air-conditioning compressor clutch and coolant fan begins. Another input signal to the control module from the brake switch on automatic transaxle-equipped vehicles with power brakes causes the module to disable the air-conditioning compressor clutch circuit and engine cooling-fan circuit for about 3 to 5 seconds. This helps prevent stalling.

Fuel Injectors

Two fuel injectors are energized or pulsed simultaneously once for every revolution of the crankshaft except during cranking. When the injectors are pulsed, it causes the pintle valve to raise off its seat against spring pressure and allow fuel to be sprayed into the intake valve port. The amount of time the injectors are pulsed determines how much fuel

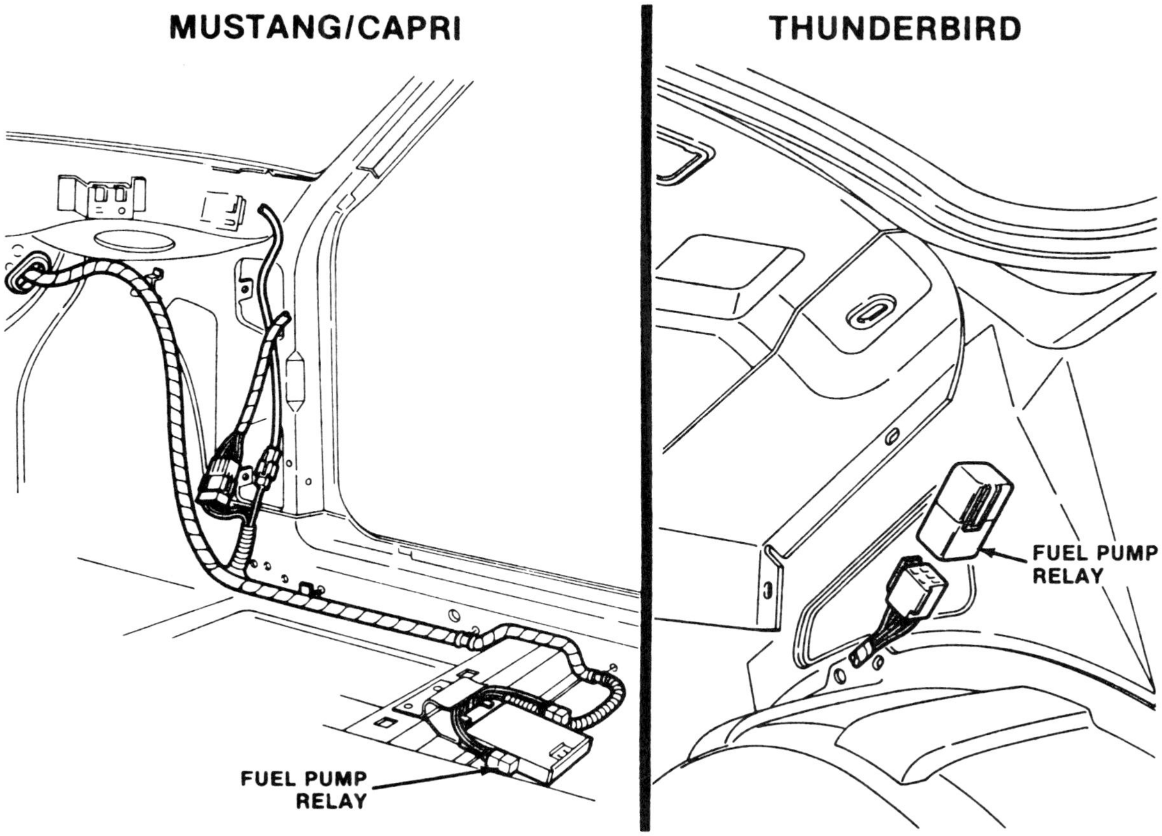

FIGURE 11-25 Fuel pump relay installation *(Courtesy of Ford Motor Company).*

is injected. This duration is known as fuel injector pulse width.

Fuel Pump Relay

As was discussed in the fuel delivery section of this chapter, the ECA controls the fuel pump relay (Fig. 11-25).

ENGINE OPERATING MODES

The multipoint fuel injection/electronic engine control-IV system has seven modes of engine operation, which include cold startup, warmup driveaway, wide-open throttle, hot curb idle, hot cruise, deceleration, and limited operating strategy. In addition, the system can operate in either open or closed loop.

Cold Startup Mode

The cold startup mode of operation begins when the driver turns on the ignition switch. The ECA receives power from its power relay and also energizes the fuel pump relay to pressurize and purge air from the fuel system. If an engine cranking signal is not received by the ECA within 1 second, the fuel pump relay will be deactivated. Fuel is not delivered until a cranking signal is received, as the timing of fuel injection is coordinated with the data from the profile ignition pickup. The ECA commands the throttle air bypass valve to open for increased idle speed. The amount of increase is based on engine coolant temperature sensor data. Idle gradually decreases as engine coolant temperature increases. Air/fuel ratio and fuel injector pulse width are determined by engine coolant temperature and engine load (airflow). As the engine coolant temperature rises, fuel injector pulse width decreases. The EGR and purge solenoids are off until the engine is warm and operating in closed loop. Spark advance is determined by engine rpm, load, air temperature, coolant temperature, and EGR flow rate. The preheated air system provides warmed air for the engine.

Warmup Driveaway Mode

During the warmup driveaway mode, additional enrichment is provided for acceleration. The amount of enrichment is determined by engine coolant temperature and engine load. The preheated air system is providing a blend of heated and cool air. The EGR and purge solenoids are still off. Spark advance is based on the same factors as before. The throttle air bypass valve closes slightly in response to rising coolant temperatures.

Wide-open-throttle Mode

The wide-open-throttle mode is activated by the throttle position sensor. The ECA reenters the open-loop mode in order to provide fuel enrichment for acceleration. The EGR and purge solenoids are off. Additional spark advance is commanded according to the same factors as discussed earlier. If the air-conditioning system is operating, it will be shut off. If the engine cooling fan is operating, it will also be shut off. The preheated air system provides fresh air.

Hot Curb Idle

When the engine is at hot curb idle, throttle closed, fuel delivery is still controlled by the ECA. The EEC-IV system will be operating in the closed-loop mode where the exhaust gas oxygen sensor signals are used for determining pulse width. During open-loop operation, the EGO sensor signals are not sufficient to be recognized by the ECA. If the ECA loses the EGO signal for longer than 15 seconds, it enters the open-loop mode. When at least two EGO sensor signals are reestablished, the ECA reenters the closed-loop mode. The preheated air system provides fresh air. Spark advance is still controlled by the ECA. The throttle air bypass valve may continually change its opening to reflect air temperature, air/fuel ratio, engine speed, and engine load. The EGR and purge solenoids are off.

Hot Cruise Mode

The hot cruise mode fuel delivery is controlled by the ECA. The EGR and purge solenoids are on. Spark advance is controlled by the ECA. The throttle air bypass valve opening is reduced by the ECA. The preheated air system provides fresh air.

Deceleration Mode

When decelerating, the EGR and purge solenoids are off. The ECA calculates an injector pulse width based on engine coolant temperature, engine speed, vane meter airflow, and EGO sensor signals. When the calculated pulse width is less than 2 milliseconds and the engine speed is greater than 1500 rpm, the ECA shuts off the injectors. They remain off until the engine speed drops to 1200 rpm, at which time the ECA again pulses the injectors. Spark advance and throttle air bypass are controlled by the ECA in accordance with input data.

Limited Operation Strategy Mode

The limited operation strategy mode is entered when certain circuit malfunctions occur. While in this mode, the EGR and purge solenoids are off, the throttle air bypass valve remains in a fixed position, spark advance remains at base timing, and the fuel injector pulse width is fixed. As a result, the engine runs and performs poorly but can be driven in spite of system problems.

SYSTEM SERVICE

Fuel Pressure Tests

Fuel system pressure tests can be made by connecting a fuel injection pressure gauge to the Schrader fitting on the fuel rail on 1.6-liter engines or on the fuel supply line on 2.3-liter engines (Fig. 11-26). The special fuel injection pressure gauge can also be used to bleed pressure from the system before servicing any part of the fuel system. The service manual provides detailed information helpful when diagnosing fuel system pressure symptoms.

Idle Speed

Curb idle is controlled by the ECA and throttle air bypass valve. Minimum idle airflow is adjusted with the TAB valve disconnected (Fig. 11-27). This enables the service technician to calibrate the position of the throttle plate. Procedures in the service manual should be closely followed and adjustments should be made at the throttle plate screw. This adjustment is necessary to allow correct operation of the TAB valve.

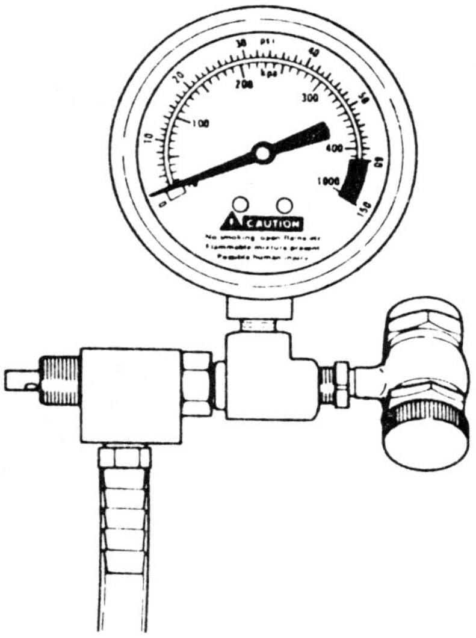

FIGURE 11-26 Fuel injection pressure gauge (T80L-9974-A) *(Courtesy of Ford Motor Company).*

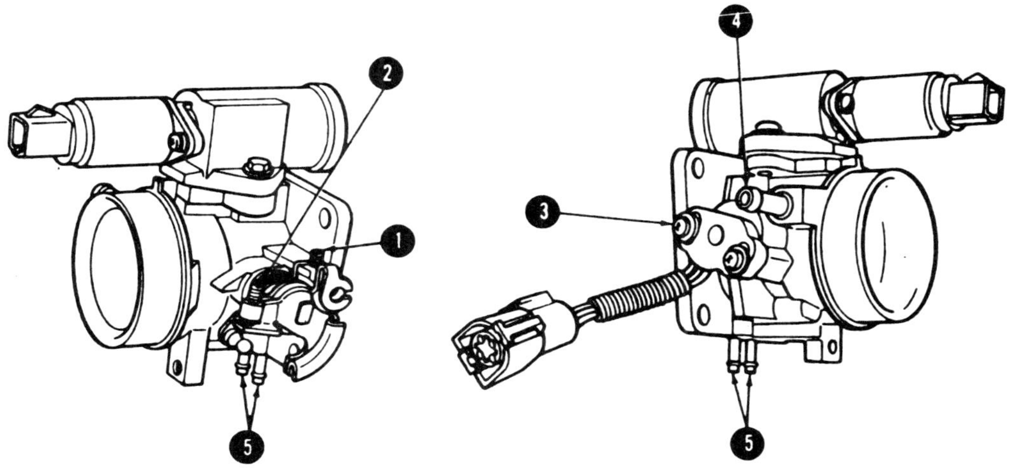

1. AN ADJUSTMENT SCREW TO SET THE THROTTLE PLATE AT A MINIMUM IDLE AIRFLOW POSITION.
2. A PRE-SET STOP TO LOCATE THE WIDE-OPEN-THROTTLE (WOT) POSITION.
3. A THROTTLE BODY MOUNTED THROTTLE POSITION SENSOR.
4. A PVC FRESH AIR SOURCE UPSTREAM OF THE THROTTLE PLATE (SOME APPLICATIONS.
5. INDIVIDUALVACUUMTAPS (SOME APPLICATIONS) FOR EGR AND SPARK CONTROL SIGNALS.

FIGURE 11-27 Air throttle body assembly *(Courtesy of Ford Motor Company).*

Ignition Timing

Ignition timing adjustment procedures are provided in the service manual. Either a magnetic probe timing meter or an inductive timing light may be used. This adjustment is very important as it affects the calibration of the ECA control of timing.

Fuel Filter

The fuel filter is serviced by replacement at the recommended change interval. The fuel system pressure must be relieved before servicing any fuel system component. The arrow on the filter points in the direction of the fuel flow.

Electrical and Electronic Components

Testing of electrical and electronic components is outlined in the service manual. A high-impedance digital multimeter is necessary for these tests.

SYSTEM DIAGNOSIS

The EEC-IV system has a self-diagnostic capability by which the ECA monitors selected circuits and stores codes representing various malfunctions when they occur.

Quick Tests

Testing of the EEC-IV system should begin with the three phases of the quick test: key on, engine off; engine running; and continuous testing. Quick test procedures vary by year and system; therefore, the appropriate service manual should be used. When quick test results indicate a problem, pinpoint tests can be performed on specific circuits. Once the problem cause has been repaired, the quick tests should be done again to assure correct system operation.

Special Tools

Special tools and equipment needed for testing and diagnosing the EEC-IV system include a jumper wire, vacuum gauge, hand vacuum pump, tachometer, spark tester, inductive timing light, fuel injection pressure gauge, and a 10-megohm, high-impedance digital multimeter (Fig. 11-28). In addition, either an analog multimeter or a self-test automatic readout (STAR) tester is necessary to access stored service codes (Fig. 11-29).

Quick Test Example

The quick test for the 2.3-liter turbocharged engine consists of ten steps. The 1.6-liter engine requires only seven steps. For purposes of discussion, the 2.3-liter turbocharged engine quick test will be described here. The quick test evaluates the functional ability of the EEC-IV system. The key on, engine off and engine running tests are used to find hard codes. Continuous testing is used to find intermittent codes. Hard codes represent faults that are currently happening, whereas intermittent codes represent faults that may come and go under varying conditions.

The quick test begins with a visual inspection and

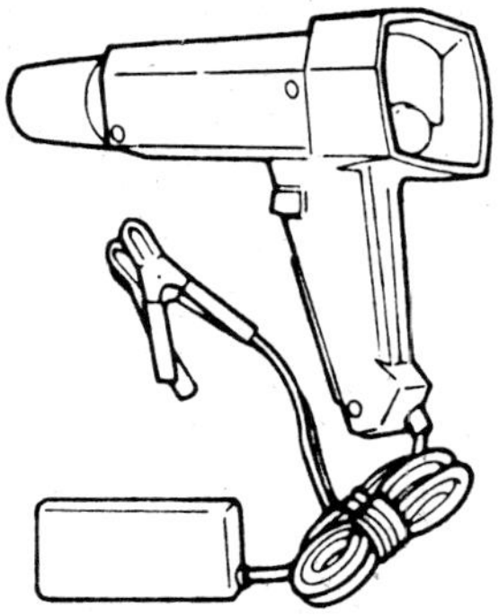

TIMING LIGHT — ROTUNDA 27-0001
OR EQUIVALENT

(a)

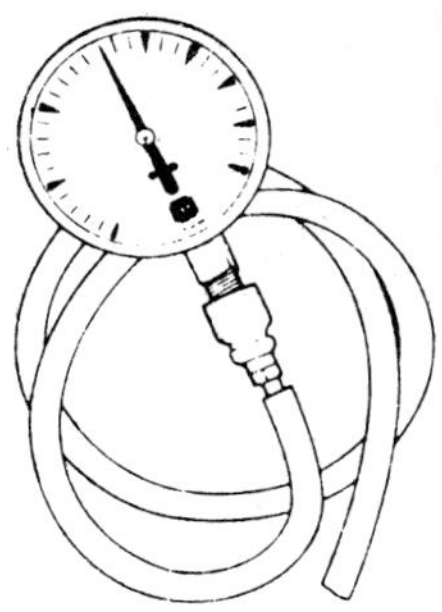

VACUUM GAUGE

(b)

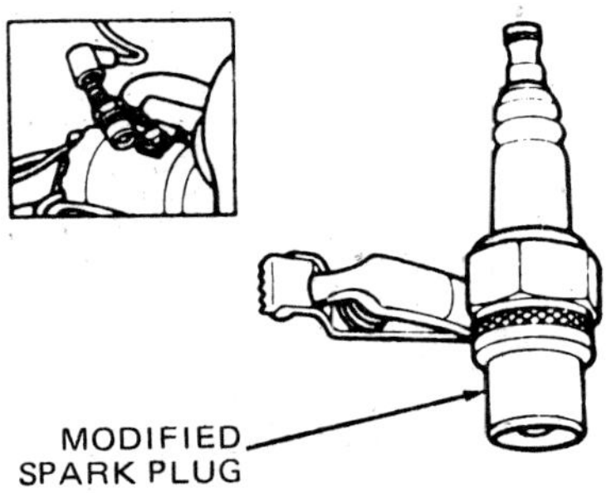

SPARK TESTER

(c)

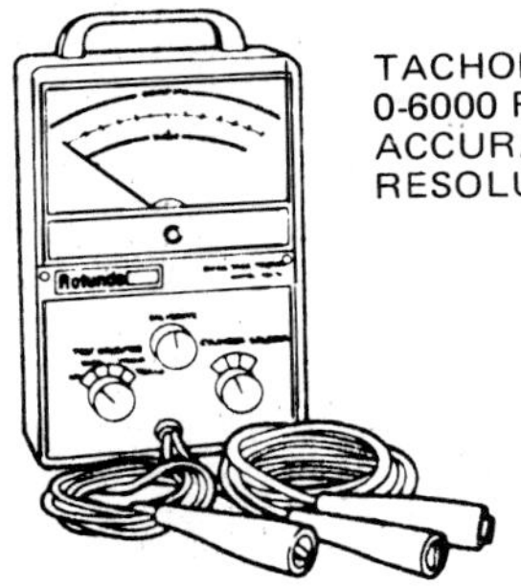

TACHOMETER

(d)

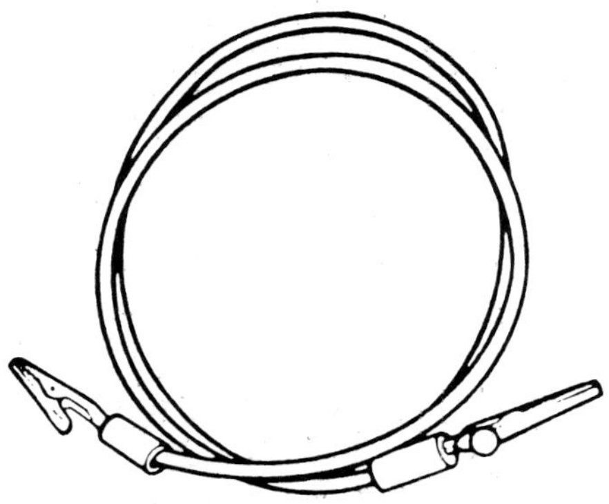

JUMPER WIRE

(e)

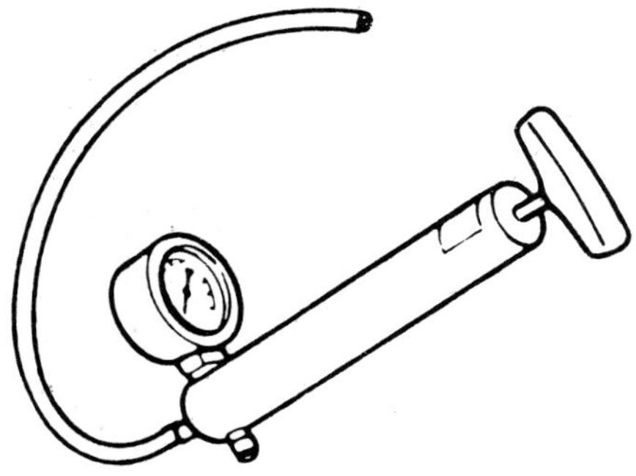

ROTUNDA PUMP 21-0014 OR EQUIVALENT

(f)

FIGURE 11-28 (a) Timing light, Rotunda 27-0001 or equivalent. (b) Vacuum gauge. (c) Spark tester. (d) Tachometer. (e) Jumper wire. (f) Rotunda pump 21-0014 or equivalent *(Courtesy of Ford Motor Company).*

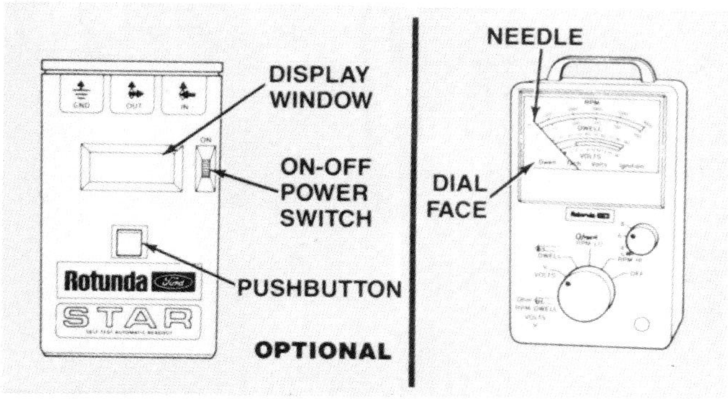

FIGURE 11-29 Analog voltmeter and STAR tester *(Courtesy of Ford Motor Company).*

preparation for testing. The next step involves connecting testing equipment. Step 3 provides access to service codes. The key on, engine off self-test is step 4; step 5 is the continuous self-test. Step 6 is known as the output cycling test, step 7 is the computed timing check, and step 8 is the engine running test. A wiggle test is included as step 9, and step 10 is where pinpoint testing routines begin.

The service technician should perform step 1 by making a visual inspection of wiring harnesses and connections. Once testing equipment is hooked up in step 2, the technician can access the service codes in the ECA memory. Initially, fast codes are displayed. These codes are used by personnel at the assembly plant and serve no purpose in the service field. Some testers may not respond to the fast codes because of the high rate of transmission. Engine identification codes are one-digit numbers that are equal to one-half the number of cylinders, such as two pulses equals a four-cylinder engine. These codes are used for assembly purposes and do not serve any function in the service field. Engine service codes representing circuit malfunctions are displayed next. There are 24 codes that can be stored in memory. They are described in the appropriate service manual. Continuous codes are displayed after the service codes. Continuous codes represent intermittent circuit faults. The key on, engine off self-test checks ECA inputs against calibrated sensor values. The continuous self-test checks sensor inputs for shorts and opens. During the output cycling test, the ECA activates all its outputs or controlled devices. The computed timing check tests the system's ability to provide 30 degrees of advanced spark timing. The engine running test checks operation of sensors and actuators while the engine is under actual operating conditions. The wiggle test can be done to find intermittent connections, opens, or shorts by observing the analog multimeter while moving, tapping, or wiggling the wiring harness or its connections. The pinpoint testing routines are used once the functional test has detected a specific circuit problem. These routines are provided in the service manual.

Service Codes

Service codes can be read by using either the analog multimeter or the STAR tester. Codes are read on the analog multimeter by counting the pulses or sweeps of the meter needle. The STAR tester provides a digital readout of the service codes. The service manual shows how to connect either of these testers to the computer system (Fig. 11-30).

Appropriate service manuals are essential to accurate diagnosis of the EEC-IV system.

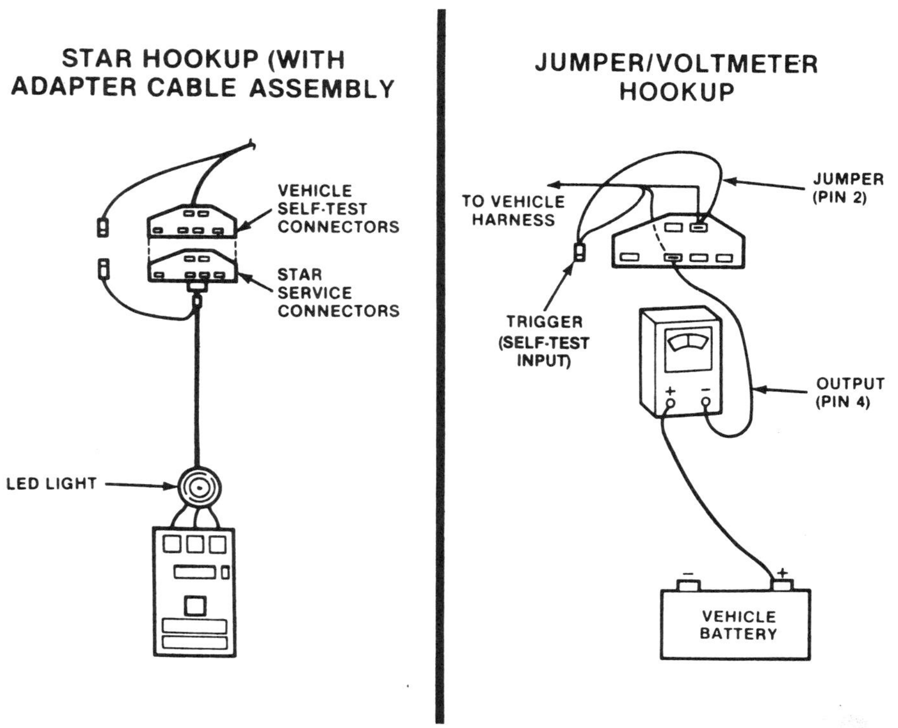

FIGURE 11-30 Equipment hookup *(Courtesy of Ford Motor Company).*

ABBREVIATIONS

ACC A/C clutch
BP Barometric pressure
DOS Distributor output signal
ECT Engine coolant temperature
EGO Exhaust gas oxygen sensor
EGR S/O EGR shut off
FP Fuel pump control
IMS Ignition module signal
ISC Idle speed control
KAM Keep alive memory
KS Knock sensor
LOS Limited operation strategy
TPS Throttle position sensor
VAF Vane air flow
VAT Vane air temperature
V Ref. Voltage reference

UNIT III
Chrysler Corporation Systems

12 Electronic Fuel Injection, Fuel Bar System, 1981–1983, Imperial V-8

OVERVIEW

The Chrysler electronic fuel injection (EFI) system is a single-point, continuous-flow, electronically controlled fuel delivery system. The system uses a computer to electronically monitor and control the air/fuel ratio, ignition timing, and emission controls. The operation of the computer is based on sensor inputs. The computer is programmed to respond to the sensor inputs, changing engine operation, and environmental conditions, and when certain sensors are inoperative or out of calibration, the computer program enables the system to compensate for them.

Because the EFI system is electronically controlled, it is not subject to the same airflow problems as is the venturi-controlled metering of fuel in the carburetor. Fuel atomization is improved even at low airflow rates through the use of injector of spray bars. This helps each cylinder get a more consistent air/fuel mixture.

The Chrysler EFI system is comprised of a fuel supply system, fuel control system, air supply system, air control system, sensors, combustion control computer and support modules, EFI circuit, autocalibration circuit, electronic spark advance circuit, power module, and automatic shutdown module. Each of these subsystems will be described in detail.

Fuel Supply System

The fuel supply system includes the fuel tank and cap, an in-tank electric fuel pump, fuel delivery line, two parallel in-line filters, control pump housing, pressure regulator and bypass orifice, fuel return line, and return line check valve (Fig. 12-1).

Fuel Tank

The EFI fuel tank is a standard stamped steel tank except for a sump area where the in-tank electric fuel pump is located. The sump area is surrounded by sheet metal baffles that prevent the fuel from running away from the pump during severe maneuvering conditions. The filler cap is a standard pressure-vacuum relief emission control cap.

In-tank Electric Fuel Pump

The in-tank electric fuel pump is a rotary, centrifugal type (Fig. 12-2). It is driven by a two-speed electric motor. A woven plastic filter is located on the intake side of the pump. This filter will not allow water to be drawn into the pump. A check valve is located between the pump outlet and the fuel delivery line. It prevents fuel from draining back into the tank when the pump is not operating. During engine operation, the fuel pump runs at an intermediate

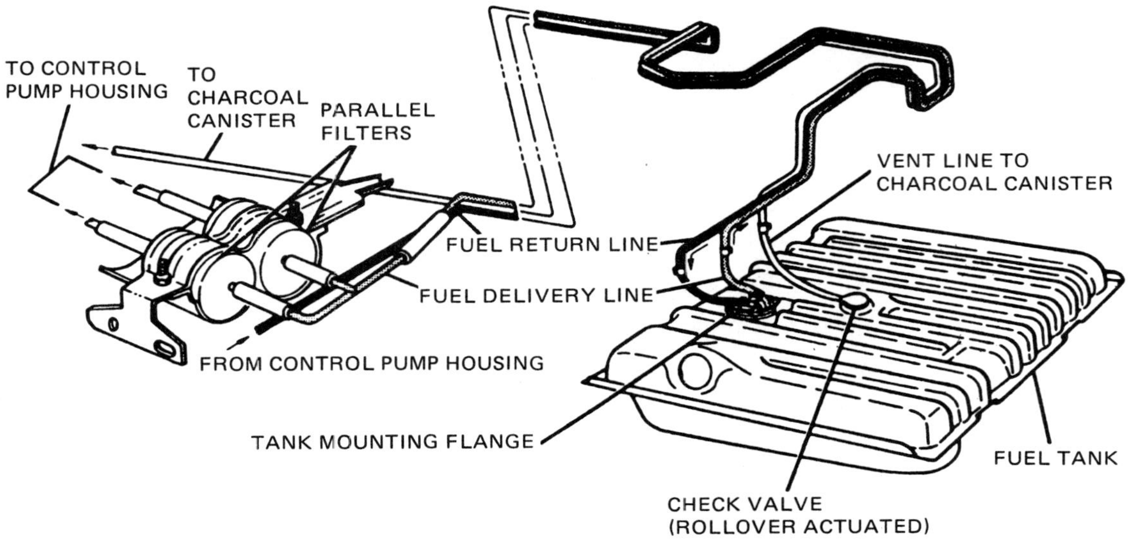

FIGURE 12-1 Electronic fuel injection fuel supply subsystem *(Courtesy of Chrysler Training Program).*

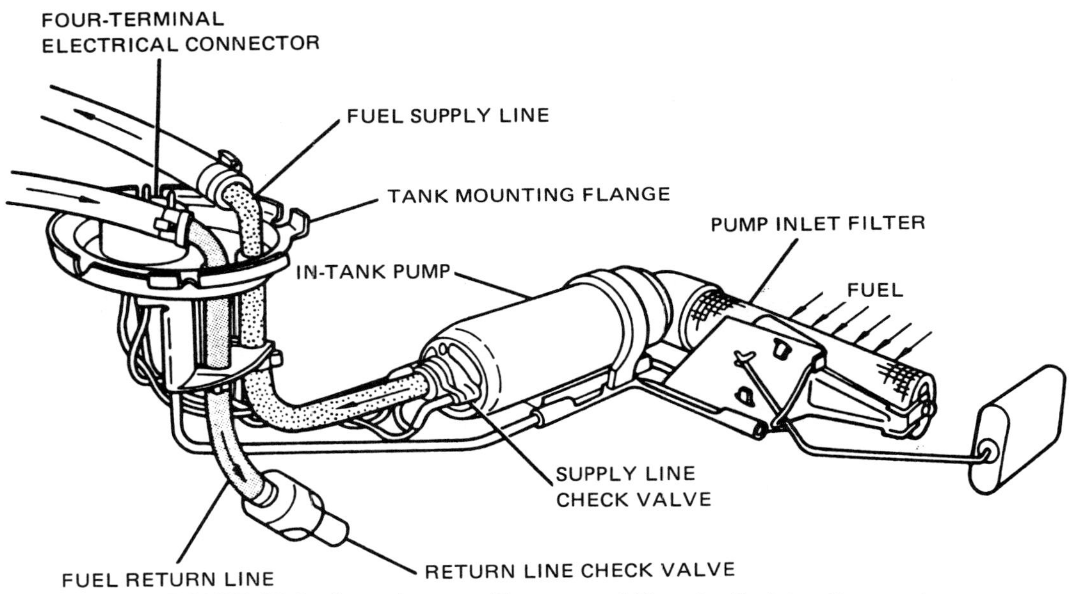

FIGURE 12-2 In-tank pump *(Courtesy of Chrysler Training Program).*

speed. This prolongs pump life. A resistor in the pump electrical circuit drops supply voltage and limits the pump speed. During cranking, a relay in the pump circuit allows full system voltage to bypass the resistor. This drives the pump at full speed, which enables the pump to fill and pressurize the fuel delivery line and the control pump housing to about 11 to 13 psi. Any fuel vapor bubbles in the lines are purged when this activity occurs.

Fuel Delivery Line

The fuel delivery line is a combination of standard steel fuel line with double flared ends and a special high-pressure fuel injection hose. Rubber hoses connect the steel line to the in-tank fuel pump, filters, and control pump housing. Standard fuel line hose is made of a neoprene rubber and should not be used as replacements. High-pressure fuel injection hose is made of a fluoroelastomer rubber and is the only acceptable replacement. This hose is resistant not only to higher pressures, but also to contaminants and additives found in today's fuels.

Fuel Filters

Two parallel in-line filters are used to filter the fuel before it reaches the control pump housing (Fig. 12-3). The filters are mounted on a common bracket fastened to the frame rail of the vehicle. The two filters cause a smaller pressure drop than a single filter and also provide increased flow capability.

Control Pump Housing

The control pump housing is actually a reservoir for the control pump (Fig. 12-4). It serves as a common junction between the fuel supply circuit and the fuel return circuit. A check valve is located between the fuel delivery line and the control pump housing. This valve prevents the control

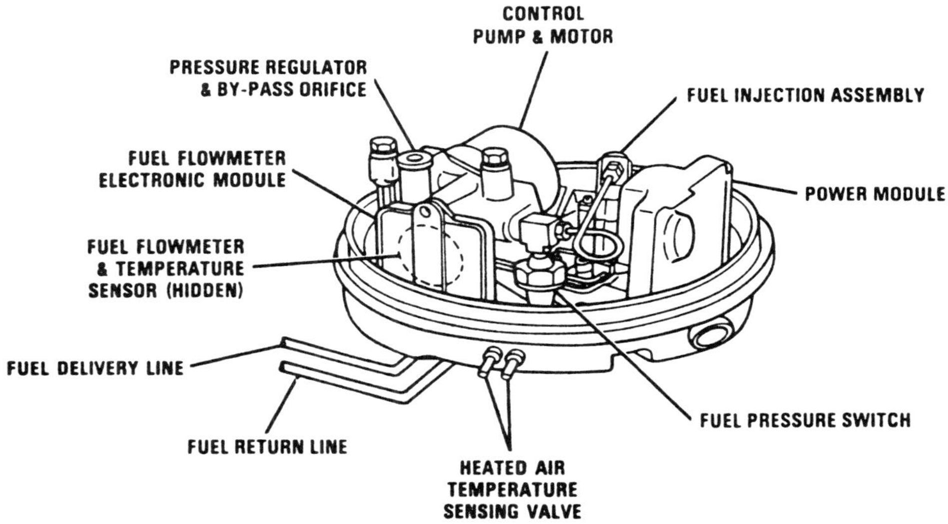

FIGURE 12-3 Hydraulic support plate assembly, fuel control subsystem *(Courtesy of Chrysler Training Program).*

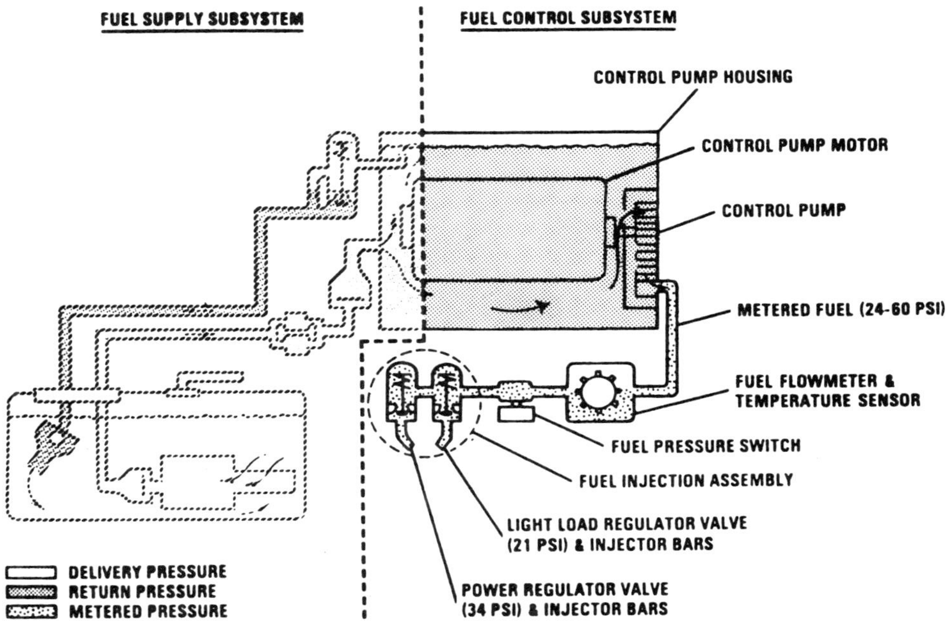

FIGURE 12-4 Fuel flow, control circuit *(Courtesy of Chrysler Training Program).*

housing fuel from draining back to the fuel tank when the system is not operating.

Pressure Regulator Valve

A pressure regulator valve maintains constant pressure in the control pump housing. It consists of a spring-loaded diaphragm and valve. Control pump housing pressure acts on one side of the diaphragm, while spring tension acts on the other side. When fuel pressure overcomes the calibrated spring tension, the diaphragm moves and raises a valve off its seat. Excess fuel is allowed to pass into the return line.

In addition, the pressure regulator valve seat has an orifice or small hole drilled through it. This allows fuel to constantly bypass the valve and seat; however, the amount of fuel passing through this orifice does not affect the operation of the system. The function of the orifice is to vent pressure that builds when the engine is turned off. Heat from the engine would otherwise cause a pressure buildup and a potential vapor-lock condition. The orifice vents this pressure into the fuel return circuit.

Fuel Return Line

The fuel return line is made of the same materials as the fuel supply line. It carries excess fuel back to the fuel tank. System air bubbles and vapors are purged through the return line. The end of the return line in the fuel tank is fitted with a return line check valve. This valve maintains about 1 to 2 psi in the return line. It also prevents draining of the fuel tank into the engine compartment when the vehicle is involved in a rollover accident.

FUEL CONTROL SYSTEM

The fuel control system includes a control pump, fuel flow meter, fuel temperature sensor, fuel pressure switch, and a fuel injection assembly (Fig. 12-4).

Control Pump

The control pump is a slipper-type positive-displacement pump that is driven by a variable-speed electric motor. This pump boosts the pressure of the in-tank electric fuel pump. The amount of boost is determined by the speed of the control pump.

Fuel Flow Meter

Fuel flowing from the control pump passes through a fuel flow meter (Fig. 12-4). The flow meter is a wheel with vanes attached to it. Passing fuel causes the vanes and wheel to turn. A light-emitting diode (LED) is mounted on one side of the vanes and a phototransistor is mounted opposite the LED. As the wheel and vanes turn, the vanes interrupt the light beam between the LED and phototransistor. When light strikes the phototransistor, a voltage signal is produced. If the light beam is interrupted, the voltage signal also is interrupted. The result is a pulsed signal from the phototransistor. As the pulsed signal frequency increases, the fuel flow rate must have increased. This signal is sent to the fuel flow meter electronic module where it is converted into a fuel flow rate value for the combustion control computer.

The fuel flow meter also has a fuel temperature sensor as an integral part. This sensor is a semiconductor thermistor whose resistance value varies inversely with temperature. As the temperature rises, its resistance decreases. As the temperature falls, its resistance increases. The voltage drop across this resistance is interpreted by the fuel flow meter electronic module as a specific fuel temperature. This value and the fuel flow rate signal are used by the combustion control computer to regulate the speed of the control pump motor.

Fuel Pressure Switch

The fuel pressure switch is a normally closed switch located between the fuel flow meter and the fuel injection assembly. Fuel pressure causes this switch to open. When the ignition switch is turned on, the fuel pressure switch is closed due to a lack of fuel pressure. This causes the control pump to run at full speed, which purges the system of air and vapors and also pressurizes the system. When the system is pressurized, the switch opens and the control pump speed is regulated by the fuel flow meter. This priming process happens in the same amount of time that it takes the crankshaft to make one revolution.

Fuel Injection Assembly

The fuel injection assembly receives fuel from the control pump (Fig. 12-5). This fuel enters two pressure-regulating valves. Each valve regulates pressure to a U-shaped fuel injection bar on top of the throttle body assembly. The light-load regulator valve opens at about 21 psi and then supplies fuel to the light-load injector bar. Four tiny holes in the bottom of the injector bar spray fuel onto crescent-shaped ridges on the throttle plates. Nozzles with an airfoil shape surround the injector holes and help shape the spray patterns. The light-load injector bar provides all the fuel the engine needs when the fuel metering pressure is between 21 and 34 psi. During starting or heavy load conditions, metering pressure goes higher than 34 psi. At this point, the other pressure regulator valve, or power regulator valve, opens. This allows fuel to be sprayed from the other U-shaped injector bar or power injector bar. Both

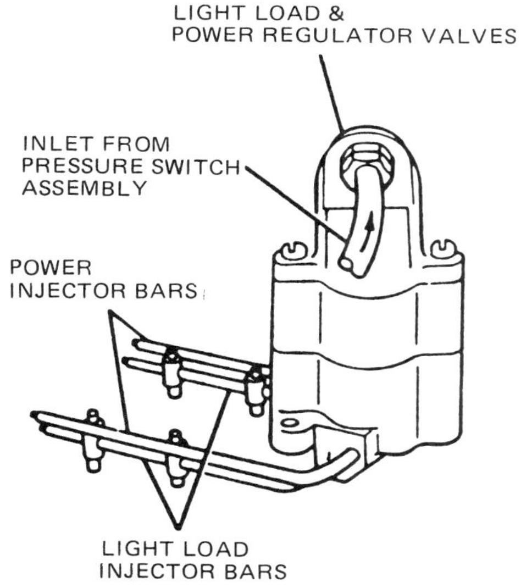

FIGURE 12-5 Fuel injection assembly *(Courtesy of Chrysler Training Program).*

light-load and power injector bars are providing fuel for the engine. The power injector bar does not have nozzles for spray pattern control. At higher engine speeds, the nozzles are not necessary.

AIR SUPPLY SYSTEM

A fresh air and heated air mixing assembly, an airflow sensor assembly, and an air cleaner assembly are components of the air supply system (Fig. 12-6).

Fresh and Heated Air Mixing Assembly

The fresh air and heated air mixing assembly functions the same as that used on carbureted engines (Fig. 12-7). Unheated air enters through fresh air ducting. Heated air enters the heat stove or sheet metal shrouding surrounding

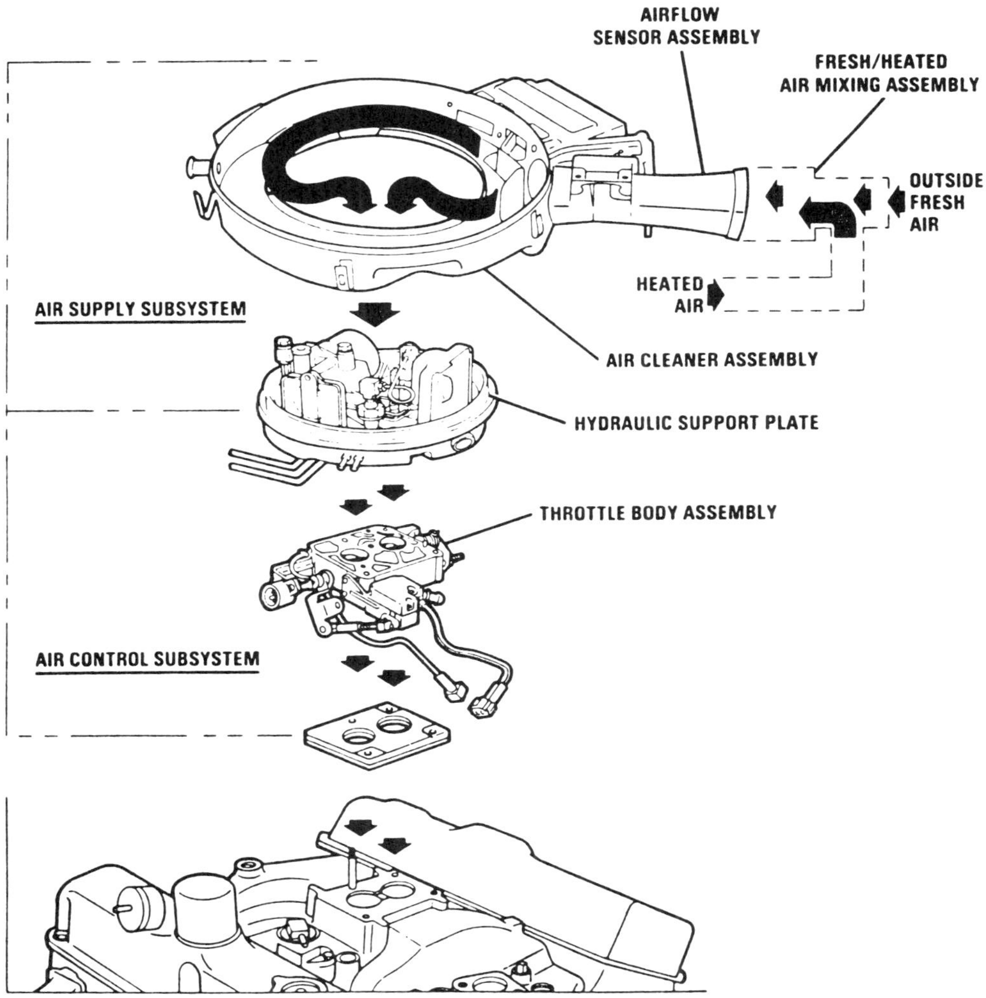

FIGURE 12-6 Electronic fuel injection air induction system *(Courtesy of Chrysler Training Program).*

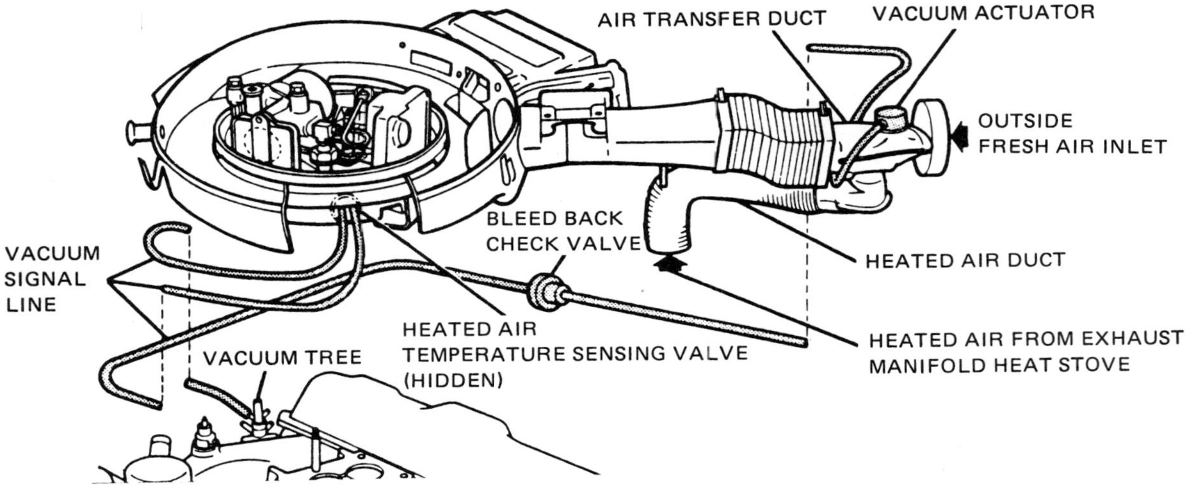

FIGURE 12-7 Fresh and heated air mixing assembly *(Courtesy of Chrysler Training Program).*

the exhaust manifold. A temperature-sensing vacuum valve in the air cleaner housing senses the temperature of the air passing into the throttle body. When the air temperature is below 70°F, the temperature sensing vacuum valve allows intake manifold vacuum to operate the vacuum actuator diaphragm on the snorkel of the air cleaner assembly. The actuator diaphragm moves upward and causes a flap valve inside the snorkel to block off unheated air. Air is then drawn through the heat stove and ducting.

When the temperature sensing vacuum valve warms, it begins to open the vacuum line. This reduces vacuum available at the actuator diaphragm, which then moves down. The flap valve moves to an intermediate position and fresh air is mixed with heated air. If the temperature-sensing vacuum valve gets warm enough, it will bleed more vacuum so that the vacuum available to the actuator diaphragm will be minimal. At this point the diaphragm moves down, the flap valve blocks the heated air passage, and air entering the air cleaner assembly will be unheated. In this way, air entering the throttle body can be maintained at a constant specified temperature regardless of operating conditions. This system provides better cold engine performance, lower overall emissions, and better fuel economy.

Airflow Sensor

The airflow sensor assembly is located between the air mixing assembly and the air cleaner housing (Fig. 12-8). It measures the volume of air entering the engine at any instant. This information, when combined with the fuel flow meter information, enables the combustion control computer to precisely control the air/fuel mixture. The airflow sensor is a hollow tube containing vanes that cause the air to swirl as it passes through the tube. The vanes are at the

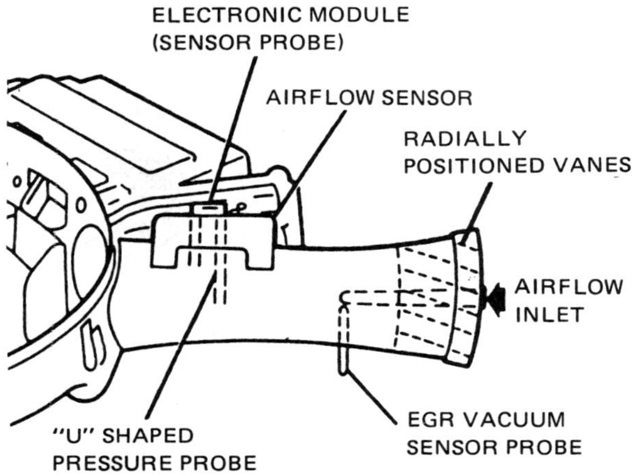

FIGURE 12-8 Airflow sensor assembly *(Courtesy of Chrysler Training Program).*

inlet end of the sensor. A U-shaped pressure-sensing tube is located at the outlet end. This pressure-sensing tube is connected, on the outside of the sensor, to an electronic module.

When air enters the sensor, the vanes cause it to swirl in a clockwise direction. This creates a small vortex or tornado inside the sensor. At the center of the vortex is a low-pressure area. Near the sensor inlet, the low-pressure area is centered within the sensor tube. As the low-pressure area approaches the outlet end of the sensor, it begins to whip off center. In this airflow sensor design, the speed or frequency at which the low-pressure area whips within the airflow sensor tube is directly related to the volume of air passing through the sensor. This speed or frequency is measured by the U-shaped tube near the outlet end of the sensor. The ends of this pressure-sensing tube extend into the path of the whipping low-pressure area. Because the ends of the pressure-sensing tube are 180 degrees apart, when low pressure is at one end, higher pressure will be at the other end. These circumstances will reverse and continue to alternate. This produces pulses of air in the pressure-sensing tube, which passes through the electronic module mounted on the outside of the airflow sensor. The electronic module contains an electrically heated silicon detector. As each pulse of air passes the detector, it momentarily cools the detector. This reduces the resistance of the detector and a minute increase in current flow is produced. The speed or frequency of the increases in current flow is interpreted by the electronic module as being an indication of airflow rate. This information is sent to the combustion control computer.

During cranking conditions, airflow is too small to be sensed. The combustion control computer is programmed to disregard the airflow sensor signal and instead use engine rpm, throttle position, and temperature sensor signals to compute airflow. The airflow sensor also provides the vacuum signal for the exhaust gas recirculation valve.

Air Cleaner

The air cleaner assembly consists of a sheet metal housing and a filtering element. Air enters the housing, passes through the filter, and is delivered to the hydraulic support plate at the top of the throttle body. The combustion control computer is mounted on the outside of the air cleaner housing.

AIR CONTROL SYSTEM

The throttle body and blade assembly, throttle position potentiometer, closed-throttle switch, and automatic idle speed motor make up the air control system. All these system functions take place in the throttle body assembly (Fig. 12-6).

Throttle Body

The throttle body and blade assembly includes two throttle plates mounted on a common shaft and a crescent-shaped ridge at the leading edge of each throttle plate. The fuel spray from the fuel injection bars is directed toward the crescents for better air/fuel mixing (Fig. 12-9).

Throttle Position Potentiometer

The throttle position potentiometer is mounted on one end of the throttle shaft (Fig. 12-9). It is a variable resistor that uses a 23-volt source and a ground. A third terminal connects to a sliding or wiping arm that sweeps across the resistive coil when the throttle shaft moves. Voltage drops across the sliding arm in accordance with the position of the arm on the resistive coil. The remaining voltage can be measured at the third terminal as being an indication of throttle position. This signal is sent to the combustion control computer and enables the computer to adjust the air/fuel ratio as engine and vehicle speed change, acceleration or deceleration is demanded by the driver, or as the engine is starting. When the throttle is opened beyond a specified

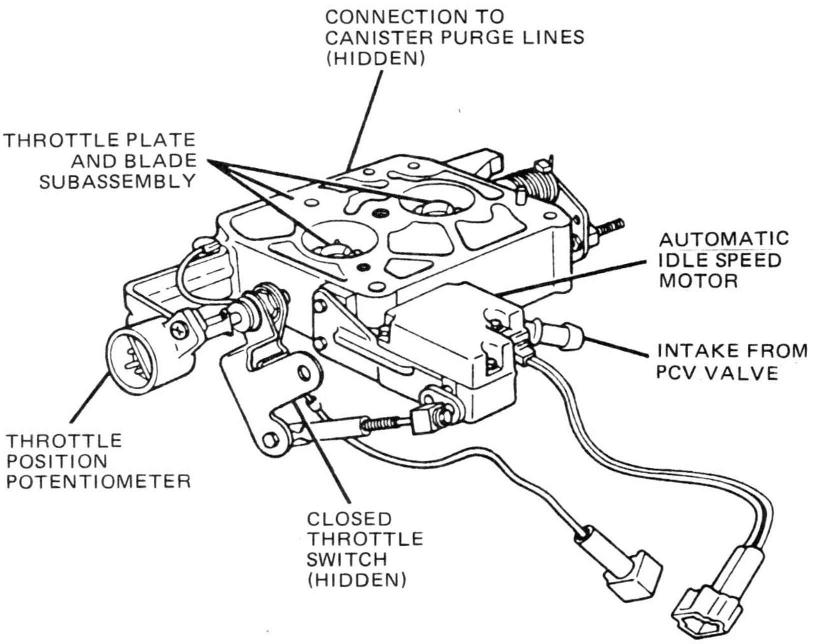

FIGURE 12-9 Throttle body assembly *(Courtesy of Chrysler Training Program).*

angle, the throttle position potentiometer signal causes the control pump to be shut off. This allows a flooded engine to be started.

Closed-throttle Switch

Located at the idle stop is a closed-throttle switch (Fig. 12-9). The switch is closed whenever the throttle blades are closed. This activates the automatic idle speed motor to take control of the idle speed. It also causes the ignition timing to return to its basic or minimum advance setting, sometimes referred to as initial or base timing. When the throttle blades are opened, the throttle switch opens and deactivates the automatic idle speed motor. The ignition timing also advances about 5 degrees. The brake switch is connected in parallel with the throttle switch and acts as a backup circuit when a malfunction occurs in the throttle switch. Either releasing the accelerator pedal or depressing the brake pedal activates the automatic idle speed motor and retards the ignition timing. Because these switches are in parallel, the automatic idle speed motor can be deactivated, and the ignition timing can be advanced only when both the throttle switch and brake switch are open.

Automatic Idle Speed Motor

The automatic idle speed motor is a reversing motor (Fig. 12-9). Idle speed is controlled by mechanically moving the throttle linkage. When the motor is deactivated by the closed-throttle switch, the motor holds its last position.

FUEL, AIR, AND IGNITION FUNCTIONS

The Chrysler electronic fuel injection system utilizes a combustion control computer to control fuel metering, ignition timing, idle speed, and fuel flow shut off under specific circumstances. The combustion control computer does not control these functions alone. It uses two modules, a power module and an automatic shutdown module, to aid in controlling engine circuits. To achieve a high degree of accuracy, the combustion control computer must rely on input information from various sensors. This information enables the computer to be aware of the constantly changing conditions of engine operation. Additional information is provided by various switches.

SENSORS AND SENSING

Information provided to the combustion control computer includes airflow, air temperature, fuel flow meter, exhaust gas oxygen content, coolant temperature, fuel temperature, fuel pressure, throttle position, closed throttle, closed-throttle backup circuit, time, air conditioner ON, ignition, engine load, ignition switch, and engine speed.

Airflow sensor

Since the airflow sensor was discussed in detail earlier in this chapter, there is no need to repeat the information here. It is located on the air cleaner assembly.

Air Temperature Sensor

The air temperature sensor is a thermistor whose resistance value decreases as the air temperature increases. It is located inside the combustion control computer, which has an air duct leading to the air cleaner housing. The temperature of the air, when combined with data from the airflow sensor, provides a more accurate measurement of the mass of the air entering the engine. This enables the combustion control computer to more precisely control the fuel mixture.

Fuel Flow Meter

The fuel flow meter was discussed earlier in this chapter and will not be repeated here. It is located on the hydraulic support plate between the control pump outlet and the fuel pressure switch. It measures the quantity of fuel delivered to the injector assembly in a given amount of time.

Oxygen Sensor

An oxygen sensor measures the exhaust gas oxygen content. The amount of oxygen in the exhaust gas is an indication of how rich or lean the air/fuel mixture entering the cylinders must be. When the mixture is too lean, the oxygen sensor generates a low-voltage signal indicating high oxygen content in the exhaust. When the mixture is rich, the oxygen sensor generates a high-voltage signal indicating low oxygen content in the exhaust. The varying oxygen sensor voltage is supplied to the combustion control computer so that it can change the amount of fuel injected. This activity occurs rapidly, so the system is very responsive and accurate in controlling the air/fuel ratio. The ideal air/fuel ratio of 14.7 to 1 can be achieved for almost all engine operating conditions. This is known as a stoichiometric mixture.

Engine Coolant Temperature Sensor

The engine coolant temperature sensor is a nickel-wire variable resistor whose resistance value increases as temperature increases. When temperature is high, the sensor resistance is high and the sensor signal is approximately 8 volts. When engine coolant temperature is low, the sensor provides a higher voltage signal of approximately 12 volts to the combustion control computer. The computer provides a richer fuel mixture and, at the same time, a higher idle speed as long as the throttle is in an idling position. As the coolant temperature rises, the sensor causes the computer to gradually lean the mixture and lower the idle speed. The coolant sensor acts as an electronic "choke," much the same as a choke on a carburetor but with much more precise control. The coolant temperature sensor is located at the left front corner of the intake manifold.

Fuel Temperature Sensor

The fuel temperature sensor is located in the fuel flow meter (Fig. 12-4). Because fuel vaporization is related to its temperature, cooler fuel would mean that a richer mixture is needed to obtain good results. This temperature signal is sent to the combustion control computer so that it can adjust the air/fuel mixture accordingly. The fuel temperature sensor is a thermistor whose resistance value decreases as fuel temperature increases. As fuel temperature lowers, the sensor resistance increases. The fuel temperature sensor signal is generally between 16 and 21 volts depending on the temperature.

Fuel Pressure Switch

A fuel pressure switch is located on the hydraulic support plate between the fuel flow meter and the fuel injection assembly. The signal from this switch is used by the combustion control computer to override the fuel flow meter signal during startup conditions. This allows for a fast purge of the fuel system. As was stated earlier in this chapter, this switch is normally closed when the system is at rest. When pressure is available, the switch opens and the signal it provides changes. This signal indicates that the system has enough pressure for starting the engine. If the switch closes as a result of a drop in fuel pressure, its signal will cause the control pump to run at full speed. The power feed line to the fuel pressure switch should have high voltage when the switch is open, system pressurized. With the system at rest, the switch should be closed and the power feed line to the switch should have low or zero voltage.

Throttle Position Potentiometer

The combustion control computer receives data concerning throttle position from the throttle position potentiometer (Fig. 12-9). It is mounted on the throttle body at the end of the throttle shaft. This sensor provides a voltage signal that is proportional to throttle blade position. It is a variable resistor whose resistance value changes as throttle position changes. The changing resistance causes changes in the voltage available at the signal line. When the driver depresses the accelerator pedal, the throttle position potentiometer is the first to indicate this demand to the combustion control computer. The airflow sensor also indicates throttle opening, but it will be slightly slower to respond. The throttle position potentiometer provides lead time for the computer to make air/fuel ratio adjustments just ahead of the actual increase in airflow.

The throttle position potentiometer also provides a means of starting a flooded engine. By depressing the accelerator pedal to open the throttle blades at least 60 degrees, the throttle position potentiometer signal, in combination with a cranking signal, causes the combustion control computer to shut off fuel flow. As long as the throttle is held open at least 60 degrees while cranking, no fuel is delivered.

Closed-throttle Switch

A closed-throttle switch is mounted on the throttle stop of the throttle body assembly (Fig. 12-9). This switch indicates to the combustion control computer whether or not it should control idle speed with the automatic idle speed motor. When the throttle is closed, so is the closed throttle switch. The voltage signal to the combustion control computer will be low. Whenever the throttle is opened to an off-idle position, the closed throttle switch opens and provides a higher-voltage signal to the computer. The computer uses this signal to determine whether or not to activate the automatic idle speed motor.

Closed-throttle Switch Backup Circuit

The closed-throttle switch backup circuit serves the same function as the closed-throttle switch. It is a switch in the brake light circuit that also indicates the need for idle speed control to the combustion control computer. Because this switch is connected in parallel to the closed-throttle switch, either switch has the ability to signal the computer when idle speed control is needed. When the brake pedal is depressed, the closed-throttle switch backup circuit closes and the signal to the combustion control computer is a low voltage. A higher-voltage signal is provided when the brake pedal is released and the switch is open.

Combustion Control Computer Timer Circuit

The combustion control computer has the ability to sense the passage of time through a timer circuit. This factor is used to help issue commands at the precise moment. It also is used to prevent problems. For example, when cranking for a longer than normal period of time, the computer will shut off fuel flow to prevent a hydrostatic lock condition.

Air-conditioning Signal

An air conditioner ON signal is provided to the combustion control computer so that it can anticipate the added load of the air-conditioning compressor by controlling the automatic idle speed motor. In this way, idle speed is maintained and the tendency to stall is prevented. When the air-conditioning system is off, a low-voltage signal is sent to the computer. When the air-conditioning system is activated, a higher voltage is sent to the computer, which, in turn, commands a signal to the automatic idle speed motor.

Ignition Switch Signal

A signal from the ignition switch is sent to the automatic shutdown module whenever the switch is moved to the start or run position. If, for some reason, this power signal is lost, the automatic shutdown module will shut off fuel flow to the engine. A possible flooding or fire hazard can be avoided with this operation. When the ignition switch is in the start or run position, a higher voltage is sent to the automatic shutdown module. If this voltage is not available to the module, the module will shut down the fuel supply.

Engine Load Computation

The load placed on an engine is an important factor to consider when trying to compute an air/fuel ratio. The combustion control computer calculates engine load from the airflow sensor and engine speed signals.

Start/Run Signal

The ignition switch, in addition to providing an ignition power signal as discussed earlier, provides a signal to the combustion control computer that indicates the position of the switch. This allows the computer to distinguish between start and run, which are both ON signals.

Engine Speed Computation

Engine speed data are obtained by the computer by counting the pulses of the ignition primary circuit. This speed signal input is one of the most important signals as it is used in all modes of operation. It is used for ignition timing, idle speed, and fuel metering.

This completes coverage of combustion control computer input data. The computer uses these data and operates in one of two modes; open feedback loop or closed feedback loop.

COMBUSTION CONTROL COMPUTER

The combustion control computer (CCC) is not a single component but is instead made up of four electronic circuits. The electronic fuel injection circuit controls air/fuel ratio; the autocalibration circuit monitors the oxygen sensor and fine tunes the EFI circuit; the electronic spark advance circuit controls the ignition power and timing; and the automatic idle speed circuit controls engine idle speed when the throttle is closed or the brake pedal is depressed.

Closed Feedback Loop

The CCC operates with a feedback feature. The feedback circuit enables the computer to monitor the results of its commands and make additional command corrections if necessary. For example, the CCC commands a specific air/fuel ratio, the oxygen sensor indicates the results of that commanded ratio, and the CCC readjusts the air/fuel ratio command if necessary. This feature allows for very precise control of the fuel mixture. The feedback feature is used only when the CCC is in the closed-loop mode of operation (Fig. 12-10).

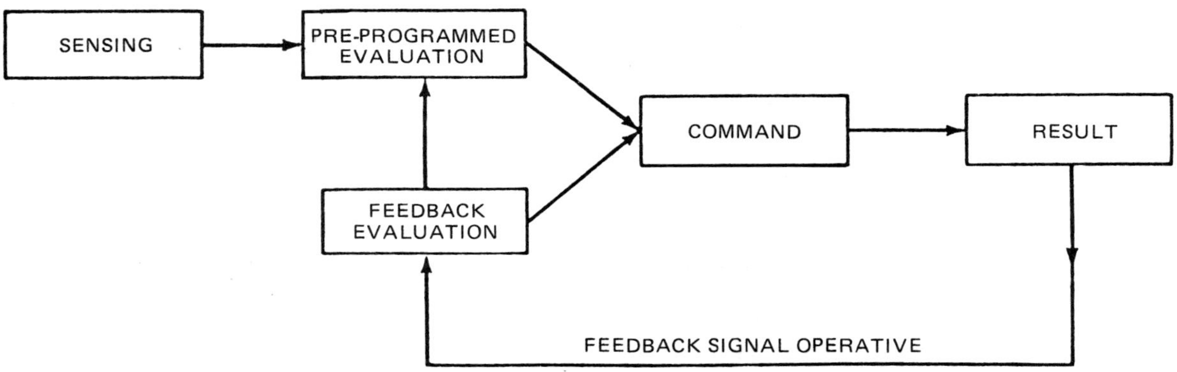

FIGURE 12-10 Closed-loop command operation *(Courtesy of Chrysler Training Program)*.

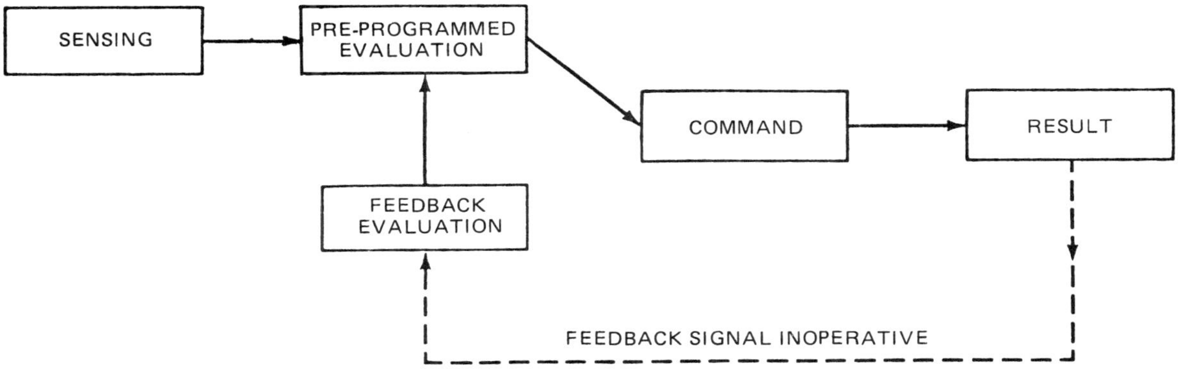

FIGURE 12-11 Open-loop command operation *(Courtesy of Chrysler Training Program).*

Open Feedback Loop

Until the oxygen sensor reaches a temperature level necessary for its operation, the computer and circuitry will operate in an open-loop mode (Fig. 12-11). The air/fuel ratio is determined by the computer program and engine sensor data while in this mode of operation. If the computer program does not provide the correct air/fuel ratios, the program can be changed by means of the automatic calibration process to some extent.

When the oxygen sensor reaches a high enough exhaust temperature level for recognizable voltage production, the CCC system enters the closed-loop mode. Air/fuel ratios are now determined by engine sensor data and the computer program, but mostly by the oxygen sensor. The system remains in the closed-loop mode unless the oxygen sensor cools to a temperature below that required for accurate operation or under heavy acceleration where power enrichment is needed. During closed-loop operation, when the oxygen sensor senses high oxygen content in the exhaust stream, it produces a low voltage that indicates to the computer that the air/fuel mixture is too lean. Conversely, low oxygen content produces a high-voltage signal, which indicates a rich mixture. The computer responds by issuing a command to the fuel injector to correct the mixture. This constant feedback of information enables the computer to attain an accurate 14.7 to 1 air/fuel ratio or stoichiometric mixture.

ELECTRONIC FUEL INJECTION CIRCUIT

The electronic fuel injection circuit (EFI) controls the fuel control pump during all modes of system operation except initial startup. It controls the air/fuel ratio by controlling the pump speed.

During initial startup, when the ignition switch is in the START position, the fuel pressure switch closes and causes the fuel control pump to run at full speed. As discussed earlier, this purges the system of air bubbles and also primes the fuel flow sensor. During this period of time, the EFI circuit has no control over the fuel control pump. When the system is purged and pressurized, the fuel pressure switch opens. The EFI circuit takes control of the fuel control pump and the CCC system is in the open-loop mode. As the oxygen sensor reaches its operation temperature, the EFI circuit will control the fuel control pump and the CCC system will be in its closed-loop mode.

While in open loop, the EFI circuit relies on the system program and engine sensor data to provide the correct air/fuel ratio. Because the EFI circuit program is volatile (loss of power results in loss of memory), the memory circuits have been designed to ensure that power to the memory is not interrupted. Power loss causes a loss of the program information.

Closed-loop operation occurs when the EFI circuit relies on its memory program, engine sensor data, and also the oxygen sensor for control of the air/fuel ratio. The EFI circuit responds primarily to the oxygen sensor, as described earlier.

AUTOCALIBRATION CIRCUIT

When the computer program does not provide the correct air/fuel ratio, the program can be changed. The correction of the EFI circuit memory is done by the autocalibration circuit. This is necessary for delivery of correct air/fuel ratios for the open-loop mode of operation. The autocalibration circuit, in conjunction with the oxygen sensor, reprograms the EFI memory when the system enters the closed-loop mode of operation. This is the only time reprogramming can occur. Recalibration takes place at steady cruising speed, about 55 mph, or at idle.

If recalibration occurs at a steady cruising speed of about 55 mph, the varying oxygen sensor voltage will be very slight. At the instant of the change in sensor voltage, the autocalibration circuit, which is monitoring the oxygen sensor voltage, compares this voltage change to the EFI memory program to see if it is in agreement. If there is no agreement, the autocalibration circuit will initiate a change in the EFI memory program. The autocalibration circuit is able to correct for any variations in the EFI program, including altitude or barometric pressure.

Calibration may need to be done at idle speed because of the inaccuracy of the fuel flow meter at this very low fuel flow speed. The recalibration is accomplished in the same manner as at cruising speed.

ELECTRONIC SPARK ADVANCE CIRCUIT

The electronic spark advance circuit (ESA) controls ignition timing through its analog circuitry. The ESA circuit monitors engine speed, engine load, and the throttle switch. As in traditional systems, any spark advance provided is in addition to initial timing, which is set mechanically.

During engine startup and idling conditions, the ESA circuit provides no advance. Timing should remain at the initial or base timing setting.

In accordance with engine speed, the ESA circuit provides timing advance. Engine speed data are provided to the ESA circuit by the ignition distributor pickup coil. No advance is provided until the engine speed reaches 1000 rpm.

The ESA circuit advances ignition timing as engine load increases. Engine load data are obtained by comparing airflow sensor data to engine speed data. In this way, no vacuum sensors are required.

When the throttle switch opens, the ESA circuit advances ignition timing in accordance with the specifications just discussed. The open throttle switch enables the ESA circuit to control ignition timing.

AUTOMATIC IDLE SPEED CIRCUIT

When the throttle switch is closed, a signal is sent to the automatic idle speed circuit (AIS) to control the engine idle speed. This enables the AIS circuit to operate a reversible dc motor, which moves the throttle linkage. To select the correct throttle position or idle speed, the AIS circuit monitors engine coolant temperature, the throttle switch, the air-conditioning circuit, and the ignition switch.

When engine coolant temperatures are low, the AIS circuit causes the idle speed motor to increase the idle speed. As coolant temperature rises, the AIS circuit gradually decreases idle speed until a programmed speed is reached. The idle speed should not go below this level.

If the ignition switch is in the ON or START position and the throttle switch is closed at the same time, the AIS circuit will position the idle speed motor so that idle speed is below specification. Once the engine starts, the engine speed signal causes the AIS circuit to initially produce an idle speed flare and then a gradual slowdown to the programmed idle speed. This flare is more noticeable when the engine is cold than when it is hot as the flare is influenced by coolant temperature. In any case, the idle speed flares and returns to the programmed level before the driver selects a transmission gear position.

Anytime the air-conditioning system cycles ON, the AIS circuit will maintain idle speed by causing the idle speed motor to open the throttle blades. This offsets the additional engine load caused by the air-conditioning compressor. It also prevents idle speed from oscillating as the air-conditioning system cycles.

POWER MODULE AND AUTOMATIC SHUTDOWN MODULE

As discussed earlier, the combustion control computer utilizes two separate modules, the power module and the automatic shutdown module, to help it control the fuel injection system.

Power Module

The power module is located on the hydraulic support plate (Fig. 12-3). It supports only the EFI system by transforming 12 volts dc into 23 volts dc. The power module converter accomplishes this and supplies the 23 volts to the entire EFI system. It also contains a driver circuit that amplifies the EFI system commands to a level high enough to control the fuel control pump. The power module also relays the fuel flow meter signal to the combustion control computer.

Automatic Shutdown Module

The automatic shutdown module (ASD) is located on the right front cowling under the hood or on the right front inner fender under the hood. All operating power for the EFI system is supplied by the ASD module. Should an accident occur, this module automatically deactivates the EFI system.

When the ignition switch is turned on, the ASD module activates the EFI system for about 500 milliseconds unless an rpm signal is also present. With no rpm signal, the EFI system is deactivated, preventing a potential safety hazard. When the engine speed signal is about 150 rpm, the ASD module keeps the EFI system activated.

When the ignition switch is in the cranking position, the ASD module prevents hydrostatic lock. During extended cranking or stalling, the ASD module interrupts power to the entire EFI system. Cranking period of longer than 10 to 20 seconds causes the ASD module to shut down the system. The ASD module must be manually reset by turning the ignition switch off and then back to start. If the fuel control pump runs at full speed during the second attempt to start the engine, the ASD module will interrupt power to the EFI system in about 3 to 7 seconds. At this point, the module must be manually reset again.

SYSTEM SERVICE AND DIAGNOSIS

Verify Problem

After becoming thoroughly familiar with the Chrysler EFI system operation and also after reviewing the electrical schematics in the service manual, the technician should verify the customer complaint. The conditions surrounding the complaint should be defined. As much information as possible about the complaint should be gathered before testing is begun.

Visual Checks

Visual checks of the system should be done first. Loose connectors, hose clamps, or cracked vacuum or fuel lines should be repaired before further tests are performed. Other visual checks include those inside the air cleaner and outside the air cleaner.

Inside Air Cleaner Checks

Inside the air cleaner are three connectors: the power module to control pump circuit, the fuel pressure switch to power module circuit, and the fuel flow meter to power module circuit. The power module is grounded to the hydraulic support plate assembly screws. Fuel lines should be checked for leaks. All other screws and bolts should be checked for tightness. Finally, all wires should be checked for pinching, routing, and cuts.

Outside Air Cleaner Checks

Outside the air cleaner checks include harness plugs and connectors, throttle body ground connection, vacuum hose condition and connections, air cleaner housing gasket condition, fuel lines and filters, and the fuel pump fuse in the fuse block.

Driveability Symptoms

Test sequences for no start, starts but stalls, and driveability symptoms are provided in the manufacturer service manual.

Throttle Position Potentiometer Service

Replacing or adjusting the throttle position potentiometer requires following a detailed procedure and the use of a voltmeter. This procedure is given in the service manual.

Combustion Control Computer Service

Whenever the combustion control computer is replaced, an autocalibration procedure must be followed and performed. This is also provided in the service manual.

A calibration verification test is described in the service manual to ensure correct system calibration. This test requires an emissions analyzer, a tachometer, and the detailed procedure steps.

Automatic Idle Speed Motor Service

The automatic idle speed system can be checked for correct operation by following the two steps outlined in the service manual. Incorrect minimum idle speed can be corrected by adjusting the AIS motor-adjusting screw after following the required steps.

Not following prescribed procedures will cause the system to be out of calibration with the computer program and also to be out of synchronization.

Fuel Filter Service

Fuel filter replacement requires normal tools; however, fuel pressure should be carefully relieved before disconnecting clamps or hoses. After replacement, always pressurize the system and check for leaks.

Fuel Pressure Tests

Fuel pressure tests can be made on the fuel control pump and the in-tank fuel pump. A fitting on the hydraulic support plate enables the service technician to connect a special fuel injection pressure gauge to the fuel control pump. Pressure should be at least 8 psi. Connecting a fuel pressure gauge to the outlet line of the in-tank electric fuel pump should provide a pressure reading of about 7.5 to 11.5 psi.

Special tools required for service are shown in Fig. 12-12.

The Chrysler EFI system is unique in design. Its overall operation is similar to other domestic fuel injection systems.

ABBREVIATIONS

AIS automatic idle speed motor
ASD automatic shutdown module

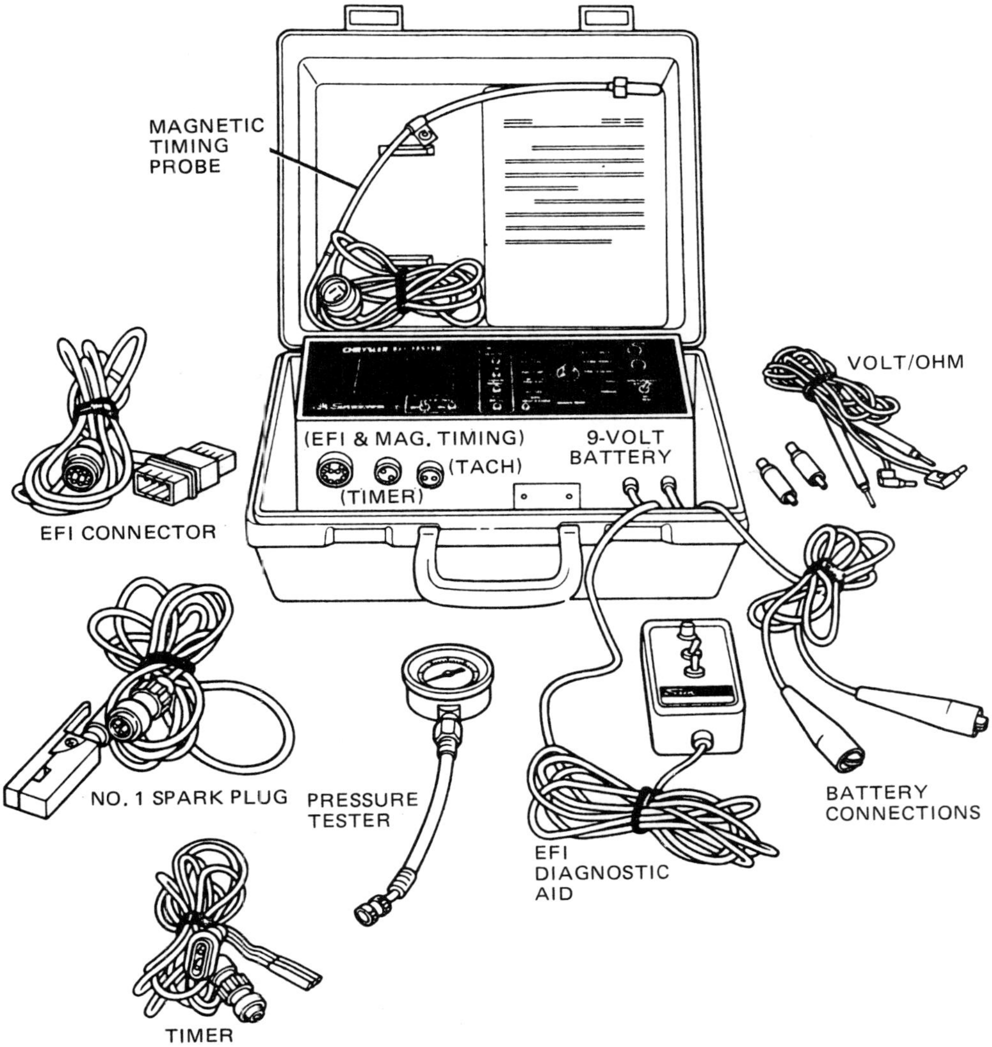

FIGURE 12-12 Electronic fuel injection tester *(Courtesy of Chrysler Training Program).*

CCC Combustion control computer
ECS Evaporation control system
EFI Electronic fuel injection
EGR Exhaust gas recirculation
ESA Electronic spark advance
FCA Fuel control assembly
PCV Positive crankcase ventilation
TPP Throttle position potentiometer

UNIT III
Chrysler Corporation Systems

13 Throttle Body Fuel Injection 1983–1985

OVERVIEW

The Chrysler electronic fuel injection system is a single-point, throttle body, pulsating, computer-controlled fuel system. It attempts to provide an ideal air/fuel mixture for all engine operating conditions. The computer is called a logic module. This module performs calculations for control of ignition timing, air/fuel ratio, emission controls, and idle speed. The logic module also has the capability of learning by updating its program and includes a self-diagnostic mode.

The logic module relies on engine sensors and switches for input data to make high-speed computations and issue output commands. The output commands are sent to a power module that controls the injector or ignition timing. The logic module also has the capability of performing self-diagnostic tests on its circuits. A fault in one of its critical sensor circuits causes a code to be set in its memory and may cause the POWER LOSS lamp to light on the instrument panel if the fault is serious enough. The codes can be accessed at a later time by the service technician as an aid in diagnosing the problem.

The Chrysler EFI system provides more accurate fuel metering and control of ignition timing, idle speed, and emission control, and uses many sensor inputs to make this control more precise. The system is simpler than a carbureted system in that many mechanical parts are eliminated.

The EFI system consists of subsystems including fuel delivery, air induction, sensors and switches, logic and power modules, and emission controls.

FUEL DELIVERY

The fuel delivery system consists of the fuel tank and cap, an in-tank electric fuel pump and inlet filter, fuel supply line, in-line fuel filter, fuel injector, fuel pressure regulator, fuel return line, and return line check valve (Fig. 13-1).

Fuel Tank

The fuel tank is a standard stamped steel tank except it has two openings for the fuel level sending unit and the in-tank electric fuel pump (Fig. 13-2). It also contains a sump or reservoir area where the electric fuel pump is located. This area is surrounded by baffles that prevent fuel from running away from the pump during low-fuel-level severe maneuvering conditions (Fig. 13-3). The fuel tank cap is a standard vacuum-pressure relief emission control cap.

In-tank Electric Fuel Pump

An in-tank electric fuel pump supplies the EFI system with pressurized fuel (Fig. 13-4). The pump is a positive-dis-

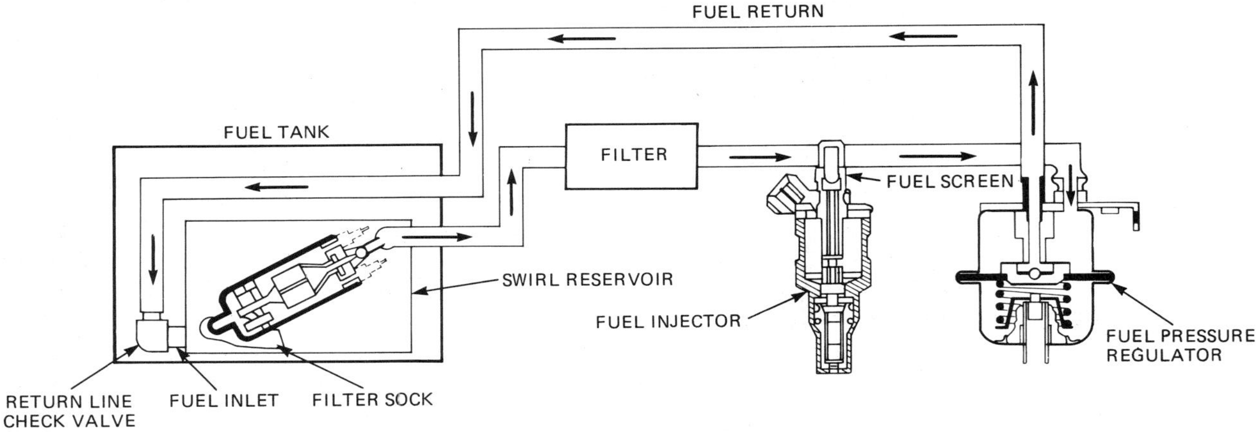

FIGURE 13-1 Electronic fuel injection system *(Courtesy of Chrysler Training Program).*

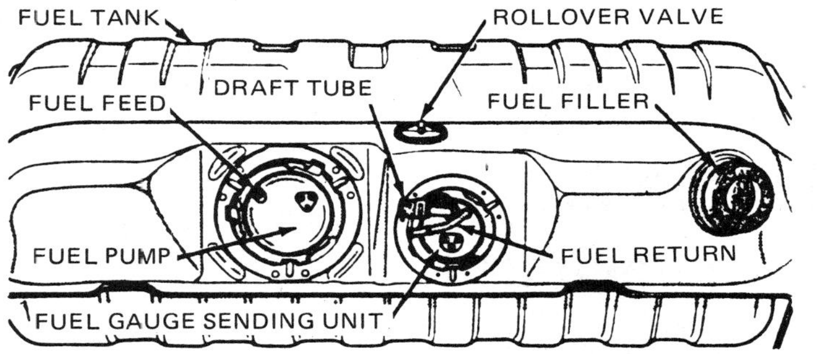

FIGURE 13-2 Fuel tank components *(Courtesy of Chrysler Training Program).*

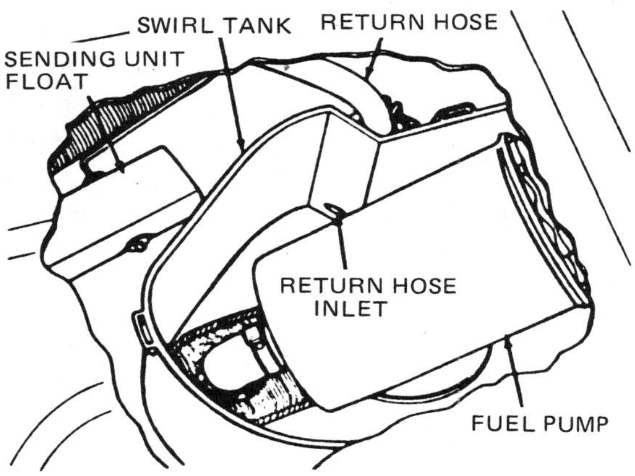

FIGURE 13-3 Fuel tank and reservoir chamber*(Courtesy of Chrysler Training Program).*

placement roller vane type that is driven by a permanent-magnet type of electric motor. Fuel is drawn into the pump through a woven plastic sock type of filter on the pump inlet. This filter is capable of removing contaminants as small as 70 microns or about 0.003 inch. It also prevents water from entering the system. The fuel pump is capable of providing about 120 psi unregulated pressure. This pres-

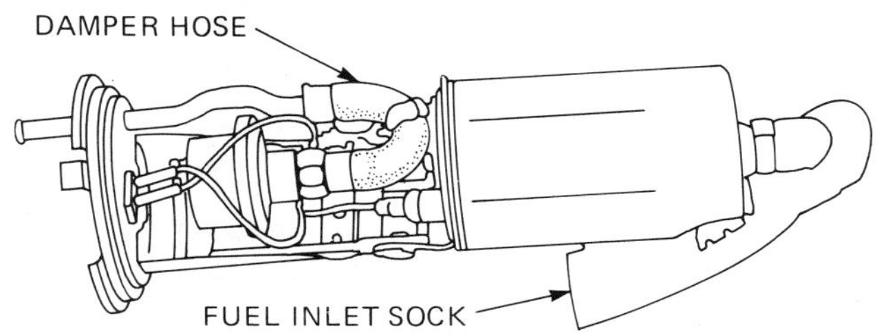

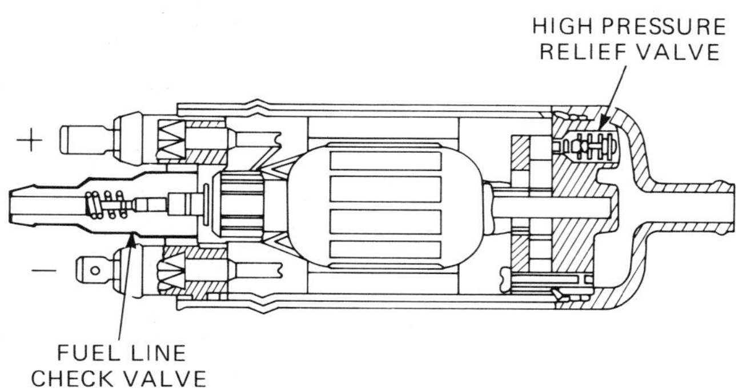

FIGURE 13-4 In-tank fuel pump *(Courtesy of Chrysler Training Program).*

sure is determined by a high-pressure relief valve inside the pump. The valve is located on the inlet side of the pump. A second check valve is located on the outlet side of the fuel pump. Its purpose is to prevent movement of fuel in either direction when the pump is shut off. The fuel pump is provided voltage for operation through the automatic shutdown (ASD) relay. When the ASD relay contacts close, the fuel pump should run.

Fuel Supply Line

The fuel supply line is made of a combination of steel tubing and special high-pressure hose material. Standard fuel line hose should not be used. Standard fuel line hose is neoprene rubber material, whereas fuel injection fuel line hose is fluoroelastomer rubber. A rubber hose connects the in-tank electric fuel pump to the steel fuel line that runs under the floor pan of the vehicle. Another rubber hose connects the steel line to the in-line filter and throttle body. Care should be taken when tightening fuel line hose clamps so that the clamps do not cut into the hose ends.

Fuel Filter

An in-line fuel filter is located in the engine compartment on some models and at the rear of the vehicle near the fuel tank on others (Fig. 13-5). This filter is a 15-micron element that will remove contaminants as small as 0.000596 inch. It is serviced by replacement at the prescribed maintenance intervals.

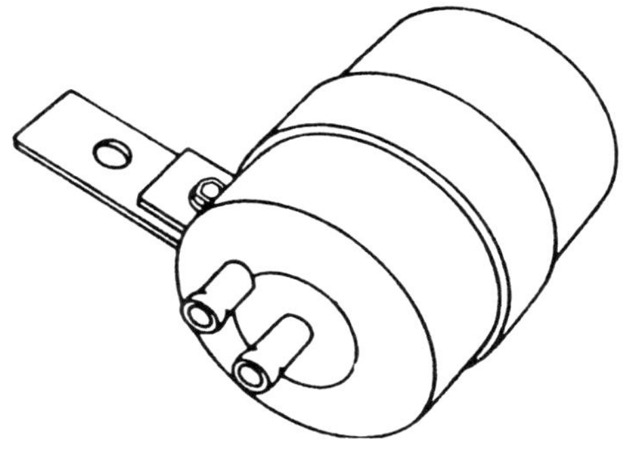

FIGURE 13-5 Fuel filtering system *(Courtesy of Chrysler Training Program).*

Fuel Injector

The fuel injector is mounted in the throttle body (Fig. 13-6). It is an electric solenoid that, when energized, raises a pintle valve against spring tension (Fig. 13-7). When de-energized, the spring returns the pintle valve against its seat. In this way, fuel is injected or sprayed into the throttle body whenever the injector is pulsed or energized. The power module controls the fuel injector. The length of time that the injector is pulsed determines how much fuel is injected. This pulse time or pulse width is calculated by the logic module and is based on input data from sensors and switches. The logic module controls the power module, which controls the fuel injector. The injector fuel inlet contains a fine-mesh stainless steel screen as a final filter. When pulsed, the injector allows fuel flow past the un-

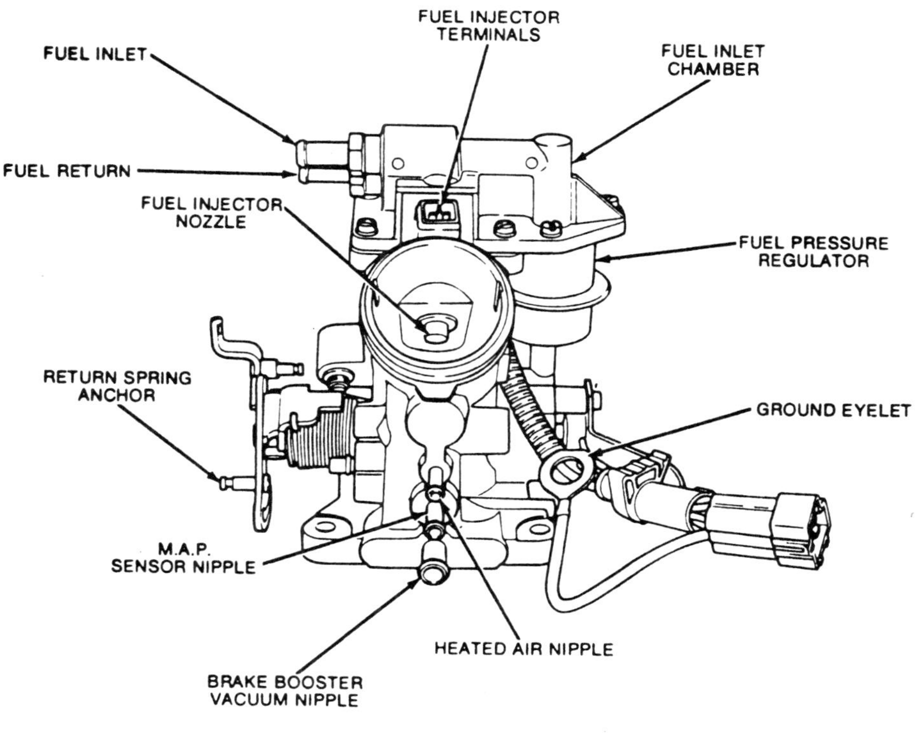

FIGURE 13-6 Throttle body assembly *(Courtesy of Chrysler Training Program).*

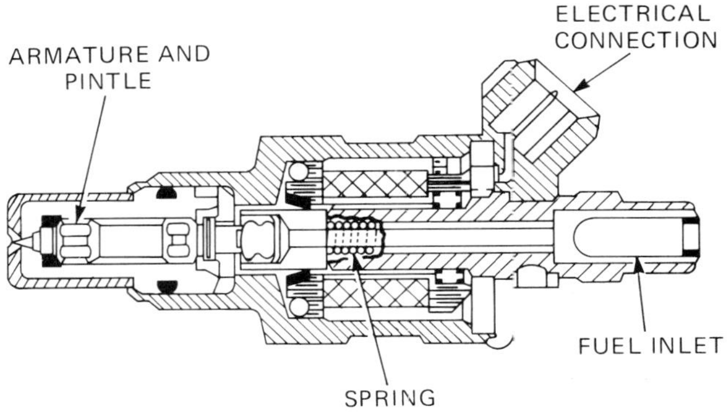

FIGURE 13-7 Fuel injector *(Courtesy of Chrysler Training Program)*.

seated pintle valve. This causes a fine, hollow, cone-shaped spray to be emitted at the injector tip. The atomized fuel mixes with incoming air at this point.

Fuel Pressure Regulator

A fuel pressure regulator maintains 38 to 40 psi fuel pressure to the fuel injector (Fig. 13-8). It is located on the return line side of the throttle body. It consists of a fuel chamber and a vacuum/spring chamber separated by a rubber diaphragm. The diaphragm is connected to a valve and seat assembly. When fuel pressure exceeds about 40 psi, it causes the diaphragm to move against spring tension and unseat the valve. This action allows the excess pressure to escape into the fuel return line. When the fuel pressure drops, the calibrated spring returns the diaphragm and valve to its seated position. A vacuum line is connected to the

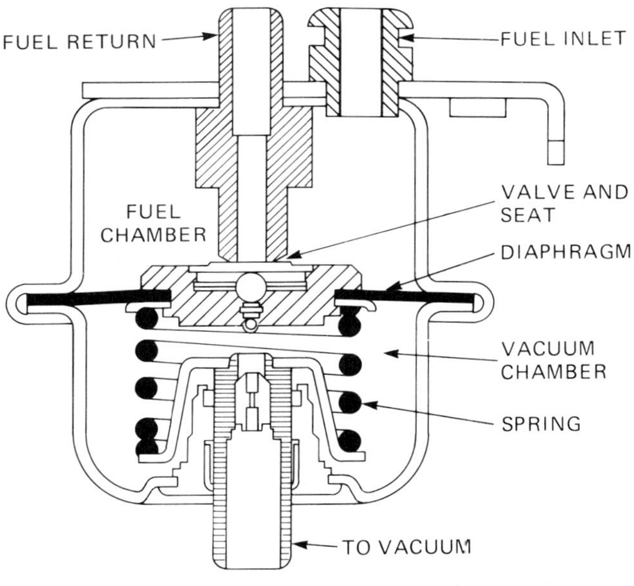

FIGURE 13-8 Fuel pressure regulator valve *(Courtesy of Chrysler Training Program)*.

vacuum chamber of the regulator. The vacuum line is connected to a port on the throttle body that is located above the throttle plate. As air flows through the throttle body, the vacuum signal increases. This causes the regulator diaphragm to be moved against spring tension, which opens the valve and allows fuel to enter the return line. The combination of fuel pressure, spring calibration, and vacuum acting on the pressure regulator fine tunes the fuel pressure for all operating conditions.

Fuel Return Line

The fuel return line begins at the fuel pressure regulator. A rubber hose connects the regulator to the steel return line, which runs under the vehicle floor pan to the rear. Another rubber hose connects the steel line to the fuel level sending unit. The return line on the sending unit empties into the fuel tank reservoir where the electric fuel pump is located. The return line fuel passes through an orifice at the reservoir, which produces a low-pressure area. This causes fuel from the tank to enter the reservoir. Return line fuel in combination with tank fuel ensures an adequate supply of fuel for the pump. The return line also has a check valve located in the reservoir. The check valve prevents reverse fuel flow. If the vehicle is involved in a rollover accident, this valve will prevent the fuel tank from draining through the fuel line.

AIR INDUCTION

The EFI air induction system consists of an air filter assembly, heated air system, throttle body, intake manifold, and the ducting necessary for connecting components.

Air Filter Assembly

The air filter assembly consists of the housing and a replaceable paper filtering element (Fig. 13-9). The housing has provisions for connecting the power module, heated air system, and throttle body to the airflow. Air is drawn through the power module as a cooling medium, is filtered, and then is drawn into the throttle body.

Heated Air System

The heated air system maintains inlet air temperature at a specified level when ambient air temperatures are low (Fig. 13-9). It also prevents icing of the throttle blade. When ambient air temperature is low, air is drawn from a sheet metal heat stove surrounding the exhaust manifold. This warmed air is blended with incoming air and results in an overall warmer air temperature. Use of this system enables

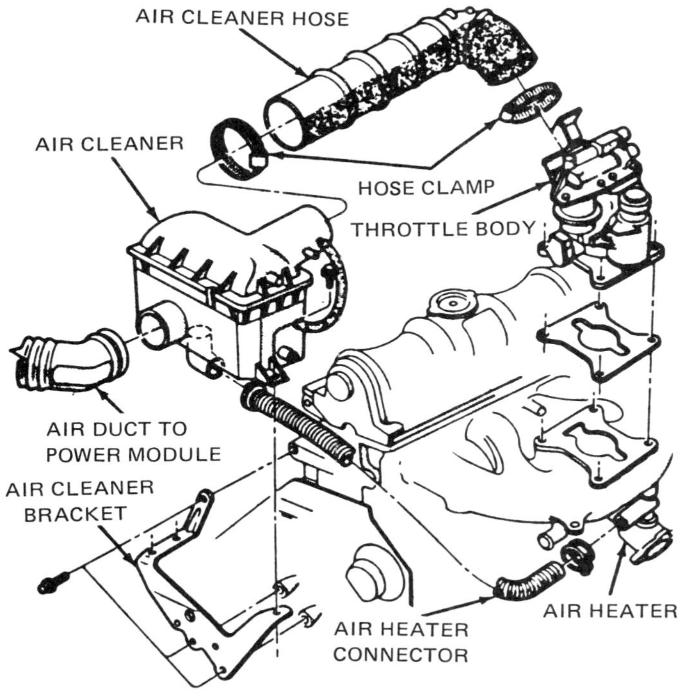

FIGURE 13-9 Heated air system *(Courtesy of Chrysler Corporation).*

the EFI system to be calibrated leaner for better fuel economy and lower emissions.

Throttle Body

A throttle body is mounted where the conventional carburetor would be located. It contains the fuel injector, fuel pressure regulator, throttle position sensor, automatic idle speed motor, and throttle blade (Fig. 13-10). The fuel injector and pressure regulator have already been discussed earlier in this chapter. The throttle position sensor will be discussed with other sensors and switches.

The automatic idle speed (AIS) motor and the throttle blade control incoming air volume. The automatic idle speed motor is operated by the logic module to control idle speed. The logic module uses data from the throttle position sensor, coolant temperature sensor, speed sensor, electric backlite switch, air-conditioning switch, park/neutral switch, and brake switch to control idle speed. The AIS motor is mounted in an air bypass passage and has the ability to either close or open the bypass passage. The throttle blade is closed at idle; therefore, air must flow through the bypass for idling. The position of the AIS motor piston restricts or opens to control how much air bypasses the throttle blade and, as a result, the idle speed. The throttle blade is attached to the throttle shaft and accelerator pedal linkage.

Intake Manifold

The intake manifold is of standard design with runners connecting the throttle body to the intake valve ports in the cylinder head.

Air Ducting

Flexible ducting is used to connect the power module to the air cleaner and the air cleaner to the throttle body.

SENSORS AND SWITCHES

Various engine sensors and selected switches are used to provide data input to the logic module. This input, along with the logic module program, is used to determine output control commands.

Sensor inputs include manifold absolute pressure, throttle position, engine speed, vehicle speed, exhaust gas oxygen content, and engine coolant temperature.

Switch inputs include park/neutral, electric backlite or heated rear window, brake, air conditioning, and air-conditioning clutch.

Manifold Absolute Pressure Sensor

The manifold absolute pressure (MAP) sensor is located in the passenger compartment above the logic module (Fig. 13-11). When the ignition switch is turned on, the sensor makes a barometric pressure reading through its piezoelectric diaphragm. This reading is used by the logic module as a reference point upon which it bases the manifold absolute pressure readings. The sensor is connected to the intake manifold by a special size and material vacuum line. It is connected electrically to the logic module (Fig. 13-12). The logic module powers the MAP sensor with 5 volts. The resistance of the sensor diaphragm changes with intake manifold pressure. As intake manifold pressure increases (vacuum decreases), the diaphragm resistance decreases and the return signal line voltage from the sensor to the logic module rises to about 4.9 volts at atmospheric pressure. At minimum intake manifold pressure (maximum vacuum), the diaphragm resistance increases and causes the return signal line voltage to fall to about 300 millivolts (0.3 volt). The MAP sensor signal is used by the logic module as an indication of engine load. When intake manifold pressure increases, more fuel is needed and the logic module commands the power module to pulse the injector longer. The MAP sensor signal is also used by the logic module to control ignition timing and the deceleration mode.

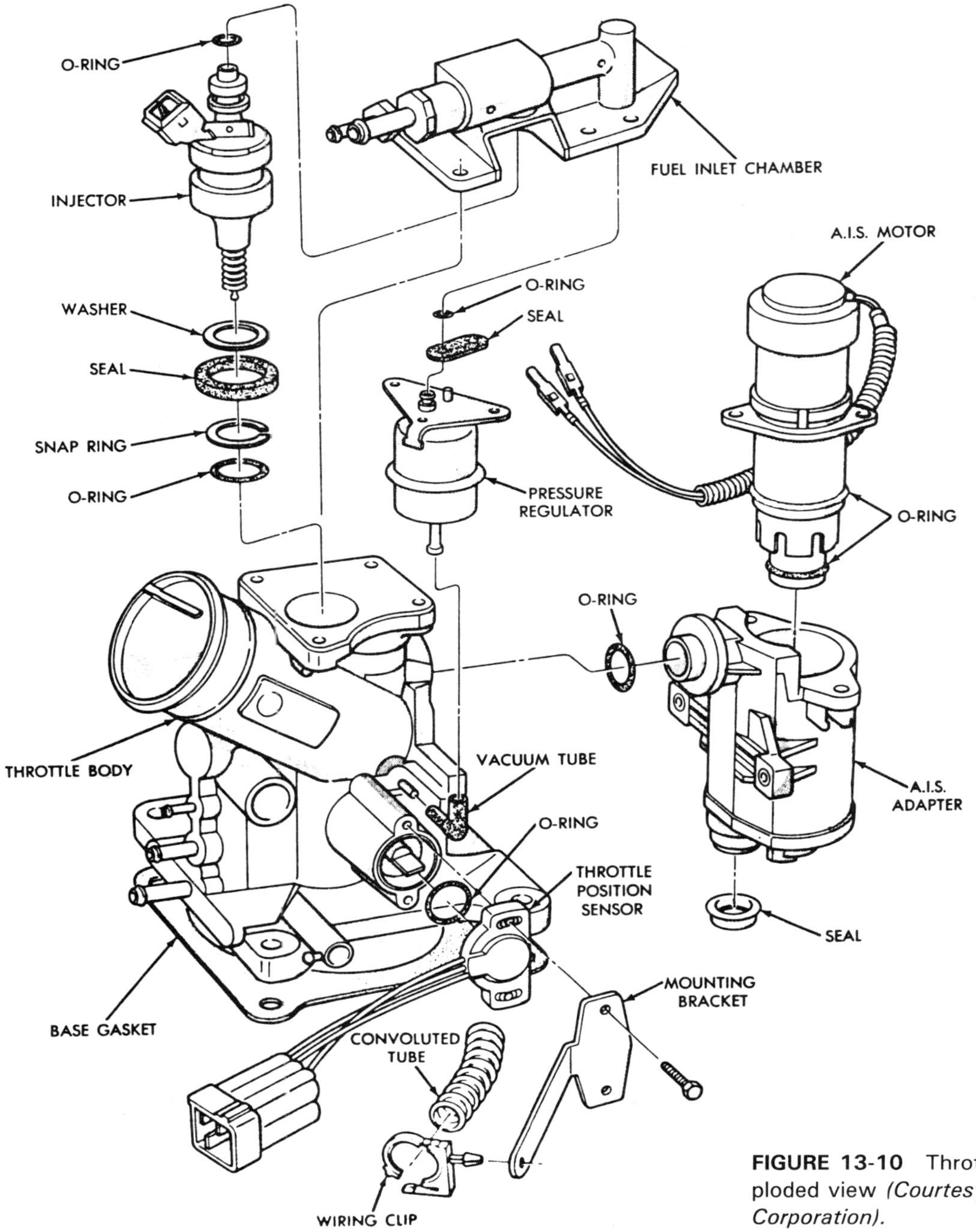

FIGURE 13-10 Throttle body exploded view *(Courtesy of Chrysler Corporation).*

FIGURE 13-11 Vacuum connectors at right-side cowl panel *(Courtesy of Chrysler Corporation).*

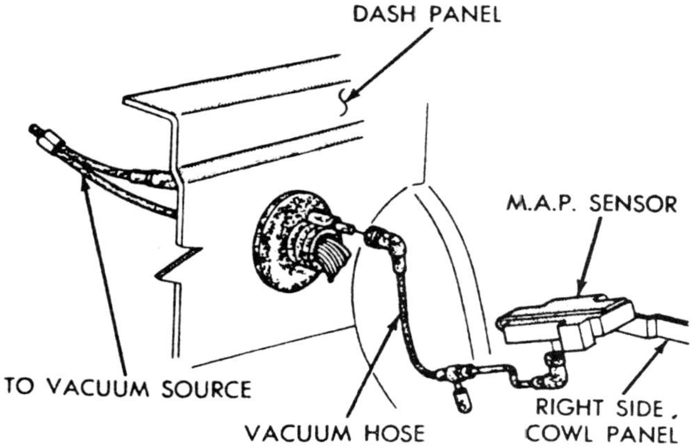

FIGURE 13-12 Manifold absolute pressure sensor, 1984 *(Courtesy of Chrysler Training Program).*

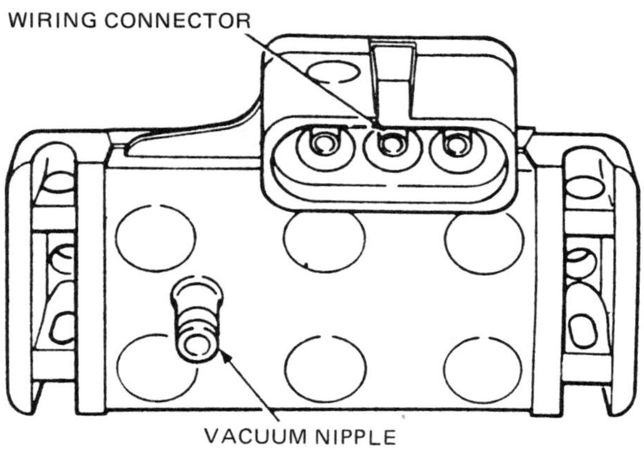

Throttle Position Sensor

A throttle position sensor is mounted on the throttle body at the end of the throttle shaft (Fig. 13-10). It is a variable resistor that is supplied 5 volts by the logic module. It senses throttle position by varying resistance as the throttle angle changes. At closed throttle, the sensor resistance is high and the return signal line voltage is about 0.5 to 1.5 volts. At wide-open throttle, the sensor resistance decreases and the return signal line voltage is about 4.7 volts. The throttle position sensor signal is used by the logic module to control the air/fuel mixture during accelerating, decelerating, idle, and wide-open throttle conditions.

Engine Speed Sensor

The engine speed sensor is an integral part of the ignition distributor (Fig. 13-13). A Hall-effect switch mounted on the pickup plate of the distributor produces a magnetic field between a permanent magnet and a transducer. Four blades extend downward from the rotor and spin with the distributor shaft and rotor. As one of the blades passes between the permanent magnet and transducer, the magnetic field is interrupted. This causes the transducer to produce an on–off signal each time the blades pass by. The signal is interpreted by the logic module to be engine speed and crankshaft position. The engine speed sensor signal is also sent to the power module for control of the automatic shutdown relay, which controls the electric fuel pump. The logic module uses the engine speed sensor signal to control ignition timing and also to pulse the injector at the right time in relation to piston position.

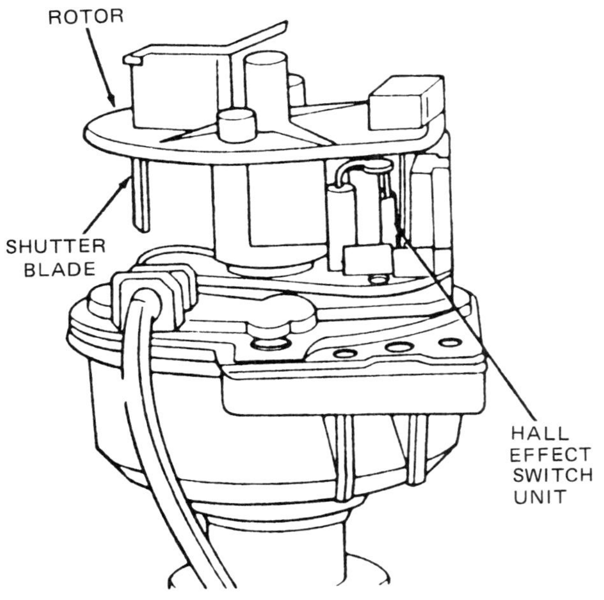

FIGURE 13-13 Hall-effect distributor pickup *(Courtesy of Chrysler Training Program).*

FIGURE 13-14 Vehicle speed sensor *(Courtesy of Chrysler Training Program).*

Vehicle Speed Sensor

The vehicle speed sensor is located in the speedometer cable at the transmission (Fig. 13-14). The sensor provides an on–off signal at a frequency of eight pulses per speedometer cable revolution. This signal is sent to the logic module where it is used to distinguish between a closed throttle deceleration and closed throttle idle. Under deceleration conditions, the logic module controls the automatic idle speed motor to achieve a desired throttle position, based on the MAP sensor value, to prevent stalling. At idle, the logic module controls the automatic idle speed motor and ignition timing to achieve a desirable idle speed.

Oxygen Sensor

Exhaust gas oxygen content is measured by a galvanic sensor called an oxygen (O_2) sensor (Fig. 13-15). The oxygen sensor is mounted in the exhaust manifold and extends into the exhaust stream. The sensor consists of two plates made

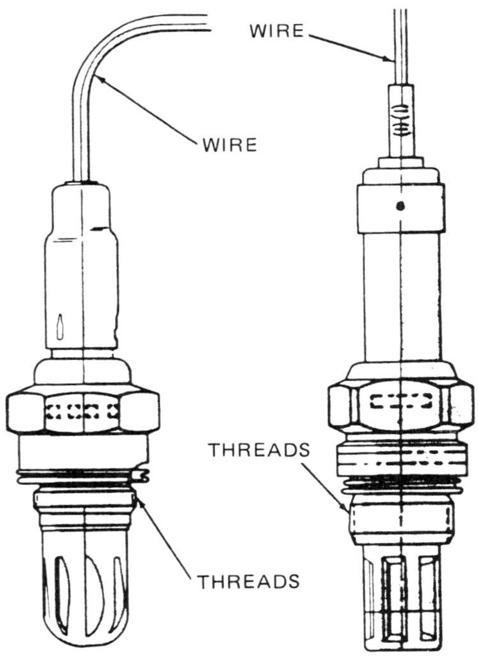

FIGURE 13-15 Typical oxygen sensors *(Courtesy of Chrysler Corporation).*

of zirconia oxide coated with platinum. One plate is exposed to fresh air and the other is exposed to the exhaust gases. When the plates are heated to about 600°F, they react to oxygen by producing a minute voltage. The difference in the two voltages produced by the plates is used by the logic module as an indication of exhaust gas oxygen content. When exhaust oxygen is high, indicating a lean air/fuel mixture, the voltage signal is low. When exhaust oxygen is low, indicating a rich air/fuel mixture, the voltage signal is high. Based on this signal, the logic module can adjust the air/fuel mixture to achieve an ideal or stoichiometric mixture for any engine operation condition. During heavy acceleration or rapid deceleration, the MAP and TPS signals may override the oxygen sensor signal to provide a rich or lean mixture.

Engine Coolant Temperature Sensor

An engine coolant temperature sensor is located in the thermostat housing (Fig. 13-16). It is a thermistor whose resistance increases as temperature decreases. The sensor resistance is high at low temperature and is low at high temperature. The coolant temperature sensor signal is used by the logic module to control idle speed, air/fuel ratio, ignition timing. Because the resistance of the sensor gradually changes as the engine warms up, the idle speed, air/fuel ratio, and ignition timing also gradually change. When at engine operating temperature, the coolant temperature sensor enables the logic module to prevent stalling, maintain an ideal or stoichiometric air/fuel mixture, and control ignition timing for best fuel economy, lowest emissions, and best performance. The coolant temperature sensor has the capability of measuring coolant temperature within the range of -60° to 270°F.

The coolant temperature sensor used on 1984 models is different from that used on 1985 models. The following chart applies to 1984 coolant temperature sensors:

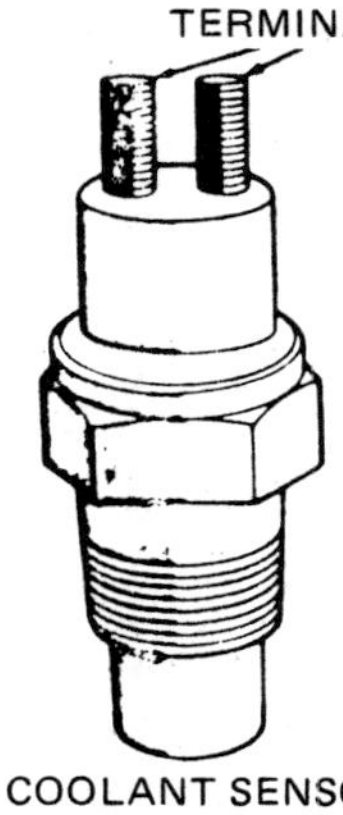

FIGURE 13-16 Coolant sensor, 1984 *(Courtesy of Chrysler Corporation).*

Temperature (°F)	*Resistance (ohms)*
32	10,000
77	5,500
190	900

The following chart applies to 1985 coolant temperature sensors:

Temperature (°F)	*Resistance (ohms)*
32	33,000
77	10,000
190	900

Park/Neutral Switch

The park/neutral switch indicates to the logic module the position of the gearshift selector. This enables the logic module to control idle speed and, on 1984 models, also open- or closed-loop operation. The logic module program for idle speed control is different for each gear position.

Electric Backlite Switch

The electric backlite or heated rear window switch on 1984 models enables the logic module to compensate for additional engine load when the system is activated. The extra load placed on the generator would normally cause idle speed to lower; however, the logic module can maintain idle speed when the heated rear window is energized by pulsing the automatic idle speed motor. In 1985 models a built-in voltage regulator provides idle compensation.

Brake Switch

The brake switch acts as a backup for the throttle position sensor for idle speed control. The TPS provides a signal to the logic module when the throttle is closed. This signal causes the logic module to control the engine idle. If the TPS malfunctions and does not provide this signal, the brake switch will provide a substitute signal when engine rpm is below 1500.

Air-conditioning Switch

To provide a stable idle speed, the logic module must know when the air-conditioning system is activated. The air-conditioning clutch cycling switch provides a signal to the logic module for controlling and maintaining idle speed due to additional engine load caused by the compressor. As the air-conditioning system cycles the compressor clutch, the compressor clutch switch causes the logic module to adjust

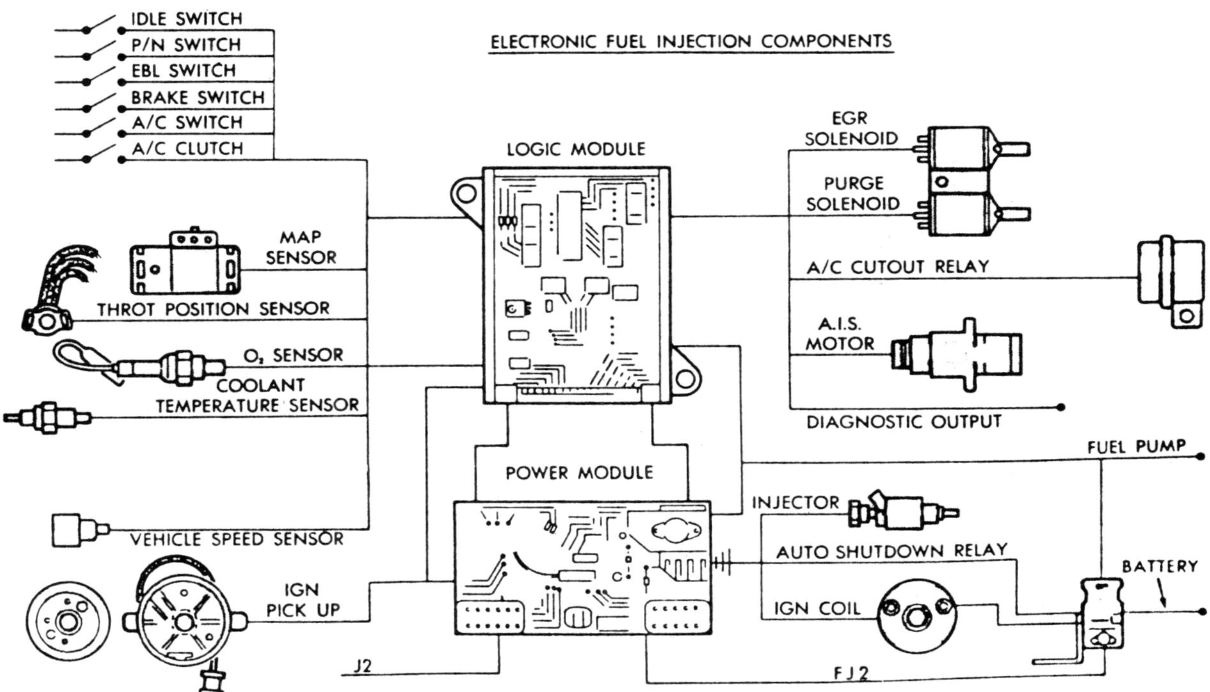

FIGURE 13-17 Electronic fuel injection components *(Courtesy of Chrysler Corporation).*

the automatic idle speed motor. This maintains idle speed as the load presented by the air-conditioning compressor changes with cycling.

LOGIC AND POWER MODULES

The logic module is a digital computer that receives sensor and switch data, makes high-speed calculations based on these data and its program, and issues commands to various motors, solenoids, or relays (Fig. 13-17). To control heavy-current devices such as motors, solenoids, or relays, the logic module uses the power module. The logic module circuitry is electronic and capable of handling only very small currents. The power module is a heavier switching, current handling module that is controlled by the logic module.

Logic Module

The logic module is located in the passenger compartment behind the right side kick panel (Fig. 13-18). The logic module provides a 5-volt reference source to the coolant temperature, throttle position, and manifold absolute pressure. The return signal from each sensor is determined by its resistance. The logic module uses the sensor data and controls the ignition timing and air/fuel ratio through the power module. It controls the exhaust gas recirculation and purge solenoids and the automatic idle speed motor directly. The logic module also contains circuitry for self-diagnosis. This enables the module to monitor and diagnose its own circuits. If a malfunction or out of range condition exists, a fault code is set and stored in a memory.

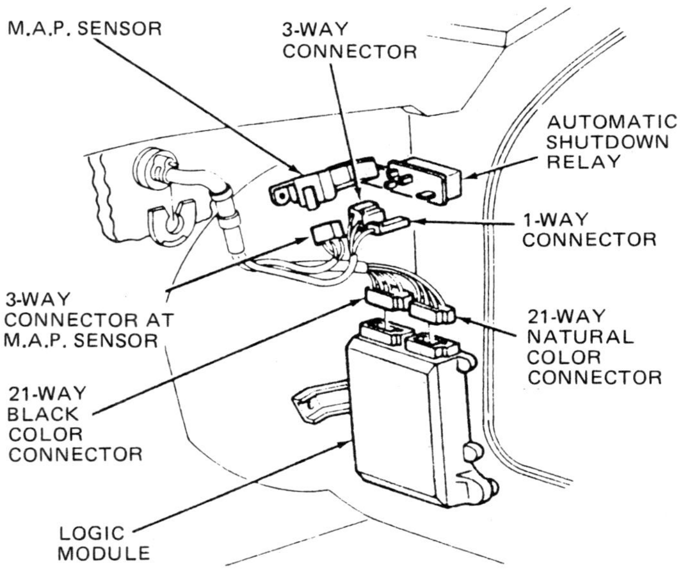

FIGURE 13-18 Cowl side area components, 1984 *(Courtesy of Chrysler Corporation).*

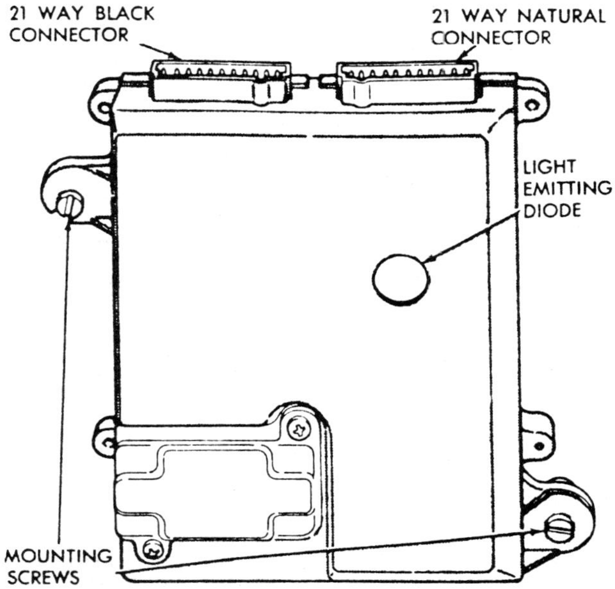

FIGURE 13-19 Logic module, 1984 *(Courtesy of Chrysler Corporation).*

The service technician can gain access to this memory by reading the flashes of the POWER LOSS lamp on the instrument panel, by reading the flashes of a light-emitting diode in the logic module (Fig. 13-19), or by using a diagnostic readout tester. This function of the logic module assists the technician in accurately diagnosing problems. The logic module can also provide a limp-in mode of operation when normal control circuits fail. This prevents vehicle failure when computer circuits malfunction such as manifold absolute pressure, throttle position, or coolant temperature sensors. During vehicle operation, the POWER LOSS light illuminates when a problem occurs with these circuits.

Power Module

The power module is located in the left front fender well behind the battery (Fig. 13-20). This location provides necessary airflow for cooling (Fig. 13-21). It grounds the ignition coil and fuel injector on circuits that are separate from other circuits connected to the logic module. In this way, interference or noise is reduced, which otherwise may affect the logic module. The power module also controls the automatic shutdown relay and provides an 8- and 12-volt power supply for operating the logic module. The automatic shutdown (ASD) relay provides power for the ignition coil and electric fuel pump. The engine speed signal from the distributor is sent first to the power module, where it controls the ASD relay, and then to the logic module. If the engine speed signal is not present or is lost during operation, the ASD relay shuts the ignition coil, fuel pump,

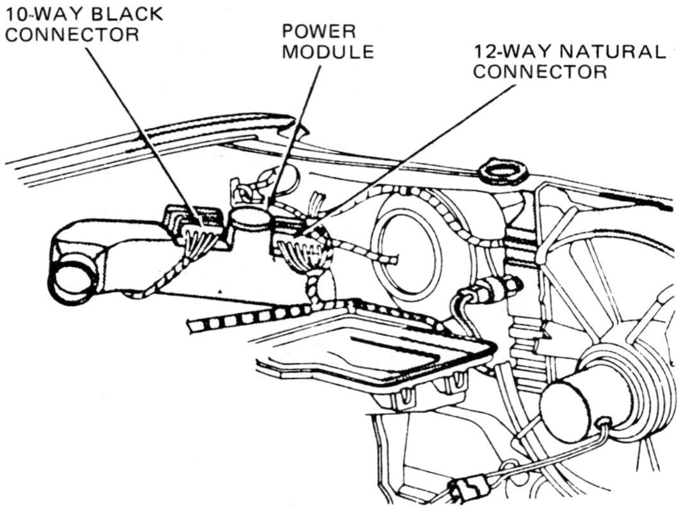

FIGURE 13-20 Power module location *(Courtesy of Chrysler Corporation).*

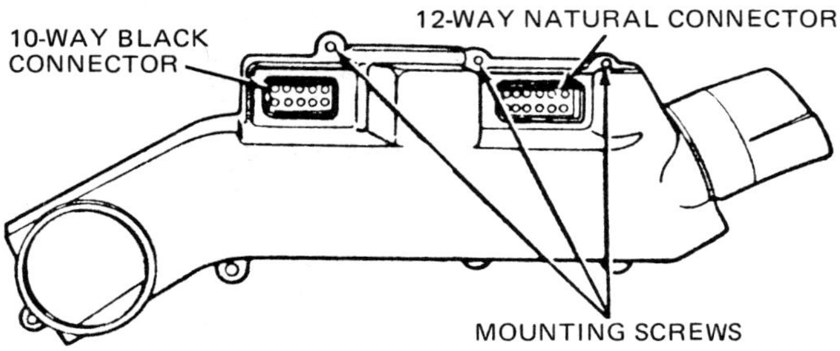

FIGURE 13-21 Power module *(Courtesy of Chrysler Corporation).*

and power module off. The 1984 model ASD relay is located in the passenger compartment in the right front kick panel above the logic module. The 1985 model ASD relay is located in the power module and is not serviceable separately.

EMISSION CONTROLS

The Chrysler EFI system also controls certain emission controls and circuits that have effects on emission levels. Emission control systems operated by the logic module include exhaust gas recirculation and evaporative emission control. In addition, the air-conditioning system is controlled by the logic module under specific operating conditions.

Exhaust Gas Recirculation Solenoid

The exhaust gas recirculation (EGR) solenoid on 1984 models is activated by the logic module when engine coolant temperature is below 70°F (Fig. 13-22). This action

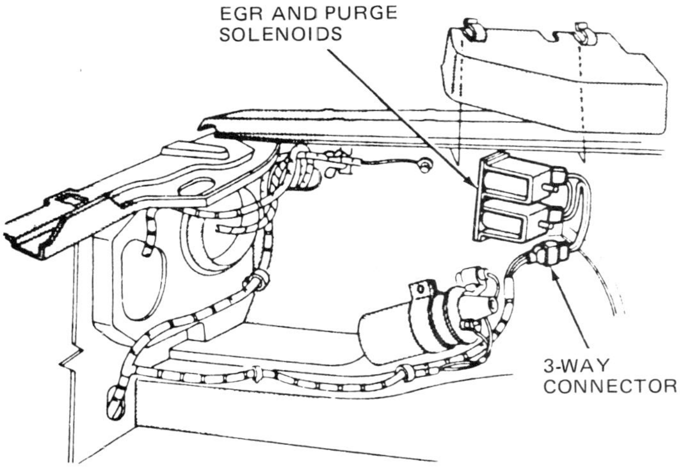

FIGURE 13-22 Electrical connector exhaust gas recirculation and purge solenoids, 1984 *(Courtesy of Chrysler Corporation).*

blocks vacuum to the EGR valve transducer, making the EGR valve inoperative. When engine coolant temperature rises above 70°F, the logic module de-energizes the EGR solenoid, which allows vacuum to the transducer valve. The EGR valve can operate as a standard back-pressure EGR valve. The EGR valve on 1985 models is a full-time operation system.

Evaporative Emission Control Purge Solenoid

The evaporative emission control system stores fuel vapors and burns them by recycling into the intake manifold. A purge solenoid, controlled by the logic module, controls when the purging occurs (Fig. 13-22). When engine coolant temperature is below 160°F, the logic module energizes the purge solenoid, which blocks vacuum to the charcoal canister. As the engine coolant temperature rises to above 160°F, the logic module de-energizes the solenoid, which allows vapors to be purged from the charcoal canister.

Air-conditioning Cut-out Relay

The air-conditioning cut-out relay is normally closed and allows operation of the air-conditioning system. When the logic module receives either an rpm signal that is less than 500 rpm or a wide-open throttle signal from the throttle position sensor, it energizes the air-conditioning cut-out relay. This shuts off the air-conditioning system and prevents stalling or lugging the engine.

SYSTEM OPERATION

The EFI system operates in modes that correspond with ignition switch position.

Ignition Switch Off

When the ignition switch is off, battery voltage is removed from the power module, automatic shutdown relay, in-tank electric fuel pump, logic module, and ignition coil.

Ignition Switch On

In the ignition switch ON position, battery voltage is available to the power module, which, in turn, provides 8 volts to the distributor pickup coil and logic module. The logic module also receives a full time 12-volts source and ignition switch ON 12-volts in addition to this 8-volt source. The logic module provides 5 volts to the charge temperature, coolant temperature, throttle position, and manifold absolute pressure sensors. The power module also provides voltage to the automatic shutdown relay. The ASD relay activates the electric fuel pump for 500 milliseconds. If the power module does not receive a distributor reference pulse within this time or if the reference pulses it does receive are less than 60 (120 rpm), it will de-energize the ASD relay and fuel pump. This acts as a safety feature to prevent potentially hazardous situations.

Ignition Switch START

With the ignition switch in the START position, distributor reference pulses are sent to both the power and logic modules. The ASD relay is activated and the fuel pump operates. The ASD relay also provides power to the logic module, ignition coil, and the injector and coil driver circuits of the power module. The logic module begins to receive data from the coolant temperature, throttle position, and MAP sensors. The logic module also causes the automatic idle speed motor to open and provide fast idle. It sends commands to the power module to regulate ignition timing and injector pulse width or air/fuel ratio.

Ignition Switch RUN

After starting, the ignition switch is released to the RUN position. For the engine to continue to run, the logic module must provide a ground for the sensors, receive an 8-volt power input, provide a 5-volt output to the sensors, provide injector and ignition control output signals, and provide a distributor signal output. The power module must provide ignition coil negative circuitry, receive battery voltage input, provide 8-volt power output, provide injector driver signals, provide injector return circuitry, provide ASD relay ground, and receive power input on circuit J2. The power module must also receive input commands from the logic module for control of the injector and ignition timing. The automatic shutdown relay must receive battery voltage to be energized so that the same battery voltage can be provided to the fuel pump. A ground for the ASD relay coil must also be provided by the power module. The distributor pickup must have an 8-volt input, a ground, and an output signal. Without all of these activities, the engine will not run.

Injector Pulse Width

The EFI system processes input signals from sensors and switches to determine the fuel injector pulse width or ON time. This is measured in milliseconds and represents with air/fuel ratio necessary to provide a stoichiometric mixture for optimum performance, economy, and emissions. A speed/density factor is used to help determine pulse width. The MAP and rpm signals are used to calculate speed/density, which is then combined with other sensor and switch data to arrive at a pulse width.

Computer Program Updating

The EFI system is also capable of correcting for production tolerance and barometric pressure changes. This is accomplished by using 12 memory cells that are activated by battery voltage but are nonvolatile memories. Ten of the cells are based on the MAP and rpm signals, while the remaining two are based on engine idle condition and gear position (P/N or drive). These cells are updated by the oxygen sensor signal whenever its signal has changed from a previous level for a specified period of time. This updating takes place only while the EFI system is in the closed-loop mode of operation.

Open- and Closed-loop Mode

Whenever the oxygen signal is sufficient to cause the logic module to regulate the air/fuel ratio, the system is in the closed-loop mode of operation. When the oxygen sensor signal is weak, while the engine is warming up, the system is in the open-loop mode of operation. During this time, the air/fuel ratio is based on engine coolant temperature and the logic module program. Once the system enters the closed-loop mode, it may reenter the open-loop mode if the oxygen sensor cools down, the throttle position sensor indicates wide-open-throttle operation, or rapid deceleration occurs. To enter the closed-loop mode, a specific coolant temperature must be reached, the oxygen sensor must be at operating temperature, and a logic module timer circuit must cycle through a period of time.

The logic module normally pulses the fuel injector two times per crankshaft revolution. During cranking, the injector is pulsed four times per crankshaft revolution. This doubling of the pulse width continues until a specified rpm is attained for a programmed period of time. During engine warmup, the injector pulse width is determined by engine coolant temperature until the oxygen sensor begins generating a signal. The pulse width is greater during this period and gradually reduces as coolant temperature rises. During closed-throttle deceleration, the logic module decreases injector pulse width to provide a lean air/fuel mixture. This improves economy and emissions. When accelerating, the logic module widens the injector pulse width to provide a rich air/fuel mixture. Deceleration and acceleration pulse width modifications are caused by the MAP, RPM, and TPS sensors. During wide-open-throttle conditions, the MAP sensor signal will not provide a stoichiometric mixture. This is necessary to prevent engine-damaging combustion temperatures.

At idle, the engine speed signal is used by the logic module to sense the results of its control of the automatic idle speed motor. During idle conditions, ignition timing is limited to a variation of no greater than 12 degrees. Idle speed control while in DRIVE is the same except for ignition timing control. Timing correction programming is not used. The logic module uses the vehicle speed sensor signal in conjunction with the automatic idle speed motor to prevent stalling during deceleration. The AIS motor is opened as vehicle speed drops to zero.

Self-diagnostic Ability

In addition to these features, the EFI system has on-board diagnostic capability. Sensors are constantly checked for correct operating range. If an input or output signal is absent or just out of range, the logic module considers this to be a real problem. It stores a code representing the problem in its memory. Problems that occur and then disappear are transient or are not considered real problems and will not cause codes to be stored. When a real problem is repaired or if it eventually goes away by itself (an intermittent connection), the code is automatically erased from memory after 30 cycles of the ignition switch. The memory can also be erased or cleared by disconnecting the quick disconnect at the battery positive terminal. The stored codes can be accessed by the service technician as an aid in diagnosing problems. Code descriptions are provided in the service manual. The codes relate to a problem in a specific circuit and do not indicate faulty components. The entire circuit may have to be diagnosed before the correct repair can be made. There are many codes that can be stored in memory.

Self-diagnostic Mode

To cause the system to enter the self-diagnostic mode, the service technician can connect a special diagnostic readout tester to a connector in the engine compartment (Fig.

C-3292

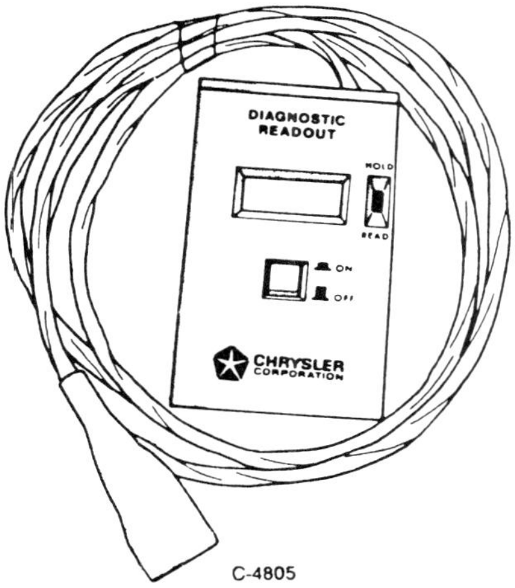

C-4805

SELF-DIAGNOSTIC READOUT TOOL

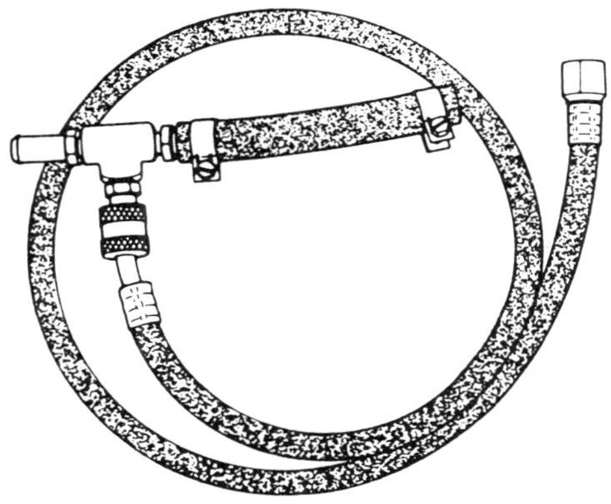

C-4799

FUEL PRESSURE GAUGE AND HOSE

FIGURE 13-23 Electronic fuel injection special tools *(Courtesy of Chrysler Corporation).*

13-23). This connector is on the right front fender well. In lieu of the tester, the technician can watch the flashes of the POWER LOSS light on the instrument panel. Cycling the ignition switch on, off, on, off, on causes the system to enter the self-diagnostic mode. The tester displays digital codes. The codes can also be read by counting flashes of the instrument panel telltale light. When code 55 appears, the system is ready to provide additional testing. Either switch tests or the actuator test mode can be chosen. The switches that are tested by the system include the brake switch, gear shift park–reverse–park switch, air-conditioning switch, speed sensor, and electric backlite switch. Switch operation is correct if the power loss lamp lights each time a switch is activated. The actuator test mode checks the operation of the ignition coil, automatic idle speed motor, and fuel injector. The power loss lamp can also indicate the presence of a distributor signal while cranking if it is lit.

Limp-in Mode

The limp-in mode of operation is entered when signals from either the MAP, TPS, or CTS sensors are incorrect. This condition sets a code in memory and cause the limp-in mode to be activated. Engine performance and economy suffer while in this mode. If the sensor signal is restored during operation, the POWER LOSS lamp will remain on and the system will stay in the limp-in mode. When the ignition switch is cycled off and then on for restarting, the system returns to normal operation. The power loss lamp will be out, but a code will remain in memory.

SYSTEM SERVICE

The first step in servicing the EFI system involves defining the problem. A check for codes should be made and a visual inspection of wiring harnesses and connections should be completed.

Ignition Timing

To set ignition timing, a tachometer and timing light are necessary. The engine should be at operating temperature. Disconnecting the coolant temperature sensor causes the system to enter the limp-in mode where ignition timing is at base specification. Check to see if the POWER LOSS lamp is illuminated as an indication of limp-in mode operation. Reconnect the coolant temperature sensor and set ignition timing. Shut the engine off and then restart it. The POWER LOSS lamp should be off. This indicates that the logic module is out of the limp-in mode and functioning normally.

Fuel Pressure Relief

Fuel system pressure can be relieved by first removing the fuel tank cap to relieve any tank pressure. Remove the fuel injector wiring harness connector from the injector and connect a jumper wire from one injector terminal to ground. Connect a second jumper wire to the other injector terminal and battery positive for no longer than 10 seconds. This pulses the injector and relieves fuel pressure. Do not connect the jumper for longer than 10 seconds as it may cause the injector winding to burn up.

Fuel Pressure Test

A special fuel injection pressure gauge is needed to check system pressure (Fig. 13-23). The gauge should be connected in the fuel line between the fuel filter and the throttle body. After starting the engine, the pressure should be 36 psi ± 2 psi.

Fuel Filter

Changing the fuel filter requires that system pressure be relieved as discussed earlier. Remove the filter and discard it and its connecting hoses. Install new hoses, clamps, and the filter. Cycle the ignition switch several times to pressurize the system and check for leaks.

The Chrysler throttle body EFI system is not unlike other systems. Component names are different and service procedures and specifications are different, but general operation is similar.

ABBREVIATIONS

AIS Automatic idle speed
ASD Automatic shutdown (relay)
CTS Coolant temperature sensor
EGR Exhaust gas recirculation
MAP Manifold absolute pressure
O_2 Oxygen (sensor)
PCV Positive crankcase ventilation
P/N Park/Neutral
rpm Revolutions per minute
TPS Throttle position sensor

UNIT III
Chrysler Corporation Systems

14
Multipoint Fuel Injection, 1984–1985

OVERVIEW

The Chrysler electronic fuel injection system used on turbocharged engines is a multipoint, computer-controlled system that provides a precise air/fuel ratio over a full range of engine operating conditions. The logic module is a digital computer that controls the air/fuel ratio, ignition timing, idle speed, and emission controls. It is the center of the electronic system. The logic module has the ability to update its program through its memory cells. Engine sensors and switches provide input data to the logic module, which uses these data along with its program to issue commands to controlled devices through the power module (Fig. 14-1). The power module acts as an electronic relay for the logic module. It controls devices operating at high currents. The logic module also monitors its own circuits and can warn the driver and aid the service technician when problems occur.

The combination of turbocharging and multipoint fuel injection results in increased horsepower, improved fuel economy, more accurate fuel metering, and reduced emissions levels.

The Chrysler electronic fuel injection system consists of subsystems that include fuel delivery, air induction, sensors and switches, logic and power modules, and emission controls.

FUEL DELIVERY SYSTEM

The fuel delivery system consists of the fuel tank and cap, in-tank electric fuel pump, fuel delivery line, in-line fuel filter, fuel rail, fuel injectors, fuel pressure regulator, and fuel return line and check valve (Fig. 14-2).

The fuel tank is a standard stamped steel tank. It has two openings for the fuel level sending unit and the in-tank electric fuel pump. A sump or reservoir area is incorporated into the tank (Fig. 14-3). The electric fuel pump is mounted in the sump, which consists of baffles. This area prevents fuel from running away from the pump during severe maneuvering conditions and low fuel level. The fuel tank cap is a standard pressure-vacuum relief emission control cap.

In-tank Electric Fuel Pump

An in-tank electric fuel pump provides the multipoint fuel injection system with pressurized fuel (Fig. 14-4). The pump is a positive-displacement roller vane type of design that contains two check valves. One check valve is located on the inlet side and limits maximum pressure to 120 psi. The second check valve is located on the outlet side and prevents movement of fuel in either direction when the system is not operating. The inlet of the fuel pump is fitted

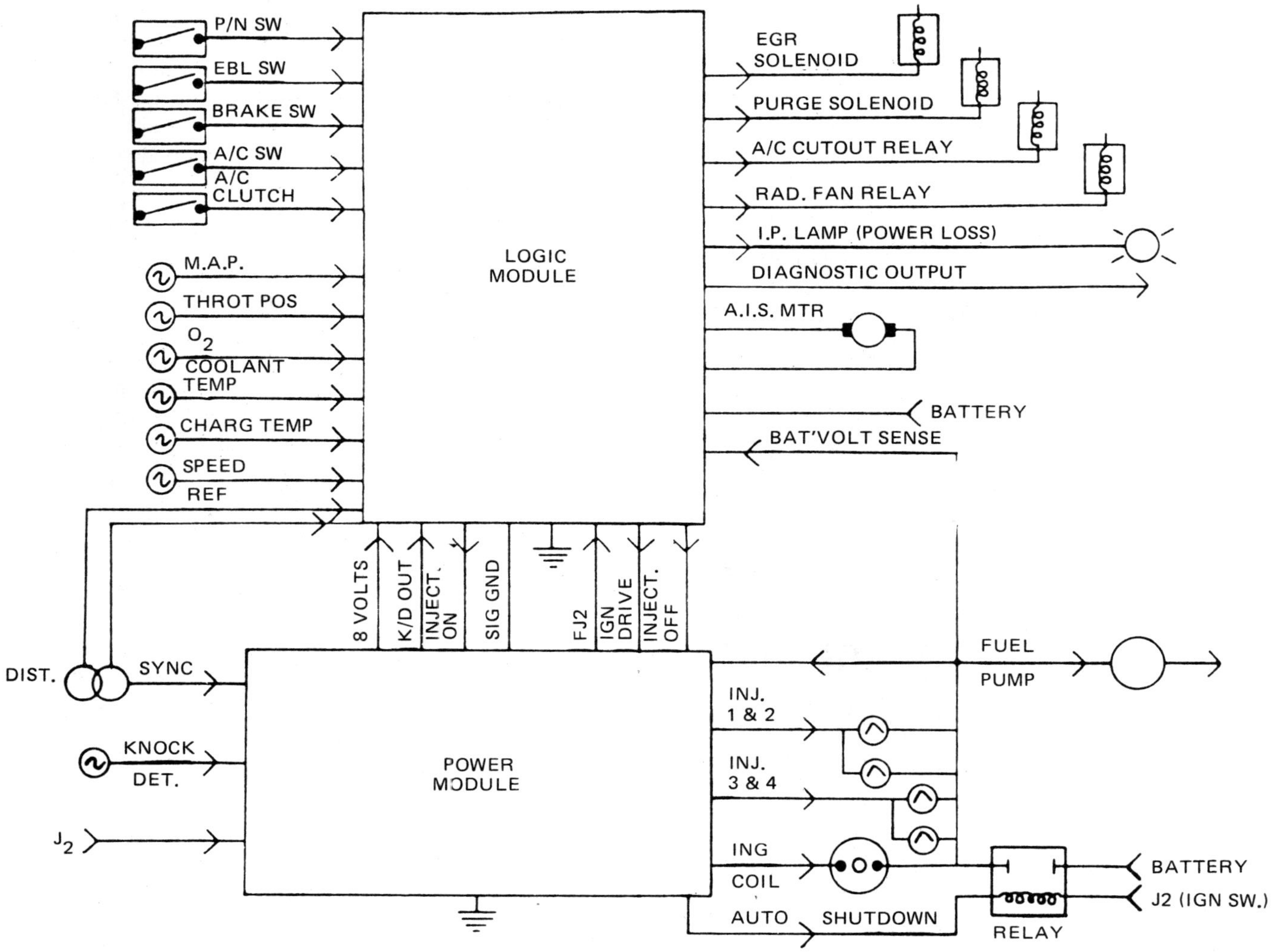

FIGURE 14-1 The 2.2-liter turbo electronic fuel injection *(Courtesy of Chrysler Training Program).*

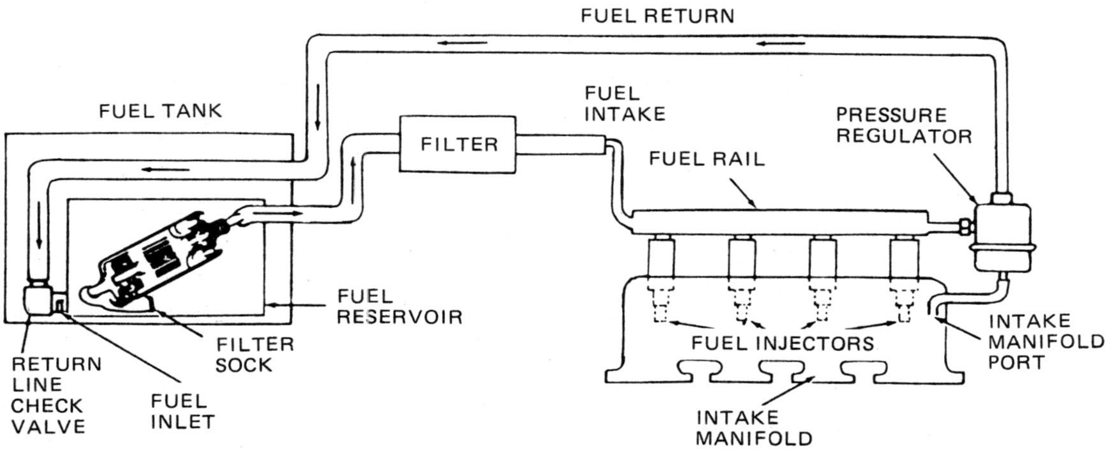

FIGURE 14-2 Electronic fuel injection fuel system *(Courtesy of Chrysler Training Program).*

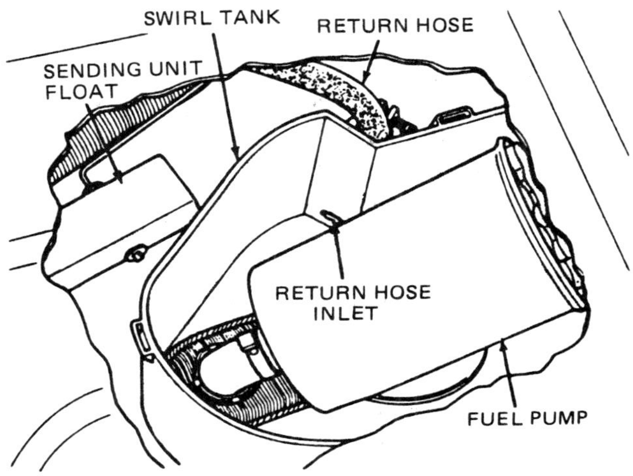

FIGURE 14-3 Fuel tank internal view *(Courtesy of Chrysler Corporation).*

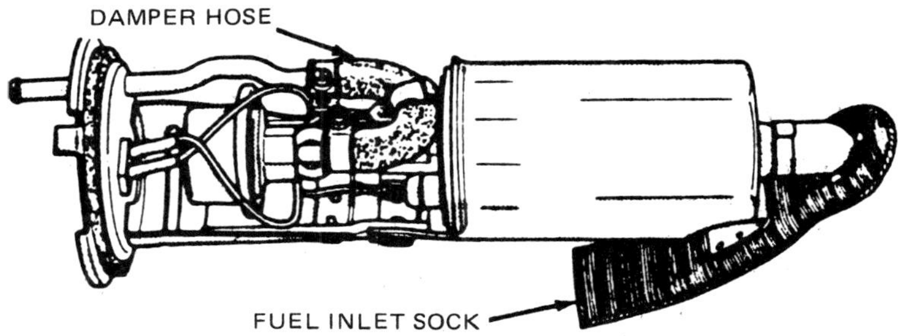

FIGURE 14-4 Fuel pump assembly *(Courtesy of Chrysler Corporation).*

with a woven plastic filter that is capable of preventing water and contaminants as small as 70 microns or 0.003 inch from passing into the pump. The pump is controlled by the automatic shutdown (ASD) relay and power module. The pump runs when the ASD relay is energized.

Fuel Delivery Line

The fuel delivery line is fuel injection rubber hose and steel material. A rubber hose connects the fuel pump outlet to a steel fuel line. Another rubber hose connects the steel line to the in-line fuel filter and the filter to the front supply line. A third rubber hose connects the fuel line to the fuel rail, which supplies fuel to the injectors. All rubber hoses are special electronic fuel injection hose material. Standard fuel line hose, which is neoprene rubber, is not an acceptable replacement. A new fluorelastomer rubber material is not only suitable for the higher pressures but also is more resistant to contaminants and additives found in today's fuels.

The in-line fuel filter is a disposable type and is located in the rear of the vehicle under the floorpan near the fuel tank (Fig. 14-5). It contains an element capable of removing contaminants as small as 15 microns or 0.0006 inch.

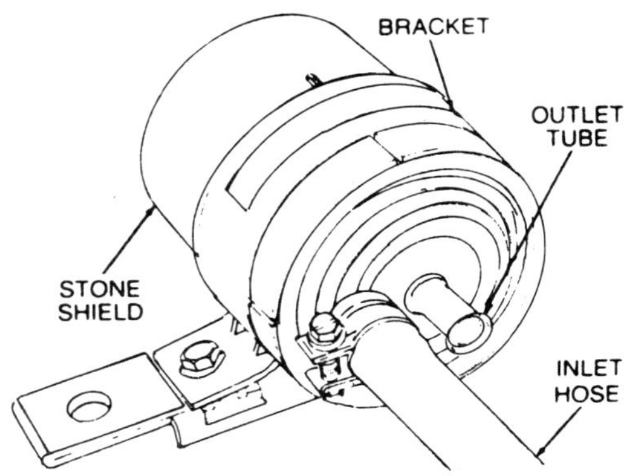

FIGURE 14-5 Fuel filtering system *(Courtesy of Chrysler Training Program).*

Fuel Rail

The fuel rail is mounted on the engine and supplies the fuel injectors with pressurized fuel from the pump. The fuel supply line is connected to one end of the rail, and the fuel pressure regulator is connected to the other end. The injectors are sealed to the rail by a rubber O-ring seal. This seal prevents fuel leakage. The injector tip is sealed to the intake manifold by a second O-ring seal. This seal prevents vacuum leaks. The fuel rail has a Schrader fitting for attachment of a fuel injection pressure gauge as an aid in diagnosing fuel supply problems.

Fuel Injectors

The fuel injectors are electric solenoids that, when energized, raise a pintle valve off its seat and allow fuel to spray into the intake valve port (Fig. 14-6). The amount of time the injector is energized determines how much fuel is injected. Battery voltage is supplied to the injectors, and the logic module provides a ground through the power module to energize the injectors. The injectors are pulsed in pairs. Injectors 1 and 2 and injectors 3 and 4 are pulsed together.

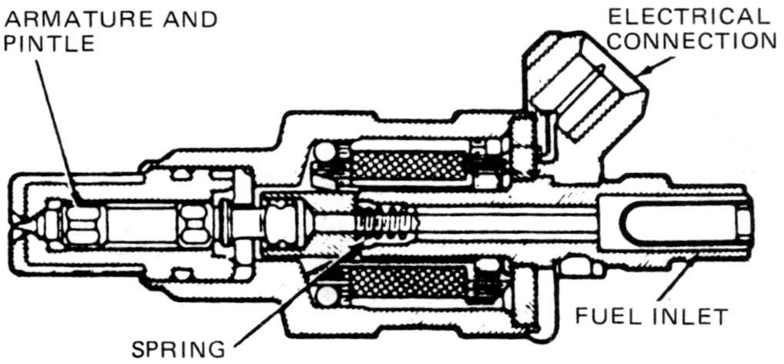

FIGURE 14-6 Fuel injector *(Courtesy of Chrysler Training Program).*

Fuel Pressure Regulator

A fuel pressure regulator is mounted at one end of the fuel rail. The regulator maintains about 55 psi pressure to the injectors. It consists of a vacuum-actuated, spring-loaded diaphragm that controls the position of a ball valve (Fig. 14-7). When fuel pressure builds to a level of about 55 psi, the pressure forces the diaphragm to move against spring tension and unseat the ball valve. Excess pressure is allowed to escape into the fuel return line. When fuel pressure drops below 55 psi, the diaphragm returns and the ball valve seats. Through this action, pressure remains constant.

A vacuum port is located on the spring side of the diaphragm. The port is connected to intake manifold vacuum. As engine operating conditions change, the vacuum assists fuel pressure in overcoming the diaphragm spring. The vacuum causes fuel pressure to be responsive to engine load; when accelerating, engine vacuum reduces and fuel pressure increases. When the turbocharger begins to provide boost, instead of vacuum operating the regulator, boost pressure will cause an even higher fuel pressure. Fuel pressures in the rail will generally vary between 42 and 62 psi. At idle, with the vacuum hose at the regulator disconnected, the rail pressure should be about 55 psi.

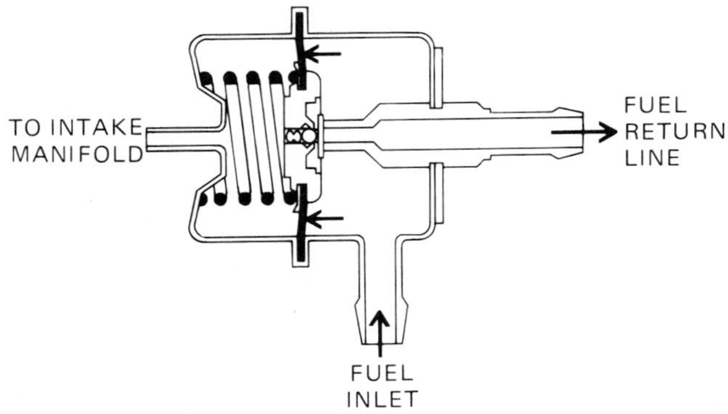

FIGURE 14-7 Fuel pressure regulator valve *(Courtesy of Chrysler Training Program).*

Fuel Return Line

The fuel return line begins at the fuel pressure regulator with a rubber hose connecting it to the steel return line. Another rubber hose connects the steel line to the return line in the fuel tank. A check valve is located on the end of the return line. It prevents reverse flow and, should the vehicle be involved in a rollover accident, it also prevents draining of the tank through the return line. Another check valve is located in the top of the fuel tank. It also prevents draining of the tank during rollover. It is located at the end of the vapor line for the charcoal canister.

AIR INDUCTION SYSTEM

A remotely mounted air cleaner assembly, air cleaner duct, throttle body, turbocharger, and intake manifold comprise the air induction system.

Air Cleaner Assembly

The air cleaner assembly is mounted in the engine compartment but not in the traditional spot on top of the engine. A flexible hose feeds air into the housing where it is filtered by a disposable paper element.

Air Ducting

Air cleaner ducting channels the filtered air to an adapter on top of the throttle body. The ducting is fastened to the housing and adapter by clamps. The adapter is sealed to the throttle body by a gasket.

Throttle Body

The throttle body is mounted by an adapter to the intake manifold. Air passes through the throttle body into the turbocharger compressor. The throttle body regulates airflow

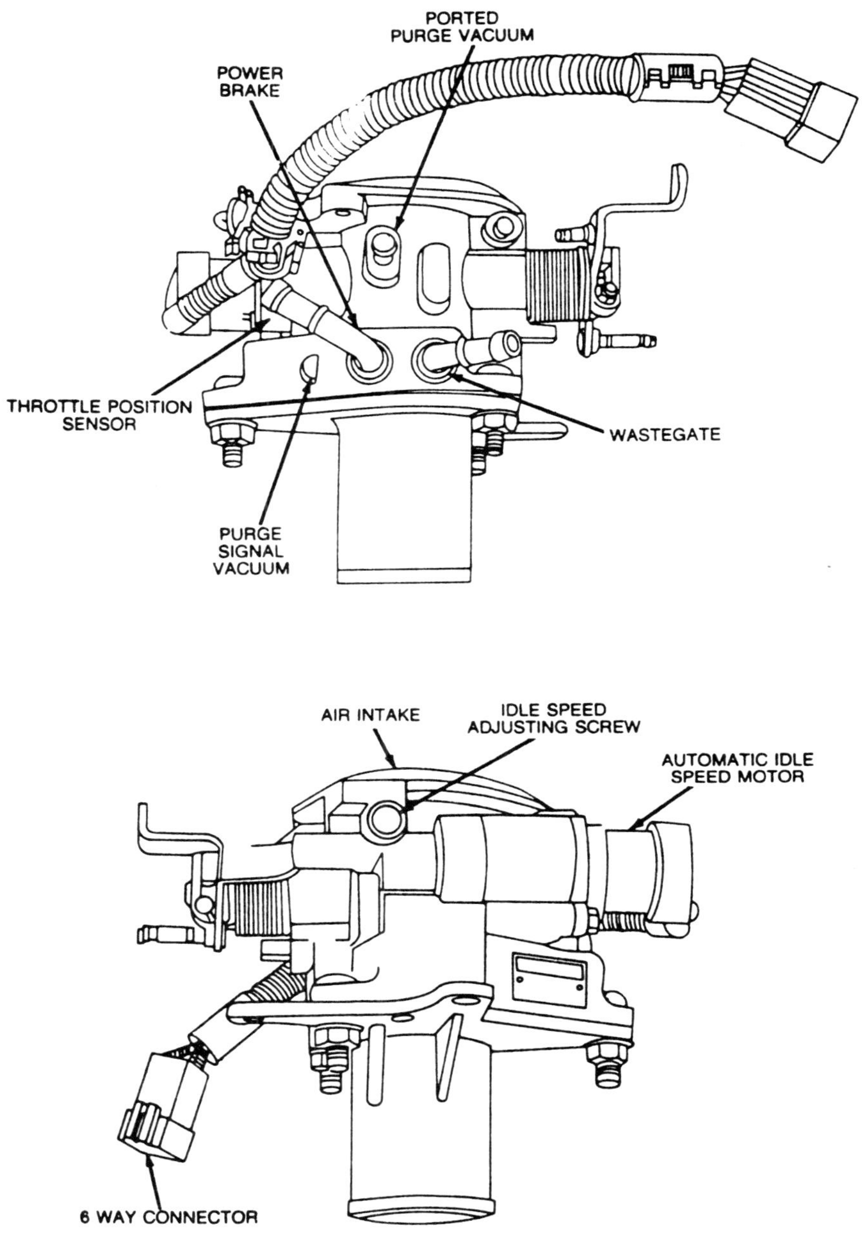

FIGURE 14-8 Throttle body assembly *(Courtesy of Chrysler Training Program).*

into the engine and also is where the throttle position sensor and automatic idle speed motor are located (Fig. 14-8). The throttle body also provides ports for all vacuum hose connections. The throttle plate is mounted on the throttle shaft and is actuated by a cable-type linkage connected to the accelerator pedal.

Turbocharger

The turbocharger consists of the housing, turbine, compressor, shaft and support bearings, wastegate, and lubrication oil passage (Fig. 14-9). The housing contains all the other parts. The turbine is driven by exhaust gases that are directed through the turbine housing. The turbine is mounted on a shaft. The compressor is mounted on the opposite end of the same shaft so that, whenever exhaust gases drive the turbine, the turbine will drive the compressor. The compressor housing directs air from the throttle body, (pressurizes or boosts it) into the intake manifold. The shaft is supported by bearings that are lubricated by engine lubrication system oil. The turbocharger can provide as much as 13 pounds of boost or a total of about 28 psi pressure. This does not use any horsepower to produce as exhaust gases are being used to drive the unit. Exhaust

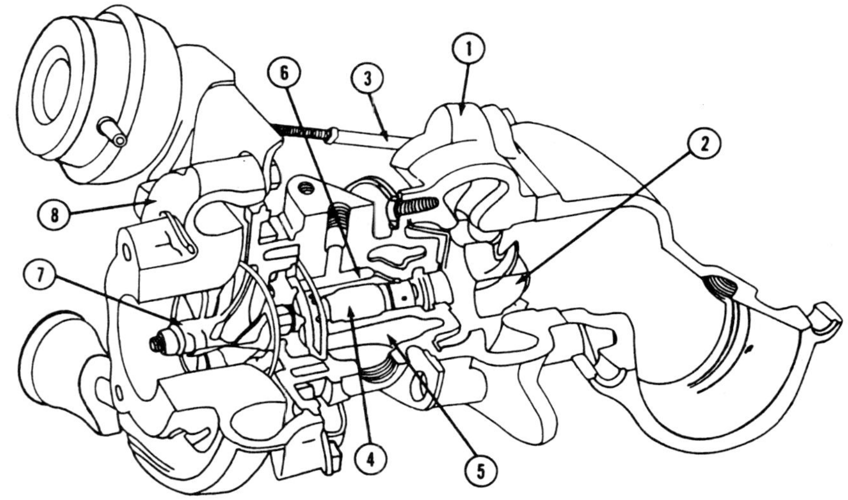

COMPONENTS OF A TURBOCHARGER

Turbine Housing: (Hot Side)(1)
Directs exhaust gages to turbine wheel.

Turbine Wheel: (Hot side)(2)
Driven by exhaust gages which drives the compressor through a shaft.

Waste Gate:(3)
Limits amount of boost pressure.

Shaft Wheel Assembly:(4)
Supported by bearings that float in oil.

Water Passage:(5)
Water circulation helps cool hot side of turbocharger.

Oil Passage:(6)
Lubricates and helps cool turbocharger.

Compressor Wheel:(7)
Compresses air to be forced into intake manifold.

Compressor Housing:(8)
Collects and directs pressurized air to intake manifold.

FIGURE 14-9 Components of a turbocharger *(Courtesy of Chrysler Training Program).*

inlet and outlet temperature at the compressor may differ by as much as 300°F. The average rotating speed of the turbocharger is 100,000 to 120,000 rpm.

The boost pressure of the unit must be limited; otherwise, engine damage will result. The wastegate serves this purpose. The wastegate is a valve in the exhaust system that is moved by an actuator. The actuator is a vacuum-pressure-operated diaphragm. The actuator causes the wastegate valve to remain closed until a specified boost pressure is reached. When this pressure is reached, intake manifold pressure causes the actuator diaphragm to move against spring tension and open the wastegate valve. Vacuum from a throttle body port is applied to the opposite side of the diaphragm and assists the boost pressure in moving the diaphragm. When the wastegate valve opens, exhaust gases are allowed to bypass the turbine and directly enter the exhaust system. The 1984 models provide about 10 pounds of boost for a total of 25-psi pressure. The 1985 models provide about 13 pounds of boost for a total of about 28 psi pressure.

A second control is used to limit boost of the turbocharger. The logic module uses the manifold absolute pressure sensor signal as an indication of intake manifold pressure. When this pressure reaches about 10 psi, the logic module shuts off fuel delivery until the boost pressure lowers. Ignition spark, however, continues.

A third control used to limit boost involves the engine speed sensor. The logic module monitors engine speed. When this speed reaches 6650 rpm, the logic module shuts off fuel delivery until 6100 rpm is reached. Ignition spark, however, continues. On 1985 models, the logic module controls a wastegate solenoid to limit boost. This is accomplished according to the MAP sensor signal.

Intake Manifold

The intake manifold is designed to accommodate the turbocharger and injectors; otherwise, it is of standard design.

SENSORS AND SWITCHES

The logic module relies on data input from engine sensors and various system switches to help it calculate correct air/fuel ratios.

Sensor input data are provided by a throttle position sensor, (TPS), exhaust gas oxygen sensor (O_2), manifold absolute pressure sensor, (MAP), engine coolant temperature sensor (CTS), vehicle speed sensor (VSS), engine speed sensor, detonation sensor, and air charge temperature sensor.

Switch inputs include the park/neutral switch, electric backlight or heated rear window switch, brake switch, air-conditioning switch, and air-conditioning clutch switch.

Throttle Position Sensor

The throttle position sensor is mounted on the end of the throttle shaft on the throttle body (Fig. 14-10). It is a potentiometer that receives a 5-volt source from the logic module, varies in resistance as throttle angle changes, and sends a varying return signal to the logic module as an indication of throttle blade position. The logic module also uses the TPS signal to determine how fast the throttle plate angle is changing. The TPS signal is used by the logic module to control automatic idle speed and the air/fuel ratio.

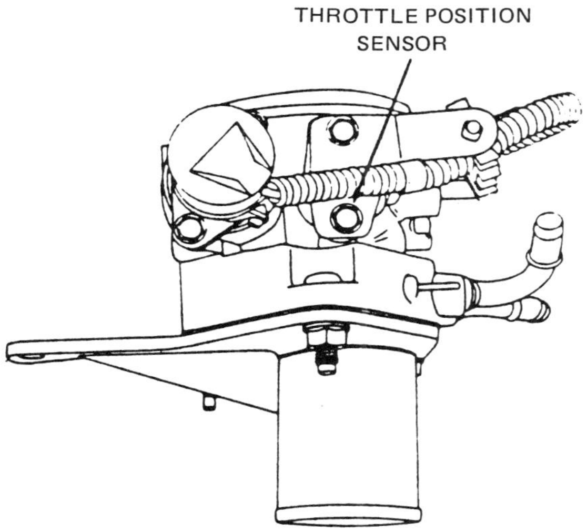

FIGURE 14-10 Throttle position sensor *(Courtesy of Chrysler Training Program).*

Exhaust Gas Oxygen Sensor

The exhaust gas oxygen sensor is mounted in the exhaust manifold and extends into the exhaust stream (Fig. 14-11). The sensor consists of an electrolyte element made of zirconia oxide coated with platinum. The element is cylindrical so that the outside surface is exposed to exhaust gas oxygen while the inside surface is exposed to fresh air oxygen. When oxygen comes in contact with zirconium oxide, a voltage is produced. The platinum coating protects the element against heat erosion. The voltage generated may vary between 100 and 900 millivolts.

When exhaust gas oxygen content is high, resulting from a lean mixture, the voltage generated is low. When exhaust gas oxygen content is low, resulting from a rich mixture, the voltage generated is high. This generated voltage signal, as compared to that produced by the oxygen in the fresh air, is used by the logic module to adjust the air/fuel ratio. To generate any voltage, the sensor element must reach an operating temperature of about 600°F. Until the sensor reaches this temperature, the logic module uses signals from the coolant temperature sensor to make air/fuel ratio adjustments. This is known as *open-loop* operation. When the oxygen sensor generates a voltage signal used by the logic module, it is known as *closed-loop* operation. During heavy acceleration, rapid deceleration, and with certain changes in the throttle position and MAP sensor signals, the oxygen sensor signal may be overridden so that the logic module can provide air/fuel ratios for performance and lowest emissions.

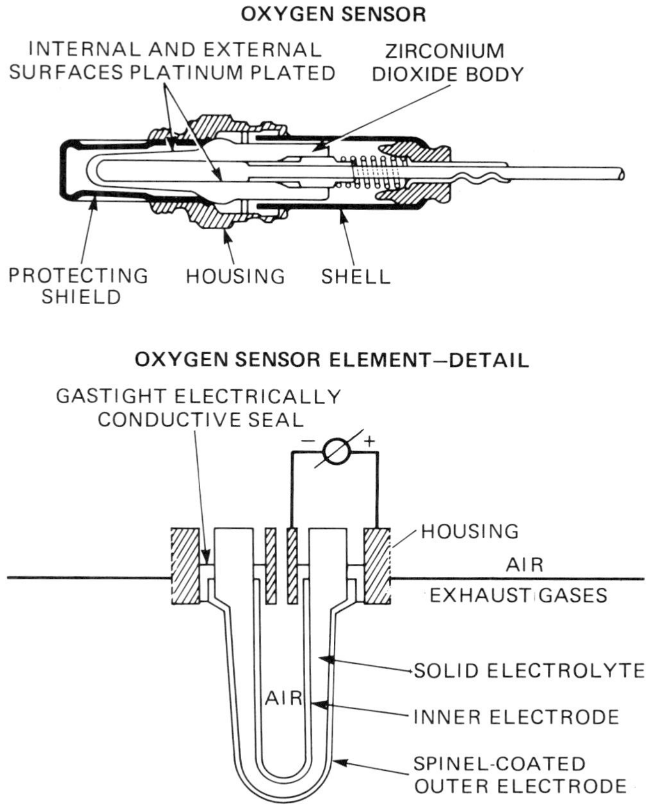

FIGURE 14-11 Oxygen sensor *(Courtesy of Chrysler Training Program).*

Manifold Absolute Pressure Sensor

A manifold absolute pressure sensor is located in the passenger compartment behind the right side kick panel above the logic module (Fig. 14-12) and is mounted on the logic

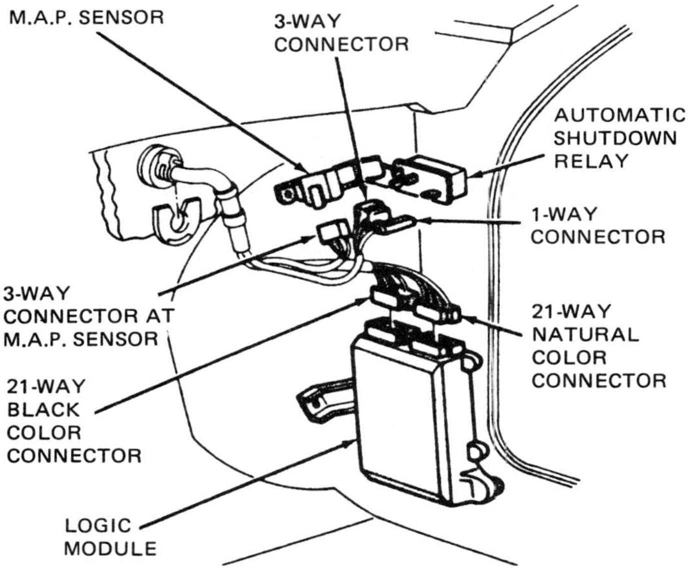

FIGURE 14-12 Electrical connectors right-side cowl panel, 1984 *(Courtesy of Chrysler Corporation).*

module in 1985 models. The MAP sensor is a piezoelectric-device that has a diaphragm that varies resistance with differences in pressure (Fig. 14-13). The logic module provides the sensor with a 5-volt source, and the sensor sends a return signal that is determined by its resistance and the engine vacuum. At atmospheric pressure or zero vacuum, the MAP sensor signal is about 2.4 volts, while at maximum vacuum the signal voltage is about 800 (0.8-volt) millivolts. The MAP sensor signal is used by the logic module to help determine the air/fuel ratio. Whenever intake manifold pressure increases (vacuum decreases), the MAP sensor signal causes the logic module to increase injector pulse width, which provides a richer mixture. As manifold pressure decreases, pulse width decreases. The MAP sensor signal is also used to determine correct spark advance. In this respect, it performs a function similar to that of the outdated vacuum advance diaphragm.

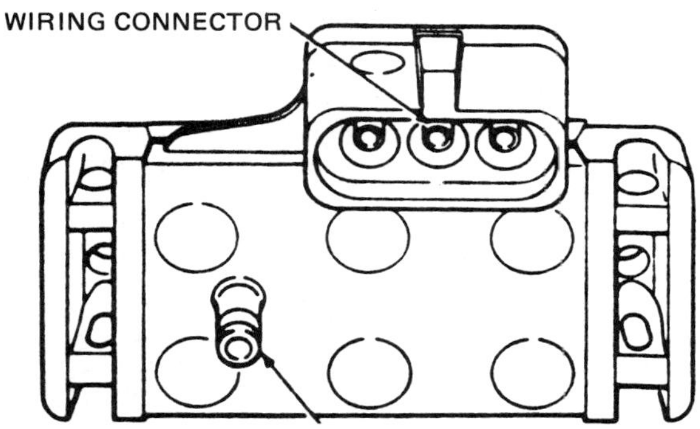

FIGURE 14-13 Manifold absolute pressure sensor, 1984 *(Courtesy of Chrysler Training Program).*

Engine Coolant Temperature Sensor

The engine coolant temperature sensor is located in the thermostat housing and senses coolant temperature. The sensor is a thermistor whose resistance value decreases as temperature increases. The coolant sensors used in 1984 models have the following resistances:

Resistance (ohms)	*Temperature (°F)*
10,000	32
5,500	77
900	190

The coolant sensors used in 1985 models have the following resistance:

Resistance (ohms)	*Temperature (°F)*
33,000	32
10,000	77
900	190

The coolant sensor signal to the logic module is used to calculate idle speed, injector pulse width, and ignition timing. The logic module supplies a 5-volt source to the sensor, which sends a return signal based on its resistance to the logic module.

Vehicle Speed Sensor

The vehicle speed sensor is located in the speedometer cable near the transmission (Fig. 14-14). It is essentially an on–off switch that supplies eight pulses per revolution of the speedometer cable. The vehicle speed sensor signal, in

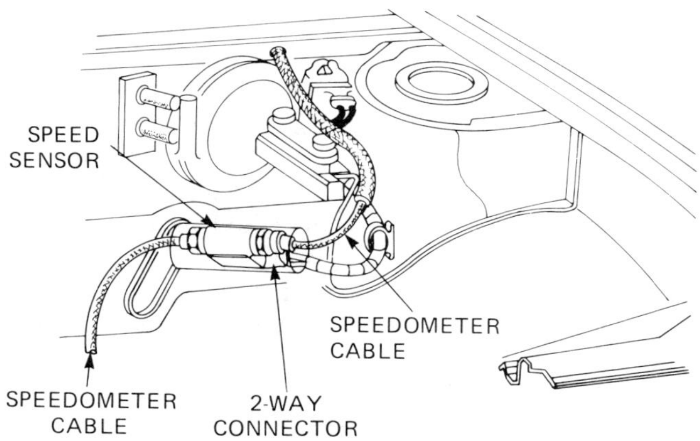

FIGURE 14-14 Speed sensor *(Courtesy of Chrysler Training Program).*

combination with the throttle position sensor signal, is used by the logic module for control of engine idle speed during deceleration and at a standstill. When decelerating, the logic module controls the automatic idle speed motor to obtain a specified manifold absolute pressure sensor signal. This prevents a rise in emission levels. At a standstill, the logic module controls the automatic idle speed motor to obtain a specified engine speed.

Engine Speed Sensor

The engine speed sensor is located in the distributor (Fig. 14-15). It is a Hall-effect switch that consists of a permanent magnet, shutter blades, and a transducer-type pickup assembly. The distributor rotor has four shutter blades attached to its underside. These blades extend downward so that, as the rotor turns, each blade will pass between the permanent magnet and pickup assembly. As this happens, the magnetic field is interrupted and a pulse is produced in the pickup assembly. This pulse is sent to the logic module, where it is interpreted as engine rpm for control of idle speed, ignition advance, and triggering of the fuel injectors through the power module.

A "synch" pickup signal is provided by a semicircular blade. It is located under the distributor breaker plate, and its signal is sent to the logic module and power module where it is used to control the injector driver circuits. The leading and trailing edge of the blade determines which pair of injectors, 1 and 2 or 3 and 4, is pulsed or energized.

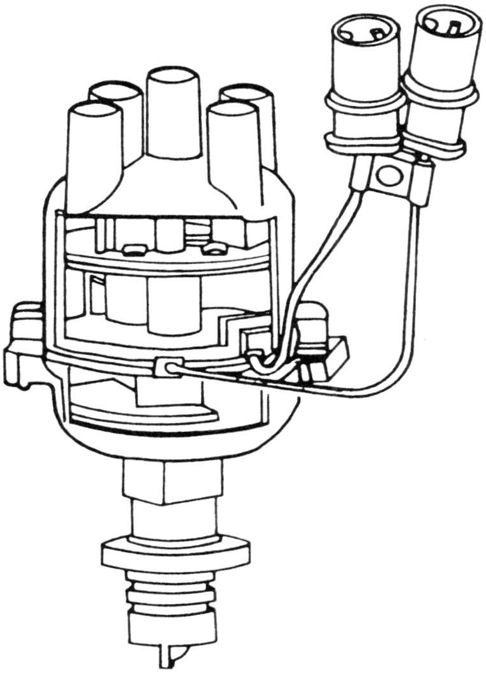

FIGURE 14-15 Hall-effect dual pickup distributor *(Courtesy of Chrysler Corporation).*

The synch signal is also used to control the automatic shutdown relay. The power module must receive an injector pulse signal from the logic module within 500 milliseconds of cranking, and the injector driver circuits must be switching in the correct sequence or the power module will remove the ground for the automatic shutdown relay coil, which deactivates the in-tank pump, logic module, ignition coil, and driver circuits in the power module.

A detonation sensor is used to detect knock or pinging. It is a piezoelectric sensor that produces a signal as a result of vibration within the engine casting. Detonation can occur as a result of overadvanced timing or the fuel octane rating being too low. The detonation sensor signal is sent to the logic module, which retards ignition timing in 5 degree increments until the detonation stops. It will, however, retard the ignition timing a maximum of 15 degrees. Timing is advanced again automatically in 2 degree increments per second. The detonation sensor circuit is ignored at engine speeds of less than 1000 rpm or if engine vacuum is high or if there is no boost from the turbocharger. It cannot be tested by tapping the engine block with a hammer as on other systems.

The air charge temperature sensor (Fig. 14-16) is a variable resistor whose resistance changes as air temperature in the intake manifold changes. As air temperature rises, the sensor resistance decreases. The sensor resistance is about 11,000 ohms at minus 4°F and is about 800 ohms at 195°F. The logic module uses the air charge temperature sensor signal when the coolant temperature sensor circuit malfunctions. The air charge temperature sensor is a backup for the coolant temperature sensor. Refer to the resistance and temperature specifications for the coolant temperature sensor provided earlier in this chapter in order to test the air charge temperature sensor. The two sensors for 1984 models have the same specifications but are not interchangeable. The two sensors for 1985 models have the

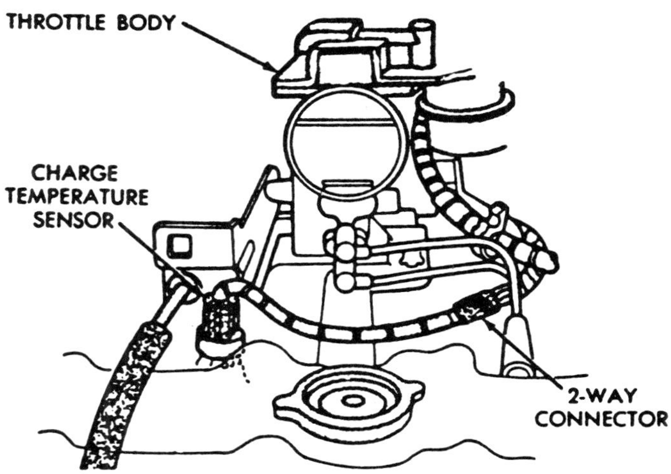

FIGURE 14-16 Charge temperature sensor *(Courtesy of Chrysler Corporation).*

same specifications but are not interchangeable. Driveability symptoms appear when they malfunction.

Park/Neutral Switch

To accurately control idle speed for either park/neutral or drive gear positions, the logic module relies on the park/neutral switch. This switch provides a signal representing gear position so that the logic module can maintain a quality idle speed.

Heated Backlite Switch

Vehicles equipped with a heated backlite or rear window also have a switch input to the logic module to indicate activation of the heated rear window circuit. This enables the logic module to compensate for the additional electrical and engine load on 1984 models by raising idle speed. The engine speed will not actually increase, but will remain stable even with the additional load. A built-in voltage regulator compensates for this load on 1985 models.

Brake Switch

The brake switch acts as a backup circuit for the throttle position sensor. One function of the TPS is to provide a signal that causes the logic module to control idle speed. If this signal is not provided, the brake switch signal input will serve the same purpose.

Air-conditioning Switch

An air-conditioning switch signal is provided to the logic module for idle speed control. The additional load of the air-conditioning compressor is offset by the automatic idle speed motor when the logic module commands it to allow more air to bypass the throttle blade. Idle speed does not increase but does remain stable, even with the additional load of the compressor. Because the air-conditioning system is a cycling clutch system, an air-conditioning clutch switch signal is needed to enable the logic module to maintain a stable idle speed as the compressor clutch cycles on and off.

LOGIC AND POWER MODULES

The logic and power modules work together to control the fuel injection system. The logic module is the decision-making component, and the power module is an electronic relay that controls many of the system circuits.

Logic Module

The logic module is located inside the passenger compartment behind the right front kick panel (Fig. 14-12). It is a digital computer that provides 5 volts to the manifold absolute pressure sensor, coolant temperature sensor, throttle position sensor, and air charge temperature sensor. The logic module uses input data from these sensors and an engine speed signal from the distributor to control ignition timing, injector pulse width, exhaust gas recirculation valve operation, evaporative emission control purge valve operation, and idle speed; it also contains circuitry that enables it to monitor and diagnose its own circuits. This self-diagnostic feature is known as on-board diagnostics. For example, if a sensor circuit develops a malfunction, the logic module will wait a programmed amount of time and then set a code in its memory. The service technician can gain access to these codes at a later time by counting the flashes of the POWER LOSS telltale on the instrument panel or by using a special diagnostic readout tester. If the input data from any of the sensors are lost or out of range, the logic module will enter the limp-in mode of operation and light the POWER LOSS telltale.

Power Module

The power module is an electronic relay controlled in part by the logic module (Fig. 14-17). It is located in the left front fender well behind the battery. Fresh air is drawn through it by the air induction system. This provides necessary cooling. Internal circuitry for control of the fuel injectors and ignition coil is separate from other circuitry to help reduce electrical noise and interference. The power module contains a voltage converter that supplies an 8-volt source for the logic module and pickup in the distributor. It also provides the ground for the coil in the automatic shutdown relay. The voltage source provided by the power

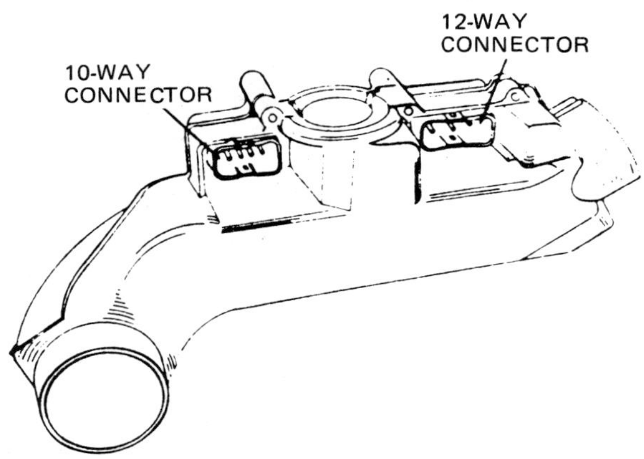

FIGURE 14-17 Power module *(Courtesy of Chrysler Training Program).*

module for the logic module and automatic shutdown relay is fused on the printed circuit board. This protects components against reversed battery polarity. Reversed battery polarity causes the fuse to blow. When this happens, power is provided through an internal diode when battery polarity is corrected. If battery polarity is reversed again, the diode will protect the components. The power module provides the ground for the injector and ignition coil circuits in accordance with the pulses from the distributor synch signal. The logic module commands the power module to either make or break the ground for these circuits.

EMISSION CONTROLS

The Chrysler EFI system operates emission controls in addition to the functions discussed so far. The emission controls used with the EFI system include a three-way catalytic converter, evaporative emission control, exhaust gas recirculation, and positive crankcase ventilation system.

Three-way Catalytic Converter

The three-way catalytic converter reduces hydrocarbon, carbon monoxide, and oxides of nitrogen emissions.

Evaporative Emission Control System

The evaporative emission control system gathers and recycles unburned hydrocarbon emissions from the fuel tank. These vapors are stored in a charcoal canister until they can be fed into the intake manifold. The EFI system logic module controls a purge valve that regulates when vapors are allowed to be drawn from the canister into the incoming air charge. The logic module controls the purge valve by means of a solenoid. When the solenoid is energized, it blocks vacuum to the purge valve, which prevents purging of the vapors. The solenoid is energized whenever the engine coolant temperature is below 180°F. When this temperature is reached, the logic module de-energizes the solenoid and allows vapors to be purged from the canister.

Positive Crankcase Ventilation System

The positive crankcase ventilation system is the same as used on carbureted engines except for the calibration of the PCV valve.

Exhaust Gas Recirculation System

The exhaust gas recirculation system is back pressure modulated. This means that both a vacuum signal and exhaust system back pressure are needed for normal operation. System back pressure is monitored by a transducer that controls the vacuum source for the EGR valve. In addition, the logic module controls a solenoid that can either block or allow vacuum to the transducer. When engine coolant temperature is below 70°F, the logic module energizes the solenoid, which blocks vacuum to the transducer and EGR valve. When engine coolant temperatures rise to about 70°F, the solenoid is de-energized. This allows normal operation of the back-pressure-modulated EGR valve. The logic module will also energize the solenoid if engine speed is below 1200 rpm or at wide-open throttle.

SYSTEM OPERATION

The multipoint EFI system operates in modes that correspond to the positions of the ignition switch.

Ignition Switch ON

When the ignition switch is in the ON position, fused power is supplied to the power module; the power module supplies 12 volts to the automatic shutdown relay, the logic module, and the distributor pickups; the logic module supplies 5 volts to its four sensors; and the automatic shutdown relay energizes the fuel pump, ignition coil, and injector circuits.

Ignition Switch CRANK

In the CRANK position, the distributor begins sending rpm pulses to the power and logic modules. The power module, at the command of the logic module, will keep the automatic shut-down relay energized, which keeps the fuel pump, injector circuits, and ignition coil active. The logic module begins receiving input data from the engine coolant, air charge, manifold absolute pressure, and throttle position sensors. The automatic idle speed motor is driven open and the injector and ignition coil circuits are commanded by the logic module through the power module. The power module must receive an ON injector command signal from the logic module within 500 milliseconds of cranking, and both injector driver circuits must be switching in the correct sequence or the automatic shutdown relay will shut the system off.

Ignition Switch RUN

When the engine starts and the ignition switch is allowed to return to the RUN position, there must be various input and output signals available at the logic and power modules, automatic shutdown relay, and distributor pickups; otherwise, the EFI system will stop. These inputs and outputs are provided in the service manual.

Ignition Switch OFF

Turning the ignition switch to OFF removes power from the power module, ASD relay, in-tank pump, logic module, and ignition coil.

Injector Pulse Width

The EFI system receives input data, processes it, and, in accordance with its computer program, issues commands to controlled devices. The injector pulse width or ON time is one of the calculations accomplished by the logic module. The pulse width is measured in milliseconds and determines the air/fuel ratio. The computer program uses the speed/density method of determining the air/fuel ratio for a given set of circumstances. The logic module relies on signals from the engine speed sensor in the distributor and the manifold absolute pressure sensor to arrive at an airflow volume and density calculation. Based on this, mass fuel flow can be calculated and the injectors can be pulsed to provide the correct mass fuel flow.

Computer Program Updating

The EFI system computer also has the capability of adjusting its program to offset variations in production tolerances or in extraneous operating parameters such as barometric pressure and altitude. This is accomplished in 12 memory cells in the logic module. Ten of these cells are based on engine speed and manifold absolute pressure, while the remaining two cells are based on idle condition in park or neutral and drive. These memory cells are nonvolatile, which means they do not lose their memory when the ignition switch is turned off. The cells are updated and revised by the oxygen sensor when the system is in closed-loop operation. If the oxygen sensor signal changes from a previous level for a specified period of time, the memory cells will be updated.

Open- and Closed-loop Mode

The EFI system can operate in two modes. The open-loop mode is entered whenever the oxygen sensor signal is not available or is not of sufficient value to be considered by the logic module. When first started, the engine exhaust is not hot enough to cause the oxygen sensor to generate voltage levels recognizable by the logic module. The system, as a result, operates in the open-loop mode. As the temperature of the oxygen sensor rises (and engine temperature also rises), the voltage levels generated increase. Eventually, the logic module recognizes these signals and the system enters the closed-loop mode of operation. The open-loop mode is reentered during wide-open-throttle or deceleration conditions and if the oxygen sensor signal stops varying for a specified amount of time. Open loop is also reentered when 1 pound of boost from the turbocharger is available. To enter the closed-loop mode, coolant temperature must reach a specified value, a timer circuit in the logic module must expire, and the oxygen sensor must be producing a valid signal.

During engine starting, the fuel injectors are pulsed two times for each revolution of the crankshaft. Coolant temperature and time are the pulse width determining factors under these conditions. This programmed pulse width is longer for a cold engine as compared to a hot restart.

If the engine coolant temperature sensor signal indicates a temperature below normal operating levels, the cold enrichment for injector pulse width is determined by the coolant temperature sensor signal, coolant temperature sensor signal at instant of startup, and engine load or MAP sensor signal. The result of these measurements is an enrichment multiplier that provides best economy and lowest emissions for warmup.

When the throttle is closed and the MAP sensor signal is below a programmed value, the logic module adjusts the injector pulse width according to engine speed. As engine rpm increases, the logic module decreases the injector pulse width.

Sudden decreases in MAP sensor signals causes the logic module to momentarily shut off the injectors to lean the air/fuel ratio for economy and emissions reasons. This is known as *deceleration leanout.*

When accelerating, the throttle position and MAP sensors signals increase. This causes the logic module to increase injector pulse width to attain good performance levels without developing driveability symptoms.

During wide-open-throttle operation, the air/fuel ratio is not stoichiometric in order to increase power production and prevent engine damage due to high combustion temperatures caused by lean mixtures. The MAP sensor signal is used to determine pulse width under these conditions.

At idle, the automatic idle speed motor opens and closes the air bypass passage to control idle speed. The logic module can modify ignition spark to achieve a specified rpm value. If the idle speed cannot be brought within specification in this manner, the logic module can pulse the AIS motor with either small or large pulses depending on how far out of specification the idle speed is at the time.

The vehicle speed sensor signal prevents complete closing of the AIS motor during closed-throttle deceleration until vehicle speed reaches zero. When the throttle is opened, the logic module pulses the AIS motor an extra ''kick'' in anticipation of the next closed-throttle deceleration. This helps reduce hydrocarbon emissions and stalling during ''panic stops.''

Self-diagnostic Capability

The on-board diagnostic feature of the EFI system enables it to diagnose many of its own circuits. If an input or output signal is malfunctioning often enough, the logic module considers this to be a "real problem." It will set a fault code representing the problem circuit in its memory. Circuit problems that occur and then disappear, such as loose connections, are not considered real problems and do not set fault codes. The codes are retained in memory until erased or cleared by the service technician or by the ignition switch being cycled on and off 30 times. The codes are two digit and are defined in the service manual. There are many codes that can be set in memory.

The fault codes can be accessed by the service technician by using a special diagnostic readout tester or by utilizing the POWER LOSS telltale on the instrument panel.

The diagnostic readout tester is connected to the EFI system at a diagnostic connector on the right front fender well in the engine compartment. The tester provides a direct readout for fault codes, distributor signal, and switch tests. It has a hold button to stop the readout display at any point. An actuator test can be made to check operation of the ignition system, AIS motor, and fuel injectors.

Self-diagnostic Mode

The flashing of the POWER LOSS telltale can also be used to gain access to fault codes. To activate the telltale and diagnostic mode, the ignition switch must be turned on, off, on, off, on within 5 seconds. The POWER LOSS telltale will light as a bulb check and then begin flashing the codes. There are short and long pauses between flashes so the codes can be distinguished from each other. When code 55 appears, the system is ready for further tests. Either "switch tests" or "actuator test mode" can be selected. The switches tested by the system include the brake switch, gear shift park-reverse-park switch, air-conditioning switch, speed sensor, and electric backlite switch. Switch operation is correct if the power loss telltale lights each time a switch is activated.

The actuator test mode checks operation of the ignition coil, AIS motor, and fuel injectors. The actuator test mode can be activated without the special diagnostic readout tester by connecting a jumper lead across the terminals in the distributor connector on the harness side with the ignition switch on. Make and break the jumper connection quickly to make the logic module think it is receiving a 120 rpm or 60 pulses per minute signal. The logic module then operates the ignition coil, AIS motor, and fuel injectors. Follow specific service manual instructions to operate the diagnostic mode. Once codes are obtained, the service manual diagnostic charts should be used to check individual circuit components.

Limp-in Mode

The limp-in mode of operation is entered when the signal from the MAP sensor, TPS, coolant or charge temperature sensors is too high, too low, or changes too fast to be real. Any of these conditions causes a code to be set and the system enters the limp-in mode. Engine performance and economy suffer while operating in this mode. If the sensor signal accuracy is restored during operation, the POWER LOSS telltale will remain on and the system will remain in the limp-in mode. When the ignition switch is cycled off and then on for restarting, the system returns to normal operation. The POWER LOSS telltale will be off, but a fault code will still be retained in memory for 30 cycles of the ignition switch.

SYSTEM SERVICE

The first step in servicing the EFI system is to define the problem according to information provided by the customer. Check for codes stored in memory next and then, depending on results so far, a visual check of wiring harnesses and connections can be made.

Distributor Positioning

To index the distributor, a timing light and a tachometer are required. The engine should be at operating temperature. Disconnecting the coolant temperature sensor causes the logic module to enter the limp-in mode where the ignition system operates at base timing. Check to see if the power loss lamp is illuminated. This indicates limp-in mode operation. Reconnect the coolant temperature sensor and set ignition timing. Shut the engine off and then restart it. The power loss lamp should be off. This indicates that the logic module is out of the limp-in mode and functioning normally.

Idle Speed

When diagnosing idle speed problems, vacuum hoses and connections should be checked for leaks. Ignition timing should be checked and corrected if necessary. The coolant temperature sensor should be checked for correct operation as it will affect idle speed. The detailed procedure provided in the service manual should be followed closely. Performance, economy, and driveability suffer if idle speed is not adjusted correctly. A tachometer is needed to make this adjustment. The throttle position sensor should be checked for correct calibration whenever idle speed is adjusted.

Fuel Pressure Relief

Fuel system pressure can be relieved by grounding the ignition coil secondary wire with a jumper and then cranking the engine. This triggers the fuel injectors to relieve pressure in the system. Fuel system service can be safely performed once pressure has been bled off.

Fuel Pressure Test

A special fuel injection pressure gauge and adaptors are needed to check system pressure. System pressure is normally 48 to 52 psi; however, the Chrysler port EFI turbocharged engine system may go as high as 75 psi when operating under boost conditions.

Fuel Filter

When servicing the in-line fuel filter, fuel pressure must be relieved as discussed earlier. Remove the filter and its connecting hoses and discard them. Install the new filter, hoses, and clamps. Cycle the ignition switch several times to pressurize the system. Check for leaks, and if the system is not leaking, start the engine.

The Chrysler multipoint EFI system is similar in many respects to other port fuel injection systems. There are, however, enough differences to justify its study as an individual system.

SUMMARY

As electronic systems use increases, the need to understand automotive computer systems will become even greater than it is today. The information provided in this text is intended to meet part of this necessary understanding. A background in automotive electronics would also help the technician prepare for today's systems and future systems.

ABBREVIATIONS

AIS Automatic idle speed
ASD Automatic shutdown (relay)
CTS Coolant temperature sensor
EGR Exhaust gas recirculation
MAP Manifold absolute pressure
O_2 Oxygen (sensor)
PCV Positive crankcase ventilation
P/N Park/Neutral
rpm Revolutions per minute
TPS Throttle position sensor

Index

C

D

E

F

G

H

I

K

L

M

N

O

P

R

S

T

V

W